Biofuels: Renewable Energy for a Sustainable Future

Biofuels: Renewable Energy for a Sustainable Future

Editor: Nina Goldman

RCALLISTO
REFERENCE

www.callistoreference.com

Callisto Reference,
118-35 Queens Blvd., Suite 400,
Forest Hills, NY 11375, USA

Visit us on the World Wide Web at:
www.callistoreference.com

ISBN: 978-1-64116-166-4 (Hardback)

Cataloging-in-Publication Data

Biofuels : renewable energy for a sustainable future / edited by Nina Goldman.
    p. cm.
Includes bibliographical references and index.
ISBN 978-1-64116-166-4
1. Biomass energy. 2. Renewable energy sources. 3. Sustainability. I. Goldman, Nina.
TP339 .B56 2019
662.88--dc23

Table of Contents

Preface

Renewable energy is replenished naturally on a human timescale. It can replace conventional fuels in the major areas of electricity generation, transportation, off-grid energy services and heating. Biofuel is a renewable fuel that is derived from plants or from agricultural, domestic, commercial and industrial wastes. Biofuels can be in a solid, liquid or gaseous form. Gaseous biofuels include biogas and landfill gas, while liquid biofuels include different bioalcohols and oils. Certain solid biomass fuels are wood, agricultural waste and dried manure. Using first, second, third and fourth generation biofuel production procedures, ethanol, propanol, butanol, biodiesel, methanol, green diesel, biofuel gasoline, etc. can be developed for varied fuel applications. Bioethanol and biodiesel have potential applications as a transportation fuel. This book contains some path-breaking studies in biofuels. There has been rapid progress in this field and its applications are finding their way across multiple industries. With state-of-the-art inputs by acclaimed experts of this field, this book targets students and professionals.

Significant researches are present in this book. Intensive efforts have been employed by authors to make this book an outstanding discourse. This book contains the enlightening chapters which have been written on the basis of significant researches done by the experts.

Finally, I would also like to thank all the members involved in this book for being a team and meeting all the deadlines for the submission of their respective works. I would also like to thank my friends and family for being supportive in my efforts.

Editor

Water for small-scale biogas digesters in sub-Saharan Africa

VAMINI BANSAL, VIANNEY TUMWESIGE and JO U. SMITH

Institute of Biological and Environmental Science, University of Aberdeen, 23 St Machar Drive, Aberdeen AB24 3UU, UK

Abstract

Biogas could provide a more sustainable energy source than wood fuels for rural households in sub-Saharan African. However, functioning of biogas digesters can be limited in areas of low water availability. The water required is approximately 50 dm^3 day^{-1} for each cow and 10 dm^3 day^{-1} for each pig providing manure to the digester, or 25 ($\pm$6) dm^3 day^{-1} for each person in the household, using a digester volume of 1.3 ($\pm$0.3) m^3 capita^{-1}. Here, we consider the potential of domestic water recycling, rainwater harvesting, and aquaculture to supply the water needed for digestion in different countries of sub-Saharan Africa. Domestic water recycling was found to be important in every country but was usually insufficient to meet the requirements of the digester, with households in 72% of countries need to collect additional water. Rooftop rainwater harvesting also has an important role, iron roofs being more effective than thatched roofs at collecting water. However, even with an iron roof, the size of roof commonly found in sub-Saharan Africa (15 to 40 m^2) is too small to collect sufficient water, requiring an extra area (in m^2) for each person of ($R/100$) (where R is the rainfall in mm). If there is a local market for fish, stocking a pond with tilapia, fed on plankton growing on bioslurry from the digester, could provide an important source of additional income and hold the water required by the digester. In areas where rainfall is low and seasonal, the fishpond might be stocked only in the rainy season, allowing the pond to be covered during the dry period to reduce evaporation. If evaporative losses (E in mm) exceed rainfall, an extra catchment area is needed to maintain the water level in the pond, equivalent to approximately (1.5 $\times$ (($E-R$)/ R)) m^2 for each person in the household.

Keywords: aquaculture, biogas, rainwater harvesting, sub-Saharan Africa, water harvesting, water recycling

Introduction

Use of biogas digesters to provide household energy in sub-Saharan Africa

Wood, charcoal, and dung are traditional biomass fuels that currently supply over 70% of the household energy in sub-Saharan Africa (SSA) (Eleri & Eleri, 2009). Biomass fuels are often the preferred energy source in rural areas because they can usually be collected locally without incurring additional cost (Karekezi & Kithyoma, 2002). However, these sources of energy can create many problems for the environment and the people using them, especially for women and children (Bryceson & Howe, 1993; Biran *et al.*, 2004). The collection of firewood has been linked to local deforestation (Subedi *et al.*, 2014), which is currently occurring at a rate of 0.7% per annum in SSA (Eleri & Eleri, 2009). This, in turn, has a detrimental effect on soil quality and increases surface run-off (Leu *et al.*, 2010; Hallett *et al.*,

2012). Cooking on a wood fire releases carbon monoxide and particulates at levels detrimental to human health (Gordon *et al.*, 2014); poor indoor air quality has been linked to over 3.5 million premature deaths annually (Lim *et al.*, 2012) and contributes to a wide range of child and adult diseases (World Health Organization (WHO), 2014).

The UN Sustainable Energy for All Initiative (UN Department of Economics and Social Affairs, 2012) includes provision of modern cooking appliances and fuels as one of its 11 key action areas. A further international initiative, the Global Alliance for Clean Cookstoves (2014), is a public–private partnership that aims to create a global market for clean and efficient household cooking solutions. Improvement in cookstoves that use biomass fuels is critical to the reduction in demand for wood and improvement in indoor air quality (MacCarty *et al.*, 2008). However, depending on thermal efficiency, improved biomass cookstoves may provide only part of the solution; Smith *et al.* (2015) estimated that if the thermal efficiency is improved from the typical value of 17% for a three-stone fire (Omer & Fadalla, 2003) to 38–50% for a pyrolysis cookstove (Roth, 2011),

Correspondence: Jo U. Smith
email: jo.smith@abdn.ac.uk

the rate of deforestation could be reduced by 41 (±25) to 50 (±30)%, with a further 21 (±12)% reduction if suitable crop residues are used as an additional fuel source, meaning that the total potential reduction in deforestation is only ~60% to 70%. Biogas presents an important opportunity to fill this energy gap, providing a clean, cheap, and renewable additional fuel source; assuming a thermal efficiency of 75% (Zielonka et al., 2010), Smith et al. (2015) estimated that the rate of deforestation could be reduced by a further 23 (±14)%. Making use of the important opportunity presented by biogas has the potential to reduce deforestation due to wood fuel demand by a total of ~70% to 100% (Smith et al., 2015).

Biogas is produced through the anaerobic digestion of organic compounds (Hamlin, 2012). Feedstock and water are added through an inlet pipe in equal ratios by volume (Amigun & Von Blottnitz, 2010). The feedstock is then broken down in an airtight chamber by anaerobic micro-organisms to produce methane and carbon dioxide. Biogas generally consists of around 60–70% methane and 20–30% carbon dioxide and can be used as energy for household cooking or lighting (Brown, 2006). The digester also produces a slurry, 'bioslurry', that is rich in available nitrogen, phosphorus, and potassium and can be used as a fertilizer to grow crops or feed fish for aquaculture (Orskov et al., 2014). The composition of the bioslurry and biogas depends on the type of feedstock used, as different substrates contain different amounts of dry matter, nutrients, and volatile solids (Amigun & Von Blottnitz, 2010; Mao et al., 2015).

Biogas technology is most suited to rural households with a readily available source of feedstock from livestock or crop residues. In Ghana, it is estimated that in 2006, at least 3.4 million households kept livestock, sufficient to generate 350 million m³ of biogas in the year 2006 (Arthur et al., 2011). Brown (2006) suggested that 1–2 cows or 5–8 pigs should provide enough manure to run a biogas digester for a household of four people (Brown, 2006). Orskov et al. (2014) agreed that sufficient biogas is produced by two cows, but suggested that the number of pigs required is at least eight. Human faeces can also be used to supply feedstock, but this would not provide sufficient feedstock for the digester. The more commonly used substrates for anaerobic digestion in SSA are cow and pig manure. Therefore, this study only considers use of cow and pig manure; crop residues and human faeces are omitted.

Over recent years, there have been increased efforts to disseminate biogas technology in SSA; in 2011, the total number of domestic biogas digesters installed in nine countries of SSA (Rwanda, Ethiopia, Tanzania, Kenya, Uganda, Burkino Faso, Cameroon, Benin and Senegal) was 24,990 (SNV, 2013); this is small compared to the numbers of domestic digesters in China, which

reached 40 million by 2010 (Dong, 2012), but is increasing under the efforts of the African Biogas Partnership Programme, the Netherlands Development Organisation (SNV), and the Humanist Institute for Development Cooperation (HIVOS) (Africa Renewable Energy Access Program, 2011). Implementation of biogas digesters in SSA is often targeted at rural households (Amigun & Von Blottnitz, 2010). The digester size usually varies from 5 to 10 m³, depending on the energy requirements of the household and the substrate retention time (Parawira, 2009). Most rural biogas plants have no moving parts to mix the feedstock and water, so this is usually done by hand; this can take up to 30 min each day, time that would otherwise be used for other household activities (Hamlin, 2012). Further time is taken in fetching water and collecting feedstock, especially if livestock are not housed, and this must be balanced against the potential time-saving due to reduced need to collect wood (Orskov et al., 2014). The size of the digester depends on the amount of energy required by the household; typically 90–100% of the energy used by a rural off-grid household is for cooking (Karekezi & Kithyoma, 2002).

Biogas digesters are a promising option for providing household energy in rural SSA, but Mengistu et al. (2015) concluded that uptake is often limited by policies and institutional arrangements, financial constraints, lack of subsidies, availability of inputs, and consumers' awareness and attitudes to the technology. Financial constraints, particularly the inability to afford the high initial investment costs, are often considered to be the principal factor preventing uptake of digesters (Bensah & Brew-Hammond, 2010; Arthur et al., 2011). Bedi et al. (2015) suggested biogas uptake in Rwanda is constrained by long payback periods and low rates of return. Mwirigi et al. (2014) recommended standardization and quality control, integrated farming using biogas and bioslurry, mobilization of local and external funds, such as from the clean development mechanism, to overcome initial construction costs, and the formation of user and disseminator associations for joint procurement and linkage to finance. However, even when subsidies and financial structures support investment in biogas, a high proportion of digesters stop working within a few years due to technical problems or lack of essential resources, such as water, feedstock, or labour (Parawira, 2009).

Water is often the key factor limiting implementation of biogas; a survey conducted in Ethiopia showed that of 700 biogas digesters, 60% were non-operational due to lack of water or manure (Eshete et al., 2006). It is suggested that to run a biogas digester efficiently, the time taken to reach the water source should be no more than 30 min (Eshete et al., 2006), and the household should be able to collect at least 25 dm^3 water per person per day (Orskov *et al.*, 2014). In providing energy for household use and organic fertilizer for food production, but in using water, anaerobic digestion is at the centre of the 'water–energy–food nexus' in SSA (Conway *et al.*, 2015; Smith *et al.*, 2015). The aim of this study is to investigate how biogas digesters can be implemented in the often water limited conditions of rural SSA, and what methods can be used to improve water availability.

Water availability in sub-Saharan Africa

Most countries in SSA are classified as having economic water scarcity, suggesting that the available water sources are not used to their full potential due to poor governance, infrastructure, or management (Van Koppen, 2003). Water is accessed by the rural population mainly though hand pumps, boreholes, wells, water vendors, piped systems, and springs (Lockwood & Smits, 2011). Due to the poor infrastructure and management of these sources, many hand pumps are not operational. A study conducted by Water Aid in Tanzania found that 25% of hand pumps did not work 2 years after installation.

The WHO (2006) suggested that 20 dm^3 capita^{-1} day^{-1} is sufficient water to meet domestic needs in SSA. However, another study (Gleick, 1998) recommends up to 50 dm^3 day^{-1} for consumption and sanitation. The volume of water is related to the distance to the source and the household size. In most cases, water use per capita decreases with household size; this trend was observed in Malawi by Rosen & Vincent (1999) where a two-member household averaged 20 dm^3 capita^{-1} day^{-1}, but an eight-member household never exceeded 10 dm^3 capita^{-1} day^{-1}.

The time taken for water collection is the main factor controlling water available to the household. In an average year, women in SSA spend approximately 40 billion hours collecting water (Blackden & Wodon, 2006). As the distance to the water source increases, the quantity of water collected generally decreases. Sugita (2006) described this in a survey of households in Uganda, where average consumption varied from 15.6 dm^3 capita^{-1} day^{-1} for households using distant hand pumps to 155 dm^3 capita^{-1} day^{-1} for households who had piped systems.

A biogas digester requires extra water for anaerobic digestion, which may result in more trips to collect water. This could be problematic because water collection has been linked to numerous health and social problems in communities. Water collection is often done by women; the more water they have to collect, the less

time is available for other activities such as education (Pickering, 2011). Carrying heavy buckets of water over long distances can cause skeletal injuries and exposes women to risk of assault and water-based diseases (Rosen & Vincent, 1999). To avoid the problems associated with an increased water demand and in the context of limited and seasonal water access and availability, additional water demand for anaerobic digestion requires alternative techniques for water collection. Therefore, three different ways to meet water demand that are appropriate to rural households are considered in this study: recycling domestic water, harvesting rainwater, and aquaculture.

Methods to meet water demand

Recycling domestic water. Recycling domestic water is the easiest way to increase the availability of water for households that get sufficient water for domestic use. Domestic water use includes drinking, laundry, bathing, cleaning, and cooking (Nyong & Kanaroglou, 1999). Gleick (1998) recommended that around 5 dm^3 capita^{-1} day^{-1} should be used for drinking, 10 dm^3 capita^{-1} day^{-1} for cooking, 15 dm^3 capita^{-1} day^{-1} for bathing, and 20 dm^3 capita^{-1} day^{-1} for cleaning; so for a four-person household, this would come to around 180 dm^3 day^{-1}. In addition, water is required for livestock, and this usually competes with domestic water use and increases the demand for water (Rosen & Vincent, 1999). Almost all of this water can be recycled in some way; water from drinking and livestock is recycled as urine and can be used for wet fermentation in the digester (Brown, 2006). The microbial content of the wastewater may be reduced by the anaerobic digestion process (Avery *et al.*, 2014), and it is therefore considered safer for use in aquaculture or for application to crops (Ogunmokun *et al.*, 2000).

The volume of water that can be recycled is largely dependent on how much is available and is allocated to different activities. The quantity of water consumed can vary greatly in different seasons; in north-eastern Nigeria, it was found that mean domestic water consumption in the rainy season was 215 dm^3 per household, whereas it was only 125 dm^3 per household in the dry season (Nyong & Kanaroglou, 1999). This was because activities, such as laundry and bathing, were done either fully or partially in streams to reduce the amount of water that must be carried to the household. However, the opposite pattern was seen in Malawi, where total household water consumption was higher in the dry seasons than the rainy seasons due to the drier climate driving higher water consumption and most households having a relatively good water source (Mloza-Banda *et al.*, 2006).

One advantage of using wastewater for household anaerobic digestion is that it can result in better sanitation as wastewater is properly disposed of. Grey water includes all wastewater produced from domestic activities, excluding toilet waste, accounting for 50–80% of total wastewater generated (Madungwe & Sakuringwa, 2007). The biogas digester does not require clean water, so grey water can be used without pretreatment (Parawira, 2009), unless there is a high amount of detergent or disinfectant in the water which could cause the digester to stop working (Orskov et al., 2014). However, for this to become normal practice, perceptions of the use of grey water need to change, as much of the rural population do not use grey water, believing it to be dirty or unfit for use (Ogunmokun et al., 2000).

Urine from humans and livestock can also be a valuable resource for the digester. A pour flush toilet can be directly connected to the input chamber of the digester (Ogunmokun et al., 2000) where all flush water and waste goes directly into the digester, reducing the amount of additional water required. An average human produces 1 dm^3 of urine per day; if piped water is available, flush water would provide an additional 3–5 dm^{-3} (Sibisi & Green, 2005). Urine has also been shown to improve biogas production when added with cow manure and water; this is due to the nitrogen-rich urine reducing the carbon: nitrogen ratio of the slurry, which also improves the quality organic fertilizer output from the digester (Haque & Haque, 2006). Cattle urine was observed to increase gas production by 30% at a proportion by volume of 50 cattle dung: 35 urine: 15 water (Haque & Haque, 2006). When human urine was used in equal proportions to cattle manure, with no additional water, Haque & Haque (2006) observed gas production to have increased by 14%. However, as household members are often away from the household during the day, human urine is not included as a source of water in this study.

Rainwater harvesting. In 2015, Rockström & Falkenmark called for increased water harvesting in Africa, emphasizing the challenge faced by SSA in meeting water requirements for food production. Extra water demand for energy production will exacerbate this situation, and the need for water harvesting becomes even more acute. According to Siegert (1994), rainwater harvesting includes 'all small-scale schemes for concentrating, storing, and collecting surface run-off water in different mediums, for domestic or agricultural use'. Lasage & Verburg (2015) classify rainwater harvesting techniques according to their size (household or community scale), and the way in which the water is stored (container, soil, or reservoir), which has implications for evaporation and the potential uses of the water. Rockström &

Falkenmark (2015) suggest that harvesting of water stored in soils, 'green water', is needed for food and biomass production, whereas water stored in containers or reservoirs, 'blue water', is needed for energy development. Here, we consider small-scale rainwater harvesting techniques that can provide blue water to the household by collecting rainwater run-off from rooftops or ground catchments in containers or reservoirs. For effective domestic rainwater harvesting, three factors should be considered: the storage facility (above- or below-ground tanks), catchment area (rooftop or courtyard), and the target use (domestic use and/or biogas) (Mwenge Kahinda et al., 2010).

Rooftop rainwater harvesting is a popular choice for rainwater collection. The volume of water collected is determined by the surface area and run-off coefficient of the roof. The run-off coefficient is defined as the proportion of rain falling on a surface that will run off into a collection vessel (Conway et al., 2009). An iron roof has a run-off coefficient of 0.8–0.9 (Sturm et al., 2009), which provides an ideal surface for rooftop rainwater harvesting. In SSA, many rural households have thatched roofs (Mwenge Kahinda et al., 2007), which have a run-off coefficient of only 0.2 (DTU, 2002). In East Africa, Pachpute et al. (2009) reported that the roof area commonly varies from 15 to 40 m^2, but in Ghana, Issaka et al. (2012) reported roof areas up to 108 m^2. To improve the efficiency of rainwater collection from rooftops, splashguards and gutters can also be added, increasing the run-off coefficient (Sturm et al., 2009). Mati et al. (2006) suggested that areas in SSA with an annual rainfall over 200 mm have potential for rainwater harvesting. If the roof area limits the amount of water collected, ground catchments can also be used.

Ground catchments allow a larger area to be used to collect the water required compared to rooftop rainwater harvesting, but may remove areas of the holding from alternative uses. Subsurface run-off can be captured from courtyards or compacted or treated surfaces with a sufficient run-off coefficient (Mwenge Kahinda et al., 2010). The run-off coefficient is higher in concrete lined catchments than in natural or treated surfaces (Sturm et al., 2009). Cement tanks are commonly used to capture rainwater from groundwater catchments as they prevent water loss; the size of these tanks typically ranges from 20 to 50 m^3 (Pachpute et al., 2009). Water collected from ground catchments is more likely to be contaminated than water collected from a rooftop (Mwenge Kahinda et al., 2007). Contaminated wastewater can be used to feed a digester, but higher levels of sand particles may mean that the digester must be cleaned out more frequently.

A problem with implementing rainwater harvesting in SSA is that rainfall is erratic and unevenly distributed (Vörösmarty *et al.*, 2005; Mwenge Kahinda *et al.*, 2007). Annual rainfall varies from 100 to 3000 mm with the highest potential for rainwater harvesting generally observed in central and western Africa (UNEP, 2010). Currently, rainwater harvesting is underutilised in SSA, and 95% of agriculture is directly fed by rainwater (Biazin *et al.*, 2012; Rockström & Falkenmark, 2015), so in areas that are already dependent on rainfall for their livelihoods, rainwater harvesting is an important potential option to capture more of the water required.

Aquaculture. In many parts of Asia, aquaculture and biogas digesters are commonly linked to produce an integrated farming system (Chan, 1993). The effluent from the digester is used to fertilize the pond, which lowers the oxygen content allowing algae to reproduce; the algae can then be used as a feed for the fish or can be used as an additional feedstock for the digester to increase biogas production (Chan, 1993). Aquaculture has the potential to provide the water required for the digester at the same time as contributing to the food security of the household. Fish consumption is lower in Africa than in other continents, with an average of only 9.1 kg of fish consumed per capita per annum (FAO, 2012), but countries on the western coast have higher rates of fish consumption than other countries in Africa; in Ghana, Gambia, and Sierra Leone, fish contribute 50% of the total animal protein consumed (FAO, 2012). Globally, only 0.15% of total fish production from aquaculture is in SSA (Hishamunda & Ridler, 2006); aquaculture is still in its infancy in SSA (Brummett & Williams, 2000) as it was only introduced in the 1950s (Hishamunda & Ridler, 2006) and is still subject to many social and political constraints. In SSA, 31% of the region would be suitable to produce tilapia, making tilapia an ideal species for African aquaculture (Kapetsky, 1994).

There are 100 different species of tilapia, but the most common species found in SSA are the Nile tilapia (*Oreochromis niloticus*) and the Mozambique tilapia (*Oreochromis mossambicus*) (Murnyak, 2010). They have a fast reproductive rate, grow quickly into adults (Murnyak, 2010), and are well adapted to sub-Saharan climates as they reproduce best at temperatures between 28 °C and 32 °C. They are fairly resistant to disease and can adapt to poor water quality with low oxygen concentrations (Boyd, 2004). Their main benefit to a rural household with a biogas digester is that they thrive on the plankton that grows on slurry produced by the digester (Orskov *et al.*, 2014). Tilapia are usually stocked in 1-m-deep earthen ponds (Murnyak, 2010); the ponds used are shallow because plankton require sunlight and carbon dioxide for photosynthesis (Chan, 1993).

Question being addressed

The aim of this study was to investigate the potential for domestic water recycling, rainwater harvesting, and aquaculture to meet the water demand of a small-scale biogas digester in rural households in different countries of SSA. The work will answer the questions:

- What proportion of the water demands of small-scale biogas digesters in rural households of sub-Saharan Africa can be met by domestic water recycling?
- Is it feasible to supply the remaining water requirement by rooftop rainwater harvesting? and
- Can aquaculture help to ensure a sufficient supply of water to run a biogas digester throughout the year?

Materials and methods

Summary of approach

The work described in this study uses a simple approach to estimate the amount of water that can be obtained for anaerobic digestion from domestic water recycling, rainwater harvesting, and aquaculture. The approach is detailed below, and a brief summary is provided here. The water required for anaerobic digestion in typical households in different countries was estimated from the national average household size. Water available for recycling to the biogas digester was estimated from the national statistics for domestic water use; the amount of extra water needed for digestion was then obtained from the difference between the water requirement and the amount that can be recycled. The time needed to collect this extra water by hand was estimated from the time required for each trip to collect water and the amount of water that can be carried in each trip; this provides an idea of the feasibility of collecting extra water without resorting to rainwater harvesting. The water that could be provided by rainwater harvesting was estimated for different roofing materials and areas of rooftop from national rainfall and potential evaporation data. The size of pond needed to stock the fish fed on nutrients from the bioslurry was estimated from the nitrogen contained in the bioslurry, and the nitrogen requirement and normal stocking density of the fish. The amount of water held by such a pond was then compared to the extra requirement of the digester to check the consistency of the water requirement and nutrient supply.

All symbol abbreviations used in this section are given in Table 1.

Water required for anaerobic digestion

Assuming that 200 dm^3 is required for anaerobic digestion of every 10 kg of manure dry matter (Orskov *et al.*, 2014), the amount of water needed to provide the maximum potential production of biogas can be estimated from the fresh weight of manure produced by livestock, W_F (kg day^{-1}), and the percentage

Table 1 Meaning of symbols used in equations

Symbol	Definition	Units
A_c	Average area of catchment	m^2
A_p	Area of pond	m^2
A_{c_aq}	Additional catchment area needed for to harvest sufficient water for anaerobic digestion if water is stored in an open pond used for aquaculture	m^2
A_{roof}	Area of iron roof needed for each person in the household to harvest enough water to run a biogas digester with no domestic water recycling	m^2 capita^{-1}
D	Average depth of pond	m
D_{fish}	Stocking density of tilapia	fish dm^{-3}
E	Annual evaporation	mm y^{-1}
H_s	Household size	capita
K	Dimensionless run-off coefficient	
n_{fish}	Number of fish produced	
$N_{bioslurry}$	Nitrogen provided by bioslurry	g y^{-1}
N_{req}	Nitrogen requirement per fish	g dm^{-3} y^{-1}
P_{DM}	Percentage dry matter in manure	%
R	Average annual rainfall	mm y^{-1}
S	Annual rainwater supply	dm^3 y^{-1}
t	Average time required for each trip to fetch water	min
T_e	Extra time required	min day^{-1}
V_a	Volume of additional water required	dm^3 day^{-1}
V_c	Volume of water required for cooking	dm^3 capita^{-1} day^{-1}
V_d	Volume of water required for drinking	dm^3 capita^{-1} day^{-1}
V_E	Annual evaporation	mm y^{-1}
V_h	Volume of water consumed per household	dm^3 day^{-1}
V_p	Volume of pond	dm^3
V_r	Volume of water that can be recycled	dm^3 day^{-1}
V_{req}	Annual water requirement	dm^3 y^{-1}
V_t	Volume of water collected at source per trip	dm^3
V_u	Volume of water usage per capita	dm^3 capita^{-1} day^{-1}
V_w	Volume of water required for optimum anaerobic digestion conditions	dm^3 day^{-1}
W_f	Fresh weight of manure	kg day^{-1}

dry matter in the dung, P_{DM} (%).The volume of water required, V_w (dm^3 day^{-1}), was estimated as follows:

$$V_w = W_F \times \left(\frac{P_{DM}}{100}\right) \times \left(\frac{200}{10}\right) \qquad (1)$$

The values for W_F and P_{DM} were taken from Omer & Fadalla (2003) and Taiganides (1978).

Orskov et al. (2014) suggested that the manure from two cows or eight pigs would provide sufficient biogas for a four-person household in rural SSA. Therefore, the volume of water needed to run digesters in households in different countries was calculated by multiplying the volume of water needed to run a digester using manure from two cows or eight pigs for a four-person household by the national average household size, H_s (capita)/4. This provides a comparative analysis of the potential of rural households in different countries to meet the water requirements to run a small-scale biogas digester.

Volume of domestic water that can be recycled

To calculate the potential volume of domestic water that can be recycled, the volume of water used each day by the household,

V_h (dm^3 day^{-1}), was first determined. The national statistics for domestic water use, V_u (dm^3 capita^{-1} day^{-1}), were collected from Dorling (2007; data from 1987 and 2003), and the household size in each country, H_s (capita), was obtained from the World Bank (2001–2009), allowing the consumption per household to be estimated as follows:

$$V_h = V_u \times H_s \qquad (2)$$

The amount of domestic water that can be recycled was estimated from water allocation to different activities. The water used for essential activities such as drinking, V_d (dm^3 capita^{-1} day^{-1}), and cooking, V_c (dm^3 capita^{-1} day^{-1}), was assumed to be unavailable for recycling. The volume of domestic water that can be recycled, V_r (dm^3 day^{-1}), was then calculated as follows:

$$V_r = V_h - ((V_d \times H_s) + (V_c \times H_s)) \qquad (3)$$

In a survey of domestic groundwater consumption in Kisumu, Kenya, Okotto et al. (2015) found that 11.7–17.6% of household water was used for drinking, and 25.5–27.5% was used for cooking. These ranges were used to set minimum and maximum values for V_d and V_c. Any country where the average household size could not be determined was omitted from the calculations. This approach assumes that any water that is not used in drinking or cooking could be used for anaerobic digestion; this will not be the case if excessive amounts of detergents are used for cleaning, so this provides an estimate of the maximum amount of water available by recycling.

Additional water required to run digester

The additional water required to run the digester was calculated by subtracting the volume of domestic water recycled (V_r) from the water required to run the biogas digester (V_w). This gives an estimate of the amount of water that must be provided by rainwater harvesting or ponds to run the digester, V_a (dm^3 day^{-1}),

$$V_a = V_w - V_r \qquad (4)$$

Time required to collect extra water

The extra time required to collect the additional water needed for anaerobic digestion, T_e (min day^{-1}), was calculated from the time required for each trip to fetch water, t (min), and the average number of additional trips needed (calculated as V_a/V_t where V_a is the volume of additional water required (dm^3 day^{-1}) and V_t is the volume of water collected per trip (dm^3)).

$$T_e = \left(\frac{V_a}{V_t}\right) \times t \qquad (5)$$

If water is collected by hand, the volume of water collected per trip was assumed to be 20 dm^3 after Orskov et al. (2014), and the mean time required to collect water (t) was assumed to be between 19 min for a centrally located water source and 104 min for a distant source after Rosen & Vincent (1999). This was also expressed as water collected per capita by dividing by the national average household size, H_s (capita). If farmers have access to additional means of transporting water, such as by donkey or using a vehicle, clearly the amount transported each trip (V_t) and the time required for each trip (t) will be different, very much reducing the time required to collect water.

Water provided by rainwater harvesting

Two types of rainwater harvesting are commonly used: rooftop and ground catchment. It was assumed that all the rainfall collected could be used to feed the digester. Potential rainwater harvesting, S (dm^3 y^{-1}), from a rooftop or ground catchment was calculated as follows:

$$S = R \times K \times A_c \qquad (6)$$

where R is the rainfall (mm y^{-1}), K is the run-off coefficient, and A_c is the area of the catchment (m^2).

Annual precipitation data were taken from the FAO AQUA-STAT database (http://www.fao.org/nr/water/aquastat/data/query/index.html?lang=en). A variety of different roof materials were considered, with run-off coefficients that ranged from 0.2 to 0.9 (DTU, 2002). The roof catchment area

was assumed to range from 15 to 40 m^2 after the range of roof sizes reported for East Africa by Pachpute et al. (2009). The annual rainwater supply was divided by 365.25 to give the average volume of water collected each day.

Sufficient water is collected for anaerobic digestion if the water supply, S (dm^3 y^{-1}), is at least equal to the annual water requirement, V_{req} (dm^3 y^{-1}), and the water lost each year by evaporation, V_E (dm^3 y^{-1}), from the holding tank.

$$S = V_{req} + V_E \qquad (7)$$

The annual water requirement, V_{req} (dm^3 y^{-1}), is calculated from the daily water requirement, V_a (dm^3 day^{-1}) as follows:

$$V_{req} = 365.25 \times V_a \qquad (8)$$

For an open tank, the volume of water lost by evaporation (V_E) is given by the annual evaporation, E (mm y^{-1}), and the area of the water holding tank, A_t (m^2), as follows:

$$V_E = E \times A_t \qquad (9)$$

Substituting V_{req} and V_E into equations 6 and 7 for S gives:

$$S = (365 \times V_a) + (E \times A_t) = R \times K \times A_c \qquad (10)$$

Rearranging equation 10 gives an equation for the area of catchment required:

$$A_c = \frac{(365 \times V_a) + (E \times A_t)}{(K \times R)} \qquad (11)$$

For a covered holding tank, the evaporation can be assumed to be low, so Equation 11 simplifies to:

$$A_c = \frac{(365 \times V_a)}{(K \times R)} \qquad (12)$$

For open holding tanks, the evaporation rates were obtained from FAO AQUASTAT (2005). The countries without evaporation values were omitted from the calculations.

Aquaculture

The amount of water required by aquaculture is dependent on the stocking density of the fish and the number of fish that can be supported by the nutrients contained in the bioslurry. Assuming that nitrogen is limiting the growth of the fish, the number of fish that can be produced by the bioslurry, n_{fish}, can be calculated from the available nitrogen provided by the bioslurry, $N_{bioslurry}$ (kg y^{-1}), and the nitrogen requirement of each fish, N_{req} (kg fish^{-1} y^{-1}), as follows:

$$n_{fish} = \frac{N_{bioslurry}}{N_{req}} \qquad (13)$$

The nitrogen requirement (N_{req}) was estimated from the minimum (0.001 kg) and the maximum weight (0.01 kg) of a fingerling and using the FAO values for the proportion of protein required with respect to the gain in body weight of fish (35–40%) (FAO, 2013b). The available nitrogen in the bioslurry ($N_{bioslurry}$) was calculated from the total nitrogen content of the feedstock, N_{feed} (kg y^{-1}), the percentage of total nitrogen lost during anaerobic digestion, P_{Nloss} (%), and the proportion of nitrogen that is available as ammonium, $p_{NH4:totalN}$, as described by Smith et al. (2013).

$$N_{\text{bioslurry}} = N_{\text{feed}} \times \left(1 - \frac{P_{\text{Nloss}}}{100}\right) \times p_{\text{NH4:totalN}} \qquad (14)$$

The value of P_{Nloss} was set to 5 and $p_{\text{NH4:totalN}}$ to 0.5 after Schievano et al. (2011). The total nitrogen content of the feedstock (N_{feed}) was calculated as follows:

$$N_{\text{feed}} = 365.25 \times \frac{P_{\text{used}}}{100} \times \frac{P_{\text{waste}}}{100} \times n_{\text{animal}} \times M_{\text{animal}} \times p_{\text{N:TS}} \quad (15)$$

where P_{used} is the percentage of the available waste of each type that is used in the digester (assumed to be 100%), P_{waste} is the wet waste produced per animal as a percentage of its live weight (kg fresh waste day^{-1} (100 kg live weight)$^{-1}$), n_{animal} is the number of each of the different types of animals on the farm, M_{animal} is the typical live weight for the type of animal specified (kg), and $p_{\text{N:TS}}$ is the proportion of nitrogen to total solids. The values used to calculate N_{feed} are given in Table 2. The value of N_{feed} was then calculated for each country by multiplying by H_s (capita)/4.

The area of the pond required to stock this number of fish, A_p (m^2), was then obtained by dividing the number of fish by the stocking density.

$$A_P = \frac{n_{\text{fish}}}{D_{\text{fish}}} \qquad (16)$$

The stocking density (D_{fish}) was taken from Yi et al. (2008) and ranged from 0.0005 to 0.003 fish dm^{-3} (note that stocking density is typically quoted as fish per area of pond surface; this was converted using an assumed 1 m depth of the pond).

The volume of water required to fill this size of pond, V_p (dm^3), was obtained from the area (A_p) and depth, d (m), assumed to be 1 m (Murnyak, 2010).

$$V_P = A_p \times d \times 1000 \qquad (17)$$

If annual rainfall exceeds evaporation ($R > E$), the pond will increase the amount of water available for digestion and aquaculture by $A_p(R-E)$. However, if evaporation exceeds rainfall, the additional catchment area needed to harvest sufficient water, A_{c_aq} (m^2), is given by:

$$A_{c\_aq} = \frac{A_p \times (E - R)}{(R \times K)} \qquad (18)$$

Note that in practice, the fish must be harvested at a time to allow the water in the pond to be utilized for the digester. This will require careful planning to synchronize rainfall, growth of fish, and the requirement of water for the digester .

Results

Water required for anaerobic digestion

For a household in SSA, using manure from two cows or eight pigs, assuming all the manure produced can be used in the digester, the volume of water required for anaerobic digestion is between 78 and 124 dm^3 day^{-1}, with a mean volume of 101 dm^3 day^{-1}. Assuming an ideal feedstock retention time of 40 days (Price & Cheremisinoff, 1981) and a digestate to gas ratio of 6 : 1 (Smith et al., 2013), this would require a digester tank of 5 ($\pm$1) m^3. This is equivalent to a water requirement for digestion for each person in the household of 25 ($\pm$6) dm^3 capita^{-1} day^{-1}, using a digester volume of 1.3 ($\pm$0.3) m^3 capita^{-1}. The typical values for water needed to run a digester in different countries of SSA obtained from the national average household size are shown in Fig. 1, ranging from a water requirement of 73 ($\pm$17) dm^3 day^{-1} with the national average household size of 2.9 capita in Cameroon, to 247 ($\pm$56) dm^3 day^{-1} with the national average household size of 9.8 capita in Senegal (Table 3).

Domestic water recycling

The potential volume of domestic water that can be recycled ranges from 21 to 411 dm^3 household^{-1} day^{-1} (Fig. 2a) and is highly dependent on country. The amount of domestic water that could potentially be recycled in 28% of countries considered (Guinea-Bissau, Zambia, Cote d-Ivoire, Cape Verde, Mauritania, Congo, Morocco South Africa and Gabon) exceeds the water requirement for the digester (Fig. 2b). However, in the

Table 2 Data used to calculate the nitrogen content of the feedstock available from two cows or eight pigs

Type of animal	Wet waste produced per animal as a percentage of its live weight, P_{waste} (kg fresh waste day^{-1} [100 kg live weight]$^{-1}$)	Number of animals, n_{animal}	Typical live weight per head, M_{animal} (kg)	Proportion of nitrogen to total solids, $p_{\text{N:TS}}$	Nitrogen content of the feedstock, N_{feed} (kg y^{-1})
Cows					
Minimum	4.6[a]	2	170[a]	0.0095[b]	54
Maximum			270[a]		86
Pigs					
Minimum	5.1[a]	8	45[b]	0.04[b]	268
Maximum					

[a]Omer & Fadalla (2003); [b]Polprasert (2007).

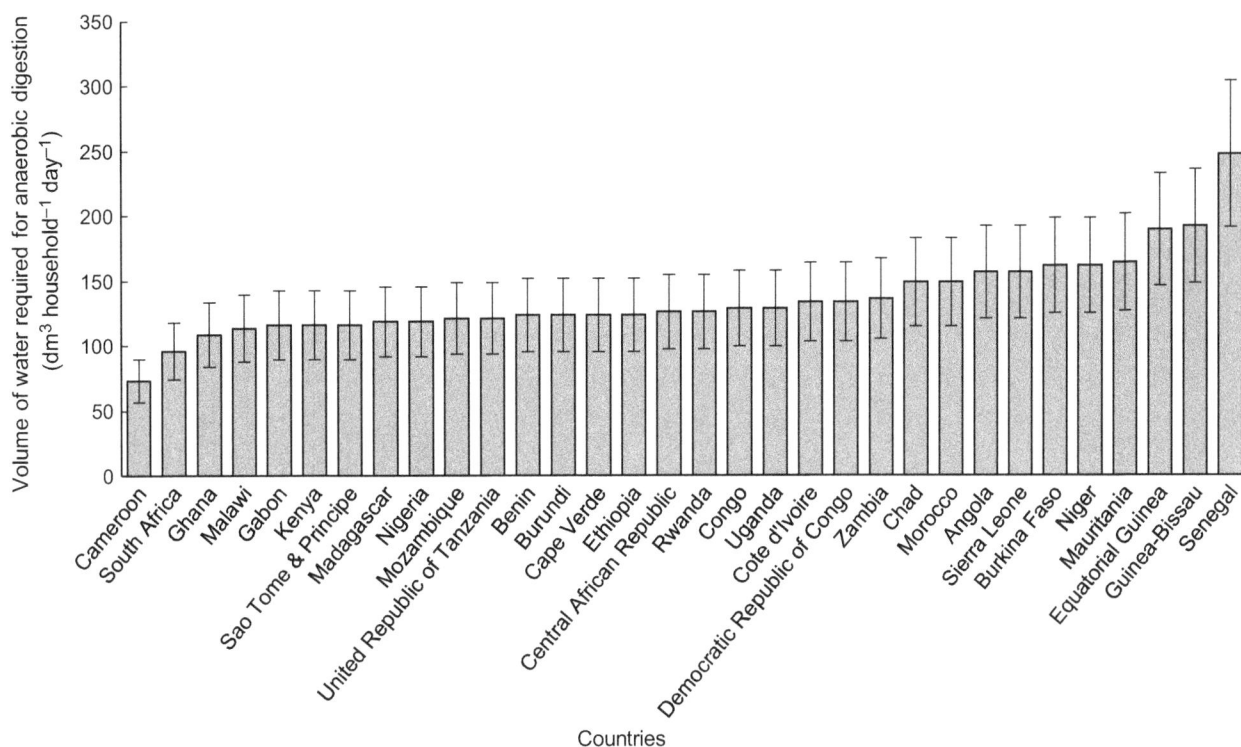

Fig. 1 Volume of water required to run the size of biogas digester needed to meet the energy demand of an average household, as per national average household statistics obtained from World Bank (2001–2009).

Table 3 Water required for anaerobic digestion of manure from two cows or eight pigs

Type of animal	[a]Wet weight of organic waste produced (% total live weight) day^{-1}	[b]Average weight of animal (kg)	Fresh weight of manure produced livestock per head, W_F (kg day^{-1})	Percentage dry matter in manure, P_{DM} (%)	Number of animals, n_{animal}	Volume of water required, V_W (dm^3 day^{-1})
Cows						
Minimum	4.6	170	7.8	25	2	78
Maximum	4.6	270	12.4	25	2	124
Pigs						
Minimum	5.1	45	2.3	25	8	92
Maximum	5.1	45	2.3	25	8	92

[a]Taiganides (1978); [b]Omer & Fadalla (2003).

remaining 72% of countries considered, recycling domestic water could only meet a proportion of the additional water required (average 44%), so installation of a biogas digester requires consideration of how the additional water needed will be accessed, either by collection from a local source or by rainwater harvesting.

Additional water requirement

The additional water required for anaerobic digestion after accounting for recycling domestic water is shown in Fig. 3. Of the countries needing extra water to run

the digester, the average additional water required is 70 ($\pm$23) dm^3 household^{-1} day^{-1}, but Senegal, with its large national average household size (9.8 capita^{-1}) and below average per capita water consumption (49% of the average of countries considered), requires over 136 ($\pm$49) dm^3 household^{-1} day^{-1}.

Time to collect additional water

The time taken to collect the additional water required for anaerobic digestion by hand from a distant water source averages 7 h household^{-1} day^{-1}, ranging from

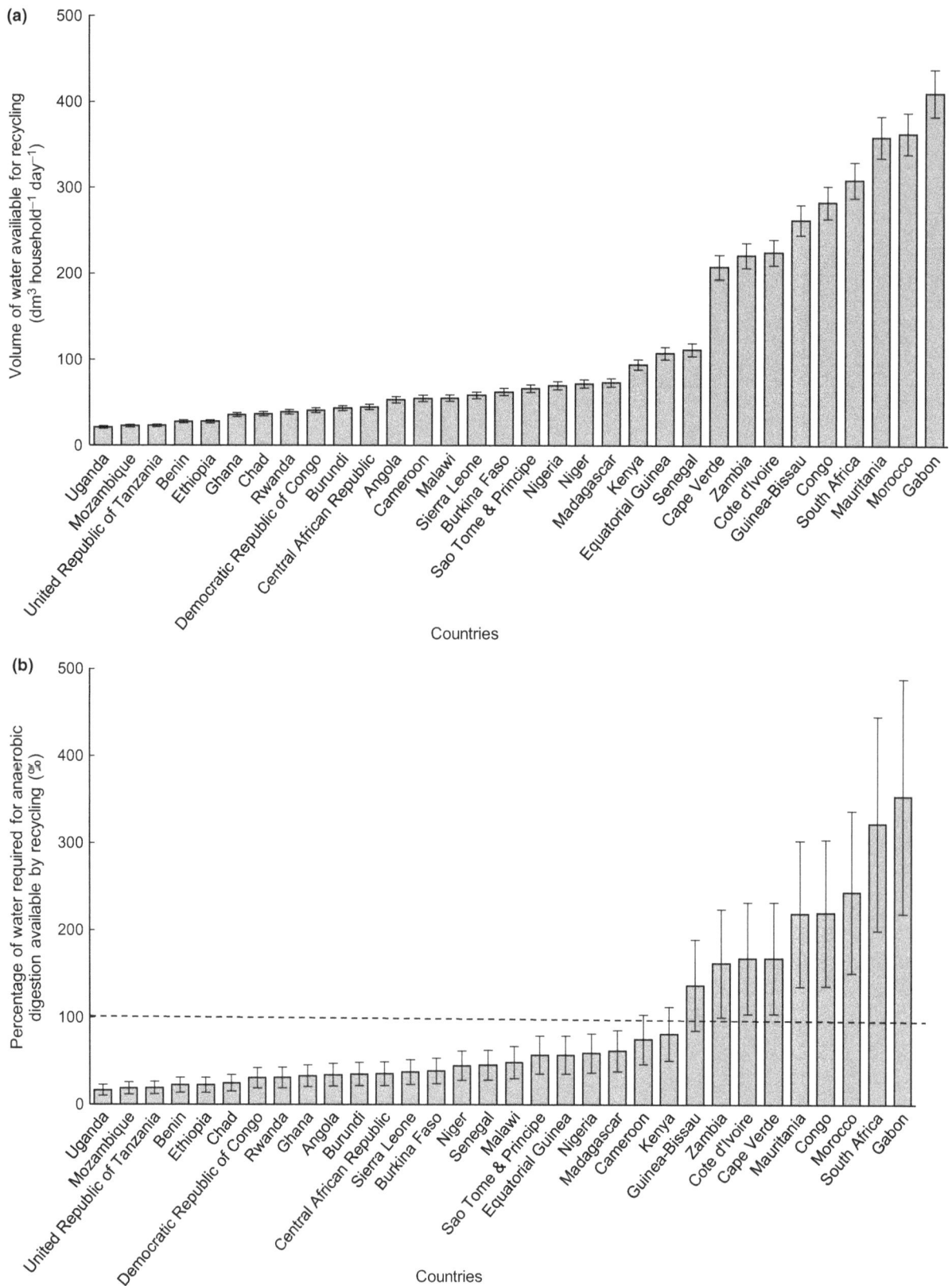

Fig. 2 Household water that could be recycled for anaerobic digestion in countries of Africa assuming domestic water use provided by Dorling (2007), household size specified by the World Bank (2001–2009), and the percentage of household water used for drinking and cooking as given by Okotto *et al.* (2015) (a) Volume of water; (b) percentage of water required for anaerobic digestion.

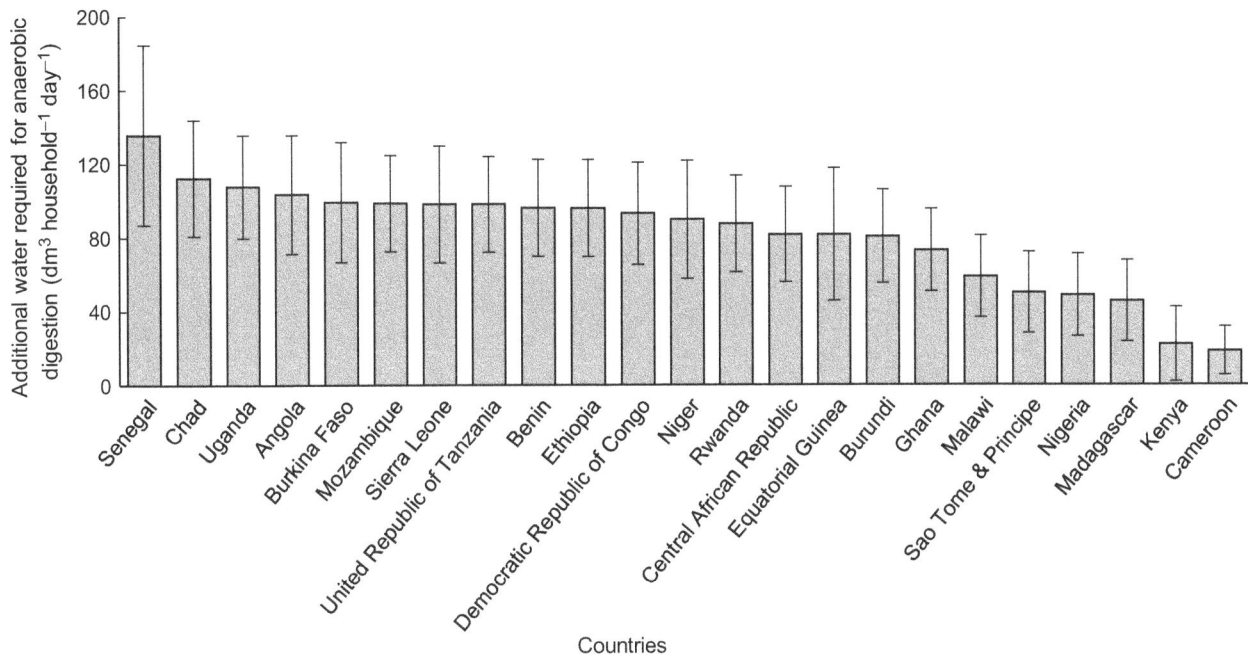

Fig. 3 Volume of additional water required to feed a biogas digester by country after accounting for recycled household water.

1.6 (± 0.4) h household^{-1} day^{-1} in Cameroon to 12 (± 2) h household^{-1} day^{-1} in Senegal (Fig. 4a). This assumes that the time for each trip to collect water from the distant water source is 104 min (Rosen & Vincent, 1999); if the time taken is less, then the time required to collect the water will be proportionally less. From a more central water source, where the time for each trip is assumed to be 19 min (Rosen & Vincent, 1999), the time required to collect the additional water by hand averages 1.3 h household^{-1} day^{-1}, ranging from 0.3 (± 0.2) h household^{-1} day^{-1} in Cameroon to 2 (± 0.8) h household^{-1} day^{-1} in Senegal. If this is re-expressed as time required for each person, this ranges from 5 (± 4) min capita^{-1} day^{-1} for a central source in Kenya to 110 (± 10) min capita^{-1} day^{-1} for a distant source in Uganda (Fig. 4b). Collection of additional water takes less than 30 min capita^{-1} day^{-1} in all countries for a central source, but in only 33% of countries if the source is distant. If no water is available from domestic water recycling, time taken to collect the additional water is even higher, between 24 (± 5) min capita^{-1} for a central source and 131 (± 30) min capita^{-1} for a distant source. Therefore, rainwater harvesting would appear to be an important adjunct to a biogas digester in most countries of SSA.

Rainwater harvesting

The rainfall collected varies greatly across different countries and different roofing materials (Fig. 5). Thatched roofs, with a run-off coefficient of 0.2, have the lowest potential for rainwater collection, while galvanized iron roofs, with a run-off coefficient of 0.9, have much higher potential. Tiled and asbestos roofs, with run-off coefficients of 0.6 and 0.8, respectively, have potential for rainwater collection in between these extremes. Figure 5 shows the potential for rainwater collection, assuming a 28 m^2 roof, the average of 15 and 40 m^2, which is the range of roof sizes reported in the literature for East Africa by Pachpute *et al.* (2009). The highest potential for rainwater harvesting is seen in Sierra Leone and Liberia, collecting over 150 (± 35) dm^3 day^{-1} with a 28 m^2 galvanized iron roof, and the lowest potential of under 25 (± 6) dm^3 day^{-1} being observed in Niger, Cape Verde, Mali, Somalia, and Chad.

Figure 6 shows the size of roof that would be required to meet the additional water requirement of the biogas digester, assuming no evaporative losses from the storage tank and all water that is not used for cooking or drinking is recycled. In all countries except Cameroon and Malawi, the area of a thatched roof required to harvest the extra water for anaerobic digestion would be outside the range of values reported in the literature (15–40 m^2 – Pachpute *et al.*, 2009). For an iron roof, the area of roof is significantly lower, and 50% of countries considered are able to harvest sufficient water with an iron roof of 40 m^2 of less. This is consistent with observations in the field; householders use iron roofs for rainwater harvesting rather than thatched roofs. A larger roof will be required if significant evaporation occurs of if less water can be recycled.

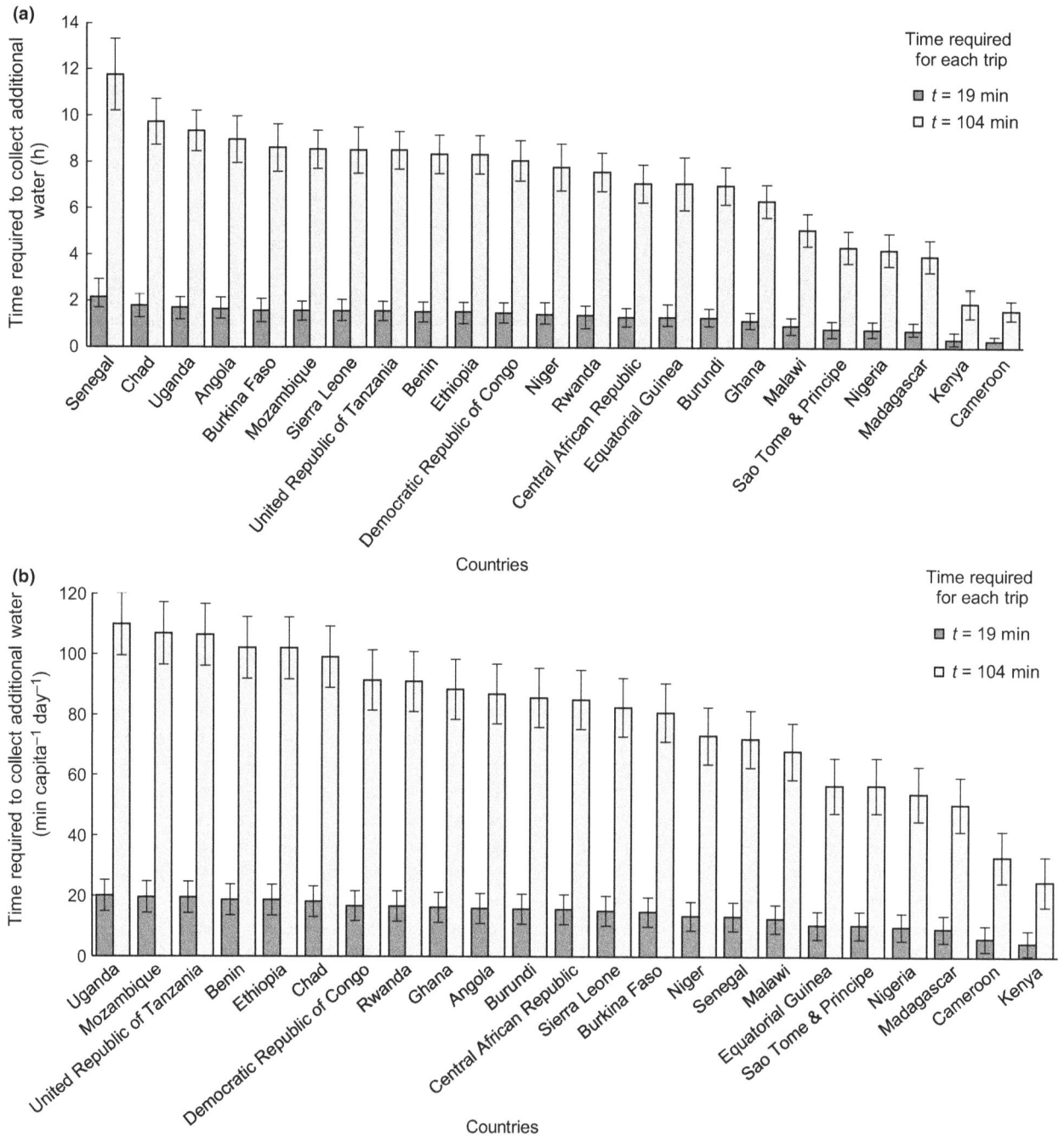

Fig. 4 Time required to collect additional water needed for anaerobic digestion. The volume of water collected per trip was assumed to be 20 dm³ after Orskov *et al.* (2014), and the mean time required to collect water was assumed to be between 19 min for a centrally located water source and 104 min for a distant source after Rosen & Vincent (1999) (a) time taken per household; (b) time taken per capita.

For the remaining countries, even using an iron roof and including domestic water recycling, an additional area is required for rainwater harvesting. If no domestic water can be recycled, the area of iron roof needed to harvest the additional water for the biogas digester with an annual rainfall of 1000 mm y⁻¹ is 10 (±2) m² capita⁻¹. This translates into a general equation for the area of iron roof

needed for rainwater harvesting if no domestic water is recycled.

$$A_{\text{roof}} = \frac{R}{100} \qquad (19)$$

where A_{roof} is the area of roof needed for each person in the household (m² capita⁻¹) and R is the annual rainfall (mm y⁻¹).

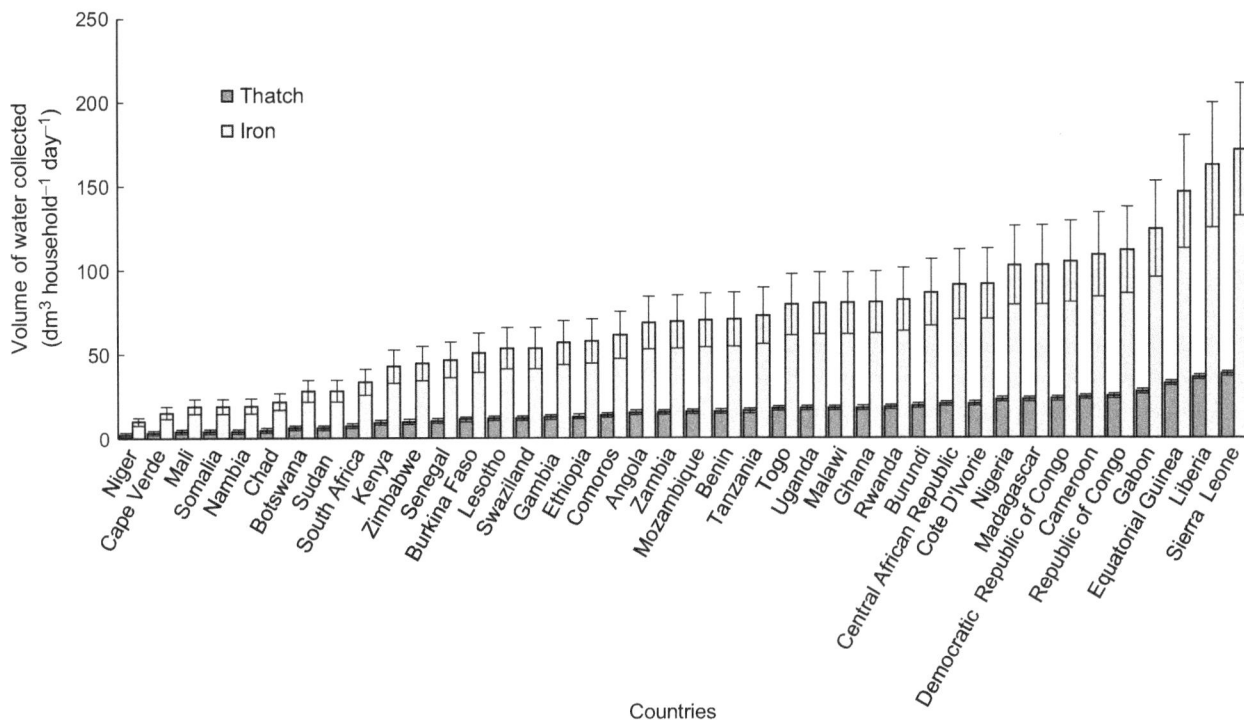

Fig. 5 Volume of water collected by rooftop rainwater harvesting. Water collection shown for a 28 m² roof, and the average of values reported in the literature of 15–40 m² (Pachpute *et al.*, 2009). Error bars show the rainwater collected at these two extremes.

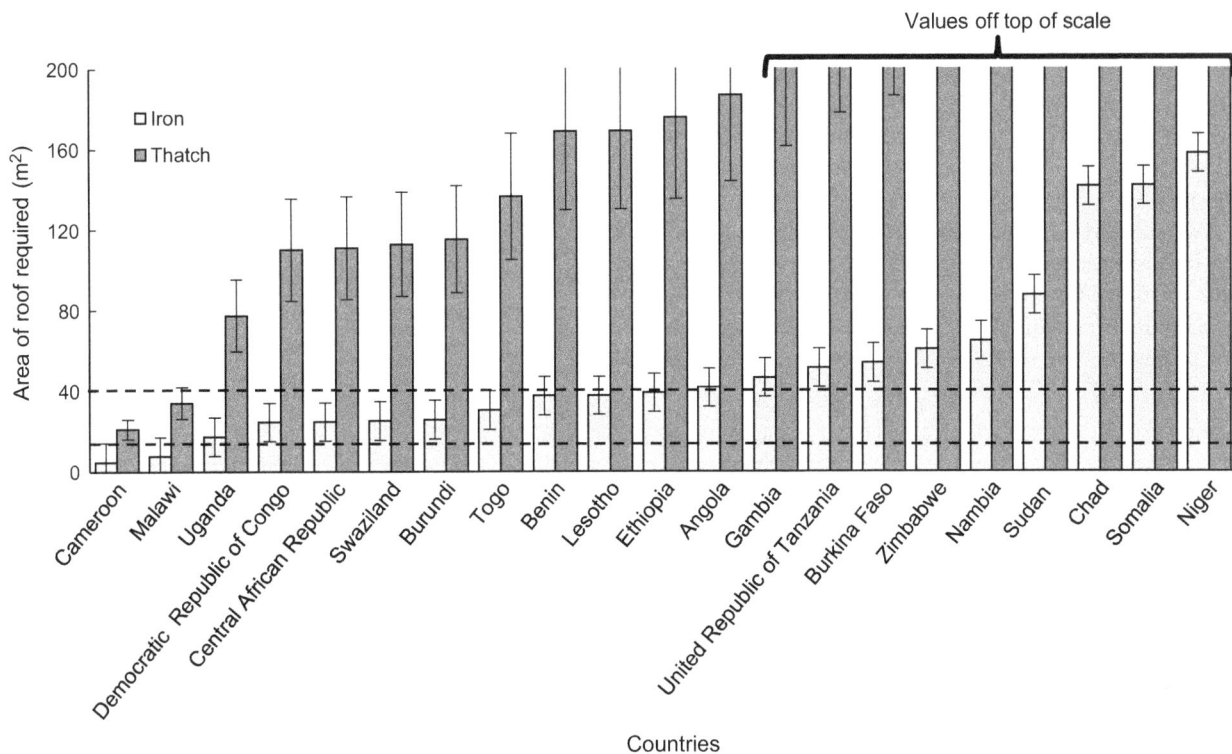

Fig. 6 Area of roof required to collect the volume of additional water needed to feed a biogas digester. Dotted lines show the normal range of roof areas reported in the literature; 15 and 40 m² (Pachpute *et al.*, 2009).

Aquaculture

Between 18 and 32 tilapia could be raised from the nitrogen in the bioslurry produced by a household with two cows. If, however, the household produces its biogas from nitrogen-rich pig manure (eight pigs), a much higher number of tilapia, between 87 and 100, can be produced (Table 4). Scaling this by $H_s/4$ gives an average over the countries considered of 117–135 tilapia. Assuming a fingerling stocking density of 0.5 to 3 fish m^{-2} (Yi *et al.*, 2008), the size of pond needed for the fish fed on the bioslurry from the pigs would be 29 to 200 m^2, equivalent to a square pond with sides 5.4 to 14.1 m. A 1-m-deep pond of this size would hold (2.9×10^4) to (2.0×10^5) dm^3 of water, providing 79–548 dm^3 day^{-1}. The distribution of pond sizes needed across countries for both pig and cow manure is given in Fig. 7.

The maximum amount of additional water required for anaerobic digestion is 136 dm^3 day^{-1} (Fig. 3). This is equivalent to an annual requirement of 5.0×10^4 dm^3. A 1-m-deep fish pond of area 50 m^2 could supply this amount of water (equivalent to a square pond with 7 m sides). This would allow a stocking density of 1.8 to 2.0 fish m^{-2} with the nitrogen available in the bioslurry produced by eight pigs, which is within the range of stocking densities given by Yi *et al.* (2008). Therefore, the size of pond needed to hold the additional water for anaerobic digestion, the stocking density of fish, and the nitrogen provided by anaerobic digestion for pig slurry are all compatible. By contrast, for the bioslurry produced by cow manure, the stocking density of 0.4 fish m^{-2} is less than the normal range, suggesting that an aquaculture/biogas digester system less viable using cow than pig manure.

As discussed above, the water required to fill the pond can be collected from rainwater harvesting from an iron roof or from an impermeable surfaced groundwater catchment surrounding the pond. However, because the pond must be uncovered to allow growth of the algae used to feed the fish, the evaporative losses from the pond may further increase the amount of water that must be collected. If annual rainfall exceeds evaporation ($R > E$), the pond will increase the amount of water available for digestion and aquaculture by $A_p(R-E)$. Assuming the average stocking density, the size of pond that can be stocked by the bioslurry produced for each household member is 23 (± 2) m^2 capita^{-1}, meaning that extra water of 3.7 (± 0.25) dm^3 day^{-1} will be provided to the digester for each 100 mm y^{-1} of hydrologically effective rainfall ($R-E$). However, if evaporation exceeds rainfall, additional catchment area is needed for rainwater harvesting. If $(R-E) = 100$ mm y^{-1} and $R = 1000$ mm y^{-1}, this comes to 1.5 (± 0.1) m^2 capita^{-1}. This can be generalized to give the extra catchment area needed, A_{c_aq} (m^2), from the household size, H_s (capita), the rainfall and evaporation as follows:

$$A_{c\_aq} = 1.5 \times H_s \times \left(\frac{(E-R)}{R} \right) \qquad (20)$$

The evaporation data available from FAO (2013a) suggest that in many cases, the potential evaporation exceeds rainfall, so water for aquaculture would need to be harvested from rooftops and a surrounding catchment.

Discussion

Water required for anaerobic digestion

The water required for anaerobic digestion was estimated from the manure provided by two cows or eight pigs (Orskov *et al.*, 2014), proportioned according to the size of household. If a different amount of biogas is needed per capita, the volume of water required would also be different; the volume required is approximately 50 dm^3 day^{-1} for each cow and 10 dm^3 day^{-1} for each pig providing manure to the digester. The uncertainty in this estimate is associated with the variation in the amount of manure produced by each animal and the

Table 4 Number of fish produced by nitrogen in bioslurry from two cows or eight pigs

Type of animal	Nitrogen content of the feedstock, N_{feed} (kg y^{-1})	Available nitrogen in the bioslurry, $N_{bioslurry}$ (kg y^{-1})	Number of fish (minimum) (N_{req} = 1.46 kg fish^{-1} y^{-1})[a]	Number of fish (maximum) (N_{req} = 1.28 kg fish^{-1})[a]
Cows				
Minimum	54	26	18	20
Maximum	86	41	28	32
Pigs				
Minimum	268	127	87	100
Maximum				

[a]FAO (2013b).

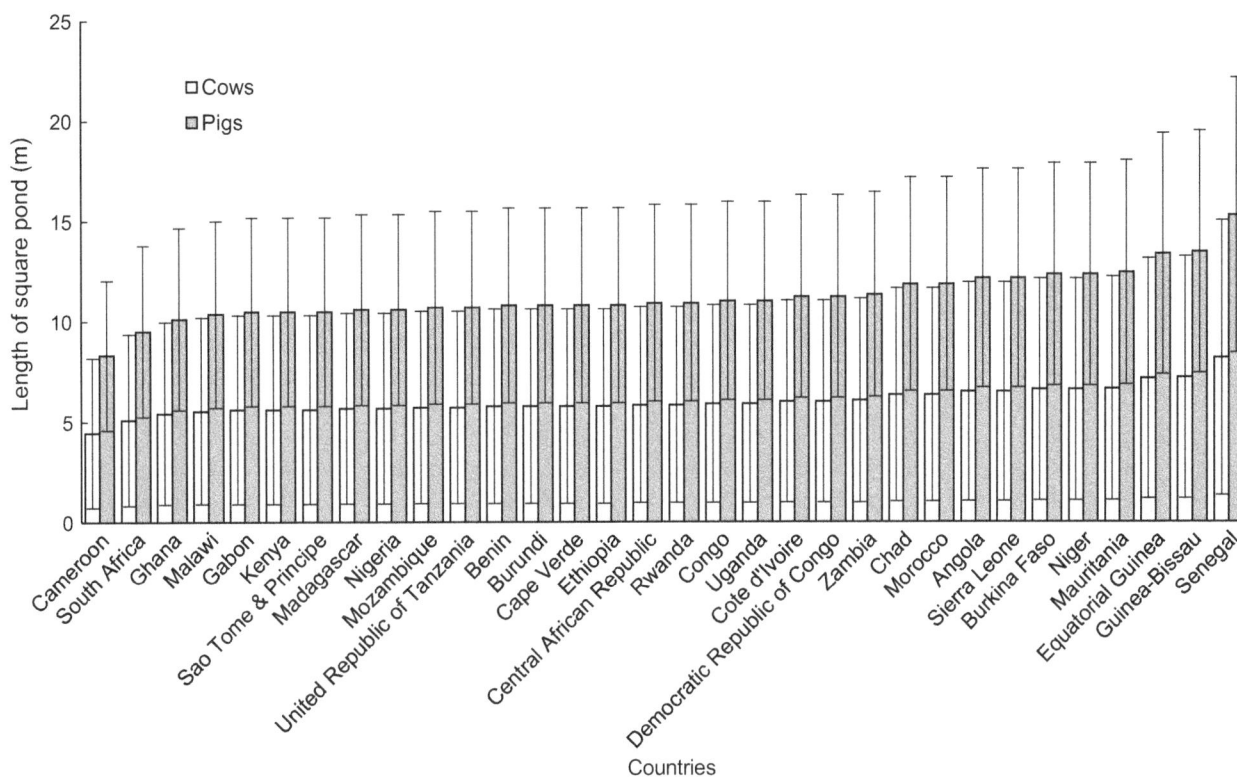

Fig. 7 The size of a square pond needed to stock the tilapia that could be raised on the household production of bioslurry. Stocking densities assumed to be 0.5 to 3 fish m^{-2} after Yi *et al.* (2008).

percentage dry matter in the manure, and is dependent on the species, breed, and diet of the livestock. For cows in Sudan, the uncertainty calculated using data provided by Omer & Fadalla (2003) was 23%; a similar level of uncertainty might be expected for pigs.

In these calculations, we estimated water requirement using the potential household requirement for biogas, rather than the potential for biogas production. In practice, the availability of feedstock to the household usually limits the biogas produced. The national potential for biogas production could be estimated using national livestock statistics, as done by Subedi *et al.* (2014). However, this potential for biogas production is unevenly distributed between households; wealthier households are likely to have more livestock and so a surfeit potential to produce biogas, while biogas production in poorer households tends to be more limited. The estimates of water required for anaerobic digestion given here represent the water requirement of households with access to sufficient feedstock to meet all of their energy needs.

Domestic water recycling

The amount of domestic water that can be recycled was calculated from national statistics, partitioned into dif-

ferent activities using data from a survey in Kisumu, Kenya. Within country and within year variation is likely to occur, depending on the accessibility of water to households in different regions and at different times of year. Furthermore, cultural differences may change the distribution of water use for different activities. For a more detailed analysis of the potential for domestic water recycling in a particular location, local surveys of water use should be done.

Domestic water use is dependent on the distance and quality of the water source, with per capita water consumption increasing if the water source is close to the household (Sugita, 2006). More economically developed countries are more likely to have water piped into households or to centrally located sources. Household in these countries are likely to be able to more easily meet the water requirements of the digester by collecting water by hand. In less economically developed countries, with less opportunity for piped water or centrally located water sources, rainwater harvesting becomes more important to the success of the digester.

Despite domestic water being available for recycling, it is often not used in the digester due to cultural perceptions that using wastewater could spread diseases (Ogunmokun *et al.*, 2000). Therefore, a successful biogas programme should include an education pro-gramme to encourage safe reuse of household wastew-ater. The use of urine from humans was not considered here. Humans can produce up to 1 dm^3 in urine every day (Sibisi & Green, 2005), which provides additional water to the digester. This was not included because it is not likely that everyone is in the house all day.

Collection of additional water required to run digester

In the majority of the countries in SSA, recycling domestic water is insufficient for anaerobic digestion and additional water collection is needed. Therefore, before a biogas digester is installed, potential sources of additional water should be surveyed, and the feasi-bility of different methods of water collection consid-ered. For many households, without access to transport for collecting water, collecting the extra water by hand is not an attractive option as it is likely to be too time-consuming. The calculations using national data suggest that, with a distant water source, only 33% of the countries considered would require less than 30 min capita^{-1} day^{-1} to collect the additional water needed. While there will be local and seasonal variation around this national norm, this result suggests that it is important to consider rainwa-ter harvesting as an adjunct to installation of a biogas digester in most conditions in SSA. Without associ-ated rainwater harvesting, in many areas of SSA, bio-gas digesters are unlikely to be successful in providing a long-term, sustainable, and widely appli-cable source of household energy; only households with a very local and reliable water source will be able to use biogas.

Rainwater harvesting

Thatched roofs are commonly used in SSA, but are not well suited to rainwater harvesting. Thatched roofs can be improved to collect water more efficiently by using polythene sheeting or by folding the roof to increase the surface area and so collect more water. However, erect-ing a roof with these features would require more labour (DTU, 2002). Alternatively, asbestos roofing could be used as it has a high run-off coefficient, but this is not recommended as the particles released from asbestos can be related to breathing problems (Worm & van Hattum, 2006). Corrugated iron and tile roofs are a more viable option, with a high run-off coefficient and producing good quality water that can also be used for human consumption. Gutters, splash guards, and pipes can also be installed on the roof edges to increase cap-ture of water, leading water straight into the inlet pipe

of the digester to avoid evaporative losses (Sturm et al., 2009).

The size of roof is critical when looking at the poten-tial for rooftop rainwater harvesting. In east Africa, it is common to find roof sizes from 15 to 40 m^2 (Pachpute et al., 2009), but even using an iron roof, in 50% of coun-tries a roof size larger than 40 m^2 would be needed to provide the water required. This could be provided by a ground catchment or open pond. The results pre-sented here are based on national data for annual rain-fall; clearly within country rainfall distribution will dictate the amount of water that can actually be har-vested in a particular household. Seasonality also impacts the amount of water that can be harvested; if a rainfall event is particularly heavy, it may be difficult to capture all of the run-off occurring in a very short per-iod of time.

Aquaculture

Aquaculture has great potential to ensure a sufficient supply of water to run a biogas digester throughout the year. In practice, some bioslurry could be used in aqua-culture, and the remainder used to fertilize crops, so the size of pond can be chosen to meet the preferences of the household. Aquaculture ponds require regular drai-nage to prevent accumulation of solids on the bottom of the pond (Boyd, 2004). This could be done by draining and refilling of the pond throughout the year. The fish-pond could be partially drained and resupplied with rainfall run-off to ensure good water quality as well as providing additional water and organic wastes to the digester (Boyd, 2004). The tilapia could be partially har-vested every few months (Murnyak, 2010), so allowing a smaller pond to be used to produce the same number of fish each year. In areas where rainfall is low and seasonal, the fishpond could be stocked with fish only in the rainy season, allowing the pond to be covered during the dry period to prevent evaporative losses (Murnyak, 2010). Seasonal use of the biogas digester might also provide a more feasible solution to house-hold energy needs, with biogas being used in the rainy season when biomass sources are wet and water is more plentiful, and biomass fuels being used in the dry sea-son when biomass is dry and easy to burn and water is scarce. The financial viability of seasonal use of biogas digesters needs further consideration; if the digester is only use for 50% of the year, then the payback period for the digester will be doubled.

Constructing the pond requires an initial input of labour, but could also provide significant advantages to the household. As well as storing water for use in the digester, aquaculture can provide an important source of income. The average market value of Nile tilapia in

SSA in 2001 was $1.27 kg^{-1} (Josupeit, 2005). The FAO Aquaculture Feed and Fertiliser Resources Information System (2013) suggests that tilapia yields of 3000 kg ha^{-1} can be sustained in a well fertilized pond. Therefore, from the 29 to 200 m^2 pond that could be fertilized using the N available in bioslurry from pigs (Table 4), the yield of tilapia would be 9–60 kg y^{-1}, with a 2001 market value of $11–$76. The annual value of the tilapia produced using the bioslurry from cows is estimated to be only $1 to $3, meaning aquaculture using the bioslurry from cows is not likely to be viable, unless the cows are fed on an unusually nitrogen-rich diet.

Recommendations

- Use of household wastewater is important to the success of the biogas digester, but is sometimes not done because it is culturally unacceptable. Therefore, an education programme should be included alongside installation of the biogas digester to encourage efficient reuse of household wastewater.

- In the majority of the countries, recycling domestic water could only meet a proportion of the additional water required for anaerobic digestion. Therefore, before the installation of the digester, methods that will be used to collect the additional water needed for the digester should be considered.

- Collection of the water needed can take a significant amount of time. Therefore, before installation of the digester, the time spent doing different activities should be budgeted to ensure that the total time spent on household activities does not significantly increase with the installation of the digester.

- In most countries, rainwater harvesting on a thatched roof cannot provide sufficient water for the digester. Therefore, if possible roofs with a higher rainwater coefficient should be used to harvest rainwater, such as iron or tile roofs.

- In 50% of countries of SSA, even an iron roof cannot harvest sufficient water for the digester. Therefore, an open pond or ground catchment should be used to collect additional water.

- If there is a local market for fish, bioslurry from pigs could be used to grow plankton to feed fish in the pond, designing the management of the pond to match local rainfall conditions.

- In countries with very low and highly seasonal rainfall, consideration should be given to the potential for limiting the use of biogas to the rainy season when water is more plentiful and the alternative biomass sources of household energy are wet and difficult to burn.

Acknowledgements

This work was part-funded by the UK Natural Environment Research Council funded ESPA project, NE/K010441/1 'ALTER – Alternative Carbon Investments in Ecosystems for Poverty Alleviation'. We are also grateful to the AUC for funding part of this work under the Afri-Flame project on 'Adaptation of small-scale biogas digesters for use in rural households in sub-Saharan Africa'.

References

Africa Renewable Energy Access Program (2011) *Energy Development for Sub-Saharan Africa: Issues and Approaches.* World Bank Group, Washington, DC, USA.

Amigun B, Von Blottnitz H (2010) Capacity-cost and location-cost analyses for biogas plants in Africa. *Resources, Conservation and Recycling*, **55**, 63–73.

Arthur R, Baidoo MF, Antwi E (2011) Biogas as a potential renewable energy source: a Ghanaian case study. *Renewable Energy*, **36**, 1510–1516.

Avery LM, Yongabi K, Tumwesige V, Strachan N, Goude PJ (2014) Potential for Pathogen reduction in anaerobic digestion and biogas generation in sub-Saharan Africa. *Biomass and Bioenergy*, **70**, 112–124.

Bedi AS, Pellegrini L, Tasciotti L (2015) The effects of Rwanda's biogas program on energy expenditure and fuel use. *World Development*, **67**, 461–474.

Bensah E, Brew-Hammond A (2010) Biogas technology dissemination in Ghana: history, current status, future prospects, and policy significance. *International Journal of Energy and Environment*, **1**, 277–294.

Biazin B, Sterk G, Temesgen M, Abdulkedir A, Stroosnijder L (2012) Rainwater harvesting and management in rainfed agricultural systems in sub-Saharan Africa – a review. *Physics and Chemistry of the Earth, Parts A/B/C*, **47**, 139–151.

Biran A, Abbot J, Mace R (2004) Families and firewood: a comparative analysis of the costs and benefits of children in firewood collection and use in two rural communities in sub-Saharan Africa. *Human Ecology*, **32**, 1–25.

Blackden CM, Wodon Q (2006) *Gender, Time Use, and Poverty in Sub-Saharan Africa*, World Bank Working Paper No. 73, pp. 1–145. World Bank, Washington, DC, USA. Available at: https://books.google.co.uk/books?hl=en&lr=&id=BA6IBjdPkcUC&oi=fnd&pg=PR3&dq=Blackden+CM,+Wodon+Q+%28eds%29+%282006%29+Gender,+time+use,+and+poverty+in+sub-Saharan+&ots=Pt-sOXBlGh&sig=wcgglanogf2rvyUCvOKTKoXRel0#v=onepage&q&f=false (accessed 21 September 2015).

Boyd CE (2004) *Farm-Level Issues in Aquaculture Certification: Tilapia*, Report Commissioned by WWF-US, pp. 1–29. WWF, Auburn, AL, USA. Available at: https://www.extension.org/mediawiki/files/e/e8/Farm-level_issues_in_Aquaculture_Certification,_Tilapia.pdf (accessed 21 September 2015).

Brown VJ (2006) Biogas: a bright idea for Africa. *Environmental Health Perspectives*, **114**, A300.

Brummett RE, Williams MJ (2000) The evolution of aquaculture in African rural and economic development. *Ecological Economics*, **33**, 193–203.

Bryceson DF, Howe J (1993) Rural household transport in Africa: reducing the burden on women? *World Development*, **21**, 1715–1728.

Chan GL (1993) Aquaculture, ecological engineering: lessons from China. *Ambio*, **22**, 491.

Conway D, Persechino A, Ardoin-Bardin S, Hamandawana H, Dieulin C, Mahé G (2009) Rainfall and water resources variability in sub-Saharan Africa during the twentieth century. *Journal of Hydrometeorology*, **10**, 41–59.

Conway D, Archer van Garderen E, Deryng D et al. (2015) Climate and southern Africa's water–energy–food nexus. *Nature Climate Change*, **5**, 837–846.

Dong L (2012) The progress of biomass energy and biogas in China, 19th Scientific Energy Management and Innovation Seminar. Available at: http://cwsolutions.hu/wp-content/uploads/2012/05/Draft-of-performance-GIEC-LiDong.pdf (accessed 12 January 2013).

Dorling D (2007) *Worldmapper Dataset 324: Domestic Water Use.* University of Sheffield. Available at: http://www.worldmapper.org/display.php?selected=324# (accessed 21 September 2015).

DTU (2002) Very-low-cost domestic roofwater harvesting in the humid tropics: existing practice, Roofwater Harvesting for Poorer Households in the Tropics – Report R1. DFID-KAR Contract R7833. Development Technology Unit, School of Engineering, University of Warwick. Available at: http://www2.warwick.ac.uk/fac/sci/eng/research/dtu/pubs/reviewed/wh/dfid/r1.pdf (accessed 21 September 2015).

Eleri A, Eleri EO (2009) Prospects for Africa – rethinking biomass energy in sub-Saharan Africa. In: *Prospects for Africa – Europe Policies*, VENRO (Association of

German Development NGOs), pp. 1–20. German NGO Forum on Environment and Development and ICEED (International Centre for Energy, Environment and Development), Bonn.

Eshete G, Sonder K, ter Heegde F (2006) Report on the feasibility study of a national programme for domestic biogas in Ethiopia. *Ethiopia: SNV.* Available at: http://www.susana.org/en/resources/library/details/491 (accessed 21 September 2015).

FAO (2012) Fisheries Department, the State of World Fisheries and Aquaculture Part 1, FAO. Available at: http://www.fao.org/docrep/016/i2727e/i2727e00.htm (accessed 21 September 2015).

FAO (2013a) AQUASTAT database, Food and Agriculture Organization of the United Nations (FAO). Available at: http://www.fao.org/nr/water/aquastat/data/query/index.html?lang=en (accessed 21 September 2015)

FAO (2013b) Aquaculture Feed and Fertiliser Resources Information System. Available at: http://www.fao.org/fishery/affris/species-profiles/nile-tilapia/nilc tila pia-home/en/ (accessed 21 September 2015)

FAO AQUASTAT (2005) Irrigation in Africa in figures. In: *AQUASTAT Survey 2005* (ed. Frenken K). FAO Land and Water Development Division, Rome. *Water Report,* Vol. **29**. Available at: http://www.fao.org/nr/water/aquastat/data/query/index.html?lang=en (accessed 21 September 2015).

Gleick PH (1998) The human right to water. *Water Policy,* **1**, 487–503.

Global Alliance for Clean Cookstoves (2014) Available at: http://www.cleancookstoves.org/ (accessed 1 September 2014).

Gordon SB, Bruce NG, Grigg J et al. (2014) Respiratory risks from household air pollution in low and middle income countries. *The Lancet Respiratory Medicine,* **2**, 823–860.

Hallett PD, Loades KW, Krümmelbein J (2012) Soil physical degradation: threats and opportunities to food security. In: *Soils and Food Security, Issues in Environmental Science and Technology* (eds Hester RE, Harrison RM), pp. 198–226. Royal Society of Chemistry, London.

Hamlin A (2012) Assessment of social and economic impacts of biogas digesters in rural Kenya, Independent Study Project (ISP) Collection, Paper 1247. Available at: http://digitalcollections.sit.edu/isp_collection/1247 (accessed 21 September 2015).

Haque MS, Haque NN (2006) Studies on the effect of urine on biogas production. *Bangladesh Journal of Scientific and Industrial Research,* **41**, 23–32.

Hishamunda N, Ridler NB (2006) Farming fish for profits: a small step towards food security in sub-Saharan Africa. *Food Policy,* **31**, 401–414.

Issaka Z, Mensah E, Agyare WA, Ofori E (2012) Appropriate rainwater harvesting storage capacity for households: a case study of central Gonja district. *World Rural Observations,* **4**, 57–63.

Josupeit H (2005) World market of tilapia. In: *Globefish Research Programme,* Vol. **79**, 28 pp. FAO, Rome.

Kapetsky JM (1994) A strategic assessment of warm-water fish farming potential in Africa, CIFA Technical Paper, No. 27, 67 pp. FAO, Rome.

Karekezi S, Kithyoma W (2002) Renewable energy strategies for rural Africa: is a PV-led renewable energy strategy the right approach for providing modern energy to the rural poor of sub-Saharan Africa? *Energy Policy,* **30**, 1071–1086.

Lasage R, Verburg PH (2015) Evaluation of small scale water harvesting techniques for semi-arid environments. *Journal of Arid Environments,* **118**, 48–57.

Leu JM, Traore S, Wang Y-M, Kan C-E (2010) The effect of organic matter amendment on soil water holding capacity change for irrigation water saving: case study in Sahelian environment of Africa. *Science Research Essays,* **5**, 3564–3571.

Lim SS, Vos T, Flaxman AD et al. (2010) A comparative risk assessment of burden of disease and injury attributable to 67 risk factors and risk factor clusters in 21 regions, 1990–2010: a systematic analysis for the Global Burden of Disease Study 2012. *Lancet,* **380**, 2224–2260.

Lockwood H, Smits S (2011) *Supporting Rural Water Supply: Moving Towards a Service Delivery Approach.* Practical Action Publishing, Rugby, UK.

MacCarty N, Ogle D, Still D, Bond T, Roden C (2008) A laboratory comparison of the global warming impact of five major types of biomass cooking stoves. *Energy and Sustainable Development,* **XII**, 56–65.

Madungwe E, Sakuringwa S (2007) Greywater reuse: a strategy for water demand management in Harare? *Physics and Chemistry of the Earth, Parts A/B/C,* **32**, 1231–1236.

Mao C, Feng Y, Wang X, Ren G (2015) Review on research achievements of biogas from anaerobic digestion. *Renewable and Sustainable Energy Reviews,* **45**, 540–555.

Mati B, De Bock T, Malesu M, Khaka E, Oduor A, Nyabenge M, Oduor V (2006) Mapping the potential of rainwater harvesting technologies in Africa. *A GIS Overview on Development Domains for the Continent and Ten Selected Countries. Technical Manual,* **6**, 126. World Agroforestry Centre (ICRAF), Nairobi, Kenya.

Mengistu MG, Simane B, Eshete G, Workneh TS (2015) A review on biogas technology and its contributions to sustainable rural livelihood in Ethiopia. *Renewable and Sustainable Energy Reviews,* **48**, 306–316.

Mloza-Banda HR, Chikuni A, Singa DD (2006) Small scale rainwater harvesting for combating water deprivation at orphan care centres in peri-urban areas of Lilongwe, Malawi, Working Paper Series No. 46. African Technology Policy Studies Network, Kenya.

Murnyak D (2010) Fish farming basics of raising tilapia & implementing aquaculture projects, ECHO Technical Note. Available at: http://c.ymcdn.com/sites/www.echocommunity.org/resource/collection/E66CDFDB-0A0D-4DDE-8AB1-74D9D8C3EDD4/Fish_Farming.pdf (accessed 21 September 2015).

Mwenge Kahinda JM, Taigbenu AE, Boroto JR (2007) Domestic rainwater harvesting to improve water supply in rural South Africa. *Physics and Chemistry of the Earth, Parts A/B/C,* **32**, 1050–1057.

Mwenge Kahinda J, Taigbenu AE, Boroto RJ (2010) Domestic rainwater harvesting as an adaptation measure to climate change in South Africa. *Physics and Chemistry of the Earth, Parts A/B/C,* **35**, 742–751.

Mwirigi J, Balana B, Mugisha J, Walekhwa P, Melamu R, Nakami S, Makenzi P. (2014) Socio-economic hurdles to widespread adoption of small-scale biogas digesters in sub-Saharan Africa: a review. *Biomass and Bioenergy,* **70**, 4–16.

Nyong AO, Kanaroglou PS (1999) Domestic water use in rural semiarid Africa: a case study of Katarko village in northeastern Nigeria. *Human Ecology,* **27**, 537–555.

Ogunmokun AA, Mwandemele OD, Dima SJ (2000) Use of recycled waste water from biogas digesters for vegetable production in the Goreangab Dam Area of Windhoek Municipality. In: *1st WARFSA/WaterNet Symposium, Maputo Mozambique,* pp. 1–2. Available at: http://www.thewaterpage.com/waternet_symposium.html (accessed 21 September 2015).

Okotto L, Okotto-Okotto J, Price H, Pedley S, Wright J (2015) Socio-economic aspects of domestic groundwater consumption, vending and use in Kisumu, Kenya. *Applied Geography,* **58**, 189–197.

Omer AM, Fadalla Y (2003) Biogas energy in Sudan. *Renewable Energy,* **28**, 499–507.

Orskov B, Yongabi K, Subedi M, Smith J (2014) Overview of holistic application of biogas for small scale farmers in sub-Saharan Africa. *Biomass and Bioenergy,* **70**, 4–16.

Pachpute JS, Tumbo SD, Sally H, Mul ML (2009) Sustainability of rainwater harvesting systems in rural catchment of sub-Saharan Africa. *Water Resources Management,* **23**, 2815–2839.

Parawira W (2009) Biogas technology in sub-Saharan Africa: status, prospects and constraints. *Reviews in Environmental Science and Bio/Technology,* **8**, 187–200.

Pickering AJ (2011) Water access, hand hygiene, and child health in sub-Saharan Africa, PhD thesis. Stanford University, Stanford, CA, USA.

Polprasert C (2007) *Organic Waste Recycling. Technology and Management* (3rd edn). IWA Publishing, London. 509pp.

Price EC, Cheremisinoff PN (1981) *Biogas: Production and Utilization,* 16 pp. Ann Arbor Science Publishers Inc., Ann Arbor, MI, USA.

Rockström J, Falkenmark M (2015) Increase water harvesting in Africa. *Nature,* **519**, 283–285.

Rosen S, Vincent JR (1999) *Household Water Resources and Rural Productivity in Sub-Saharan Africa: A Review of the Evidence,* Vol. **673**, pp. 1–52. Harvard Institute for International Development, Harvard University, Cambridge, MA, USA. Available at: http://www.cid.harvard.edu/archive/events/cidneudc/papers/rosenvincent.pdf (accessed 21 September 2015).

Roth C (2011) *Micro-Gasification: Cooking with Gas from Biomass.* GIZHERA, Berlin. Poverty-oriented Basic Energy Service. Available at: http://www.newdawnengineering.com/HERA-GIZ%20micro (accessed 1 October 2014).

Schievano A, D'Imporzano G, Salati S, Adani F (2011) On-field study of anaerobic digestion full-scale plants (Part I): an on-field methodology to determine mass, carbon and nutrients balance. *Bioresources Technology,* **102**, 7737–7744.

Sibisi NT, Green JM (2005) A floating dome biogas digester: perceptions of energising a rural school in Maphephetheni, KwaZulu-Natal. *Journal of Energy in Southern Africa,* **16**, 45–52.

Siegert K (1994) Introduction to water harvesting: some basic principles for planning, design and monitoring. In: *Water Harvesting for Improved Agricultural Production,* Proceedings of the FAO Expert Consultation, Cairo, Egypt, November 1993, Rome. FAO, Italy. Available at: http://agris.fao.org/agris-search/search.do?recordID=XF9764657.

Smith JU, Apsley A, Avery L et al. (2013) The potential of small-scale biogas digesters to improve livelihoods and long term sustainability of ecosystem services in sub-Saharan Africa. In: *Final Report,* 188 pp. University of Aberdeen, Aberdeen, UK. Available at: http://r4d.dfid.gov.uk/pdf/outputs/energy/60928-FinalReport140613.pdf (accessed 21 September 2015)

Smith JU, Fischer A, Hallett PD *et al.* (2015) Sustainable use of organic resources for bioenergy, food and water provision in rural sub-Saharan Africa. *Renewable and Sustainable Energy Reviews*, **50**, 903–917.

SNV (2013) *Domestic Biogas Newsletter*, Issue 8, February 2013. Available at: www.snvworld.org/download/publications/snv_domestic_biogas_newsletter_-_issue_8_-_february_2013.pdf (accessed 3 April 2014).

Sturm M, Zimmermann M, Schütz K, Urban W, Hartung H (2009) Rainwater harvesting as an alternative water resource in rural sites in central northern Namibia. *Physics and Chemistry of the Earth, Parts A/B/C*, **34**, 776–785.

Subedi M, Matthews R, Pogson M, Abegaz A, Balana B, Oyesiku-Blakemore J, Smith J (2014) Can biogas digesters help to reduce deforestation in Africa? *Biomass and Bioenergy*, **70**, 87–98.

Sugita EW (2006) Increasing quantity of water: perspectives from rural households in Uganda. *Water Policy*, **8**, 529–537.

Taiganides EP (1978) Energy and useful by-product recovery from animal wastes. In: *Water Pollution Control in Developing Countries* (eds Ouano EST, Lohani BN, Thanh NC), pp. 315–323. Asian Institute of Technology, Bangkok.

UN Department of Economics and Social Affairs (2012) UN sustainable energy for all (SE4All) initiative. Available at: http://www.se4all.org/ (accessed 1 September 2014).

UNEP (2010) Africa water atlas. In: *Division of Early Warning and Assessment (DEWA)*, 314 pp. United Nations Environment Programme, Nairobi, Kenya. Available at: http://www.unep.org/pdf/africa_water_atlas.pdf (accessed 21 September 2015)

Van Koppen B (2003) Water reform in sub-Saharan Africa: what is the difference? *Physics and Chemistry of the Earth, Parts A/B/C*, **28**, 1047–1053.

Vörösmarty CJ, Douglas EM, Green PA, Revenga C (2005) Geospatial indicators of emerging water stress: an application to Africa. *Ambio*, **34**, 230–236.

World Bank (2001–2009) *World Databank*, Online Database. Available at: http://databank.worldbank.org/ (accessed 21 September 2015).

World Health Organisation (2014) *Global Health Observatory (GHO), World Health Statistics 2014*. Available at: http://www.who.int/gho/publications/world_health_statistics/2014/en/ (accessed 21 September 2015).

World Health Organization (2006) *Guidelines for Drinking Water Quality, Vol. 1: Recommendations* (3rd edn). WHO Press, Geneva, Switzerland. Available at: http://www.who.int/water_sanitation_health/dwq/gdwq3rev/en/ (accessed 21 September 2015).

Worm J, van Hattum T (2006) Rainwater harvesting for domestic use. In: *Agrodoc 43*, 84 pp. Agromisa Foundation and CTA, Wageningen, The Netherlands. Available at: http://journeytoforever.org/farm_library/AD43.pdf (accessed 21 September 2015)

Yi Y, Lin CK, Diana JS (2008) A manual of fertilization and supplemental feeding strategies for small-scale Nile tilapia culture in ponds. In: *Nineteenth Annual Technical Report: Aquaculture*, 14 pp. CRSP, Oregon State University, Corvallis, OR, USA. Available at: http://pdacrsp.oregonstate.edu/pubs/featured_titles/FertilizerManual.pdf (accessed 21 September 2015)

Zielonka S, Lemmer A, Oechsner H, Jungbluth T (2010) Energy balance of a two-phase anaerobic digestion process for energy crops. *Engineering in Life Sciences*, **10**, 515–519.

2

Miscanthus as biogas substrate – cutting tolerance and potential for anaerobic digestion

ANDREAS KIESEL and IRIS LEWANDOWSKI

Department of Biobased Products and Energy Crops, Institute of Crop Science, University of Hohenheim, Stuttgart, Germany

Abstract

In the anaerobic digestion and biogas industry in Germany, the step of energy crop production accounts for a high proportion of the greenhouse gas emissions and environmental impacts. Replacing annual energy crops, for example maize, by perennial biomass crops such as miscanthus offers the potential to increase the sustainability of biogas crop production. However, the cutting tolerance of miscanthus and the mechanisms influencing it need to be investigated to assess its potential as a biogas crop. For this purpose, a field trial with different harvest regimes was conducted to identify the potential methane yield and cutting tolerance of *Miscanthus x giganteus*. Several fertilization regimes were tested under nitrogen-limited conditions in a pot trial to investigate the mechanisms behind the cutting tolerance. The refilling of carbohydrate (starch) stores in the rhizome was identified as a very important factor influencing the cutting tolerance of miscanthus, whereas the nutrient relocation appeared to be of less importance. The field trial revealed that *Miscanthus x giganteus* offers a very high methane yield potential of approx. 6000 m^3 ha^{-1} when harvested in October, which is within the range of the methane hectare yield of energy maize. The substrate-specific methane yield of *Miscanthus x giganteus* biomass decreased with later harvest dates and reached 247 ml (g oDM)$^{-1}$ in October. This harvest date delivered very high, stable yields of on average 26 t DM ha^{-1} over two years and enabled a good cutting tolerance. Green harvest in October was identified to be suitable for *Miscanthus x giganteus* and is recommended for biogas utilization. In conclusion, the perennial biomass crop *Miscanthus x giganteus* is a very promising biogas crop and offers the potential to increase the sustainability of the anaerobic digestion sector in Germany by replacing a substantial area of biogas maize cultivation.

Keywords: anaerobic digestion, biogas, carbohydrates, cutting tolerance, energy crop, green cut, *Miscanthus x giganteus* perennial, relocation of nutrients

Introduction

In Germany, almost 8.000 biogas plants with a total installed electric capacity of 3.8 GW were under operation in 2014 (FNR 2014). At the same time, 1.27 Mha or 10.5% of the total arable land in Germany (11.9 Mha) was used for the cultivation of energy crops for biogas production (FNR 2014; Statistisches Bundesamt 2014). Maize is the most important biogas energy crop making an input proportion of 73% of crop-derived biomass (FNR 2014). This is criticized because maize cultivation can be characterized as intensive due to high fertilizer demands often combined with intensive soil cultivation. Also the environmental impact can be high, due to high erosion and nitrate leaching risk and negative impacts on biodiversity caused by pesticide use when monoculture of maize prevails (Altieri, 1999; Svoboda *et al.*, 2013; Vogel *et al.*, 2016).

When considering the entire biogas value chain, the step of energy crop cultivation accounts for a high share of the environmental impact and greenhouse gas emissions (Lijó *et al.*, 2014; Pacetti *et al.*, 2015). Perennial energy crops offer the potential to reduce the environmental impacts of crop production and thereby increase the sustainability of the biogas sector. Various perennial biogas crops are currently being researched as biogas substrates, including cup plant (*Silphium perfoliatum* L.), szarvasi (*Elymus elongatus* ssp. *ponticus* cv. Szarvasi-1), energy dock (*Rumex schavnat*) and giant knotweed (*Falopia sachalinensis* var. Igniscum). However, Mast *et al.* (2014) revealed that the overall and substrate-specific methane yield of such novel energy crops is lower than for energy maize. A lower methane yield per hectare means that a larger cultivation area is required which consequently could lead to increased competition with land for food production and biodiversity conservation. To avoid such negative effects, alternative biogas crops should ideally be higher yielding than maize, should have a better environmental profile and be able to be

Correspondence: Andreas Kiesel
e-mail: akiesel@uni-hohenheim.de

grown under conditions that are marginal for food production.

Miscanthus is a rhizomatus, perennial C4 grass species, which originates from South-East Asia. The sterile clone *Miscanthus x giganteus* is a high-yielding genotype, which is currently the standard cultivar in commercial utilization. This high yield potential has led to miscanthus being identified as a promising energy crop in several studies (Lewandowski *et al.*, 2000; Clifton-Brown *et al.*, 2004). As its fertilizer and pesticide requirements are low, miscanthus can be also characterized as a low-input crop (Lewandowski & Schmidt, 2006). *Miscanthus x giganteus* has a good environmental profile with the potential to increase soil carbon, soil fertility and biodiversity and to reduce nutrient run-off and leaching (McCalmont *et al.*, 2015). Despite these benefits, miscanthus cultivation and the utilization of its biomass are still not widespread in Europe [approx. 38.300 ha in Europe (Elbersen *et al.*, 2012)]. Here, the biomass is mainly used for the low-value application of combustion, mostly for heat generation and therefore harvested in late winter or spring. Water content and concentration of critical elements are the major determinants of the combustion quality of biomass. For miscanthus, both are positively influenced by delaying the harvest to late winter or spring, which is however accompanied by biomass losses (Iqbal & Lewandowski, 2014). The low relevance of miscanthus production in Germany, and in Europe in general, has been described by McCalmont *et al.* (2015) as a 'chicken-and-egg' problem. There is no significant market for miscanthus biomass in Europe, and biomass production costs for the low-value application of heat production are still too high. Opening up the biogas sector as a new market for miscanthus biomass could encourage the introduction of this environmentally beneficial crop into European agriculture and thereby help reduce the ecological burden of biogas production.

Miscanthus is currently not used for biogas production on account of the low suitability of the winter-/spring-harvested biomass, which is characterized by high lignin and low water contents. A green cut increases both yield and suitability of the biomass as biogas substrate. However, harvesting miscanthus when it is still green is not recommended by Fritz & Formowitz (2010) as it negatively impacts biomass yields in the following years. Later studies had contradictory findings. Some, such as Mayer *et al.* (2014) and Wahid *et al.* (2015), consider green-cut miscanthus to be the most promising future biogas crop. Wahid *et al.* (2015) identified September to October as the ideal harvest time for miscanthus when its biomass is to be used for biogas production. However, neither of these studies looked into the cutting tolerance of miscanthus. Cutting tolerance has been defined by Kiesel & Lewandowski (2014)

in this context as the ability of a crop to recover from an early green harvest without yield reductions in the following year. Miscanthus recycles a large proportion of nutrients from the aboveground biomass to the rhizomes during senescence in autumn and reuses them for the production of new shoots in spring (Lewandowski *et al.*, 2003). The prevention of this nutrient relocation could be one explanation for yield losses in miscanthus in the year following a green cut, but the mechanisms influencing the cutting tolerance are still not clear and need to be explored.

The objectives of this study were to investigate the mechanisms determining the cutting tolerance of miscanthus, to identify green-cut regimes suitable for *Miscanthus x giganteus* and to quantify the biogas yield derived from these. For this purpose, a field trial with *Miscanthus x giganteus* and a pot trial with a novel *Miscanthus sacchariflorus* genotype (OPM 19) were performed. OPM 19 indicates that the *Miscanthus sacchariflorus* genotype was genotype number 19 of the OPTIMISC (EU FP7 No. 289159) genotype set. The field trial was used to analyse the cutting tolerance of *Miscanthus x giganteus* with three different green-cut regimes and two nitrogen levels and to measure the effects on dry matter (DM) and specific biogas yield. In the pot trial, the response of miscanthus to different nitrogen levels and application dates was tested under nitrogen-limited conditions. Rhizome weight and starch production were measured to help understand the mechanisms behind cutting tolerance.

Materials and methods

Field trial

The cutting tolerance field trial was performed using a *Miscanthus x giganteus* stand established in 2008 at the research station 'Ihinger Hof' in south-west Germany (48.7° latitude, 8.9° longitude, approx. 480 m a.s.l.). The soil is classified as Haplic Luvisol with a silty clay texture and an overlay of loess loam. The site is characterized by a long-term average annual air temperature and precipitation of 8.3 °C and 689 mm, respectively. The climate data relevant for the field trial (2012–2015) are shown in Table 1 on a monthly basis. As miscanthus is a perennial crop, the 2012 data are included to show that the year preceding the cutting tolerance trial was within the range of average conditions. The original planting density was three rhizomes m^{-2}, and weeding was performed during the establishment period only. The crop was fertilized annually from 2010 onwards with 80 kg N ha^{-1} a^{-1} of stabilized ammonium nitrate fertilizer ENTEC® 26 (EuroChem Agro GmbH, Mannheim, Germany). Harvests were conducted in spring, as practised in commercial miscanthus cultivation.

The cutting tolerance field trial was set up in 2013 in the mature miscanthus crop described above as a randomized

Table 1 Climate data on monthly basis at research station Ihinger Hof for years 2012–2015. Average air temperature was measured 2 m above soil surface

Month	Average air temperature (°C)				Precipitation (mm)			
	2012	2013	2014	2015	2012	2013	2014	2015
January	2.0	0.5	3.2	1.9	56.1	21.6	36.5	76.5
February	−3.5	−1.6	4.1	−0.2	8.4	54.6	43.8	13.2
March	7.3	1.4	7.2	5.4	7.7	31.0	8.4	34.3
April	8.1	8.4	10.9	9.0	42.3	60.7	49.9	31.9
May	14.3	10.8	12.1	13.0	43.1	138.6	68.2	67.7
June	16.3	15.8	16.7	16.5	116.5	82.8	24.3	75.2
July	17.3	19.8	18.4	20.8	96.0	173.4	162.0	28.9
August	19.2	17.5	15.7	20.0	39.3	69.5	142.4	75.0
September	14.0	13.7	14.6	12.6	57.2	97	77.0	36.0
October	8.8	10.9	12.1	8.4	58.3	87.1	50.1	16.0
November	5.5	4.0	6.7	7.2	133.7	60.3	59.0	69.5
December	1.8	3.0	2.8	NA	68.1	46.3	41.7	NA
Average	9.3	8.7	10.4	10.4*	NA	NA	NA	NA
Sum	NA	NA	NA	NA	726.7	922.9	763.3	524.2†

NA, not assessed.
*Preliminary average, no data from December included.
†Preliminary sum, no data from December included.

block design with three replicates. The plot size was 9 m². The central 4 m² were harvested for yield estimation. The variants included three green-harvest regimes, two nitrogen (N) fertilization levels and a winter control (Table 2). The green-harvest regimes comprised one double-cut (July/October) and two single-cut regimes, one early (August) and one late (October). Each green-harvest regime was tested at a lower (80 kg N ha^{-1} a^{-1}) and higher (140 kg N ha^{-1} a^{-1}) nitrogen fertilization level. The winter control was only fertilized with 80 kg N ha^{-1} a^{-1}. The fertilizer used was also ENTEC® 26. The 2013 N fertilization was split into two applications: 80 kg N ha^{-1} on 22 April 2013 and 60 kg N ha^{-1} on 10 June 2013 for the plots with the higher fertilization level. In 2014, the total amount was given in one application on 10 April 2014. During the course of the field trial, the mineral content of the soil (P, K and Mg) was monitored for each plot to avoid negative effects due to nutrient limitation. The plant-available mineral supply was found to be sufficient, and therefore, no mineral fertilizer other than

nitrogen was applied. The mineral nitrogen content of the soil was measured after the last green cut each year. As only very low values were detected (on average 4.4 kg N ha^{-1} in 2013 and 3.6 kg N ha^{-1} in 2014), these were neglected in the calculation of nitrogen fertilization for the following year.

The crop was harvested using a sickle bar mower at a cutting height of approx. 5 cm. The border of each plot was removed, and the central 4 m² were collected and weighed. In literature, a minimum sampling area of 3 m² is recommended (Knörzer et al., 2013). A subsample of approx. 1 kg was taken and dried at 60 °C in a drying cabinet to constant weight to establish the dry matter (DM) content. The DM yield was calculated based on the fresh matter (FM) yield and the DM content. The dried subsample was milled in a cutting mill SM 200 (Retsch, Haan) using a 1-mm sieve for chemical and biogas analyses. An aliquot of five shoots was used to establish the average dry weight per shoot and the leaf-to-stem ratio. For this purpose, the five shoots were sepa-

Table 2 Experimental treatments of the cutting tolerance field trial

No.	Harvest regime	Fertilization (kg N ha^{-1})	Harvest date Year 1 (2013)	Harvest date Year 2 (2014)
1	Double cut	80	1st cut: 18.07.13	1st cut: 28.07.14
2		140	2nd cut: 24.10.13	2nd cut: 23.10.14
3	Early single cut	80	29.08.13	28.08.14
4		140		
5	Late single cut	80	24.10.13	23.10.14
6		140		
7	Winter control	80	20.02.14*	09.03.15†

*Biomass from growing season 2013.
†Biomass from growing season 2014.

rated into leaf and stem biomass, dried at 60 °C and weighed. The leaf sheath was counted as stem biomass and therefore not removed from the stems. Flowers were only present at the harvest of the late single-cut regime in 2014 and were counted as stem biomass.

Plant measurements were taken on 14 May 2014 and 10 June 2014 and on 29 April 2015, 21 May 2015 and 23 June 2015. The measurements included shoot density, stem height and diameter. The shoot density was established by counting the shoots taller than 5 cm in the central square metre. The stem height was measured as the distance from the soil surface to the point where the last fully developed leaf projects from the stem. The stem diameter was measured 5 cm above the soil surface. To take into account the fact that stems may not be perfectly round, each stem was measured horizontally from several angles at the defined height and the largest diameter was recorded. The stem height and diameter were measured on five representative shoots (minimum 60% of mean height) from each plot.

Pot trial

The objective of the pot trial was to compare different nitrogen fertilization rates and application times. It was conducted from 29 April 2014–29 September 2014 in the greenhouse using *Miscanthus sacchariflorus* (OPM19) plantlets. The greenhouse temperature was maintained at a minimum of 20 °C during the day and 15 °C during the night with no artificial lighting. The plantlets were approx. one year old (pot volume 75 cm^3) and selected according to similar plant height and shoot number. The selected plantlets were transferred into pots with a volume of approx. 5 l several months before the beginning of the pot trial and watered carefully (no excess water). The pots were placed on saucers (height 3 cm) to avoid uncontrolled leaching of nutrients. The pots were filled with a soil mixture consisting of fertilized peat, loam and sand in a volumetric ratio of 1 : 2 : 1. The soil mixture was sterilized at 80 °C for 24 h before planting. After successful establishment, the plants were watered with excess water for several weeks to remove the remaining mineral nitrogen in the soil. The excess water was able to run off the saucers, and the water remaining in the saucers was taken up by the plants within few hours to one day. No negative effects on the plants were observed during this period, and viable roots were visible at the bottom of the saucers.

At the start of the pot trial, the plants were approx. 2 years old and the pots were placed in a randomized block design

with 4 replications. The treatments of the pot trial are shown in Table 3. Two fertilization levels [100 and 200 mg N (kg soil)$^{-1}$] were tested in two application regimes (single and split application). For the split applications, half of the fertilizer was applied at the beginning of the trial and the other half directly after the first cut. In addition, a control with no fertilization and a treatment with a reduced fertilizer application [50 mg N (kg soil)$^{-1}$] after the first cut were conducted. The first fertilizer application was on 23 May 2014. In the following weeks, the plants were watered according to their specific needs without excess water to avoid nitrate leaching. The plants were harvested on 15 July 2014 (after 13 weeks' growth) directly followed by the second fertilizer application. The second harvest was performed on 29 September 2014 (after 11 weeks' growth). Directly after the second harvest, the rhizomes were also harvested by washing off the soil and separating the rhizomes from the roots. The biomass samples from both harvests and the rhizomes were dried at 60 °C to constant weight and milled in a cutting mill SM 200 (Retsch, Haan) using a 1-mm sieve to be used for the chemical analysis.

Chemical analysis

Phosphor (P), potassium (K), magnesium (Mg) and calcium (Ca) contents of the field and pot trial samples were analysed according to DIN EN ISO 15510 and VDLUFA Method Book III, method 10.8.2 (Naumann & Bassler, 1976/2012). For this analysis, 0.5 g of each sample was diluted with 8 ml HNO$_3$ and 6 ml H$_2$O and digested in an ETHOS.lab microwave (MLS GmbH, Leutkirch, Germany). The extract was analysed by an ICP-OES from the State Institute of Agricultural Chemistry, Hohenheim. The nitrogen content was analysed according to the DUMAS principle (method EN ISO 16634/1 and VDLUFA Method Book III, method 4.1.2) using a Vario Macro Cube (Elementar Analysensysteme GmbH, Hanau, Germany). The starch content of the rhizomes was measured enzymatically using the starch analysis kit 10207748035 (Hoffmann-La Roche Ltd., Basel; R-Biopharm AG, Darmstadt, Gemany) at the State Institute of Agricultural Chemistry, Hohenheim. The principle of the starch analysis kit is also described in method 7.2.5 in the VDLUFA Method Book III.

Biogas analysis

The substrate-specific biogas and methane yields were measured in a biogas batch test under mesophilic conditions at

Table 3 Overview of the pot trial fertilizer treatments

Treatment	Description	Total fertilization In mg N (kg soil)$^{-1}$	1st application	2nd application
0/0	Control	0	0	0
100/0	Half amount, single application	100	100	0
200/0	Full amount, single application	200	200	0
0/50	Reduced single late application	50	0	50
50/50	Half amount, split application	100	50	50
100/100	Full amount, split application	200	100	100

39 °C according to VDI guideline 4630. The biogas batch test was certified by the KTBL and VDLUFA interlaboratory comparison test 2014. The fermentation period was 35 days. Four replicates of each sample were analysed. Standard maize was analysed alongside the miscanthus samples to monitor the activity of the inoculum. The inoculum originated from the fermenter of a commercial mesophilic biogas plant which uses the following substrates: maize silage, grass silage, cereal whole crop silage, liquid and solid cattle manure and small quantities of horse manure. The inoculum was sieved and diluted to 4% DM with deionized water. Various macro- and micronutrients were added according to Angelidaki *et al.* (2009). Afterwards, the inoculum was incubated at 39 °C under anaerobic conditions for 6 days.

For the biogas batch analysis, 200 mg organic dry matter (oDM) of milled sample was transferred into a 100 ml fermentation flask, 30 g inoculum was added, and the gas-containing headspace was flushed with nitrogen to attain anaerobic conditions. The oDM content of the milled samples was estimated by drying an aliquot of approx. 1 g at 105 °C in a cabinet dryer and incineration at 550 °C in a muffle kiln to constant weight. The weight was recorded before and after drying and incineration. The fermentation flasks were closed gastight by a butyl rubber stopper and an aluminium cap. The pressure increase in the fermentation flasks was measured by puncturing the butyl rubber stopper with a cannula attached to a HND-P pressure meter (Kobold Messring GmbH, Hofheim, Germany). The biogas production was calculated as dry gas (water vapour pressure was considered) from the pressure increase and was standardized to 0 °C and 1013 hPa using Formula (1) and (2). Formula (1) was used for the first measurement and takes into account pressure increase caused by warming from laboratory temperature to 39 °C and the water vapour partial pressure. Formula (2) was used for the subsequent 17 measurements, which were taken on a regular basis.

$$V_{biogas} = V_{HS} * T_S/T_F * ((P_{A1} + P_{F1}) - (P_{A0} * T_F/T_{Lab}) - P_{WP})/P_S \quad (1)$$

where V_{biogas} = volume of biogas produced

V_{HS} = volume of gas-containing headspace in the fermentation flasks

T_S = standard temperature (=273.15 K = 0 °C)

T_F = fermentation temperature (=312.15 K = 39 °C)

P_{A1} = ambient pressure at first measurement

P_{F1} = overpressure in fermentation flasks at first measurement

P_{A0} = ambient pressure at sealing of the fermentation flasks (batch test start)

T_{Lab} = laboratory temperature at sealing of the fermentation flasks (batch test start)

P_{WP} = water vapour partial pressure at 39 °C

P_S = standard pressure (1013 hPa)

$$V_{biogas} = V_{HS} * T_S/T_F * ((P_{An} + P_{Fn}) - (P_{A(n-1)} + P_{F(n-1)}))/P_S \quad (2)$$

where P_{An} = ambient pressure at each measurement

P_{Fn} = overpressure in fermentation flask at each measurement

$P_{A(n-1)}$ = ambient pressure at previous measurement

$P_{F(n-1)}$ = overpressure in the fermentation flasks at previous measurement

During the course of the biogas batch test, it was occasionally necessary to remove the produced biogas from the fermentation flasks. The overpressure in the fermentation flasks was removed using a gastight syringe once it had reached an approximate value of 500 mbar. The biogas was transferred to a gastight evacuated storage flask where it was kept until the end of the batch test. After each gas collection, the remaining overpressure in the fermentation flasks was allowed to level off to ambient pressure by injecting a blank cannula. For the subsequent measurement, $P_{F(n-1)}$ was then set to zero in formula (2). At the end of the batch test, the remaining biogas in the headspace of the fermentation flasks was removed by active extraction with a gastight syringe and also transferred into the storage flask. An aliquot of the collected biogas was used to analyse the methane content by a GC-2014 gas chromatograph (Shimadzu, Kyoto). The gas chromatograph was equipped with a thermal conductivity detector (TCD), and the detection temperature was set to 120 °C. Two columns (HayeSep and Molsieve column) were used (oven temperature 50 °C) with argon as carrier gas. The gas samples were injected using a Combi-xt PAL autosampler (CTC Analytics AG, Zwingen, Switzerland).

Statistical analysis

Statistical analysis was performed using the software SAS version 9.4 (SAS Institute Inc., Cary, NC, USA). The program 'Procmixed' was used, and a mixed model was developed for the field trial according to Formula (3). A test on homogeneity of variance and normal probability of residues was performed. The effects were tested at a level of probability of $\alpha = 0.05$.

$$y = \mu + rep + yr + rep * yr + hr + Nf + hr * Nf + yr * hr + yr * Nf + yr * hr * Nf + e \quad (3)$$

where μ = general mean effect

rep = effect of field replicate

yr = effect of year

rep * yr = effect of interaction of field replicate and year

hr = effect of harvest regime

Nf = effect of nitrogen fertilization level

hr * Nf = effect of interaction of harvest regime and nitrogen fertilization level

yr * Nf = effect of interaction of year and nitrogen fertilization level

yr * hr * Nf = effect of interaction of year, harvest regime and nitrogen fertilization

e = residual error

The effect of the nitrogen fertilization and the interactions between harvest regime and nitrogen fertilization; year and nitrogen fertilization; and year, harvest regime and nitrogen fertilization were not significant. For this reason, the model shown in Formula (3) was adapted as shown in Formula (4).

$$y = \mu + \text{rep} + \text{yr} + \text{hr} + \text{yr} * \text{hr} + e \qquad (4)$$

Field replicate and year were fixed effects, and the interaction between field replicate and year and the interaction between field replicate, harvest regime and nitrogen fertilization level were random effects.

The data from the pot trial were also analysed by SAS version 9.4, and a mixed model according to Formula (5) was applied. A test on homogeneity of variance and normal probability of residues was performed. The effects were tested at a level of probability of $\alpha = 0.05$.

$$y = \mu + \text{treat} + \text{rep} + \text{treat} * \text{rep} + e \qquad (5)$$

where μ = general mean effect
  treat = effect of treatment
  rep = effect of replicate in the greenhouse
  treat * rep = effect of interaction of treatment and replicate
  e = residual error

The treatment consisted of nitrogen fertilization level and nitrogen application regime. Treatment and replicate were fixed effects, and no random effects were allowed.

Results

Field trial

Plant measurements. In the years 2014 and 2015 – the first and second year after the cutting regimes were first applied – nitrogen fertilization had very little effect on the regrowth in each harvest regime (Figs 1 and 2). Therefore, the following results are discussed based on the harvest regimes and not on fertilization level. The double-cut and the early single-cut regime negatively influenced the regrowth of the following vegetation period (Fig. 1). The lower shoot density and the reduced stem height and diameter revealed that the plants were growing less vigorously and sprouting started later than in the late single-cut regime and the winter control. However, no overwinter plant losses or development of gaps were observed throughout the trial. In the year after the first green cut (2014), the double-cut regime showed a high number of shoots per square metre (Fig. 1a). This could indicate that the crop reacts to this harvest regime by increased shoot numbers in the second season. However, the standard deviation in 2014 was high and this effect was no longer visible in 2015 (Fig. 1b). For this reason, the identified effect may also be caused by chance.

The average dry weight per shoot increased with later harvest in 2013, but the proportion of leaf biomass decreased sharply over winter (Fig. 2). The average dry weight per shoot of the double-cut and the early single-cut regime was lower in 2014 than in 2013. The leaf-to-stem ratio of the early single-cut regime was also lower in 2014 than in 2013. In the single late-cut regime, the average dry weight per shoot was significantly higher

in 2014 than 2013, but the leaf-to-stem ratio was stable. However, the weather conditions in 2013 were not ideal for miscanthus growth (cold spring until end of May and drought during June/July), which could explain the higher average dry weight per shoot in 2014 in the late single-cut regime. In the winter control, this effect was not visible. This may be due to losses over winter.

Dry matter yield and methane yield. The double-cut regime showed a low dry matter (DM) yield in both fertilization levels and years, because the low yields of the first cut were not compensated by the second (Fig. 3). It appeared that the first cut was performed before the end of the crop's main growth phase and therefore the crop was prevented from producing a higher yield. The yield of the early single-cut regime was high in the first year (2013) in both fertilization levels. However, there was a DM yield decrease from 2013 to 2014 in the double-cut and the early single-cut regime of approx. 40% and 60%, respectively. The green cut in the first year (2013) therefore greatly influenced the yield of the second year (2014), whereas the nitrogen fertilization showed almost no effect on the DM yield in the second year. By contrast, the DM yield of the late single-cut regime was even slightly higher in 2014 than in 2013 and showed the significantly highest DM yield in both years and fertilization levels. The DM yield of the late single-cut regime was on average 39% higher than the winter control, as biomass losses occur over winter. The DM content increased with later harvest dates and was ideal for ensiling in the early single-cut regime [on average 35% of fresh matter (FM)].

The substrate-specific methane yield (SMY) decreased with later harvest dates and the significantly highest SMY was measured in the both cuts of the double-cut regime (Fig. 4). The SMY of the late single-cut regime was significantly lower than those of the early single-cut regime, but significantly higher than the winter control. The methane yield per hectare was influenced mainly by the DM yield; the SMY had only of minor influence. The DM and methane yield of the double-cut and early single-cut regime decreased sharply from 2013 to 2014. The late single-cut regime and the winter control delivered stable DM and methane yields. However, the methane yield of the late single-cut regime was about 45% higher than the winter control, due to the higher DM yield and also higher SMY.

Mineral content of and nutrient removal by the biomass. Here, 'content' refers to the concentration of the respective nutrient in the biomass (unit % DM). 'Nutrient removal' is calculated from the nutrient content and dry matter yield and expresses the amount of nutrients removed from the field (kg ha^{-1}).

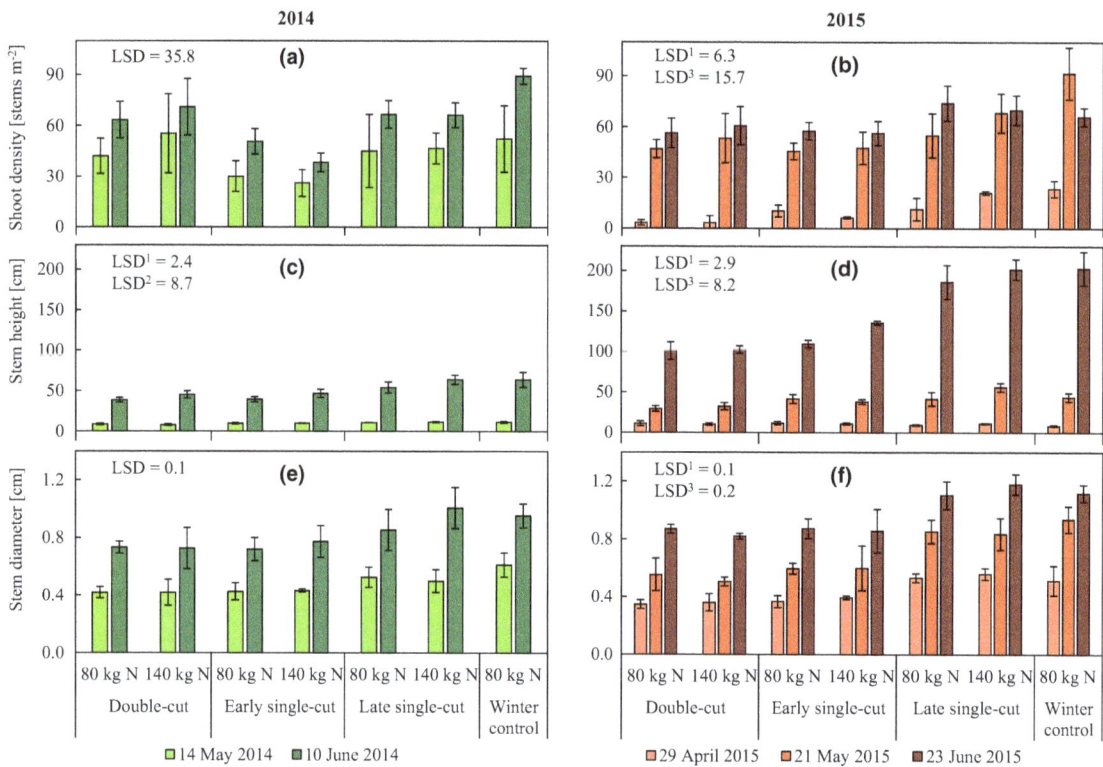

Fig. 1 Plant measurements of the regrowth of the four harvest regimes and two fertilization levels [80 and 140 kg nitrogen (N) ha^{-1} a^{-1}] in spring 2014 and 2015 in the field trial. Letters a, c and e show data from 2014 (first year after first green cut) and b, d and f from 2015 (second year after first green cut). Error bars indicate standard deviation. [1]LSD was calculated over the first measurement date in 2014 or 2015. [2]LSD was calculated over the second measurement date in 2014. [3]LSD was calculated over the second and third measurement date in 2015.

The nitrogen, phosphorus, potassium, magnesium and calcium contents of the harvested biomass are shown in Table 4. High quantities of potassium and nitrogen in particular were removed by the biomass of the high-yielding green-cut regimes, especially in 2013. In green-cut treatments where the DM yield was low, the nutrient removal was correspondingly low, for example in the double-cut and early single-cut regime in 2014. The removal of both nutrients decreased with later harvest dates, and the lowest nutrient contents were found in the biomass of the winter control. The highest nitrogen and potassium removal by the biomass was found in the late single-cut regime and the first year of the early single-cut regime, where the DM yield was high. The removal in the early and late single-cut regime was considerably lower in 2014 than in 2013. In the early single-cut regime, this was mainly influenced by the reduced DM yield and in the late single-cut regime by a lower nitrogen and potassium content. In the first year, the potassium content of the biomass from the late single-cut regime was much lower than that from the early single-cut regime, indicating that potassium was either actively relocated to

the rhizome or lost through leaf fall. In 2013, the leaf-to-stem ratio was lower at the harvest of the late single-cut regime than at the early single-cut regime (Fig. 2), but most of the dead leaves were still attached to the stem. For this reason, potassium seems to be a good indicator of how far the relocation of nutrients and carbohydrates to the rhizomes has proceeded. The nitrogen removal by the biomass in the late single-cut regime is higher than in the winter control, especially in the first year, when the plants had an oversupply of nitrogen. This can be seen from far higher nitrogen fertilization application than removal by the spring-harvested biomass in the previous year, in particular for the higher fertilization level.

Pot trial

Dry matter yield. Under conditions of limited nitrogen availability, increased nitrogen fertilization led to significantly higher biomass and rhizome production, except the treatment 0/50 (Fig. 5a–c). It is notable that the treatments with single application at the beginning of the trial (100/0 and 200/0) had a significant higher bio-

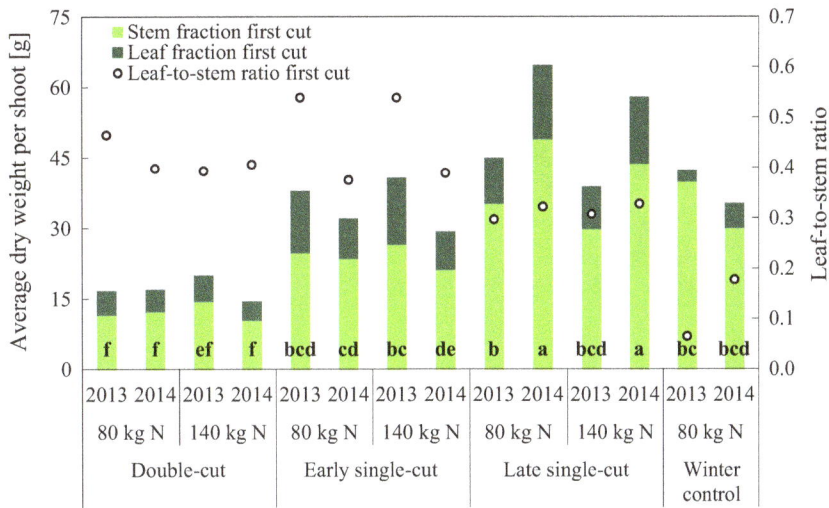

Fig. 2 Average dry weight per shoot and leaf-to-stem ratio of the first cut from the four different harvest regimes and two fertilization levels [80 and 140 kg nitrogen (N) ha^{-1} a^{-1}] in the field trial. Letter display corresponds to total average dry weight per shoot (leaf and stem biomass). The columns with different lower-case letters differ significantly from each other according to a multiple t-test $\alpha = 0.05$.

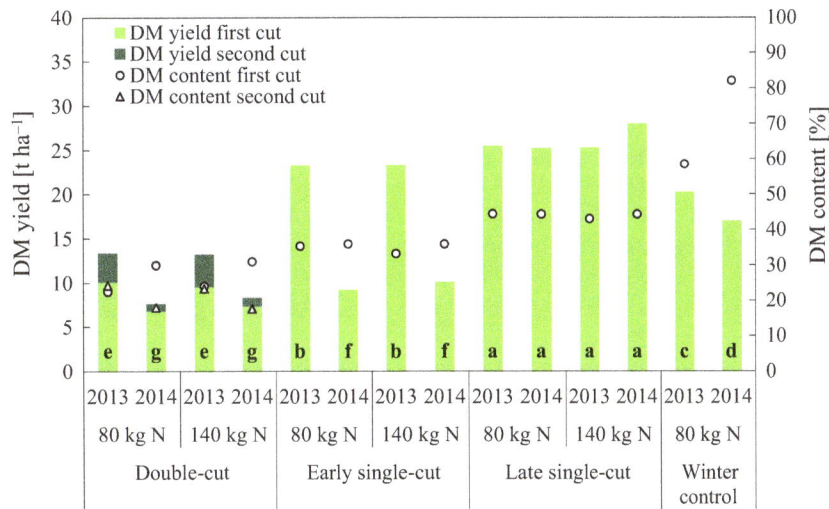

Fig. 3 Mean dry matter (DM) yield and DM content of the harvested biomass from the four different cutting regimes at two fertilization levels [80 and 140 kg nitrogen (N) ha^{-1} a^{-1}] in the field trial. Letter display corresponds to total mean DM yield (first and second cut). The columns with different lower-case letters differ significantly from each other according to a multiple t-test $\alpha = 0.05$.

mass production at the first cut than the split application treatments. The second nitrogen application after the first harvest stimulated the biomass production and led to nonsignificant differences between both application regimes of the same fertilization level. The rhizome production was not significantly influenced by the application regime, but by the fertilization level. Overall, single application of nitrogen fertilizer was advantageous, due to significantly higher biomass yield at the first cut and similar rhizome production and biomass yield at the second cut.

Nitrogen and starch content. Higher nitrogen fertilization resulted in higher nitrogen content in the biomass of both cuts and the rhizomes (Fig. 6). Nitrogen (N) removal was higher in the biomass of the fertilized treatments than in the unfertilized control due to both higher biomass and rhizome yield and higher N content of the biomass and the rhizomes. The N content of the aboveground biomass was higher when N fertilizer was applied in an early single application. These differences were not observed for the N content of rhizomes. Interestingly, the total N removal in the single application

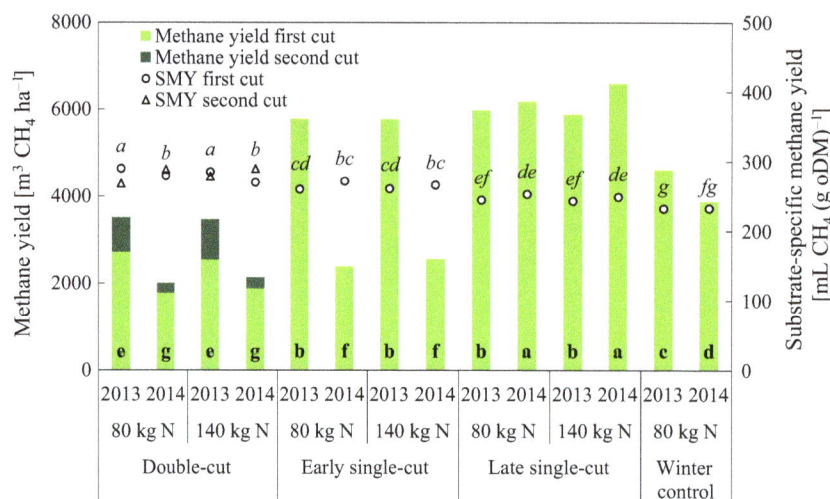

Fig. 4 Mean methane yield per hectare and substrate-specific methane yield (SMY) of the four harvest regimes at two fertilization levels [80 and 140 kg nitrogen (N) ha^{-1} a^{-1}] in the field trial. Letter display with bold lower-case letters corresponds to total mean methane yield (first and second cut). Letter display with italic lower-case letters corresponds to the SMY of the first cut. The columns with different bold or italic lower-case letters differ significantly from each other according to a multiple t-test $\alpha = 0.05$.

treatments was 792 and 1163 mg N in the treatments 100/0 and 200/0, respectively. This means more N was taken up by the crop than the applied N amount of 500 and 1000 mg N in the treatments 100/0 and 200/0, respectively. This could be due to residual nitrogen and mineralization in the soil or the nitrogen content of the rhizomes at the start of the trial. The split application treatments had a lower total N removal than the single application treatments. However, root biomass and soil were not analysed, so it is possible that a larger proportion of nitrogen was still attached to the roots or in the soil in the split application treatments.

Higher nitrogen fertilization resulted in lower starch content of the rhizomes (Fig. 7), due to the dilution effect of the higher biomass and protein production. The nitrogen fertilization increased growth of above- and belowground biomass production and in particular the amount of starch in the rhizome. The application of fertilizer after the first cut seemed to positively influence the amount of starch in the rhizome, whereas full application at the beginning of the trial seemed to positively influence the aboveground biomass production. However, the quantity of rhizome starch in the 200/0 treatment was not significantly lower than those of the 100/100 treatment.

Discussion

For a perennial crop, such as miscanthus, long-term productivity is the key factor for economically viable crop production. In the literature, the long-term productivity of *Miscanthus x giganteus* harvested in late winter

is reported to be relatively stable with annual fluctuations due to seasonal effects (Christian *et al.*, 2008; Gauder *et al.*, 2012). It is also reported to be more stable than other perennial energy grasses such as switchgrass (Iqbal *et al.*, 2015). Long-term productivity of green-harvested miscanthus is, however, much more critical on account of the early harvest and very much depends on the cutting tolerance of the crop (Fritz & Formowitz, 2010). The results of this study indicate that *Miscanthus x giganteus* tolerates harvest in October. However, detailed knowledge of the mechanisms influencing the cutting tolerance is necessary to avoid damaging the crop and to identify optimal harvest regimes. The mechanisms driving cutting tolerance of miscanthus are discussed here, and recommendations for optimal management of miscanthus as a biogas crop are elaborated.

Mechanisms influencing cutting tolerance of miscanthus

The field and pot trial in this study were performed based on the hypothesis that nitrogen fertilization and harvest time interactively affect the cutting tolerance of miscanthus by influencing processes determining cutting tolerance. Under conditions of nitrogen limitation, as here in the pot trial, a positive effect of nitrogen fertilization on the biomass yield, biomass regrowth after the first cut and rhizome weight was indeed recorded. However, under conditions where nitrogen supply is not limited, as in the field trial, only very little effect of nitrogen fertilization on cutting tolerance was observed. The amount of nutrients removed by the harvested bio-

Table 4 Nitrogen, phosphorus, potassium, magnesium and calcium content of and removal by harvested biomass. Values in parentheses indicate standard deviation

Harvest regime	Nitrogen fertilizer level kg ha⁻¹ a⁻¹	Year	Nitrogen		Phosphorus		Potassium		Magnesium		Calcium	
			Content % DM	Removal kg ha⁻¹	Content % DM	Removal kg ha⁻¹	Content % DM	Removal kg ha⁻¹	Content % DM	Removal kg ha⁻¹	Content % DM	Removal kg ha⁻¹
Double cut–first cut	80	2013	1.0 (0.1)	103 (8)	0.2 (0.0)	19 (1)	2.4 (0.1)	243 (12)	0.1 (0.0)	12 (1)	0.3 (0.0)	27 (2)
		2014	0.6 (0.1)	44 (11)	0.1 (0.0)	9 (2)	1.3 (0.2)	92 (28)	0.1 (0.0)	6 (1)	0.3 (0.1)	21 (8)
	140	2013	1.1 (0.2)	103 (20)	0.2 (0.0)	18 (3)	2.3 (0.2)	219 (45)	0.1 (0.0)	12 (2)	0.3 (0.0)	25 (3)
		2014	0.7 (0.0)	51 (5)	0.1 (0.0)	9 (0)	1.3 (0.1)	94 (3)	0.1 (0.0)	7 (1)	0.3 (0.1)	22 (7)
Double cut–second cut	80	2013	1.0 (0.1)	32 (3)	0.2 (0.0)	6 (1)	1.7 (0.2)	54 (11)	0.2 (0.0)	5 (1)	0.4 (0.0)	14 (3)
		2014	1.6 (0.0)	14 (3)	0.4 (0.0)	3 (0)	2.9 (0.4)	24 (4)	0.2 (0.0)	2 (0)	0.7 (0.1)	6 (2)
	140	2013	1.1 (0.2)	39 (11)	0.2 (0.0)	6 (2)	1.8 (0.3)	64 (14)	0.2 (0.0)	6 (2)	0.4 (0.1)	16 (6)
		2014	1.6 (0.1)	16 (0)	0.4 (0.0)	4 (0)	2.9 (0.1)	29 (2)	0.2 (0.0)	2 (0)	0.6 (0.0)	6 (1)
Early single cut	80	2013	0.6 (0.0)	132 (3)	0.1 (0.0)	28 (2)	1.4 (0.1)	319 (12)	0.1 (0.0)	18 (3)	0.2 (0.0)	44 (6)
		2014	0.5 (0.1)	47 (1)	0.1 (0.0)	12 (0)	1.0 (0.1)	96 (5)	0.1 (0.0)	7 (1)	0.3 (0.0)	27 (3)
	140	2013	0.7 (0.2)	164 (39)	0.1 (0.0)	31 (4)	1.5 (0.2)	355 (39)	0.1 (0.0)	26 (6)	0.2 (0.1)	57 (14)
		2014	0.6 (0.0)	60 (7)	0.1 (0.0)	12 (1)	1.0 (0.1)	105 (8)	0.1 (0.0)	10 (1)	0.3 (0.0)	31 (4)
Late single cut	80	2013	0.6 (0.1)	135 (22)	0.1 (0.0)	21 (4)	0.8 (0.1)	192 (44)	0.1 (0.0)	19 (3)	0.2 (0.0)	52 (6)
		2014	0.4 (0.1)	100 (24)	0.1 (0.0)	23 (4)	0.6 (0.1)	162 (32)	0.1 (0.0)	19 (2)	0.2 (0.0)	59 (14)
	140	2013	0.7 (0.2)	184 (46)	0.1 (0.0)	22 (3)	1.0 (0.2)	237 (46)	0.1 (0.0)	23 (3)	0.2 (0.0)	58 (8)
		2014	0.5 (0.1)	121 (37)	0.1 (0.0)	20 (3)	0.6 (0.1)	171 (32)	0.1 (0.0)	24 (4)	0.2 (0.0)	66 (10)
Winter control	80	2013	0.3 (0.0)	52 (12)	0.1 (0.0)	12 (3)	0.5 (0.1)	103 (34)	0.1 (0.0)	10 (3)	0.1 (0.0)	26 (8)
		2014	0.4 (0.1)	61 (17)	0.0 (0.0)	7 (1)	0.3 (0.1)	49 (9)	0.0 (0.0)	7 (1)	0.1 (0.0)	18 (5)

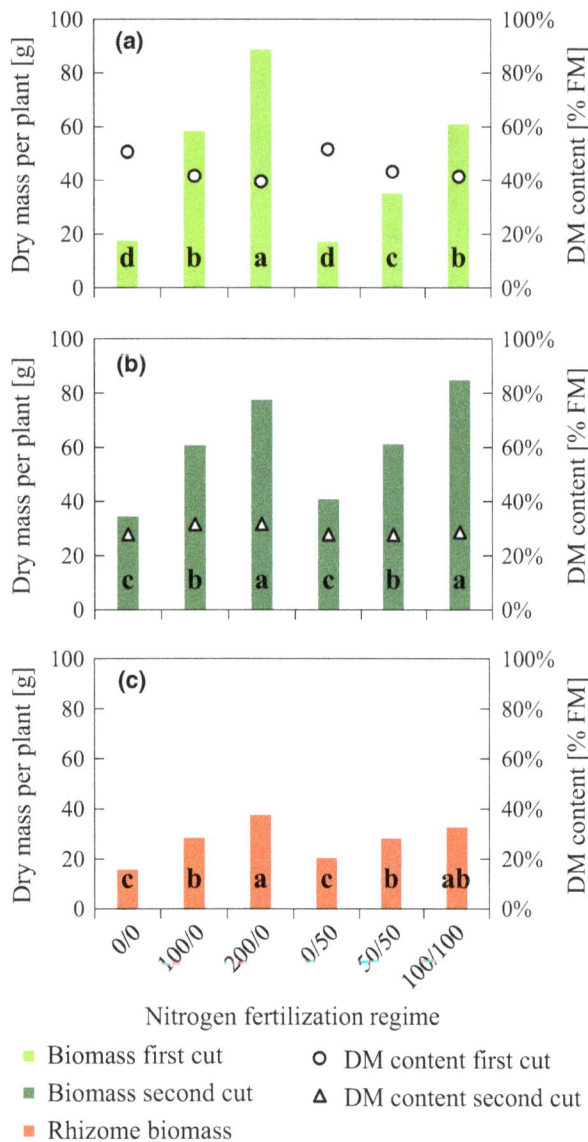

Fig. 5 Mean dry matter (DM) production and DM content of the first and second cut and the rhizomes in the pot trial treatments. The treatments comprised two nitrogen fertilization levels [100 and 200 mg N (kg soil)$^{-1}$] in two application regimes [single (full fertilizer amount at beginning) and split application (half at beginning and other half after first cut)]. Additionally, a low nitrogen application after the first cut was tested [50 mg N (kg soil)$^{-1}$]. The labelling of the treatments refers to the application rate of nitrogen [0, 50, 100 or 200 mg N (kg soil)$^{-1}$] and the fertilizer regime (fertilizer application at beginning/fertilizer application after first cut). Letter display corresponds to mean DM production. Columns with different lower-case letters differ significantly from each other according to a multiple t-test $\alpha = 0.05$.

mass does not seem critical for the cutting tolerance of miscanthus, as removed nutrients can easily be replaced by fertilization and taken up by the efficient, deep root-

ing system characteristics of miscanthus (Strullu *et al.*, 2011; Cadoux *et al.*, 2012). From these observations, it is concluded that the relevance of nitrogen fertilization increases with nitrogen limitation, for example on soils poor in nutrients or organic matter.

Nitrogen fertilization also positively affects starch production in the rhizomes, as was measured in the pot trial. This is likely to be the effect of more photosynthetically active biomass present when nitrogen fertilizer is applied. Purdy *et al.* (2015), and also the observations made in both the pot and the field trial of this study, indicate that carbohydrate – in this case starch – reserves in the rhizomes play an important role in the cutting tolerance of the crop. In the first year, the yield of the late single-cut regime was only slightly higher than that of the early single-cut regime, although the crop had two months more time for yield formation. This leads to the hypothesis that the crop invested a large proportion of the biomass accumulation into belowground biomass to refill the carbohydrate stores in the rhizomes with starch. Purdy *et al.* (2015) found that starch concentration in miscanthus rhizomes is likely to reach a peak in late autumn (November) and concluded that an earlier harvest could have a negative effect on the crop. The optimal harvest date for a green cut would therefore be at the time of maximum starch content in the rhizomes. Mutoh *et al.* (1968) revealed that, under Japanese growth conditions, the largest proportion of the net production was used to build up rhizomes, roots and reserve material in August and September. Later studies found that the starch content in the rhizomes increases until the end of the vegetation period and peaks in late autumn (Masuzawa & Hogetsu, 1977; de Souza *et al.*, 2013). Our study found the optimal harvest date for a green cut of *Miscanthus x giganteus* to be October and indicated that potassium seems to be a good indicator of how far the relocation of nutrients and carbohydrates to the rhizomes has proceeded. Himken *et al.* (1997) revealed that the content of potassium and nitrogen in the rhizome increased from June to late autumn, which supports this hypothesis. However, senescence and starch storage in the rhizomes are influenced by climatic conditions and accelerated by low daily minimum temperatures (Purdy *et al.*, 2015). In 2013, the first frost occurred at the beginning of October and several cold nights with temperatures around 0 °C were recorded before harvest. Both could have triggered carbohydrate transport to and starch formation in the rhizomes (Purdy *et al.*, 2015). This could explain why *Miscanthus x giganteus* tolerated cutting as early as October in this study and suggests that not only date, but also daily minimum temperature or first frost should be considered when determining the optimal harvest time for a green cut.

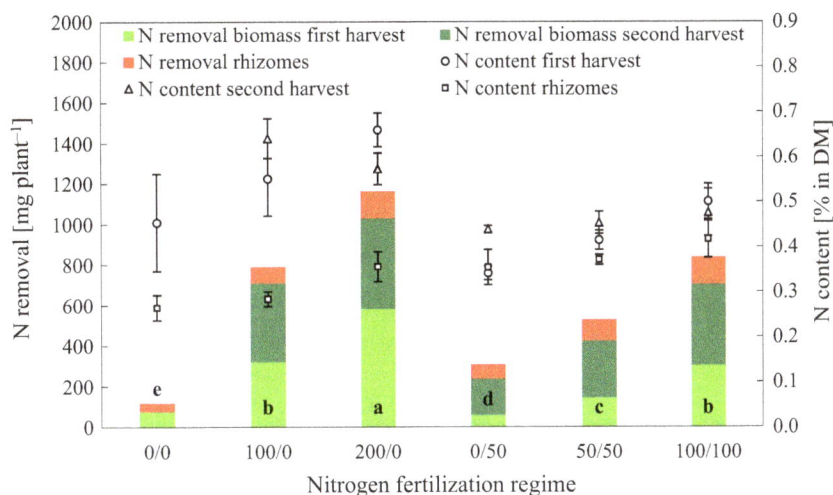

Fig. 6 Average nitrogen content and removal of the first and second cut and the rhizomes in the pot trial treatments. The treatments comprised two nitrogen fertilization levels [100 and 200 mg N (kg soil)$^{-1}$] in two application regimes [single (full fertilizer amount at beginning) and split application (half at beginning and other half after first cut)]. Additionally, a low nitrogen application after the first cut was tested [50 mg N (kg soil)$^{-1}$]. The labelling of the treatments refers to the application rate of nitrogen [0, 50, 100 or 200 mg N (kg soil)$^{-1}$] and the fertilizer regime (fertilizer application at beginning/fertilizer application after first cut). Figure display corresponds to total nitrogen removal (by biomass of first and second cut and rhizomes). Columns with different lower-case letters differ significantly from each other according to a multiple t-test $\alpha = 0.05$. Error bars indicate standard deviation of the respective nitrogen content.

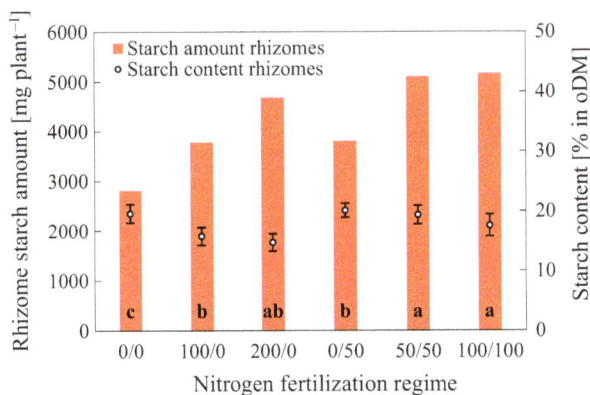

Fig. 7 Average quantity and content of starch in the rhizomes of the pot trial treatments. The treatments comprised two nitrogen fertilization levels [100 and 200 mg N (kg soil)$^{-1}$] in two application regimes [single (full fertilizer amount at beginning) and split application (half at beginning and other half after first cut)]. Additionally, a low nitrogen application after the first cut was tested [50 mg N (kg soil)$^{-1}$]. The labelling of the treatments refers to the application rate of nitrogen [0, 50, 100 or 200 mg N (kg soil)$^{-1}$] and the fertilizer regime (fertilizer application at beginning/fertilizer application after first cut). Figure display corresponds to average quantity of starch. Columns with different lower-case letters differ significantly from each other according to a multiple t-test $\alpha = 0.05$. Error bars indicate standard deviation of the respective starch content.

The results of Purdy *et al.* (2015) indicate that the seasonal fluctuations in rhizome starch content also differ between genotypes and the starch content of the rhizomes increases with ongoing senescence. Therefore, the potential of other genotypes that senesce earlier than *Miscanthus x giganteus* should be explored for biogas production. These genotypes may be characterized by earlier flowering. The carbohydrate sink in such genotypes may switch earlier from stems to rhizomes, and consequently, the starch stores in the rhizomes are refilled earlier. Such genotypes may tolerate earlier cutting than *Miscanthus x giganteus*. The advantage of genetic variation in ripening would be the option of broadening the harvest window of miscanthus to enable harvest under ideal climatic and soil conditions.

Cutting tolerance and yield of Miscanthus x giganteus

In the field trial, a significant yield decline was observed in the year after the early single-cut and the double-cut regime with harvests in July and August, indicating that these regimes are not sustainable. Both cutting regimes affected the regrowth of the crop in the second year negatively. This can be seen in a reduced leaf-to-stem ratio and average dry weight per shoot, especially in the early single-cut regime. Negative effects of an early green cut in August on the yield the following year have also been reported by Fritz & Formowitz (2010). The double-cut regime showed noteworthy regrowth after the first cut in 2013, but considerably less than in the late single-cut regime. In the field trial, nitrogen fertilizer was applied in spring and no fertilizer was applied after the first cut of the double-cut regime. In the pot trial, the split nitrogen application mainly increased the biomass yield of the second cut, but also promoted rhizome starch production. Therefore, we suggest investigating the effect of nitrogen fertilization on the yield of the second cut and the cutting tolerance of the double-cut regime when applied directly after the first cut under field conditions. However, the genotype used in the pot trial was not *Miscanthus x giganteus*, as in the field trial. Therefore, it cannot be ruled out that the *Miscanthus sacchariflorus* genotype used starts to relocate carbohydrates to the rhizomes earlier than *Miscanthus x giganteus*. In addition, the conditions in the greenhouse (e.g. higher temperatures) may have affected the generative development of the plants and thereby promoted the carbohydrate relocation to the rhizome.

In contrast to the early single- and double-cut regime, the late single-cut regime with harvest in October showed a stable or even slightly increased biomass yield from 2013 to 2014. Here, the slightly higher biomass yield and the higher average dry weight per shoot in 2014 may have been influenced by better weather conditions in 2014 than in 2013. Yield stability, with no negative effects on the yield in the following year, has also been reported for October harvest regimes by Mayer *et al.* (2014) and Yates *et al.* (2015). Based on these findings, it appears that a green-harvest regime with harvest in October is possible for *Miscanthus x giganteus*. However, as discussed above, the interactions between the processes of senescence and rhizome starch production with temperature should be further investigated. Such investigations should also include locations characterized by climates different from those for which an October harvest was examined in this study.

It is concluded that, with proper nutrient management and harvest timing, *Miscanthus x giganteus* tolerates green cutting in October. However, long-term studies are required to assess the cutting tolerance over a longer period and identify the fertilization requirements. As the harvest time is determined here by cutting tolerance, the biomass quality for biogas production may not be ideal. Therefore, the overall suitability of *Miscanthus x giganteus* for biogas production will be discussed in the following section.

Potential of Miscanthus x giganteus for biogas utilization

For combustion, *Miscanthus x giganteus* is conventionally harvested in late winter with dry biomass. A major advantage of a green harvest is the higher biomass yield. The average dry matter yield of the winter control (18.7 t DM ha^{-1}) was about 28% lower than the yield of the late green harvest in October (26.0 t DM ha^{-1}). Similar biomass losses over winter have been reported in the literature (Lewandowski *et al.*, 2003; Cadoux *et al.*, 2012). Therefore, the utilization of green biomass has the potential to substantially increase the biomass yield per unit area and to exceed that of maize.

The results presented confirm that *Miscanthus x giganteus* has a high potential for biogas utilization. The methane yield of the October harvest was very high (on average 6153 m^3 ha^{-1}) and within the range of the methane hectare yield of energy maize (6008 m^3 ha^{-1}) (Mast *et al.*, 2014). The ability of miscanthus to compete with maize has also been reported by Mayer *et al.* (2014). *Miscanthus x giganteus* could thus replace a significant share of maize cultivation for biogas production without increasing the cultivation area required. Schorling *et al.* (2015) revealed a total potential for *Miscanthus x giganteus* cultivation in Germany of 4 million ha. This large potential cultivation area indicates that miscanthus could make a significant contribution to substrate production for anaerobic digestion and help increase the sustainability of the biogas sector.

Before using *Miscanthus x giganteus* biomass in commercial biogas plants, its performance and substrate-specific methane yield (SMY) needs to be further investigated on a larger scale. As its SMY was analysed here in a batch test using milled biomass, the particle size may have positively influenced the SMY and in particular the rate of biogas production. For this reason, further research with commercially chopped biomass is required, also to assess the risk of floating layers forming in wet fermentation plants. As *Miscanthus x giganteus* biomass has a lower rate of biogas production than maize in anaerobic digestion (data not shown), it may require larger digester volumes or additional pretreatment. For commercial application, the biomass needs to be preserved by ensiling and the high dry matter content in October (on average 44% of fresh matter) may be problematic. However, the successful ensiling of miscanthus biomass has been reported in several studies (Huisman & Kortleve, 1994; Klimiuk *et al.*, 2010; Mayer *et al.*, 2014). On a commercial scale, additional silage additives can be used for efficient ensiling of *Miscanthus x giganteus* biomass or mixed ensiling with biomass of other crops, such as ryegrass or maize, may be performed.

The SMY decreased with later harvest dates and reached the lowest values at winter harvest [on average 233 ml (g oDM)$^{-1}$]. This is due to the effect of progressing lignification, relocation of easily degradable carbohydrates to the rhizomes and losses of faster degradable leaves over winter. As early green cuts are not tolerated by the crop, the SMY of the October harvest [on average 247 ml (g oDM)$^{-1}$] is suggested here as a reference SMY of *Miscanthus x giganteus* biomass. There are diverging findings for SMY of *Miscanthus x giganteus* biomass in previous studies. Wahid *et al.* (2015) and Mayer *et al.* (2014) found similar values for biomass harvested in autumn as in this study. Menardo *et al.* (2013) measured the SMY of pretreated miscanthus biomass and reported a very low SMY of 84 ml (g oDM)$^{-1}$ for the untreated, winter-harvested control. Klimiuk *et al.* (2010) analysed ensiled *Miscanthus x giganteus* biomass in a continuously operated digester and measured only 100 ml (g oDM)$^{-1}$. These lower yields may be an effect of the fermentation technology applied by Klimiuk *et al.* (2010). It is assumed here that *Miscanthus x giganteus* biomass requires longer retention times in continuously operated digesters, as the biogas production rate of this lignified biomass is comparatively low.

For anaerobic digestion, a harvest before winter is favourable, but the nutrient removal, especially for nitrogen and potassium, is significantly higher than after winter [see also Cadoux *et al.* (2012)]. The maximum removal was obtained in August and decreased until winter harvest. As too early harvest resulted in yield decline, only the late single-cut regime (harvest in October) is discussed here. A higher nitrogen fertilization resulted in a slightly higher yield, but also in a higher nitrogen removal. The nitrogen removal in 2013 was higher than in 2014. This can be explained as oversupply, due to lower removal of nitrogen in the year before the trial started. In 2014, the nitrogen removal was slightly higher than the fertilization in the 80 kg ha^{-1} treatment and slightly lower than that of the 140 kg N ha^{-1} treatment. Both nitrogen levels delivered high yields, and the yield response to the higher nitrogen level was low. Therefore, the ideal fertilization level for long-term yield stability is considered to be between 80 and 140 kg N ha^{-1} a^{-1}. However, deposition from the air and soil fertility should also be taken into account when estimating nutrient requirements. Long-term observations are required to analyse which nitrogen level is sufficient for a steady green harvest of the crop. Low nitrogen fertilization requirements are seen as an important advantage of perennial crops, especially in terms of reducing the environmental impacts of biomass production (McCalmont *et al.*, 2015). Early harvesting of miscanthus decreases this benefit because larger amounts of nutrients are withdrawn and need to be replaced by fertilization. However, in the case of biomass for biogas production, the largest part of these nutrients can be recycled by the application of biogas digestate. This is common practice in commercial biogas production. Direct emissions from nitrogen fertilizer or digestate application increase the global warming potential of crop cultivation, but only low-yield increases of 0.26 to 2.54 t DM ha^{-1} are required to offset these (Roth *et al.*, 2015).

In conclusion, *Miscanthus x giganteus* can be used for anaerobic digestion when harvested in October, but long-term effects of the green harvest on the productivity need to be assessed. The removed nutrients need to be replaced to ensure long-term productivity, but recycling of digestates should be sufficient. The methane yield and behaviour of the biomass in large-scale digesters need to be further researched. Due to the slower rate of biogas production, additional pretreatment or larger digester volumes may be required. Breeding and selection of new miscanthus genotypes for biogas production should focus on development of genotypes with higher substrate-specific methane yield in October (e.g. less lignified) and earlier refilling of rhizome starch stores, to allow a broader harvest window. The replacement of biogas maize by miscanthus offers great potential for reducing the environmental impacts of biogas production without increasing land-use competition.

Acknowledgements

The research leading to these results was performed in the OPTIMISC project and received funding from the European Union Seventh Framework Programme (FP7/2007–2013) under grant agreement No. 289159. The authors are grateful to Rene Pfitzer, Dagmar Mezger and Martin Zahner for their support in performing the laboratory analyses and to the staff of the research station Ihinger Hof, especially Thomas Trucksses, for maintaining and managing the field trial. Particular thanks go to Nicole Gaudet for editing the manuscript and to Karin Hartung for her support in the statistical analysis.

References

Altieri MA (1999) The ecological role of biodiversity in agroecosystems. *Agriculture, Ecosystems & Environment*, **74**, 19–31.

Angelidaki I, Alves M, Bolzonella D *et al.* (2009) Defining the biomethane potential (BMP) of solid organic wastes and energy crops: a proposed protocol for batch assays. *Water science and technology*, **59**, 927–934.

Cadoux S, Riche AB, Yates NE, Machet J (2012) Nutrient requirements of *Miscanthus x giganteus*. *Biomass and Bioenergy*, **38**, 14–22.

Christian DG, Riche AB, Yates NE (2008) Growth, yield and mineral content of *Miscanthus×giganteus* grown as a biofuel for 14 successive harvests. *Industrial Crops and Products*, **28**, 320–327.

Clifton-Brown J, Stampfl PF, Jones MB (2004) Miscanthus biomass production for energy in Europe and its potential contribution to decreasing fossil fuel carbon emissions. *Global Change Biology*, **10**, 509–518.

Elbersen B, Startisky I, Hengeveld G, Schelhaas M, Naeff H, Böttcher H (2012) Atlas of EU biomass potentials - Spatially detailed and quantified overview of EU biomass potential taking into account the main criteria determining biomass availability from different sources.

FNR (2014) Basisdaten Bioenergie Deutschland 2014: Festbrennstoffe, Biokraftstoffe, Biogas, Gülzow.

Fritz M, Formowitz B (2010) Eignet sich Miscanthus als Biogassubstrat?, Freising.

Gauder M, Graeff-Hönninger S, Lewandowski I, Claupein W (2012) Long-term yield and performance of 15 different Miscanthus genotypes in southwest Germany. *Annals of Applied Biology*, **160**, 126–136.

Himken M, Lammel J, Neukirchen D, Czypionka-Krause U, Olfs H (1997) Cultivation of Miscanthus under West European conditions: seasonal changes in dry matter production, nutrient uptake and remobilization. *Plant and Soil*, **189**, 117–126.

Huisman W, Kortleve WJ (1994) Mechanization of crop establishment, harvest, and post-harvest conservation of Miscanthus sinensis Giganteus. *Industrial Crops and Products*, **2**, 289–297.

Iqbal Y, Lewandowski I (2014) Inter-annual variation in biomass combustion quality traits over five years in fifteen Miscanthus genotypes in south Germany. *Fuel Processing Technology*, **121**, 47–55.

Iqbal Y, Gauder M, Claupein W, Graeff-Hönninger S, Lewandowski I (2015) Yield and quality development comparison between Miscanthus and switchgrass over a period of 10 years. *Energy*, **89**, 268–276.

Kiesel A, Lewandowski I (2014) Miscanthus as Biogas Substrate. ETA-Florence Renewable Energies. doi: 10.5071/22ndEUBCE2014-1BO.10.5

Klimiuk E, Pokój T, Budzyński W, Dubis B (2010) Theoretical and observed biogas production from plant biomass of different fibre contents. *Bioresource Technology*, **101**, 9527–9535.

Knörzer H, Hartung K, Piepho H, Lewandowski I (2013) Assessment of variability in biomass yield and quality. *GCB Bioenergy*, **5**, 572–579.

Lewandowski I, Schmidt U (2006) Nitrogen, energy and land use efficiencies of Miscanthus, reed canary grass and triticale as determined by the boundary line approach. *Agriculture, Ecosystems & Environment*, **112**, 335–346.

Lewandowski I, Clifton-Brown J, Scurlock J, Huisman W (2000) Miscanthus. *Biomass and Bioenergy*, **19**, 209–227.

Lewandowski I, Clifton-Brown JC, Andersson B *et al.* (2003) Environment and harvest time affects the combustion qualities of genotypes. *Agronomy Journal*, **95**, 1274.

Lijó L, González-García S, Bacenetti J, Fiala M, Feijoo G, Moreira MT (2014) Assuring the sustainable production of biogas from anaerobic mono-digestion. *Journal of Cleaner Production*, **72**, 23–34.

Mast B, Lemmer A, Oechsner H, Reinhardt-Hanisch A, Claupein W, Graeff-Hönninger S (2014) Methane yield potential of novel perennial biogas crops influenced by harvest date. *Industrial Crops and Products*, **58**, 194–203.

Masuzawa T, Hogetsu K (1977) Seasonal changes in the amount of carbohydrate and crude protein in the rhizome of Miscanthus sacchariflorus. *The Botanical Magazine Tokyo*, **90**, 181–191.

Mayer F, Gerin PA, Noo A *et al.* (2014) Assessment of energy crops alternative to maize for biogas production in the Greater Region. *Bioresource Technology*, **166**, 358–367.

McCalmont JP, Hastings A, McNamara NP, Richter GM, Robson P, Donnison IS, Clifton-Brown J (2015) Environmental costs and benefits of growing Miscanthus for bioenergy in the UK. *GCB Bioenergy*. doi:10.1111/gcbb.12294.

Menardo S, Bauer A, Theuretzbacher F *et al.* (2013) biogas production from steam-exploded Miscanthus and utilization of biogas energy and CO2 in greenhouses. *BioEnergy Research*, **6**, 620–630.

Mutoh N, Yoshida K, Yokoi Y, Kimura M, Hogetsu K (1968) Studies on the production processes and net production of *Miscanthus sacchariflorus* community. *Japanese Journal of Botany*, **20**, 67–92.

Naumann C, Bassler R (1976/2012) Die chemische Untersuchung von Futtermitteln. Methodenbuch / Verband Deutscher Landwirtschaftlicher Untersuchungs- und Forschungsanstalten, Bd. 3. VDLUFA-Verl., Darmstadt.

Pacetti T, Lombardi L, Federici G (2015) Water–energy nexus. *Journal of Cleaner Production*, **101**, 278–291.

Purdy SJ, Cunniff J, Maddison AL *et al.* (2015) Seasonal carbohydrate dynamics and climatic regulation of senescence in the perennial grass, Miscanthus. *BioEnergy Research*, **8**, 28–41.

Roth B, Finnan JM, Jones MB, Burke JI, Williams ML (2015) Are the benefits of yield responses to nitrogen fertilizer application in the bioenergy crop *Miscanthus × giganteus* offset by increased soil emissions of nitrous oxide? *GCB Bioenergy*, **7**, 145–152.

Schorling M, Enders C, Voigt CA (2015) Assessing the cultivation potential of the energy crop *Miscanthus × giganteus* for Germany. *GCB Bioenergy*, **7**, 763–773.

de Souza AP, Arundale RA, Dohleman FG, Long SP, Buckeridge MS (2013) Will the exceptional productivity of *Miscanthus x giganteus* increase further under rising atmospheric CO₂? *Agricultural and Forest Meteorology*, **171**, 82–92.

Statistisches Bundesamt (2014) Statistisches Jahrbuch Deutschland 2014, 1., Auflage. Statistisches Bundesamt, Wiesbaden.

Strullu L, Cadoux S, Preudhomme M, Jeuffroy M, Beaudoin N (2011) Biomass production and nitrogen accumulation and remobilisation by *Miscanthus×giganteus* as influenced by nitrogen stocks in belowground organs. *Field Crops Research*, **121**, 381–391.

Svoboda N, Taube F, Wienforth B, Kluß C, Kage H, Herrmann A (2013) Nitrogen leaching losses after biogas residue application to maize. *Soil and Tillage Research*, **130**, 69–80.

Vogel E, Deumlich D, Kaupenjohann M (2016) Bioenergy maize and soil erosion — Risk assessment and erosion control concepts. *Geoderma*, **261**, 80–92.

Wahid R, Nielsen SF, Hernandez VM, Ward AJ, Gislum R, Jørgensen U, Møller HB (2015) Methane production potential from *Miscanthus* sp. *Biosystems Engineering*, **133**, 71–80.

Yates N, Riche AB, Shield I, Zapater M, Ferchaud F, Ragaglini G, Roncucci N (eds) (2015) Investigating the Longterm Biomass Yield of *Miscanthus Giganteus* and Switchgrass when Harvested as a Green Energy Feedstock. ETA-Florence Renewable Energies.

3

Cellulosic feedstock production on Conservation Reserve Program land: potential yields and environmental effects

STEPHEN D. LEDUC[1], XUESONG ZHANG[2,3], CHRISTOPHER M. CLARK[1] and R. CÉSAR IZAURRALDE[3,4,5]

[1]National Center for Environmental Assessment, U.S. Environmental Protection Agency, 1200 Pennsylvania Ave., NW, 8623P, Washington, DC 20460, USA, [2]Joint Global Change Research Institute, Pacific Northwest National Lab, 5825 University Research Court, Suite 1200, College Park, MD 20740, USA, [3]Great Lakes Bioenergy Research Center, Michigan State University, East Lansing, MI 48824, USA, [4]Department of Geographical Sciences, University of Maryland, College Park, MD 20742, USA, [5]Texas AgriLife Research, Texas A&M University, Temple, TX 76502, USA

Abstract

Producing biofuel feedstocks on current agricultural land raises questions of a 'food-vs.-fuel' trade-off. The use of current or former Conservation Reserve Program (CRP) land offers an alternative; yet the volumes of ethanol that could be produced and the potential environmental impacts of such a policy are unclear. Here, we applied the Environmental Policy Integrated Climate model to a US Department of Agriculture database of over 200 000 CRP polygons in Iowa, USA, as a case study. We simulated yields and environmental impacts of growing three cellulosic biofuel feedstocks on CRP land: (i) an Alamo-variety switchgrass (*Panicum virgatum* L.); (ii) a generalized mixture of C4 and C3 grasses; (iii) and no-till corn (*Zea mays* L.) with residue removal. We simulated yields, soil erosion, and soil carbon (C) and nitrogen (N) stocks and fluxes. We found that although no-till corn with residue removal produced approximately 2.6–4.4 times more ethanol per area compared to switchgrass and the grass mixture, it also led to 3.9–4.5 times more erosion, 4.4–5.2 times more cumulative N loss, and a 10% reduction in total soil carbon as opposed to a 6–11% increase. Switchgrass resulted in the best environmental outcomes even when expressed on a per liter ethanol basis. Our results suggest planting no-till corn with residue removal should only be done on low slope soils to minimize environmental concerns. Overall, this analysis provides additional information to policy makers on the potential outcome and effects of producing biofuel feedstocks on current or former conservation lands.

Keywords: biofuel, biomass, carbon, Conservation Reserve Program, erosion, nitrogen, no-till corn, residue removal, switchgrass

Introduction

In December 2007, the US Congress enacted the Energy Independence and Security Act (EISA), mandating an increase in annual renewable fuel volumes from ca. 34 billion L in 2008 to 136 billion L by 2022 (U.S. Congress, 2007). Of this 136 billion L total, ca. 60 billion L are required to come from cellulosic feedstocks. Despite this mandated increase, it is uncertain where these cellulosic feedstocks will be cultivated. Agro-economic modeling suggests that cellulosic feedstocks could replace current commodity crops and pastureland (U.S. DOE, 2011), potentially exacerbating the 'fuel vs. food' dilemma (Tilman *et al.*, 2009).

US Conservation Reserve Program (CRP) land could offer an alternative land-use type for cultivation of these feedstocks. Established by the 1985 Farm Bill and administered by the US Department of Agriculture (USDA) Farm Service Agency (FSA), the CRP is the largest agricultural land-retirement program in the US (Stubbs, 2013). The program offers annual payments to agricultural landowners in exchange for the establishment of perennial cover, providing environmental benefits, including increased carbon (C) sequestration, wildlife habitat, and reductions in soil erosion and nutrient runoff (Allen & Vandever, 2012). Despite these benefits, the amount of CRP land has steadily declined since reaching a peak enrollment of 14.9 million ha in 2007. In the 2008 Farm Bill, the US Congress capped the program at 12.9 million ha (U.S. Congress, 2008) and again reduced the enrollment cap to 9.7 million ha in 2014 (U.S. Congress, 2014). Growing cellulosic feedstocks on CRP land might offer a means to both produce ethanol and maintain CRP land under perennial cover. This raises, however, important questions

Correspondence: Stephen D. LeDuc

e-mail: leduc.stephen@epa.gov

regarding whether environmental benefits could be maintained under cellulosic feedstocks.

Here, using a modeling approach, we explored two timely questions: (i) how productive is CRP land for cellulosic feedstock cultivation? And (ii) what are the environmental effects of this production? To answer these questions, we employed the Environmental Policy Integrated Climate (EPIC) model in a spatially explicit framework and USDA-FSA field-level CRP polygons for Iowa, USA. We simulated the cultivation of three potential biofuel feedstocks on CRP land: (i) Alamo-variety switchgrass (*Panicum virgatum* L.); (ii) a generalized mixture of C4 and C3 grasses; (iii) and no-till corn (*Zea mays* L.) with residue removal. We simulated yields, soil erosion, and soil C and nitrogen (N) stocks and fluxes, with comparisons made to an unharvested grass cover.

Materials and methods

We used EPIC, a comprehensive biophysical and biogeochemical process-based model, to simulate the production capacity and environmental effects of using CRP land located in the state of Iowa. The EPIC model has been extensively tested for many agricultural cropping systems, and applied to understand agronomic and environmental impacts of alternative management practices and climate change (Wang *et al.*, 2012). The crop growth and soil organic carbon modules of EPIC have been examined against field observations from numerous sites across the world (Wang *et al.*, 2005; He *et al.*, 2006; Izaurralde *et al.*, 2006, 2007; Causarano *et al.*, 2007, 2008; Apezteguía *et al.*, 2009; Schwalm *et al.*, 2010; Zhang *et al.*, 2013), including on marginal and CRP land (Izaurralde *et al.*, 2006; Gelfand *et al.*, 2013). This has made it a useful tool for assessing conservation effects of the CRP (FAPRI, 2007; USDA-FSA, 2008, 2010). Key components of EPIC include plant growth and productivity, C and nutrient cycling, soil erosion, and greenhouse-gas emissions. More information about the methods used in the EPIC simulations is provided in Data S1.

We applied a spatially explicit integrative modeling framework for EPIC, developed by Zhang *et al.* (2010, 2015), whereby we combined multiple data layers to define modeling units. We were granted provisional access to a 2008 USDA-FSA database of farm-level CRP polygons for Iowa, and in adherence to stipulations, all results shown in this paper are rescaled-up to the county or state-level. Maps of CRP parcels for Iowa, soils [from the Soil Survey Geographic (SSURGO) database], and county boundaries were discretized to raster format with a grid resolution of 30 m, and were further combined to define over 200 000 homogeneous spatial modeling units with a total area of ca. 290 000 ha. For each modeling unit, we further derived elevation and climate information from the Shuttle Radar Topography Mission digital elevation model (Farr *et al.*, 2007) and re-analysis North-American Land Data Assimilation System 2 (NLDAS-2; ldas.gsfc.nasa.gov/nldas/), respectively. More information about the spatial data used is provided in Data S2. We employed the Python-based parallel computing software by Zhang *et al.* (2013) to execute EPIC in parallel on the Department of Energy Evergreen cluster, and compiled spatially explicit modeling results into relational databases linked to GIS maps for geospatial analysis and presentation.

Using this framework, we modeled the effects of using CRP land for cellulosic feedstock production. We restricted our simulations to the potential conversion of CRP under grass cover, specifically three conservation practices (CP): CP1 (introduced grasses), CP2 (native grasses), and CP10 (established grasses) covering ca. 80, 58, and 152 thousand ha, respectively, according to the USDA-FSA database (Fig. 1). To initialize the soil C and N pools prior to the simulated treatments, we ran the model for 30 years with the generic C3/C4 grass mixture as land-cover and using the NLDAS-2 data from 1979 through 2008. We then simulated three treatments harvested annually: (i) switchgrass; (ii) the generalized C4/C3 grass mixture; and (iii) no-till corn with residue removal (hereafter, these feedstocks are referred to as 'switchgrass', 'grass mixture', and 'no-till corn', respectively). Although, the no-till corn yielded both cellulosic and grain ethanol, we were primarily interested in cellulosic ethanol feedstocks, and therefore chose to simulate continuous corn with residue removal and did not simulate it with a soybean rotation.

We simulated each treatment for 30 years, again using the 1979 through 2008 NLDAS-2 data. For the establishment of these three treatments, land-use conversion from the C3/C4

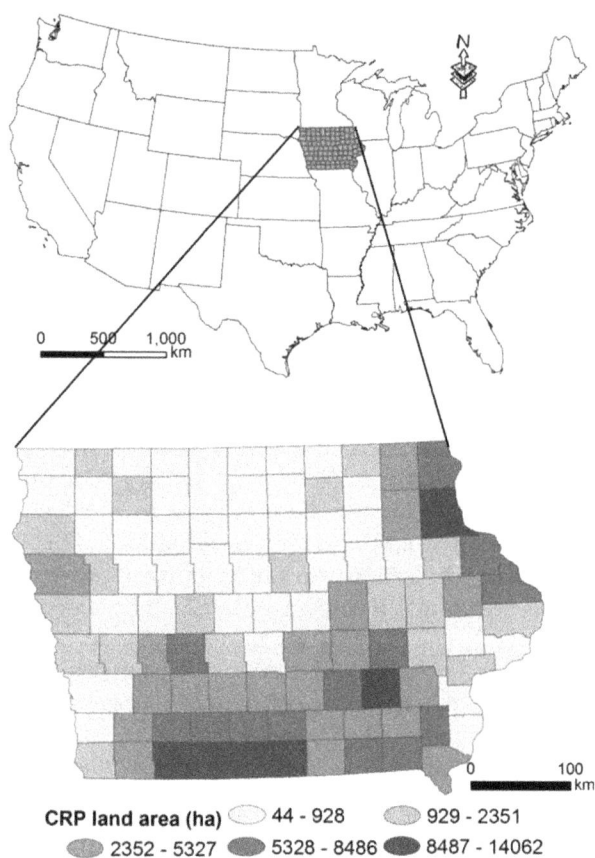

Fig. 1 Map of simulated Conservation Reserve Program (CRP) area by county in Iowa, USA.

baseline was simulated without tillage as suggested by Gelfand *et al.* (2011). The fertilization levels for perennials were chosen to be consistent with previous studies (Schmer *et al.*, 2008; Gelfand *et al.*, 2013), with switchgrass having a N application rate of 60 kg N ha^{-1} yr^{-1} while C3/C4 grasses receiving a one-time application of 40 kg N ha^{-1} in the first establishment year. The no-till corn received 140 kg N ha^{-1} yr^{-1}, slightly less than the average N application rate for corn between 2001 and 2010 in Iowa (ca. 147 kg N ha^{-1} yr^{-1}; USDA-ERS, 2015). We simulated a moderate corn stover removal rate of 50% (Karlen *et al.*, 2014). All results were compared to the generic C3/C4 mixture without harvesting, which was re-established at year zero, but received no fertilizer amendment.

Outputs modeled included: aboveground biomass and grain yield, erosion, and soil C and N stocks and fluxes. The latter included CO_2 emissions as well as N runoff and leaching. To gain insights into variation in simulated yields and environmental impacts, we tested for correlations with slope and land-capability class. Land-capability classes (1 through 8) are assigned by the USDA-Natural Resources Conservation Service to indicate the suitability of soils to grow crops, with higher numbers indicating greater limitations to crop yields. We used a Spearman's Rank Correlation Test to test for correlations since residuals from regression analyses maintained a non-normal distribution, despite data transformations. Mean output values were calculated across all modeling units. All statistical analyses were conducted in R 3.2.1.

Results

Biomass and fuel production

In our simulations, switchgrass produced the most cellulosic biomass per ha, with an average yield of 4.4 Mg ha^{-1} annually. The grass mixture produced the least and no-till corn was intermediate (Table 1). When both grain and cellulosic ethanol are considered, the model simulations suggest no-till corn could produce approximately 2.5 and 4 times more ethanol per area than switchgrass and the grass mixture, respectively (Table 1). Yields for all feedstocks, including grain yields for no-till corn, were negatively correlated with slope and land-capability class ($P < 0.001$; P-values ranged from -0.490 for switchgrass yields and slope to -0.618 for no-till corn-grain yields and slope).

Soil and water quality impacts

No-till corn resulted in greater simulated soil and water quality impacts than the other two feedstocks. Soil erosion values were 4–5 times higher under no-till corn than for switchgrass and the grass mixture (Table 2). No-till corn resulted in a net loss of total soil C over time compared to the unharvested grass mixture; by contrast, switchgrass and the grass mixture increased soil C (Table 2; Fig. 2). Carbon lost in erosion and a reduction in C inputs in crop residues accounted predominantly for the decline in total soil C under no-till corn simulations. Respiration also declined under no-till corn, but this did not balance out the losses of C (Table 2). Larger C inputs from litter under the perennial feedstocks (mostly arising from belowground root litter) compared to the no-till corn was another factor explaining the divergent patterns in soil C dynamics (Table 2; Fig. 2). Nitrogen losses, particularly associated with sediment, were higher in the no-till corn vs. the other feedstocks (Table 2; Fig. 3).

When considered on a per unit ethanol basis (including both grain and cellulosic), no-till corn's soil and water quality impacts were closer to those of the other feedstocks given the greater amount of ethanol produced (Fig. 4). Both no-till corn and mixed grasses had similar

Table 1 Model outputs for simulations of feedstock production on Conservation Reserve Program land in Iowa. Shown are annual biomass, ethanol production per ha, and total cellulosic ethanol production, expressed in absolute values and as a percentage of the annual cellulosic ethanol target volume for the year 2022 in the Energy Independence and Security Act (EISA) of 2007. Results listed by feedstock type

Model outputs	Switchgrass	Grass mixture	No-till corn with residue removal
Cellulosic feedstock yield, mean (5th and 95th percentile) (Mg ha^{-1} yr^{-1})	4.3 (2.0, 9.8)	2.6 (1.2, 6.9)	3.6 (3.0, 4.4)
Grain yield, mean (5th, 95th percentiles) (Mg ha^{-1} yr^{-1})	NA	NA	7.2 (5.9, 8.6)
Cellulosic ethanol, mean (L ha^{-1} yr^{-1})*	1653	970	1379
Grain ethanol, mean (L ha^{-1} yr^{-1})†	NA	NA	2903
Grain + cellulosic ethanol, mean (L ha^{-1} yr^{-1})	1653	970	4282
Total cellulosic ethanol produced over the 287 000 + ha simulated area (millions L yr^{-1})	474.9	278.7	396.4
Percent of cellulosic EISA target volume (ca. 60 billion L) in 2022	0.78	0.46	0.65

*Calculated using a conversion rate of 0.38 L of ethanol per kg biomass (Schmer *et al.*, 2008).

†Calculated using a conversion rate of 402.38 L of ethanol per Mg biomass (derived from Rendleman & Shapouri, 2007).

Table 2 Model outputs for simulations of feedstock production on Conservation Reserve Program land in Iowa. Shown are annual average soil erosion, and soil carbon (C) and nitrogen (N) stocks and fluxes by feedstocks and unharvested grass mixture

Model outputs	Unharvested grass mixture	Switchgrass	Grass mixture	No-till corn with residue removal
Soil erosion (Mg soil ha^{-1} yr^{-1})	0.2	3.1	3.6	14.1
Total soil C (Mg C ha^{-1})	139.9	155.6	147.9	126.0
C input from above- and belowground litter (Mg C ha^{-1} yr^{-1})	4.5	5.9	4.9	3.9
CO_2 heterotrophic respiration (Mg C ha^{-1} yr^{-1})	4.5	5.3	4.6	4.0
C loss with sediment (kg C ha^{-1} yr^{-1})	4.4	91.2	77.0	317.8
Cumulative N loss (kg N ha^{-1} yr^{-1})	0.8	13.3	11.3	58.2
Sediment N loss (kg N ha^{-1} yr^{-1})	0.5	11.7	10.2	40.3
Surface N runoff (kg N ha^{-1} yr^{-1})	0.2	1.3	0.9	4.4
Nitrate leaching (kg N ha^{-1} yr^{-1})	0.1	0.2	0.1	13.5

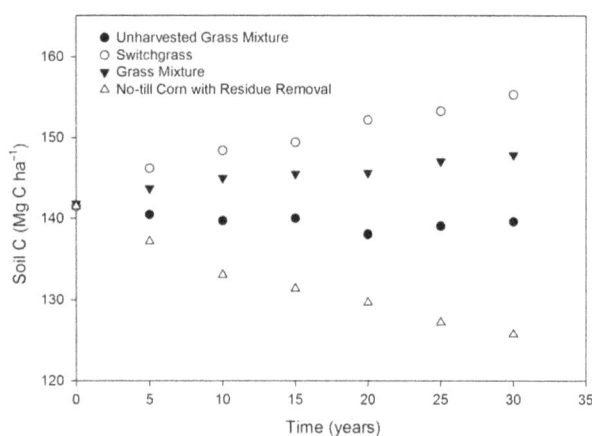

Fig. 2 Soil carbon (C) stock (to a depth of ~1.5 m) by feedstock and unharvested grass mixture during the simulation period.

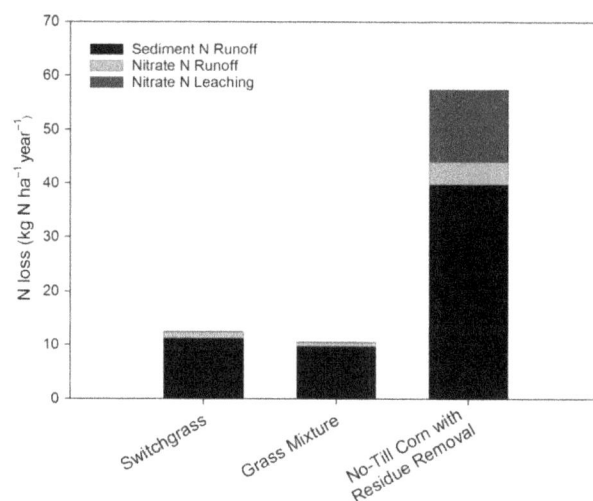

Fig. 3 Annual soil nitrogen (N) loss by feedstock type, expressed relative to the unharvested grass mixture.

values for soil erosion and N loss on a per unit ethanol basis, although no-till corn still resulted in a loss of total soil C. Of the feedstocks, switchgrass produced the best environmental outcomes per unit ethanol.

Lastly, erosion relative to the unharvested grass mixture increased with slope and land-capability class under all feedstocks (Table 3; Fig. 5a). Nitrogen losses were also strongly correlated with slope and land-capability class for switchgrass and the grass mixture, but were only weakly associated with slope and land-capability class for no-till corn (Table 3, Fig. 5b). Changes in total soil C relative to the unharvested grass mixture were not strongly associated with slope or land-capability class for any of the feedstocks (Table 3).

Discussion

In this paper, we simulated the effects of producing cellulosic feedstocks for over 200 000 individual mod-

eling units, making up ca. 290 000 ha of CRP – either currently in the program or recently exited. These simulations suggest that CRP land can be used to produce cellulosic biofuel feedstocks, but that careful consideration of environmental vs. energy trade-offs are warranted.

Biomass and ethanol yields

The yields simulated here by EPIC are generally within the range of published field trials in the Midwestern US Tilman *et al.* (2006) reported annual yields of prairie grass of approximately 3.7 Mg ha^{-1} in study plots on degraded soils in Minnesota. Average switchgrass annual yields of 5.2–11.1 Mg ha^{-1} have been observed in established fields in the Great Plains (Schmer *et al.*, 2008), and in a recent study, Bonin & Lal (2014) found

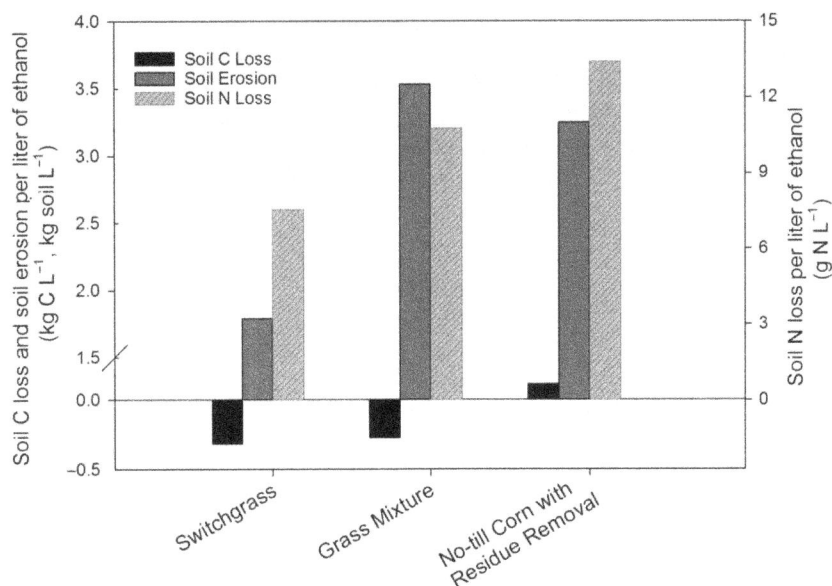

Fig. 4 Soil carbon (C) and nitrogen (N) loss and erosion by feedstock type relative to the unharvested grass mixture, expressed on a per-unit ethanol basis. Note: negative soil C loss values denote a net increase in soil carbon.

Table 3 Spearman's Rank Correlation Test *P*-values for correlation between soil and water quality impacts of each feedstock relative to the unharvested grass mixture, and slope and land-capability class (note: *P*-values in this test range from 0 to 1, with stronger relationships indicated by numbers closer to 1). All correlations were significantly positive ($P \leq 0.001$)

	Soil erosion	Cumulative N loss	Total soil C
Land-capability class			
Switchgrass	0.774	0.702	0.104
Grass mixture	0.748	0.736	0.206
No-till corn with residue removal	0.765	0.076	0.294
Slope			
Switchgrass	0.913	0.832	0.115
Grass mixture	0.892	0.895	0.236
No-till corn with residue removal	0.893	0.155	0.322

average corn stover yields of 4.1 and 13.1 Mg ha^{-1} at two sites in Ohio. Nationwide, 2015 corn-grain yields were estimated by the USDA to average approximately 8.9 Mg ha^{-1} after accounting for moisture content (15.5%) (USDA-NASS, 2015). In all these cases, the averages reported are well within our simulated range (Table 1), except for the high corn stover yield in Ohio (Bonin & Lal, 2014) and the nationwide corn-grain yield average (USDA-NASS, 2015). Conservation Reserve Program land is typically less productive than prime farm land, likely explaining the lower simulated corn yields

compared to a highly productive site and the nationwide average.

It is constructive to consider that at these simulated yields, there is not enough CRP land nationwide to meet the cellulosic goals of EISA. Volumes of cellulosic biofuels have been lagging behind EISA targets – ca. 125 million L of cellulosic biofuel were produced in 2014, far smaller than the EISA set-goals of 6.6 and 60 billion L annually for 2014 and 2022, respectively (US EPA 2010, 2015). According to the model, switchgrass had the highest yield of cellulosic ethanol per hectare (1653 L ha^{-1}) of all the feedstocks. If that yield is extrapolated, by multiplying it by all the area currently under CRP nationally (9.5 million ha; USDA-FSA, 2015), the volume of cellulosic ethanol produced (ca. 15.8 billion L) would still only yield about 25% of the 60 billion L cellulosic ethanol volume EISA requires annually. Expressed another way, it would require more than 3.5 times the current enrollment of CRP nationally to meet the EISA target. Further, CRP land includes tree plantings, windbreaks, wetland restorations, and other land-covers less suitable for farming than the CRP grasslands studied here. Thus, these results suggest CRP-produced cellulosic ethanol could not by itself meet targets set by Congress; rather, it would need to be a contributing piece, amongst many other land-use types and feedstocks.

Environmental impacts

Our results indicate that if only total ethanol production is considered, corn would be the preferred feedstock of the three simulated due to its high starch and cellulosic

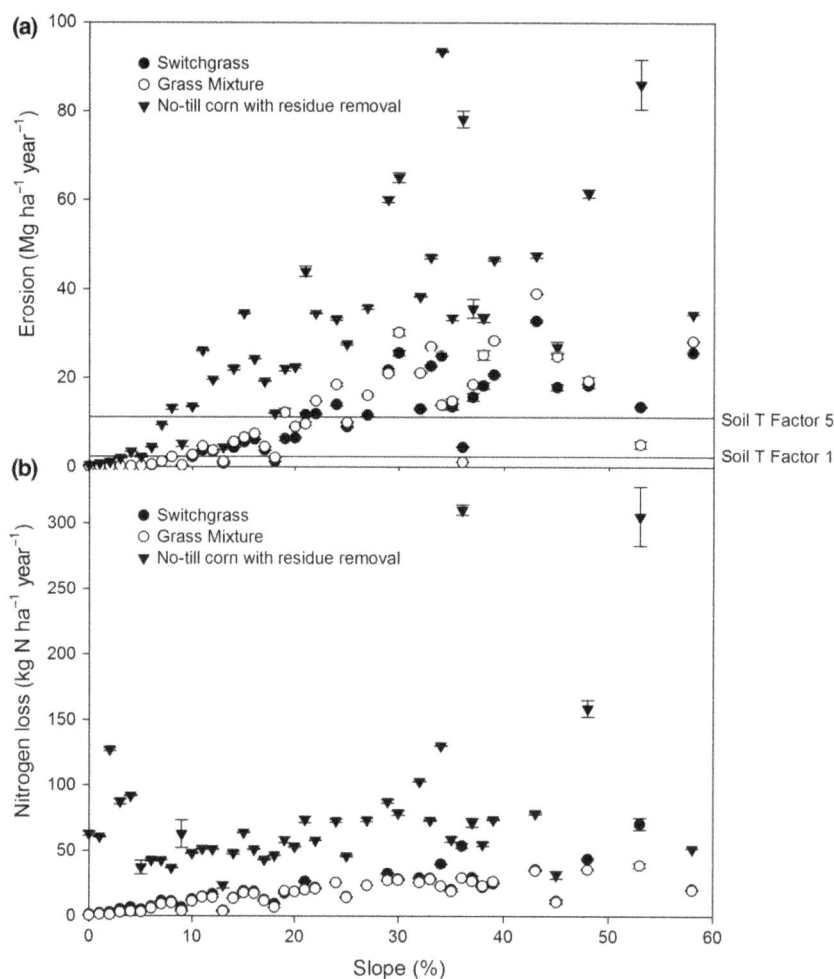

Fig. 5 Soil erosion (a) and nitrogen loss (b) by slope for the three feedstock types relative to the unharvested grass mixture. Symbols represent mean values for all modeling units with a given slope; error bars equal ±1 SE. Horizontal lines depict soil loss tolerances for deep soils (Soil T Factor 5) and for shallow or otherwise fragile soils (Soil T Factor 1).

output for ethanol production. However, if judged on environmental effects, our findings suggest no-till corn with residue removal would cause much greater negative environmental effects than perennial grasses. When expressed on a per-unit ethanol basis, the erosion and N loss of no-till corn are similar to the grass mixture (though not for carbon loss), but even then, switchgrass would still yield better environmental outcomes (Fig. 4).

These findings are consistent with other studies. Gelfand *et al.* (2011) found CRP grassland conversion to no-till continuous corn (without residue removal) could result in a total carbon debt of 68 Mg CO_2 equivalents ha^{-1}, requiring 40 years to repay with the C savings from the displacement of gasoline. In a meta-analysis of literature values, Anderson-Teixeira *et al.* (2009) concluded corn residue harvesting resulted in large average soil organic C losses even in no-till systems, predicting losses of 3–8 Mg ha^{-1} over 10 years' time under 25%

and 100% residue removal. Here, soil C losses averaged 6.6 Mg ha^{-1} over the first 10 years of the simulations, and continued over the entire 30 year period (Fig. 2). Conversely, growing switchgrass and mixed prairie grasses resulted in a simulated increase in soil C, 8.7 and 5.3 Mg ha^{-1} over the first 10 years, respectively, and 15.7 and 8.3 Mg ha^{-1} over the entire 30 years, respectively. The effects of N amendments and higher residue inputs, particularly under switchgrass, likely explain this increase in soil C (Liebig *et al.*, 2008). This also suggests these CRP soils may not have reached C equilibrium even after years of perennial cover, a finding supported by field-based evidence (Gebhart *et al.*, 1994; Gelfand *et al.*, 2011; O'Connell *et al.*, 2016)).

No-till corn also resulted in much higher soil erosion and N loss relative to the perennial grasses (Table 1). Simulated erosion for no-till corn averaged 14.1 Mg ha^{-1}, above the average water erosion rate of

12.2 Mg ha^{-1} on cultivated land in Iowa (USDA-NRCS, 2013); the soil loss tolerance for deep soils (11.2 Mg ha^{-1}); and much higher than the soil loss tolerance for most shallow or otherwise fragile soils (2.2 Mg ha^{-1}) (USDA-NRCS, 2015a). These soil loss tolerance levels are the maximum rates of annual soil erosion permitting sustainable crop productivity (USDA-NRCS, 2015a). Strikingly, the perennial grasses minimized, but did not eliminate, erosion according to the model. EPIC uses a modified version of the universal soil loss equation (USLE; see Data S1) to estimate erosion. Harvesting biomass increases erosion by reducing surface cover, which affects the cover-management factor in the USLE. In our simulations, switchgrass and the grass mixture had erosion rates (3.1 and 3.6 Mg ha^{-1}, respectively) marginally higher than the soil loss tolerance for the most shallow or otherwise fragile soils, and much lower than the tolerance classes for other soils. The erosion rates simulated were also similar to those observed on noncultivated or pastureland in Iowa (USDA-NRCS, 2013), and were almost entirely driven by erosion on steep slopes (above 20%; further discussed below). Field studies of erosion rates under harvested perennial grass systems are generally lacking, limiting direct comparisons (Blanco-Canqui, 2010). Most of the research to date has focused on the erosion reduction benefits of planting perennial grasses in buffer strips, which have been shown to substantially reduce, but not always eliminate erosion (Blanco-Canqui, 2010). Field trials incorporating the harvesting of perennial grasses for biomass on steeper, erosion-prone slopes are needed to more fully evaluate this finding.

We found that all of these biofuel types are likely to contribute to N loadings, especially no-till corn. The perennial grasses greatly reduced N loss relative to the no-till corn (Fig. 3), but did increase N loss relative to the unharvested grass mixture. The greatest loss of N was associated with sediment erosion under the no-till corn treatment. The N losses under the no-till corn are similar in magnitude to those found in field observations. Zhu & Fox (2003) for instance, observed annual nitrate (NO$_3^-$) leaching averages of 20 and 41 kg N ha^{-1} in a no-till corn system (without residue removal) receiving 100 kg N ha^{-1}, while we simulated an annual loss of 13.5 kg N ha^{-1} via NO$_3^-$ leaching and a total N loss of 58.2 kg N ha^{-1} under a similar fertilization rate. EPIC is an edge-of-field model and so N lost may not necessarily enter waterways, but it certainly suggests these biofuel types could contribute to N loadings, especially if no-till corn were grown on CRP land. If widespread across the Mississippi River Basin, increased biofuel production on CRP land would be in conflict with the goal of reducing nutrient loads and shrinking

the hypoxic zone in the Gulf of Mexico (Gulf Coast Ecosystem Restoration Task Force 2011).

Our simulations predict most environmental impacts would arise from producing these feedstocks on steeper sloped CRP land. Under no-till corn, most simulated erosion above soil loss tolerance levels occurred on lands with a slope of 10% or greater (Fig. 5a). For the perennial grasses, this occurred on lands with a slope of 20% or more (Fig. 5a). Likewise, N loss was significantly correlated with slope, with much of these losses associated with erosion (Table 3; Fig. 5b). We also found a decline in yields for all feedstocks with slope. This occurred despite higher sloped soils being a small fraction of the area modeled. More than half of the area modeled (ca. 148 000 ha) contained soils with slopes of less than 10%, and over 95% of the area contained soils with slopes of less than 20% (274 000 ha). These findings are echoed by other studies. Secchi *et al.* (2009) used EPIC to simulate the effects of converting CRP to corn-ethanol production and found greater erosion and N loss rates at higher slopes under continuous corn production. Likewise, in field experiments, Blanco-Canqui & Lal (2007, 2008) found corn stover removal negatively impacted soil nutrients, C, and grain and stover yields, particularly on steeper soils.

It may be that a lower corn stover removal rate than the one simulated here (50%) would have reduced some of the environmental effects on the steep sloped soils. Graham *et al.* (2007) suggested a national stover removal rate of 30% to keep erosion below soil tolerance loss levels. Blanco-Canqui & Lal (2007, 2008) concluded that a removal rate of 25% or less may be sustainable, a finding based on soils with slopes of ≤10%, much less than some of the slopes simulated here. At these low removal rates, however, it might not be economically profitable to remove stover. Given the environmental concerns, reductions in yield, and the difficulties farming these steep soils, it may be that producers avoid these soils altogether. Overall, our results suggest it makes best economic and environmental sense to preferentially use flatter, deeper CRP soils and not steeper sloped CRP land.

Notably, areas in CRP have been declining nationally, with much of this land going back to agriculture (USDA-NRCS, 2015b). Broader than CRP, recent studies have shown large-scale conversions of grasslands to agriculture in the US, predominantly to corn and soybeans (Wright & Wimberly, 2013; Lark *et al.*, 2015). Here, we compared our results to a baseline of an unharvested grass mixture, and we did not explicitly compare to row-crop agriculture. Nevertheless, growing perennial grasses for feedstocks is likely to provide better environmental outcomes than annual row crops (Ranney & Mann, 1994). Thus, against this backdrop, a cellulosic perennial grass industry might afford a means to keep CRP land under perennial cover, and provide more environmental benefits than the current status quo of converting it back to intensive row-crop agriculture. By contrast, it is possible that a corn stover industry could further incentive an expansion of corn and corn stover removal onto marginal lands, including CRP land. Our study suggests the environmental outcome of this would be worse than growing perennial grasses for biofuel feedstocks.

When evaluating the results of this study, it is important to note what was not considered in our modeling approach. This research study focused on select environmental end-points, namely soil C, N, and erosion. Conservation Reserve Program land, and other set-aside areas, however, provide environmental benefits not considered here – for example, wildlife habitat. Moreover, we did not consider any potential effects from the production of feedstocks on the economic benefits of these lands, such as hunting and recreation. We also focused on Iowa and three particular types of feedstocks. Other feedstocks in other areas may present alternative benefits or impacts. Finally, we did not include social drivers in our analysis. Recent research, for instance, suggests that landowner preferences may pose a barrier to growing cellulosic biomass on marginal lands (Skevas et al., 2016).

Conclusions

In this study, we employed a CRP field-level database and modeling framework to simulate the effects of producing biofuel feedstocks on over 200 000 individual modeling units, representing 290 000 ha of former or current CRP grasslands. We conclude that CRP land can be used to produce substantial volumes of grain and cellulosic ethanol. Ignoring environmental effects, corn would be the preferred feedstock due to its high grain and cellulosic output for ethanol production, producing approximately 2.6–4.4 times more ethanol per area compared to switchgrass and the grass mixture. However, when environmental considerations are included, the balance changes. No-till corn led to 3.9–4.5 times more erosion, 4.4–5.2 times more cumulative N losses, and a 10% reduction in total soil carbon as opposed to a 6–11% increase. Switchgrass resulted in the best environmental outcomes in absolute terms and on a per unit ethanol basis. Our simulations further suggest planting no-till corn with residue removal should only be done on the lower slope soils (with slopes less than 10%) in order to avoid some of the most negative effects. Relevant to all feedstock types simulated here, we conclude CRP land can only meet a small fraction of the cellulosic volume goal set by the 2007 EISA if these results from Iowa hold for the nation. Thus, growing cellulosic biomass on CRP land

would likely need to be a smaller part of a larger portfolio of renewable energy, rather than viewed as a 'silver bullet'. Overall, this analysis provides additional information to policy makers on the potential outcomes and effects of producing feedstocks on current or former conservation areas.

Acknowledgements

Funding for this work was provided by US EPA Biofuel Research Initiative, US DOE Great Lakes Bioenergy Research Center (DOE BER Office of Science DE-FC02-07ER64494, DOE BER Office of Science KP1601050, DOE EERE OBP 20469-19145), and NASA (NNH12AU03I and NNH13ZDA001N). We thank Ellen Cooter and Mark Johnson of the US EPA, and three anonymous reviewers, whose suggestions improved the quality of this manuscript. We also thank the USDA Farm Service Agency (FSA), particularly Rich Iovanna, for providing the CRP data. These data were obtained from the FSA under a Non-Disclosure Agreement (NDA 525277) signed with Battelle Memorial Institute, Pacific Northwest Division. Disclosure: The views expressed in this paper are those of the authors and do not necessarily represent the views or policies of the US Environmental Protection Agency.

References

Allen AW, Vandever MW (2012) Conservation Reserve Program (CRP) contributions to wildlife habitat, management issues, challenges and policy choices – an annotated bibliography: U.S. Geological Survey Scientific Investigations Report 2012-5066.

Anderson-Teixeira KJ, Davis SC, Masters MD, Delucia EH (2009) Changes in soil organic carbon under biofuel crops. Global Change Biology Bioenergy, 1, 75–96.

Apezteguía HP, Izaurralde RC, Sereno R (2009) Simulation study of soil organic matter dynamics as affected by land use and agricultural practices in semiarid Córdoba, Argentina. Soil and Tillage Research, 102, 101–108.

Blanco-Canqui H (2010) Energy crops and their implications on soil and environment. Agronomy Journal, 102, 403–419.

Blanco-Canqui H, Lal R (2007) Soil and crop response to harvesting corn residues for biofuel production. Geoderma, 141, 355–362.

Blanco-Canqui H, Lal R (2008) No-tillage and soil-profile carbon sequestration: an on-farm assessment. Soil Science Society of America Journal, 72, 693–701.

Bonin CL, Lal R (2014) Aboveground productivity and soil carbon storage of biofuel crops in Ohio. Global Change Biology Bioenergy, 6, 67–75.

Causarano HJ, Shaw JN, Franzluebbers AJ et al. (2007) Simulating field-scale soil organic carbon dynamics using EPIC. Soil Science Society of America Journal, 71, 1174–1185.

Causarano HJ, Doraiswamy PC, Mccarty GW, Hatfield JL, Milak S, Stern A (2008) EPIC modeling of soil organic carbon sequestration in croplands of Iowa. Journal of Environmental Quality, 37, 1345–1353.

FAPRI (Food and Agricultural Policy Research Institute) (2007) Estimating water quality, air quality, and soil carbon benefits of the Conservation Reserve Program, FAPRI-UMC Report #01-07. University of Missouri, Columbia, Missouri.

Farr TG, Rosen PA, Caro E et al. (2007) The shuttle radar topography mission. Reviews of Geophysics, 45, 33.

Gebhart DL, Johnson HB, Mayeux HS, Polley HW (1994) The CRP increases soil organic-carbon. Journal of Soil and Water Conservation, 49, 488–492.

Gelfand I, Zenone T, Jasrotia P, Chen JQ, Hamilton SK, Robertson GP (2011) Carbon debt of Conservation Reserve Program (CRP) grasslands converted to bioenergy production. Proceedings of the National Academy of Sciences of the United States of America, 108, 13864–13869.

Gelfand I, Sahajpal R, Zhang XS, Izaurralde RC, Gross KL, Robertson GP (2013) Sustainable bioenergy production from marginal lands in the US Midwest. Nature, 493, 514–517.

Graham RL, Nelson R, Sheehan J, Perlack RD, Wright LL (2007) Current and potential US corn stover supplies. Agronomy Journal, 99, 1–11.

Gulf Coast Ecosystem Restoration Task Force (2011) Gulf of Mexico Regional Ecosystem Restoration Strategy. Available at: http://archive.epa.gov/gulfcoasttaskforce/web/pdf/gulfcoastreport_full_12-04_508-1.pdf (accessed 7 March 2016).

He X, Izaurralde R, Vanotti MB, Williams JR, Thomson AM (2006) Simulating long-term and residual effects of nitrogen fertilization on corn yields, soil carbon sequestration, and soil nitrogen dynamics. *Journal of Environmental Quality*, **35**, 1608–1619.

Izaurralde RC, Williams JR, Mcgill WB, Rosenberg NJ, Jakas MCQ (2006) Simulating soil C dynamics with EPIC: model description and testing against long-term data. *Ecological Modelling*, **192**, 362–384.

Izaurralde R, Williams JR, Post WM, Thomson AM, Mcgill WB, Owens L, Lal R (2007) Long-term modeling of soil C erosion and sequestration at the small watershed scale. *Climatic Change*, **80**, 73–90.

Karlen DL, Birrell SJ, Johnson JMF et al. (2014) Multilocation corn stover harvest effects on crop yields and nutrient removal. *Bioenergy Research*, **7**, 528–539.

Lark TJ, Salmon JM, Gibbs HK (2015) Cropland expansion outpaces agricultural and biofuel policies in the United States. *Environmental Research Letters*, **10**, 044003.

Liebig MA, Schmer MR, Vogel KP, Mitchell RB (2008) Soil carbon storage by switchgrass grown for bioenergy. *Bioenergy Research*, **1**, 215–222.

O'Connell JL, Daniel DW, Mcmurry ST, Smith LM (2016) Soil organic carbon in playas and adjacent prairies, cropland, and Conservation Reserve Program land of the High Plains, USA. *Soil & Tillage Research*, **156**, 16–24.

Ranney JW, Mann LK (1994) Environmental considerations in energy crop production. *Biomass and Bioenergy*, **6**, 211–228.

Rendleman CM, Shapouri H (2007) New technologies in ethanol production. US Department of Agriculture-Office of the Chief Economist. Agricultural Economic Report 482. Available at: http://www.usda.gov/oce/reports/energy/aer842_ethanol.pdf (accessed 7 March 2016).

Schmer MR, Vogel KP, Mitchell RB, Perrin RK (2008) Net energy of cellulosic ethanol from switchgrass. *Proceedings of the National Academy of Sciences of the United States of America*, **105**, 464–469.

Schwalm CR, Williams CA, Schaefer K et al. (2010) A model-data intercomparison of CO_2 exchange across North America: results from the North American Carbon Program site synthesis. *Journal of Geophysical Research: Biogeosciences (2005–2012)*, **115**, G00H05.

Secchi S, Gassman PW, Williams JR, Babcock BA (2009) Corn-based ethanol production and environmental quality: a case of iowa and the conservation reserve program. *Environmental Management*, **44**, 732–744.

Skevas T, Hayden NJ, Swinton SM, Lupi F (2016) Landowner willingness to supply marginal land for bioenergy production. *Land Use Policy*, **50**, 507–517.

Stubbs M (2013) Agricultural conservation: a guide to programs. Congressional Research Service Report for Congress. R40763.

Tilman D, Hill J, Lehman C (2006) Carbon-negative biofuels from low-input high-diversity grassland biomass. *Science*, **314**, 1598–1600.

Tilman D, Socolow R, Foley JA et al. (2009) Beneficial biofuels – the food, energy, and environment trilemma. *Science*, **325**, 270–271.

U.S. Congress, 110th (2007) Pub. L. 110–140, Energy Independence and Security Act of 2007. Available at: http://www.gpo.gov/fdsys/pkg/PLAW-110publ140/html/PLAW-110publ140.htm, U.S. Government Printing Office (accessed 31 August 2015).

U.S. Congress, 110th (2008) Pub L. 110–234, Food, Conservation, and Energy Act of 2008. Available at: http://www.gpo.gov/fdsys/pkg/PLAW-110publ234/pdf/PLAW-110publ234.pdf, U.S. Government Printing Office (accessed 31 August 2015).

U.S. Congress, 113th (2014) Pub. L. 113–79, Agricultural Act of 2014. Available at: http://www.gpo.gov/fdsys/pkg/PLAW-113publ79/html/PLAW-113publ79.htm, U.S. Government Printing Office (accessed 31 August 2015).

U.S. DOE (U.S. Department of Energy) (2011) *U.S. Billion-Ton Update: Biomass Supply for a Bioenergy and Bioproducts Industry*. R.D. Perlack and B.J. Stokes (Leads), ORNL/TM-2011/224. Oak Ridge National Laboratory, Oak Ridge, Tennessee.

U.S. EPA (U.S. Environmental Protection Agency) (2010) Fact Sheet: EPA Finalizes Regulations for the National Renewable Fuel Standard Program for 2010 and Beyond. EPA-420-F-10-007. Available at: https://www.epa.gov/renewable-fuel-standard-program/renewable-fuel-standard-rfs2-final-rule-additional-resources (accessed 7 March 2016).

U.S. EPA (2015) Renewable Fuel Standard Program: Standards for 2014, 2015, and 2016 and Biomass-Based Diesel Volume for 2017; Final Rule. Available at: https://www.gpo.gov/fdsys/pkg/FR-2015-12-14/pdf/2015-30893.pdf (accessed 7 March 2016).

USDA-ERS (U.S. Department of Agriculture-Economic Research Service) (2015) Fertilizer use and price. Available at: http://www.ers.usda.gov/data-products/fertilizer-use-and-price.aspx (accessed 23 November 2015).

USDA-FSA (U.S. Department of Agriculture-Farm Services Agency) (2008) Fact sheet: Conservation Reserve Program (CRP) benefits: water quality, soil productivity and wildlife estimates. Available at: https://www.fsa.usda.gov/Internet/FSA_File/crpbennies.pdf (accessed 7 March 2016).

USDA-FSA (U.S. Department of Agriculture-Farm Services Agency) (2010) Conservation Reserve Program Annual Summary and Enrollment Statistics FY 2010. Available at: http://www.fsa.usda.gov/Internet/FSA_File/annual2010summary.pdf (accessed 7 March 2016).

USDA-FSA (U.S. Department of Agriculture-Farm Services Agency) (2015) Monthly Summary December 2015. Available at: http://www.fsa.usda.gov/Assets/USDA-FSA-Public/usdafiles/Conservation/PDF/dec2015summary.pdf (accessed 4 February 2016).

USDA-NASS (U.S. Department of Agriculture-National Agricultural Statistics Services) (2015) Available at: http://www.nass.usda.gov/Statistics_by_Subject/?sector=CROPS (accessed 14 October 2015).

USDA-NRCS (U.S. Department of Agriculture-Natural Resources Conservation Service) (2013) Summary Report: 2010 National Resources Inventory, Natural Resource Conservation Service, Washington, DC, and Center for Survey Statistics and Methodology. Iowa State University, Ames, Iowa. Available at: http://www.nrcs.usda.gov/Internet/FSE_DOCUMENTS/stelprdb1167354.pdf (accessed 7 March 2016).

USDA-NRCS (U.S. Department of Agriculture-Natural Resources Conservation Service) (2015a) National soil survey handbook, title 430-VI. Available at: http://www.nrcs.usda.gov/wps/portal/nrcs/detail/soils/ref/?cid=nrcs142p2_054223#66 (accessed 15 October 2015).

USDA-NRCS (U.S. Department of Agriculture-Natural Resources Conservation Service) (2015b) Summary Report: 2012 National Resources Inventory, Natural Resources Conservation Service, Washington, DC, and Center for Survey Statistics and Methodology. Iowa State University, Ames, Iowa. Available at: http://www.nrcs.usda.gov/Internet/FSE_DOCUMENTS/nrcseprd396218.pdf (accessed 7 March 2016).

Wang X, He X, Williams J, Izaurralde R, Atwood J (2005) Sensitivity and uncertainty analyses of crop yields and soil organic carbon simulated with EPIC. *Transactions of the ASAE*, **48**, 1041.

Wang X, Williams JR, Gassman PW, Baffaut C, Izaurralde RC, Jeong J, Kiniry JR (2012) The EPIC and APEX models: use, calibration and validation. *Transactions of the ASABE*, **55**, 1447–1462.

Wright CK, Wimberly MC (2013) Recent land use change in the Western Corn Belt threatens grasslands and wetlands. *Proceedings of the National Academy of Sciences of the United States of America*, **110**, 4134–4139.

Zhang X, Izaurralde RC, Manowitz D et al. (2010) An integrative modeling framework to evaluate the productivity and sustainability of biofuel crop production systems. *Global Change Biology Bioenergy*, **2**, 258–277.

Zhang X, Izaurralde RC, Arnold JG, Williams JR, Srinivasan R (2013) Modifying the soil and water assessment tool to simulate cropland carbon flux: model development and initial evaluation. *Science of the Total Environment*, **463**, 810–822.

Zhang X, Izaurralde RC, Manowitz DH et al. (2015) Regional scale cropland carbon budgets: evaluating a geospatial agricultural modeling system using inventory data. *Environmental Modelling & Software*, **63**, 199–216.

Zhu Y, Fox RH (2003) Corn-soybean rotation effects on nitrate leaching. *Agronomy Journal*, **95**, 1028–1033.

4

Harvest management affects biomass composition responses of C4 perennial bioenergy grasses in the humid subtropical USA

CHAE-IN NA, JEFFREY R. FEDENKO, LYNN E. SOLLENBERGER and JOHN E. ERICKSON

Agronomy Department, University of Florida, Gainesville, FL 32611, USA

Abstract

Elephantgrass (*Pennisetum purpureum* Schum.) and energycane (*Saccharum* spp. hybrid) are high-yielding C4 grasses that are attractive biofuel feedstocks in the humid subtropics. Determining appropriate harvest management practices for optimal feedstock chemical composition is an important precursor to their successful use in production systems. In this research, we have investigated the effects of harvest timing and frequency on biomass nutrient, carbohydrate and lignin composition of UF1 and cv. Merkeron elephantgrasses and cv. L 79-1002 energycane. Biomass properties under increased harvest frequency (twice per year) and delayed harvest (once per year after frost) were compared with a control (once per year prior to frost). There were no differences between elephantgrass entries in structural carbohydrates; however, elephantgrasses had greater structural hexose concentration than energycane for single-harvest treatments (avg. 398 vs. 366 mg g^{-1}), a trait that is preferred for biofuel production. Delayed harvest of energycane decreased structural hexose compared with the control (374 vs. 357 mg g^{-1}) because nonstructural components accumulated in energycane stem as harvest was delayed. Frequent defoliation (2X) increased N, P, and ash concentrations (75% for N and P and 58% for ash) in harvested biomass compared with single-harvest treatments. We conclude that multiple harvests per year increase the harvest period during which feedstock is available for processing, but they do not result in optimal feedstock composition. In contrast, extending the period of feedstock supply by delaying a single harvest to after first freeze did not negatively affect cell wall constituent properties, while it increased length of the harvest period by ~30 days in the southeast USA.

Keywords: biofuel, elephantgrass, energycane, fiber composition, lignin, nitrogen, *Pennisetum purpureum*, *Saccharum* spp., structural carbohydrates

Introduction

Plant structural biomass consists primarily of three major types of polymers: cellulose, hemicellulose, and lignin. These polymers are strongly bonded by noncovalent forces and by covalent cross-linkages (Perez *et al.*, 2002). The cell wall is composed of long crystalline cellulose microfibrils embedded in a matrix of other polysaccharides (Perez *et al.*, 2002). In grasses, the predominant hemicellulosic polysaccharides in cell walls are glucuronoarabinoxylans that have a xylose backbone with arabinose and acetyl substitutions and constituents (Vermerris, 2008). Given the abundance of cellulose and hemicellulose in cell walls of perennial grasses, they represent a major source of structural carbohydrates for conversion to bioenergy. Lignin is the third major component of plant structural biomass and limits not only

the physical accessibility of cellulose and hemicellulose, but also the activity of cellulolytic enzymes (Jung *et al.*, 2013). This occurs as a result of multiple factors, including shielding cellulose from microbial degradation by providing a surface that cellulolytic enzymes adsorb to irreversibly (Akin, 2007; Vermerris, 2008).

The proportions of cellulose and lignin in biomass affect the yield of biochemical conversion processes. For instance, due to the high lignin concentration in wood, more ethanol can be produced from switchgrass (*Panicum virgatum* L.) than that from the same weight of wood biomass (McKendry, 2002). Actual concentrations of monomers from cellulose and hemicellulose can be analyzed by procedures established by the National Renewable Energy Laboratory (NREL). These procedures allow for the quantification of sugar monomers from extractives (nonstructural carbohydrates or soluble sugars) and structural carbohydrates, and measurement of acid-soluble and insoluble lignin (Sluiter, 2008a,b).

Correspondence: Chae-In Na
e-mail: dasan00@ufl.edu

The desired composition of biomass for bioenergy is dependent upon the postharvest conversion process used, and in most processes, lower levels of N and ash in the feedstock are preferred (Lewandowski & Heinz, 2003; Waramit et al., 2011). High levels of N and/or ash can reduce thermochemical conversion efficiency and increase wear, while decreasing energy generation when used for co-firing (Shahandeh et al., 2011). Perennial C4 grasses have up to two times greater N-use efficiency than C3 plants (Jakob et al., 2009), and high biomass production with low tissue N concentration makes them important candidate bioenergy grasses.

Harvest management affects not only biomass yield but also its composition (Casler & Boe, 2003; Lewandowski & Heinz, 2003; Adler et al., 2006). Miscanthus (Miscanthus giganteus) and giant reed (Arundo donax L.) showed a gradual decline in stem N concentration with increasing maturity in the United Kingdom (Smith & Slater, 2011). The decline in plant N concentration with increasing maturity has been associated with decreasing leaf proportion and much lower N concentration in stem than in leaves. For instance, average switchgrass leaf N concentration was 13.5 mg g^{-1} compared with 5.7 mg g^{-1} for stem (Shahandeh et al., 2011). Similarly, giant reed and switchgrass leaves contained twofold or greater N when compared with the stem and showed a decline in N and P concentrations from October to December (Kering et al., 2012).

Elephantgrass (or napiergrass; Pennisetum purpureum Schum.) is indigenous to sub-Saharan Africa in areas of rainfall exceeding 1000 mm. It is in the tribe Paniceae of the Poaceae family (Hanna et al., 2004). Elephantgrass is a robust, creeping rhizomatous plant that perennates in the tropics and subtropics. Sollenberger et al. (2014) reported that elephantgrass breeding lines developed for biomass averaged 3.77 m in plant height, 271 g in tiller mass, and 7.4 mm in tiller diameter in Florida. Sugarcane (Saccharum spp.) hybrids have a high concentration of cellulose instead of sucrose, and they are a potentially valuable feedstock resource for cellulosic ethanol production. This is why high-cellulose Saccharums are called 'energycanes (Saccharum spp. hybrid)' (León et al., 2010). To better understand Saccharum spp. as a biofuel crop, it is useful to note that there are three distinctive types. These include sugarcane (primarily sugar, conventional sugarcane), type I energycane (sugar and fiber), and type II energycane (primarily fiber).

Recent studies on harvest frequency and timing for elephantgrass and energycane grown in Florida, USA, found that two harvests annually negatively affected long-term biomass yield and plant persistence (Na et al., 2015a,b). Harvest management is an important determi-

nant of chemical composition of perennial grasses used for forage (Chaparro & Sollenberger, 1997), and it is reasonable to hypothesize that it affects composition of bioenergy feedstocks as well. Limited data exist describing harvest management effects on chemical composition of perennial grasses used for biofuel feedstocks grown on sandy soils in a subtropical environment. Therefore, the objective of this research was to determine the effect of harvest frequency and timing and grass entry on concentrations of plant structural and nonstructural carbohydrates, lignin, N, P, and ash in harvested biomass.

Materials and methods

Experimental site

The experiment was conducted during 2010 and 2011 at the Plant Science Research and Education Unit (PSREU) at Citra, FL (29.41°N, 82.17°W). The soil was a well-drained Candler sand (hyperthermic, uncoated Lamellic Quartzipsamments). Initial soil characterization (0–20 cm) showed an average soil pH of 7.0, and Mehlich-1 extractable P, K, and Mg of 54, 20, and 123 mg kg^{-1}, respectively. These concentrations are considered to be high for P, very low for K, and very high for Mg (Mylavarapu et al., 2009). Monthly average, maximum, and minimum temperatures (Fig. 1) and monthly precipitation (Fig. 2) are shown for the experimental period.

Treatments and experimental design

Treatments included all factorial combinations of three grass entries and three harvest management practices. Each treatment was replicated four times in a split-plot arrangement of a randomized complete block design. Harvest treatment was the main plot and grass entry was the subplot. The three grass entries included two elephantgrasses, cv. Merkeron (Burton, 1989) and a breeding line referred to as UF1, and cv. L 79-1002 type II energycane (Bischoff et al., 2008). These two species were chosen because earlier work with biomass feedstocks identified them as having the greatest potential in this region (Woodard & Prine, 1993). Merkeron and L 79-1002 were also chosen because they were the cultivars with the largest current presence in the region. Breeding line UF1 was included because previous research had demonstrated its high yield potential and preferred morphological characteristics (Sollenberger et al., 2014; Na et al., 2015a,b).

Three harvest management treatments were implemented that included different harvest frequency and timing. These were (i) two harvests per year (late July, named 2X-July; November harvest of regrowth after 2X-July harvest, named 2X-Nov), (ii) one harvest per year in fall [at initiation of Merkeron flowering (first entry to flower) and before first freeze; named 1X-Nov], and (iii) one harvest per year in winter (within 1 week following first freeze, with a freeze defined as a temperature of less than 0 °C at 2 m above soil level resulting in

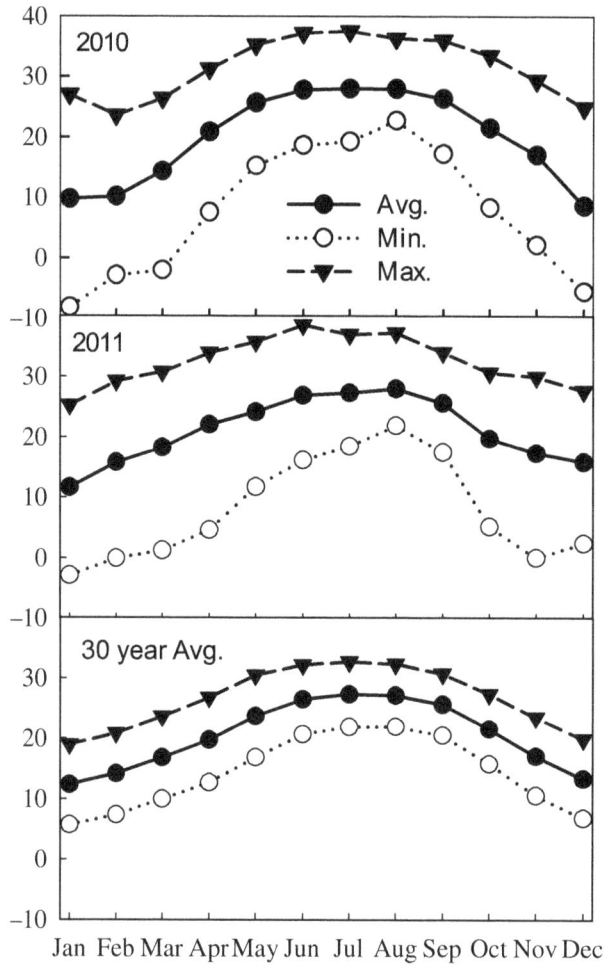

Fig. 1 Monthly average and monthly maximum and minimum air temperatures for 2010 and 2011 for the experimental location (available at Florida Automated Weather Network, http://fawn.ifas.ufl.edu), and the 30-year average for Gainesville, Florida (available at Florida Climate Center, http://climatecenter.fsu.edu).

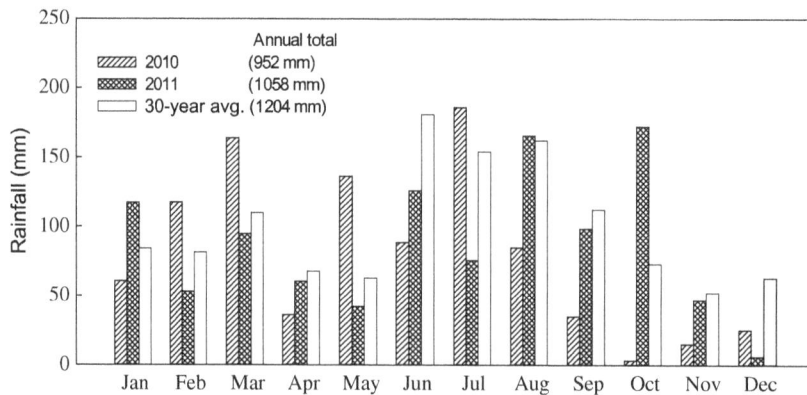

Fig. 2 Monthly rainfall for 2010 and 2011 for the experimental location (available at Florida Automated Weather Network, http://fawn.ifas.ufl.edu), and the 30-year average for Gainesville, Florida (available at Florida Climate Center, http://climatecenter.fsu.edu).

complete kill of leaves, or at full flowering of Merkeron, whichever happened first; named 1X-Dec). In 2010, harvests occurred on 30 July for 2X-July, 10 November for 1X-Nov and 2X-Nov, and 9 December for 1X-Dec. In 2011, harvest dates were 21 July for 2X-July, 8 November for 1X-Nov and 2X-Nov, and 15 December for 1X-Dec.

Plot preparation and management

Plots contained six rows, each 6 m long, with 1 m spacing between rows, and were established using aboveground stem pieces planted on December 15, 2009. Thus, the 2010 data are from the establishment year. In both years, N was applied as ammonium sulfate [$(NH_4)_2SO_4$] at a rate of 150 kg N ha^{-1}, and K was applied as muriate of potash (KCl) at a rate of 90 kg K ha^{-1}. Nutrients were split applied, with applications of 50 kg N and 45 kg K ha^{-1} in mid-April and 100 kg N and 45 kg K ha^{-1} in mid-May. No P was applied based on soil test results indicating sufficient levels. Limited irrigation was applied to the experiment only at visual signs of significant drought stress, for example, severe leaf rolling. There were five irrigation events in 2010 (total of 60 mm) and three irrigation events in 2011 (total of 50 mm) using a traveling gun irrigation system.

At harvest, four representative tillers (approximately 3 kg fresh weight) were selected from mid-row at least a meter from all borders for determination of plant part proportion, dry matter, and chemical composition of harvested feedstock. Tillers were hand-separated into leaf (blade and sheath) and stem (including inflorescence, if present) components. Leaf : stem ratio data were reported previously (Na *et al.*, 2015b). Samples were dried at 60 °C until constant weight. Stem samples were initially grounded through a hammer mill to reduce particle size. Stem and leaf samples were ground to pass a 1-mm stainless steel screen in a Wiley mill (Model 4 Thomas-Wiley Laboratory Mill, Thomas Scientific, Swedesboro, NJ, USA). Ground samples were transferred to an airtight container immediately.

Biomass fiber analysis

Each component of dried biomass (stem and leaves) was analyzed for chemical composition. A modified NREL procedure was used for compositional analysis (Sluiter, 2008a,b). In the modified NREL procedure, dried biomass was analyzed for nonstructural extractives, structural carbohydrates, and total lignin (Fedenko *et al.*, 2013). For nonstructural extractives, 100 ml of deionized (D.I.) water was added to 1 g of dried sample and autoclaved in sealed pressure tubes (ACE Glass, Inc., Vineland, NJ, USA) at 121 °C for 1 h. Samples were then vacuum-filtered through wet-strengthened 113 filter paper (Whatman, GE healthcare, Little Chalfont, UK). Filtered extractives were collected for nonstructural carbohydrate analysis. The captured structural biomass from extraction was dried at 40 °C and then weighed. A 0.3-g dry subsample was used for two-stage acid hydrolysis; the first stage was a 1-h incubation at 30 °C with concentrated sulfuric acid (72%, Fluka Alalytical; Sigma-Aldrich, St. Louis, MO, USA). In the second stage, sulfuric acid was diluted to 4% by adding D.I. water for 1-h digestion under elevated pressure and temperature (121 °C and 103 kPa). Hydrolyzed samples were vacuum-filtered through a medium-porosity filtering crucible (Coors #60531; CoorsTek, Golden, CO, USA). Two aliquots of filtrate were taken, with one analyzed in duplicate for acid-soluble lignin at a wavelength of 240 nm using a UV–vis spectrophotometer (StellarNet, Inc., Tampa, FL, USA). The second aliquot collected from hydrolyzed samples was neutralized to pH 5–7 using calcium carbonate. Insoluble lignin was determined gravimetri-

cally as total solids remaining in the crucible after vacuum filtration of hydrolyzed biomass. Total lignin was calculated as the sum of acid-soluble and insoluble lignin. Extractives and neutralized hydrolyzed samples were filtered through a 0.22-μm syringe filter and analyzed by HPLC (Perkin-Elmer Flexar system, Waltham, MA, USA) using a refractive index detector and a Bio-Rad Aminex HPX-87H column (300 × 7.8 mm) maintained at 50 °C. Sulfuric acid (HPLC grade, 4 mM) was used as the mobile phase at a flow rate of 0.4 ml min^{-1} with a 10-μl injection and 40-min run time. Perkin-Elmer's Chromera software was used to quantify peaks (Fedenko, 2011).

Extractives are considered to be the sum of all nonstructural components of plant tissue removed during extraction, of which soluble sugars are a major component. They were determined on a mass balance basis as the difference between initial sample weight and sample weight after hot water extraction (Sluiter, 2008a). Hexoses and pentoses are the sum of six-carbon (6C; glucose and mannose) and five-carbon (5C; xylose and arabinose) structural carbohydrates, respectively. Mannose was never present at minimum quantifiable levels in this study. Total lignin is the sum of acid-soluble and insoluble lignin.

Total nitrogen, phosphorus, and ash

For N and P analyses, a modified Kjeldahl procedure was conducted. Samples were digested using a modification of the aluminum block digestion procedure (Gallaher *et al.*, 1975). Sample weight was 0.25 g, catalyst used was 1.5 g of 9 : 1 K_2SO_4 : $CuSO_4$, and digestion was conducted for at least 4 h at 375 °C using 6 ml of sulfuric acid and 2 ml hydrogen peroxide. Analysis of digestate was carried out using the Technicon Autoanalyzer and semiautomated colorimetry to determine N and P in the digestate (Hambleton, 1977). Absolute dry matter was determined by oven drying at 105 °C to constant weight and then samples were ashed at 500 °C for 6 h to determine ash concentration.

Statistical analysis

Data were analyzed using mixed-model methods in PROC MIXED (SAS Institute, 2008). In all models, harvest treatment and grass entry were considered fixed effects. Year was considered a repeated measurement (fixed). Block and interactions with block were considered random effects. Because there were two harvests per year in the 2X treatment, samples from both harvests were analyzed separately in the laboratory, and the model effectively contains four levels of the harvest management treatment, that is, 2X-July, 2X-Nov, 1X-Nov, and 1X-Dec. Leaf, stem, and total aboveground biomass data were analyzed separately for each factor. Total aboveground biomass data were calculated as the weighted average of leaf and stem composition for each factor. Means were compared using the pdiff test of LSMEANS. All means reported in the text are least-square means and were considered different if $P \leq 0.05$. Because the effects of greatest interest were harvest management, entry, and their interaction, and because the harvest management × entry interaction was significant in most cases, data presented are the means for this interaction. If the interaction was not significant, only main effect means were compared.

Results

Extractives and total soluble sugars

A harvest management × entry interaction occurred ($P = 0.044$) for leaf extractives because UF1 elephantgrass showed a greater concentration than other entries in the 1X-Dec harvest, but there were no differences between elephantgrass entries for the other harvest treatments (Table 1). Single-harvest (1X-Nov and 1X-Dec) leaf-extractive concentrations were lower than either 2X-July or 2X-Nov for all grass entries. A harvest management × entry interaction occurred for stem extractives ($P < 0.001$, Table 1). L 79-1002 stem extractives were greatest in 1X-Dec (338 mg g^{-1}) followed by 1X-Nov (313 mg g^{-1}), while elephantgrass entries had greater stem extractives in 2X-July and 2X-Nov treatments (avg. 275 and 268 mg g^{-1}, respectively) than 1X-

Nov and 1X-Dec (avg. 256 and 252 mg g^{-1}, respectively). In 1X-Dec, L 79-1002 showed the greatest stem extractive concentration followed by UF1 (263 mg g^{-1}) and then Merkeron (241 mg g^{-1}). There also was a harvest management × entry interaction ($P < 0.001$) for extractive concentration in total aboveground biomass. There were no differences among entries within the 2X-Nov harvest, but L 79-1002 had greater extractive concentration than Merkeron for all other harvest treatments and greater than UF1 in both 1X-Nov and 1X-Dec.

Total soluble sugars in leaf tissue extractives showed a harvest management × entry interaction ($P = 0.044$) because there were no differences among entries within the 2X-July management, but differences among entries occurred for other harvest managements. One of the elephantgrasses always had the greatest concentration, except for 2X-July, when there were no differences among entries (Table 2). In the stem, a harvest management × entry interaction occurred ($P < 0.001$) for total soluble sugar concentrations (Table 2). For all three entries, 2X-July total soluble sugar concentration was

Table 1 Effect of grass entry and harvest management interaction on extractives in leaf ($P = 0.044$), stem ($P < 0.001$), and total above-ground biomass ($P < 0.001$). Data are means across four replicates and 2 years ($n = 8$)

Entry	Harvest management*			
	2X-July	2X-Nov	1X-Nov	1X-Dec
	mg extractives g^{-1} dry matter			
Leaf				
L 79-1002	199 a†B‡	196 aB	164 bB	173 bC
Merkeron	221 aA	226 aA	194 bA	187 bB
UF1	216 aA	225 aA	197 bA	201 bA
SE	4.0			
Stem				
L 79-1002	302 bcA	289 cA	313 bA	338 aA
Merkeron	265 aC	262 aC	250 abC	241 bC
UF1	285 aB	274 abB	262 bB	263 bB
SE	7.9			
Total				
L 79-1002	267 bA	251 cA	269 bA	289 aA
Merkeron	251 aB	248 abA	234 bcB	229 cC
UF1	262 aA	256 abA	244 bB	254 abB
SE	6.0			

*Harvest management treatments were harvested twice per year in July and November (2X-July and 2X-Nov), once per year in November (1X-Nov), and once per year after first freeze in December (1X-Dec).
†Harvest management means within an entry not followed by the same lower case letter are different ($P < 0.05$).
‡Entry means within a harvest management not followed by the same upper case letter are different ($P < 0.05$).

Table 2 Effect of grass entry and harvest management interaction on total soluble sugars concentration in leaf ($P = 0.044$), stem ($P < 0.001$), and total above-ground biomass ($P < 0.001$). Data are means across four replicates and 2 years ($n = 8$)

Entry	Harvest management*			
	2X-July	2X-Nov	1X-Nov	1X-Dec
	mg total soluble sugars g^{-1} dry matter			
Leaf				
L 79-1002	23 b†A‡	32 aB	14 cB	22 bB
Merkeron	26 bA	40 aA	24 bA	22 bB
UF1	26 bA	35 aAB	29 abA	33 abA
SE	2.9			
Stem				
L 79-1002	152 cA	165 cA	228 bA	280 aA
Merkeron	97 bB	134 aB	138 aC	142 aC
UF1	138 bA	147 bB	161 bB	189 aB
SE	11.2			
Total				
L 79-1002	109 cA	110 cA	163 bA	205 aA
Merkeron	74 cB	96 bB	106 abC	116 aC
UF1	102 cA	106 cAB	126 bB	167 aB
SE	6.9			

*Harvest management treatments were harvested twice per year in July and November (2X-July, and 2X-Nov), once per year in November (1X-Nov), and once per year after first freeze in December (1X-Dec).
†Harvest management means within an entry not followed by the same lower case letter are different ($P < 0.05$).
‡Entry means within a harvest management not followed by the same upper case letter are different ($P < 0.05$).

less than 1X-Dec. The magnitude of this difference varied among entries (84% increase in L 79-1002, 51% in Merkeron, and 36% in UF1, respectively). Across treatments, total soluble sugars in the leaves were several times lower than that in the stem (avg. 27 vs. 164 mg soluble sugars g^{-1}). In total aboveground biomass, there was a harvest management x entry interaction for soluble sugars (Table 2, $P < 0.001$). Because the concentration of leaf soluble sugar was very small and because leaves composed a small percentage of harvested biomass, particularly for 1X-Nov and 1X-Dec, total aboveground biomass soluble sugar concentrations generally followed the same pattern as stem. Concentration of total soluble sugars varied mostly among entries for 1X vs. 2X harvest treatments. Interestingly, among elephantgrass entries, UF1 showed 44% greater soluble sugar concentration in total aboveground biomass compared with Merkeron. Further investigation of UF1 soluble sugar concentration is needed to determine the specific sugars present.

Structural hexose and pentose

In the leaves, structural hexose concentration was affected by harvest management and entry ($P < 0.001$ for both). Entry L 79-1002 showed greater hexose concentration than elephantgrass entries by 7% (Table 3). Structural hexose concentrations were greater in 1X-Nov and 1X-Dec followed by 2X-July and then 2X-Nov. In stem, there was a harvest management x entry interaction ($P = 0.001$). Stems of elephantgrass entries had greater structural hexose concentration than those of L 79-1002 for all harvest treatments (Table 3). Structural hexose concentration in L 79-1002 was similar for 2X and 1X-Nov treatments, but it was least for 1X-Dec. Merkeron and UF1 had lowest structural hexose concentration in 2X-Nov, while UF1 had greater hexose concentration in 1X-Nov and then 1X-Dec, but for Merkeron, the 1X-Nov and 1X-Dec treatments were not different.

In total aboveground biomass, harvest management x entry interaction occurred ($P = 0.001$). For 2X-Nov, there were no differences among grass entries; however, entry L 79-1002 had lower structural hexose concentration than the elephantgrasses in 2X-July, 1X-Nov, and 1X-Dec (Table 3). Comparing harvest managements within an entry, both Merkeron and UF1 had greatest hexose concentration in 1X-Nov and 1X-Dec and least in 2X-Nov, while energycane hexose concentration varied relatively little and was greater in 1X-Nov than in 2X-Nov and 1X-Dec.

Concentration of structural pentose in leaves was affected by harvest management ($P = 0.001$). The 2X-July treatment had the lowest pentose concentration compared with other treatments, but there were no

Table 3 Effect of grass entry and harvest management interaction on structural hexose concentration in leaf ($P = 0.562$), stem ($P = 0.001$), and total above-ground biomass ($P = 0.001$). Data are means across four replicates and 2 years ($n = 8$)

Entry	Harvest management*				
	2X-July	2X-Nov	1X-Nov	1X-Dec	Mean
	mg hexose g^{-1} dry matter				
Leaf					
L 79-1002	371	356	404	395	381 A[†]
Merkeron	346	334	370	369	355 B
UF1	342	333	369	357	350 B
Mean	353 b[‡]	341 c	381 a	374 a	
SE	5.2				
Stem					
L 79-1002	363 a[§]B[¶]	361 aB	361 aB	342 bB	
Merkeron	400 bA	385 cA	415 aA	406 abA	
UF1	397 bA	378 cA	413 aA	394 bA	
SE	7.5				
Total					
L 79-1002	367 ab[§]B[¶]	360 bA	374 aB	357 bB	
Merkeron	382 bA	365 cA	403 aA	399 aA	
UF1	381 bA	362 cA	402 aA	389 bA	
SE	6.2				

*Harvest management treatments were harvested twice per year in July and November (2X-July, and 2X-Nov), once per year in November (1X-Nov), and once per year after first freeze in December (1X-Dec).

[†]Entry means across harvest managements not followed by the same upper case letter are different ($P < 0.05$).

[‡]Harvest management means across entries not followed by the same lower case letter are different ($P < 0.05$).

[§]Harvest management means within an entry not followed by the same lower case letter are different ($P < 0.05$).

[¶]Entry means within a harvest management not followed by the same upper case letter are different ($P < 0.05$).

differences among the other harvests. There was harvest management x entry interaction for stem pentose concentration ($P = 0.003$). The most pronounced difference was that 2X-Nov had the greatest pentose concentration among all entries. For both 2X treatments, L 79-1002 had the greatest pentose concentration compared with the two elephantgrasses (Table 4). There were no differences among entries for the 1X-Nov harvest, while for 1X-Dec, Merkeron had greater pentose concentration than L 79-1002. In total aboveground biomass, structural pentose concentration was affected by harvest management and entry ($P = 0.001$, $P < 0.001$, respectively). Entry L 79-1002 had greater pentose concentration than the elephantgrasses; however, the differences among entries were relatively small (4–5 mg of pentose g^{-1}). The 2X-Nov harvest management had greater pentose concentration than the other defoliation treatments.

Table 4 Effect of grass entry and harvest management interaction on structural pentose concentration in leaf ($P = 0.977$), stem ($P = 0.003$), and total above-ground biomass ($P = 0.139$). Data are means across four replicates and 2 years ($n = 8$)

Entry	Harvest management*				
	2X-July	2X-Nov	1X-Nov	1X-Dec	Mean
	mg pentose g^{-1} dry matter				
Leaf					
L 79-1002	252	266	273	272	266A
Merkeron	251	266	271	272	265A
UF1	254	273	279	274	270A
Mean	252 b[†]	268 a	274 a	273 a	
SE	4.0				
Stem					
L 79-1002	217 b[‡]A[§]	244 aA	209 bcA	202 cB	
Merkeron	206 bB	233 aB	209 bA	213 bA	
UF1	206 bB	230 aB	209 bA	209 bAB	
SE	3.1				
Total					
L 79-1002	229	253	228	222	233 A[¶]
Merkeron	221	246	225	225	229 B
UF1	222	246	228	218	228 B
Mean	224 b[‡]	248 a	227 b	222 b	
SE	2.9				

*Harvest management treatments were harvested twice per year in July and November (2X-July, and 2X-Nov), once per year in November (1X-Nov), and once per year after first freeze in December (1X-Dec).
[†]Harvest management means across entries not followed by the same lower case letter are different ($P < 0.05$).
[‡]Harvest management means within an entry not followed by the same lower case letter are different ($P < 0.05$).
[§]Entry means within a harvest management not followed by the same upper case letter are different ($P < 0.05$).
[¶]Entry means across harvest managements not followed by the same upper case letter are different ($P < 0.05$).

Lignin

Leaf lignin concentration was affected by harvest management ($P = 0.005$) and entry ($P < 0.001$). Entry L 79-1002 leaf lignin concentration was greater than in Merkeron and UF1 by 7–8% (218, 204, and 201 mg lignin g^{-1}, respectively, Table 5). Single-harvest management (1X-Nov and 1X-Dec; 214 and 212 mg lignin g^{-1}, respectively) had greater leaf lignin concentration than 2X harvest managements (2X-July and 2X-Nov; 204 and 202 mg lignin g^{-1}, respectively). For stem lignin concentration, there was a harvest management × entry interaction ($P = 0.002$). For all entries, 1X-Nov and 1X-Dec biomass had greater stem lignin concentration than either of the 2X harvest managements (Table 5). The interaction occurred because there were no differences between Merkeron and UF1 for 1X-Nov; in contrast,

within the other harvest management treatments, Merkeron had greater stem lignin concentration than UF1 which had a greater lignin concentration than L 79-1002. In total aboveground biomass, a harvest management x entry interaction occurred ($P = 0.002$). Similar to stem lignin concentration, both the 2X harvest managements generally had lesser lignin concentration compared with 1X-Nov and 1X-Dec. Merkeron generally had greater lignin concentration than energycane.

Nitrogen, phosphorus, and ash

Nitrogen concentration in the leaves was affected by harvest management and entry ($P < 0.001$ for both). Energycane leaves had a lower N concentration (9.9 mg g^{-1}) than elephantgrass entries (average of 11.5 mg g^{-1}, Table 6). Leaf N concentration averaged approximately 37% lower for the single-harvest treat

Table 5 Effect of grass entry and harvest management interaction on lignin concentration in leaf ($P = 0.069$), stem ($P = 0.002$), and total above-ground biomass ($P = 0.002$). Data are means across four replicates and 2 years ($n = 8$)

Entry	Harvest management*				
	2X-July	2X-Nov	1X-Nov	1X-Dec	Mean
	mg lignin g^{-1} dry matter				
Leaf					
L 79-1002	213	213	226	221	218 A[†]
Merkeron	199	197	210	211	204 B
UF1	200	197	204	202	201 C
Mean	204 b[‡]	202 b	214 a	212 a	
SE	2.8				
Stem					
L 79-1002	167 b[§]C[¶]	167 bC	182 aB	177 aC	
Merkeron	184 bA	182 bA	203 aA	206 aA	
UF1	176 bB	174 bB	199 aA	199 aB	
SE	3.5				
Total					
L 79-1002	183 c[§]B[¶]	186 bcAB	195 aB	190 abC	
Merkeron	189 bA	188 bA	204 aA	207 aA	
UF1	183 bB	183 bB	201 aA	200 aB	
SE	2.9				

*Harvest management treatments were harvested twice per year in July and November (2X-July, and 2X-Nov), once per year in November (1X-Nov), and once per year after first freeze in December (1X-Dec).
[†]Entry means across harvest managements not followed by the same upper case letter are different ($P < 0.05$).
[‡]Harvest management means across entries not followed by the same lower case letter are different ($P < 0.05$).
[§]Harvest management means within an entry not followed by the same lower case letter are different ($P < 0.05$).
[¶]Entry means within a harvest management not followed by the same upper case letter are different ($P < 0.05$).

Table 6 Effect of grass entry and harvest management interaction on nitrogen concentration in leaf ($P = 0.334$), stem ($P = 0.942$), and total above-ground biomass ($P = 0.955$). Data are means across four replicates and 2 years ($n = 8$)

	Harvest management*				
	2X-July	2X-Nov	1X-Nov	1X-Dec	Mean
Entry	mg nitrogen g^{-1} dry matter				
Leaf					
L 79-1002	12.5	12.9	7.4	6.6	9.9 B[†]
Merkeron	14.1	14.2	9.5	8.2	11.5 A
UF1	13.0	13.7	9.5	9.5	11.4 A
Mean	13.2 a[‡]	13.6 a	8.8 b	8.1 b	
SE	0.76				
Stem					
L 79-1002	5.3	4.4	3.3	3.0	4.0 B
Merkeron	7.0	5.2	4.4	3.9	5.1 A
UF1	5.5	4.9	3.4	3.4	4.3 B
Mean	5.9 a	4.9 b	3.7 c	3.5 c	
SE	0.56				
Total					
L 79-1002	7.8	7.9	4.8	4.1	6.1 B
Merkeron	9.1	8.7	5.8	4.8	7.1 A
UF1	7.6	8.0	4.6	4.2	6.1 B
Mean	8.2 a	8.2 a	5.0 b	4.4 b	
SE	0.58				

*Harvest management treatments were harvested twice per year in July and November (2X-July, and 2X-Nov), once per year in November (1X-Nov), and once per year after first freeze in December (1X-Dec).

[†]Entry means across harvest managements not followed by the same upper case letter are different ($P < 0.05$).

[‡]Harvest management means across entries not followed by the same lower case letter are different ($P < 0.05$).

ments (1X-Nov and 1X-Dec) compared with the 2X treatments.

In stem, N concentration was affected by harvest management ($P = 0.002$) and entry ($P = 0.011$). Merkeron (5.1 mg g^{-1}) N concentration was greater than other entries (4.3 mg g^{-1} in UF1 and 4.0 mg g^{-1} in L 79-1002, Table 6). The 2X-July treatment, which was harvested about 8 week after the final spring fertilizer application, had greatest N concentration, while 1X-Nov and 1X-Dec harvest management were lowest in N. Similar to stem N, total aboveground biomass N concentration was affected by harvest management ($P < 0.001$) and entry ($P = 0.004$). Merkeron had greatest N concentration. Nitrogen concentration decreased by 39% from 2X-July and 2X-Nov to 1X-Nov and 46% to 1X-Dec.

Phosphorus concentration in leaves, stem, and total aboveground biomass showed similar trends to N concentrations. Phosphorus concentration was affected by

harvest management and entry in leaf ($P = 0.002$ and 0.001, respectively), stem ($P < 0.001$ for both), and total aboveground biomass ($P = 0.001$ and <0.001, respectively). In both plant parts and in total biomass, Merkeron had the greatest overall P concentration (Table 7). In leaves, stem, and total plant biomass, P concentration was greater for the 2X treatments compared with 1X-Nov and 1X-Dec. For example, in total aboveground biomass, P concentration in 2X-Nov was approximately twice as great as the single-harvest treatment.

There were harvest management and entry effects on leaf ash concentration ($P = 0.012$ and <0.001, respectively). Entry UF1 had the greatest leaf ash concentration followed by Merkeron and then L 79-1002 (Table 8). The 1X-Nov and 1X-Dec harvested biomass had lesser ash concentration (48 and 44 mg g^{-1}, respectively) than 2X-July and 2X-Nov treatments (54 mg g^{-1} for both). In stem, ash concentration was affected by harvest management ($P < 0.001$). The average of the sin-

Table 7 Effect of grass entry harvest and management interaction on phosphorus concentration in leaf ($P = 0.139$), stem ($P = 0.360$), and total above-ground biomass ($P = 0.348$). Data are means across four replicates and 2 years ($n = 8$)

	Harvest management*				
	2X-July	2X-Nov	1X-Nov	1X-Dec	Mean
Entry	mg phosphorus g^{-1} dry matter				
Leaf					
L 79-1002	1.40	1.94	0.72	0.57	1.16 B[†]
Merkeron	1.95	2.13	1.14	1.03	1.56 A
UF1	1.32	1.51	0.88	1.00	1.17 B
Mean	1.56 a[‡]	1.86 a	0.91 b	0.87 b	
SE	0.18				
Stem					
L 79-1002	1.30	1.44	0.70	0.63	1.02 B
Merkeron	1.64	1.99	1.21	1.09	1.48 A
UF1	1.07	1.56	0.78	0.94	1.09 B
Mean	1.34 b	1.66 a	0.89 c	0.89 c	
SE	0.14				
Total					
L 79-1002	1.31	1.66	0.73	0.62	1.08 B
Merkeron	1.71	2.04	1.19	1.07	1.50 A
UF1	1.13	1.53	0.80	0.95	1.10 B
Mean	1.38 b	1.74 a	0.90 c	0.88 c	
SE	0.15				

*Harvest management treatments were harvested twice per year in July and November (2X-July, and 2X-Nov), once per year in November (1X-Nov), and once per year after first freeze in December (1X-Dec).

[†]Entry means across harvest managements not followed by the same upper case letter are different ($P < 0.05$).

[‡]Harvest management means across entries not followed by the same lower case letter are different ($P < 0.05$).

Table 8 Effect of grass entry and harvest management interaction on ash concentration in leaf ($P = 0.359$), stem ($P = 0.257$), and total above-ground biomass ($P = 0.853$). Data are means across four replicates and 2 years ($n = 8$)

	Harvest management*				
	2X-July	2X-Nov	1X-Nov	1X-Dec	Mean
Entry	mg ash g^{-1} dry matter				
Leaf					
L 79-1002	44	45	33	32	38 C[†]
Merkeron	59	57	50	43	52 B
UF1	60	60	60	56	59 A
Mean	54 a[‡]	54 ab	48 bc	44 c	
SE	3.9				
Stem					
L 79-1002	44	43	28	22	34 A
Merkeron	40	43	26	25	34 A
UF1	39	46	25	27	34 A
Mean	41 a	44 a	26 b	24 b	
SE	2.5				
Total					
L 79-1002	44	44	29	24	35 B
Merkeron	46	48	32	29	39 A
UF1	45	51	32	31	40 A
Mean	45 a	48 a	31 b	28 b	
SE	2.4				

*Harvest management treatments were harvested twice per year in July and November (2X-July, and 2X-Nov), once per year in November (1X-Nov), and once per year after first freeze in December (1X-Dec).

[†]Entry means across harvest managements not followed by the same upper case letter are different ($P < 0.05$).

[‡]Harvest management means across entries not followed by the same lower case letter are different ($P < 0.05$).

gle-harvest treatments (1X-Nov and 1X-Dec) was 41% less than the average of 2X-July and 2X-Nov (Table 8). In total aboveground biomass, ash concentration was affected by harvest management and entry ($P < 0.001$, <0.012, respectively). Elephantgrass entries had greater (up to 11%) ash concentration than energycane (Table 8). Because of generally greater ash concentration in leaves than stem and leaf senescence over the season, total ash concentration in 1X-Nov and 1X-Dec was less than in either of the 2X treatments.

Discussion

Extractives and total soluble sugars

Elephantgrass and energycane extractive concentration responses to harvest management were different. Elephantgrass biomass had greatest extractives concentrations when harvested twice per year, while energycane extractives were greatest when biomass was harvested once per year. This response was due to accumulation of extractives in energycane stems late in the growing season. Because of the preponderance of stem in harvested biomass, especially when harvested once per year, the effects of leaf extractives concentration were small. There are very limited data available for extractives concentration in perennial C4 grasses. An elephantgrass study in Vietnam reported 182 mg g^{-1} of extractives measured using a hot water extraction (Hoa *et al.*, 2008). As observed in the present study, Fedenko *et al.* (2013) found that total extractives concentration was greater for energycane than elephantgrass when harvested once per year in fall prior to a freeze event (avg. 234 vs. 188 mg g^{-1}) due to sugar accumulation in energycane. In Colorado, extractives concentration varied among entries including 164 mg g^{-1} for the C4 perennial switchgrass, 254 mg g^{-1} for the C3 perennial grass tall fescue (*Festuca arundinacea* L.), and 172 mg g^{-1} for maize stover (*Zea may* L.; Thammasouk *et al.*, 1997).

Stem was the major contributor of soluble sugars to total aboveground biomass for both grass species in the present study. Although differences existed in soluble sugar concentration in leaves, they contributed marginally to differences on a total harvested biomass basis because leaf proportion was low and leaf abscission occurred late in the growing season (Na *et al.*, 2015b). Single-harvest (1X-Nov and Dec) energycane showed greatest soluble sugar concentration in total harvested biomass because of greater accumulation in the stem, mostly of sucrose. Although energycane stem has a relatively low sugar concentration compared with sugarcane (*Saccharum* spp. hybrid), it shows measurable levels of soluble sugar accumulation late in the season (Bischoff *et al.*, 2008; Kim & Day, 2011; Tew *et al.*, 2011). Unlike structural carbohydrates, non-cell wall carbohydrates are directly fermentable; however, they are susceptible to microbial degradation during storage (Dien *et al.*, 2006). Moreover, by degradation of free sugars, cellulosic fermentation inhibitors including furfural can be produced (Tran & Chambers, 1986; Zhang *et al.*, 2011). If they cannot be used for conversion to biofuel, they will decrease feedstock quality as they reduce concentration of structural carbohydrates proportionally. In the previous research, L 79-1002 had 101% (first year) and 45% (second year) greater soluble sugar concentration than Merkeron (Fedenko *et al.*, 2013), a similar result to the present study.

Structural hexose and pentose

For lignocellulosic biomass, ethanol fermentation efficiency is determined by the quality of structural carbohydrates (Lu & Mosier, 2008), so it is critical to investigate tissue structural carbohydrates. Energycane had greater leaf structural hexose concentration than elephantgrass entries; however, elephantgrass entries showed greater hexose concentration in both stem and total aboveground biomass than energycane. This was due at least in part to an increase in total soluble sugar over time in energycane stem tissue, which resulted in a dilution of the concentration of structural components. A large amount of glucose from cell wall structural components is advantageous for ethanol production because glucose can be converted by most organisms more efficiently to ethanol than most other sugars, especially pentose sugars (Lu & Mosier, 2008). Similar to values observed in the present study, structural hexose concentrations of 347 mg g^{-1} have been reported for mixtures of C4 grasses in Minnesota, USA (Gillitzer *et al.*, 2013). Chemical composition of six-carbon structural sugars was 300–337 mg g^{-1} for switchgrass (Xu *et al.*, 2011). In the prior work at the location of the present study, structural glucose concentration in elephantgrass and energycane was 374 vs. 366 mg g^{-1} in the first year and 448 vs. 432 mg g^{-1} in the second year, respectively (Fedenko *et al.*, 2013). Late harvests and increasing maturity have been observed to increase glucose and nonglucose structural sugars in reed canarygrass (*Phalaris arundinacea* L.) (233–286 mg g^{-1}) and switchgrass (294–340 mg g^{-1}) (Dien *et al.*, 2006). This corresponds to what was observed for elephantgrass, but not for energycane, in the present study.

Unlike structural hexose results, energycane showed the greatest pentose concentration in total aboveground biomass. As pentose concentrations increased, structural hexose concentrations decreased in the 2X-Nov harvest for total biomass. As indicated earlier, because of the fermentation inefficiency of pentose by most microorganisms, it is less favorable than hexose (Lu & Mosier, 2008). In switchgrass, the concentration of five-carbon structural sugars ranged from 183 to 196 mg g^{-1} (Xu *et al.*, 2011). Structural xylose concentration in season-long growth of elephantgrass in Florida was similar to that of energycane (135 and 145 mg g^{-1}, respectively) (Fedenko, 2011). For both reed canarygrass and switchgrass, structural five-carbon sugars increased with increasing maturity (147–191 mg g^{-1} in reed canarygrass; 210–253 mg g^{-1} in switchgrass) (Dien *et al.*, 2006). This trend was not observed in the present study. Pentose concentrations of 198 and 187 mg g^{-1} from structural carbohydrates have been reported for a mixture of C4 grasses and for a mixture of C3 grasses, respectively, in Minnesota (Gillitzer *et al.*, 2013). Although there were statistical differences in pentose concentrations in the present study, unlike hexose concentrations, pentose was relatively constant across harvest managements and entries (avg. of 248 mg g^{-1}) except for 2X-Nov, for which the range was 218–229 mg g^{-1}.

Lignin

Merkeron consistently exhibited the greatest total lignin concentration, and single-harvest per year treatments (1X-Nov and 1X-Dec) showed greater lignin concentration in total aboveground biomass than 2X treatments. Total lignin concentration in the present study is similar to 204–291 mg g^{-1} reported in a previous study for full-season growth of elephantgrass and energycane (Fedenko *et al.*, 2013). The results of both studies agree that elephantgrass generally has greater lignin concentration than energycane. The reported lignin concentrations are also similar to those for three switchgrasses, which ranged from 214 to 230 mg g^{-1} for full-season growth (Xu *et al.*, 2011). Prior research with alfalfa (*Medicago sativa* L.) and pine (*Pinus* sp.) has shown that greater lignin concentrations decrease pretreatment efficiency relative to lesser concentrations and that even slight changes are sufficient to markedly affect conversion efficiency (Dixon & Chen, 2007; Studer *et al.*, 2011). Therefore, greater lignin concentrations in elephantgrass may negatively affect pretreatment and biofuel production.

Nitrogen, phosphorus, and ash

In general, 2X treatments showed greater N concentration compared with 1X treatments. Merkeron had the greatest N concentration in total aboveground biomass. Similar to responses in the present study, harvest frequency of reed canarygrass in Iowa affected N concentration. When biomass was harvested twice per year (June and fall), N concentration was 13.4 (June) and 8.8 mg g^{-1} (fall) compared with 8.3 mg g^{-1} for a single harvest in fall (Tahir *et al.*, 2011). In Japan, elephantgrass that was harvested less frequently had lower N concentrations in both leaves and stem than that harvested more frequently (Hsu *et al.*, 1990). In the same study, the difference between leaf and stem N concentration increased as harvest interval was extended. When plants reached a height of 1 m, N concentrations were 10.8 mg g^{-1} in leaf vs. 5.0 mg g^{-1} in stem, and when plants were 2 m tall, N concentrations were 8.6 mg g^{-1} in leaf vs. 2.6 mg g^{-1} in stem. Switchgrass biomass N concentration decreased over the season (Madakadze *et al.*, 1999); however, N concentration was relatively constant after September at about 5 mg g^{-1}. Miscanthus N concentration decreased until October and then remained constant for the rest of the season (Heaton *et al.*, 2009). Similarly, delaying harvest from 1X-Nov to 1X-Dec in the present study did not result in a significant decrease in either leaf or stem N, but 1X-Dec N concentration of total aboveground biomass was slightly less than 1X-Nov because of leaf senescence and the resultant decrease in leaf : stem ratio in 1X-Dec (Na *et al.*, 2015a). When grown with no fertilizer on loamy sands in Georgia, Merkeron elephantgrass had greater overall N concentration than L 79-1002 (3.8 vs. 2.7 mg g^{-1}) (Knoll *et al.*, 2012), similar to the pattern of response in the present study, but Singh *et al.* (2015) reported no difference in total aboveground biomass N concentration between L 79-1002 and Merkeron across three sites in Florida.

Similar to N concentration, Merkeron showed the greatest tissue P concentration. The 1X treatments had lesser P concentrations than either of the 2X treatments. Similar results were found for reed canarygrass in Iowa, where P concentration was greater for two harvests per year than for a single harvest (Tahir *et al.*, 2011). Switchgrass P concentration was also affected by harvest management, and decreased slightly from the first to second harvest of a two-harvest per year treatment, and was lowest in the fall harvest of a single-harvest per year treatment (1.3, 1.1, and 0.8 mg g^{-1}, respectively) (Guretzky *et al.*, 2011). Giant reed and switchgrass P concentrations decreased from October to December in Oklahoma (Kering *et al.*, 2012). In switchgrass, leaf N was much greater compared with stem, but leaf and stem P concentrations varied only slightly (Shahandeh *et al.*, 2011).

In the present study, harvest management affected ash concentration. For the 1X treatment, total aboveground elephantgrass biomass ash concentration was lower than that for either harvest of the 2X treatment. The results of the present study are similar to those reported previously for Merkeron and L 79-1002. When 4-year data were averaged, Merkeron had greater ash concentration than L 79-1002 when grown with no fertilizer on loamy sands of the coastal plain of Georgia, USA (45.9 vs. 34.4 mg g^{-1}, respectively) (Knoll *et al.*, 2012). Reed canarygrass responded differently than elephantgrass and energycane, as ash concentration increased slightly over the season (96, 106, and 107 mg g^{-1} for two harvests per year in June and October, and for one harvest in October, respectively) (Tahir *et al.*, 2011). In switchgrass, ash concentration in Iowa peaked in July (71 mg g^{-1}) and decreased until fall after which it remained relatively constant between 43 and 45 mg g^{-1} (Wilson *et al.*, 2013). Limited data for plant part ash concentration indicate that leaf was found to have greater ash than stem (Summers *et al.*, 2001; Bakker & Elbersen, 2005).

In conclusion, harvest frequency (2X vs. single) significantly affects compositional quality of perennial grasses. A single harvest in fall appears to maximize the concentration of cellulose in total biomass, and delaying harvest from 1X-Nov to 1X-Dec had little impact on most response factors. The exception to this was for extractives

and soluble sugars in energycane, which increased significantly from 1X-Nov to 1X-Dec harvest. A relatively greater concentration of soluble sugars in energycane reduced the concentrations of the structural components. A major factor affecting concentration differences due to harvest management was differences in leaf:stem ratio because leaves generally had greater N, P, and ash than stem. Later harvests were associated with lesser leaf percentage in total biomass, which caused N, P, and ash to decrease in 1X-Nov and 1X-Dec relative to 2X treatments. Total aboveground biomass N concentration in Merkeron decreased to a greater extent than the other entries because of greater leaf abscission (Na *et al.*, 2015b). The 1X-Nov and 1X-Dec treatments of all entries had lesser concentrations of N and ash, components which can negatively affect some conversion processes, but they had slightly greater lignin concentration than 2X harvests. On the basis of this study results, a single harvest in fall will result in preferred biomass compositional properties, with maximum cellulose concentration of both elephantgrass and energycane. However, unlike energycane, delaying a single harvest of elephantgrass to after a freeze will not compromise cellulose concentration. This study provides needed information on nutrient, carbohydrate, and lignin concentrations and carbohydrate composition in elephantgrass and energycane that will guide decision making regarding biomass harvest frequency and timing for perennial grasses grown for biomass in the southeastern USA.

Acknowledgements

We gratefully acknowledge the assistance of Kenneth Woodard, Dwight Thomas, Miguel Castillo, Kim Mullenix, Nick Krueger, and Marcelo Wallau for their help in data collection and field management. This study is part of the Ph.D. dissertation of the senior author and was directed by Dr. Lynn E. Sollenberger at the University of Florida. This research is funded by the Florida Department of Agriculture and Consumer Services.

References

Adler PR, Sanderson MA, Boateng AA, Weimer PJ, Jung HJ (2006) Biomass yield and biofuel quality of switchgrass harvested in fall or spring. *Agronomy Journal*, **98**, 1518–1525.

Akin DE (2007) Grass lignocellulose: strategies to overcome recalcitrance. *Applied Biochemistry and Biotechnology*, **137**, 3–15.

Bakker RR, Elbersen HW (2005) Managing ash content and quality in herbaceous biomass: an analysis from plant to product. In: *14th European Biomass Conference and Exhibition* (ed. Sjunnesson, L. *et al.*), pp. 210–213. ETA – Renewable Energies, Paris, France.

Bischoff KP, Gravois KA, Reagan TE, Hoy JW, Kimbeng CA, Laborde CM, Hawkins GL (2008) Registration of 'L 79-1002' Sugarcane. *Journal of Plant Registrations*, **2**, 211–217.

Burton GW (1989) Registration of 'Merkeron' napiergrass. *Crop Science*, **29**, 1327–1327.

Casler MD, Boe AR (2003) Cultivar X environment interactions in switchgrass. *Crop Science*, **43**, 2226–2233.

Chaparro CJ, Sollenberger LE (1997) Nutritive value of clipped 'Mott' elephantgrass herbage. *Agronomy Journal*, **89**, 789–793.

Dien BS, Jung HJ, Vogel KP *et al.* (2006) Chemical composition and response to dilute-acid pretreatment and enzymatic saccharification of alfalfa, reed canary-grass, and switchgrass. *Biomass and Bioenergy*, **30**, 880–891.

Dixon RA, Chen F (2007) Lignin modification improves fermentable sugar yields for biofuel production. *Nature Biotechnology*, **25**, 759–761.

Fedenko JR (2011) Biomass yield and composition of potential bioenergy feedstocks. Ph.D dissertation. University of Florida, Florida, USA.

Fedenko JR, Erickson JE, Woodard KR *et al.* (2013) Biomass production and composition of perennial grasses grown for bioenergy in a subtropical climate across Florida, USA. *Bioenergy Research*, **6**, 1082–1093.

Gallaher RN, Weldon CO, Futral JG (1975) An aluminum block digester for plant and soil analysis. *Soil Science Society of America Journal*, **39**, 803–806.

Gillitzer PA, Wyse DL, Sheaffer CC, Taff SJ, Lehman CC (2013) Biomass production potential of grasslands in the oak savanna region of Minnesota, USA. *Bioenergy Research*, **6**, 131–141.

Guretzky JA, Biermacher JT, Cook BJ, Kering MK, Mosali J (2011) Switchgrass for forage and bioenergy: harvest and nitrogen rate effects on biomass yields and nutrient composition. *Plant and Soil*, **339**, 69–81.

Hambleton LG (1977) Semiautomated method for simultaneous determination of phosphorus, calcium and crude protein in animal feeds. *Journal-Association of Official Analytical Chemists*, **60**, 845–852.

Hanna WW, Chaparro CJ, Mathews BW, Burns JC, Sollenberger LE, Carpenter JR (2004) Perennial Pennisetums. In: *Warm-Season (C4) Grasses* (eds Moser LE, Burson BL, Sollenberger LE), pp. 503–535. ASA-CSSA-SSSA, Madison, WI.

Heaton EA, Dohleman FG, Long SP (2009) Seasonal nitrogen dynamics of *Miscanthus × giganteus* and *Panicum virgatum*. *Global Change Biology-Bioenergy*, **1**, 297–307.

Hoa DT, Man TD, Hau NG (2008) Pretreatment of lignocellulosic biomass for enzymatic hydrolysis. *ASEAN Journal on Science and Technology for Development*, **25**, 341–346.

Hsu F, Hong K, Lee M, Lee K (1990) Effects of cutting height on forage yield, forage and silage quality of napier grasses (written in Chinese with abstract in English). *Journal of the Agricultural Association of China*, **151**, 77–88.

Jakob K, Zhou F, Paterson A (2009) Genetic improvement of C4 grasses as cellulosic biofuel feedstocks. *In Vitro Cellular & Developmental Biology*, **45**, 291–305.

Jung JH, Vermerris W, Gallo M, Fedenko JR, Erickson JE, Altpeter F (2013) RNA interference suppression of lignin biosynthesis increases fermentable sugar yields for biofuel production from field-grown sugarcane. *Plant Biotechnology Journal*, **11**, 709–716.

Kering MK, Butler TJ, Biermacher JT, Guretzky JA (2012) Biomass yield and nutrient removal rates of perennial grasses under nitrogen fertilization. *Bioenergy Research*, **5**, 61–70.

Kim M, Day D (2011) Composition of sugar cane, energy cane, and sweet sorghum suitable for ethanol production at Louisiana sugar mills. *Journal of Industrial Microbiology and Biotechnology*, **38**, 803–807.

Knoll JE, Anderson WF, Strickland TC, Hubbard RK, Malik R (2012) Low-input production of biomass from perennial grasses in the coastal plain of Georgia, USA. *Bioenergy Research*, **5**, 206–214.

León RG, Gilbert RA, Korndorfer PH, Comstock JC (2010) Selection criteria and performance of energycane clones (*Saccharum* spp. × *S. spontaneum*) for biomass production under tropical and sub-tropical conditions. *Ceiba*, **51**, 11–16.

Lewandowski I, Heinz A (2003) Delayed harvest of miscanthus – influences on biomass quality and environmental impacts of energy production. *European Journal of Agronomy*, **19**, 45–63.

Lu Y, Mosier NS (2008) Current technologies for fuel ethanol production from lignocellulosic plant biomass. In: *Genetic Improvement of Bioenergy Crops*. (ed. Vermerris W), pp. 161–182. Springer, New York.

Madakadze IC, Stewart K, Peterson PR, Coulman BE, Smith DL (1999) Switchgrass biomass and chemical composition for biofuel in eastern Canada. *Agronomy Journal*, **91**, 696–701.

McKendry P (2002) Energy production from biomass (part 1): overview of biomass. *Bioresource Technology*, **83**, 37–46.

Mylavarapu R, Wright D, Kidder G, Chambliss C (2009) *UF/IFAS Standard Fertilization Recommendations for Agronomic Crops*. Florida Cooperative Extension Service, UF/IFAS, Florida, USA.

Na C, Sollenberger LE, Erickson JE *et al.* (2015a) Management of perennial warm-season bioenergy grasses. II. Seasonal differences in elephantgrass and energycane morphological characteristics affect responses to harvest frequency and timing. *Bioenergy Research*, **8**, 618–626.

Na C, Sollenberger LE, Erickson JE, Woodard KR, Vendramini JM, Silveira M (2015b) Management of perennial warm-season bioenergy grasses. I. Biomass harvested, nutrient removal, and persistence responses of elephantgrass and energycane to harvest frequency and timing. *Bioenergy Research*, **8**, 581–589.

Perez J, Munoz-Dorado J, de la Rubia T, Martinez J (2002) Biodegradation and biological treatments of cellulose, hemicellulose and lignin. *International Microbiology*, **5**, 53–63.

SAS Institute (2008) *SAS Online DOC 9.2*. SAS Inst., Cary, NC.

Shahandeh H, Chou CY, Hons FM, Hussey MA (2011) Nutrient partitioning and carbon and nitrogen mineralization of switchgrass plant parts. *Communications in Soil Science and Plant Analysis*, **42**, 599–615.

Singh MP, Erickson JE, Sollenberger LE, Woodard KR, Vendramini JM, Gilbert RA (2015) Mineral composition and removal of six perennial grasses grown for bioenergy. *Agronomy Journal*, **107**, 466–474.

Sluiter A, Ruiz R, Scarlata C, Sluiter J, Templeton D (2008a) *Determination of Extractives in Biomass; Laboratory Analytical Procedure (LAP)*. National Renewable Energy Laboratory, Colorado, USA.

Sluiter A, Hames B, Ruiz R, Scarlata C, Sluiter J, Templeton D, Crocker D (2008b) *Determination of Structural Carbohydrates and Lignin in Biomass; Laboratory Analytical Procedure (LAP)*. National Renewable Energy Laboratory, Colorado, USA.

Smith R, Slater FM (2011) Mobilization of minerals and moisture loss during senescence of the energy crops *Miscanthus × giganteus, Arundo donax* and *Phalaris arundinacea* in Wales, UK. *Global Change Biology-Bioenergy*, **3**, 148–157.

Sollenberger LE, Woodard KR, Vendramini JM *et al.* (2014) Invasive populations of elephantgrass differ in morphological and growth characteristics from clones selected for biomass production. *Bioenergy Research*, **7**, 1382–1391.

Studer MH, Demartini JD, Davis MF *et al.* (2011) Lignin content in natural Populus variants affects sugar release. *Proceedings of the National Academy of Sciences*, **108**, 6300–6305.

Summers ME, Jenkins BM, Hyde PR, Williams JF, Scardacci SC (2001) *Properties of Rice Straw as Influenced by Variety, Season and Location*. American Society of Agricultural Engineers Annual International Meeting, California, USA.

Tahir MH, Casler MD, Moore KJ, Brummer EC (2011) Biomass yield and quality of reed canarygrass under five harvest management systems for bioenergy production. *Bioenergy Research*, **4**, 111–119.

Tew TL, Miller JD, Dufrene EO *et al.* (2011) Registration of 'HoCP 91-552' sugarcane. *Journal of Plant Registrations*, **5**, 181–190.

Thammasouk K, Tandjo D, Penner MH (1997) Influence of extractives on the analysis of herbaceous biomass. *Journal of Agricultural and Food Chemistry*, **45**, 437–443.

Tran A, Chambers R (1986) Lignin and extractives derived inhibitors in the 2,3-butanediol fermentation of mannose-rich prehydrolysates. *Applied Microbiology and Biotechnology*, **23**, 191–197.

Vermerris W (2008) Composition and biosynthesis of lignocellulosic biomass. In: *Genetic Improvement of Bioenergy Crops* (ed. Vermerris W), pp. 89–142. Springer, New York, USA.

Waramit N, Moore KJ, Heggenstaller AH (2011) Composition of native warm-season grasses for bioenergy production in response to nitrogen fertilization rate and harvest date. *Agronomy Journal*, **103**, 655–662.

Wilson DM, Dalluge DL, Rover M, Heaton EA, Brown RC (2013) Crop management impacts biofuel quality: influence of switchgrass harvest time on yield, nitrogen and ash of fast pyrolysis products. *Bioenergy Research*, **6**, 103–113.

Woodard KR, Prine GM (1993) Regional performance of tall tropical bunchgrasses in the southeastern USA. *Bioenergy Research*, **5**, 3–21.

Xu J, Chen Y, Cheng J, Sharma-Shivappa RR, Burns JC (2011) Delignification of switchgrass cultivars for bioethanol production. *Bioresource Technology*, **6**, 707–720.

Zhang K, Agrawal M, Harper J, Chen R, Koros WJ (2011) Removal of the fermentation inhibitor, furfural, using activated carbon in cellulosic-ethanol production. *Industrial & Engineering Chemistry Research*, **50**, 14055–14060.

5

Exploring soil microbial 16S rRNA sequence data to increase carbon yield and nitrogen efficiency of a bioenergy crop

LEONARDO M. PITOMBO[1,2,3], JANAÍNA B. DO CARMO[3], MATTIAS DE HOLLANDER[1], RAFFAELLA ROSSETTO[4], MARYEIMY V. LÓPEZ[5], HEITOR CANTARELLA[2] and EIKO E. KURAMAE[1]

[1]Department of Microbial Ecology, Netherlands Institute of Ecology (NIOO/KNAW), Droevendaalsesteeg 10, 6708 PB Wageningen, The Netherlands, [2]Soils and Environmental Resources Center, Agronomic Institute of Campinas (IAC), Av. Barão de Itapura 1481, 13020-902 Campinas, SP, Brazil, [3]Department of Environmental Sciences, Federal University of São Carlos (UFSCar), Rod. João Leme dos Santos Km 110, 18052-780 Sorocaba, SP, Brazil, [4]Polo Piracicaba, Agência Paulista de Tecnologia (APTA), Rodovia SP 127 km 30, 13400-970 Piracicaba, SP, Brazil, [5]Department of Soil Science, University of São Paulo (ESALQ/USP), Avenida Pádua Dias 11, CEP 13418-260 Piracicaba, SP, Brazil

Abstract

Crop residues returned to the soil are important for the preservation of soil quality, health, and biodiversity, and they increase agriculture sustainability by recycling nutrients. Sugarcane is a bioenergy crop that produces huge amounts of straw (also known as trash) every year. In addition to straw, the ethanol industry also generates large volumes of vinasse, a liquid residue of ethanol production, which is recycled in sugarcane fields as fertilizer. However, both straw and vinasse have an impact on N_2O fluxes from the soil. Nitrous oxide is a greenhouse gas that is a primary concern in biofuel sustainability. Because bacteria and archaea are the main drivers of N redox processes in soil, in this study we propose the identification of taxa related with N_2O fluxes by combining functional responses (N_2O release) and the abundance of these microorganisms in soil. Using a large-scale *in situ* experiment with ten treatments, an intensive gas monitoring approach, high-throughput sequencing of soil microbial 16S rRNA gene and powerful statistical methods, we identified microbes related to N_2O fluxes in soil with sugarcane crops. In addition to the classical denitrifiers, we identified taxa within the phylum Firmicutes and mostly uncharacterized taxa recently described as important drivers of N_2O consumption. Treatments with straw and vinasse also allowed the identification of taxa with potential biotechnological properties that might improve the sustainability of bioethanol by increasing C yields and improving N efficiency in sugarcane fields.

Keywords: Anaeromyxobacter, nitrous oxide, straw, sugarcane, sustainability, vinasse

Introduction

Bioethanol from sugarcane is becoming an increasingly important alternative energy source worldwide. The sugarcane planted area in Brazil is around 9 million-ha (CONAB, 2013) and 26 million-ha worldwide (Kao *et al.*, 2014). The environmental benefits of replacing fossil fuels by ethanol from sugarcane can be only achieved if management practices are applied which lead to a minimum of greenhouse gases (GHG) losses to the atmosphere. Approximately 45% of the total amount of methane (CH_4) and nitrous oxide (N_2O) emitted by the cultivation of sugarcane are derived from straw burning (Macedo *et al.*, 2008); this practice also releases charcoal particles, causing health problems for humans and animals. Recently, legislation was passed that restricts burning, which is being phased out. The unburned cane (green cane) leaves a thick mulch of plant material after harvest (trash or straw) ranging from 10 to 20 t ha^{-1} (dry matter) consisting of leaves and tops. The conversion from burning to green management of sugarcane will impact the biogeochemical cycling of carbon and nitrogen in the plant–soil system. Straw preservation affects the entire production process of sugarcane, influencing yields, fertilizer management and application, soil erosion, soil organic matter dynamics, and GHG emissions.

In addition to plant straw, a second by-product of the sugarcane ethanol industry is vinasse (Mutton *et al.*,

Correspondence: Eiko E. Kuramae
e-mail: e.kuramae@nioo.knaw.nl

2010). Each liter of ethanol generates 10–15 l of vinasse, which is a liquid organic residue that has a high potential for polluting water streams. For this reason, there is legislation regulating the destination of vinasse. Currently, Brazil produces approximately 337 billion liters of vinasse that is almost entirely recycled as fertilizer in sugarcane fields (Mutton *et al.*, 2010). Vinasse has a high dissolved organic carbon (DOC) content resulting in high biochemical oxygen demand (BOD) and contains several nutrients; it is particularly rich in potassium and sulfur and contains considerable amounts of phosphorus, nitrogen, calcium, magnesium, and micronutrients. Although vinasse use is of relevance for nutrient cycling in sugarcane fields its application contributes to GHG emissions especially when applied with chemical fertilizers and on the straw (Carmo *et al.*, 2013).

The major sources of atmospheric N_2O are the nitrification and denitrification processes (Bouwman *et al.*, 2013). Globally, approximately 10.9 Tg $N-N_2O$ is estimated to be lost from land surfaces every year (Galloway *et al.*, 2004). Although the metabolic pathways that result in N_2O production during nitrification are not entirely clear, some authors suggest that this process might represent an even larger source of N_2O from agricultural fields than denitrification (Khalil *et al.*, 2004; Toyoda *et al.*, 2011). Moreover, nitrification supplies oxidized N compounds that can be used as electron acceptors, subsequently resulting in N_2O release. Denitrification can also be a detoxification pathway as a result of the accumulation of NO_2^- in the system (Zhu *et al.*, 2013). Nitrogen oxidation regulates nutrient availability for plant and the element mobility on the system, making the nitrification process of agronomical and environmental interest. Nitrification is a microbial-restricted niche (Hayatsu *et al.*, 2008) while denitrification can be performed by a broad diversity of microorganisms, especially prokaryotes (Philippot *et al.*, 2007). Denitrification has being described as a modular pathway in which different cells may drive different steps to the complete reduction of nitrate to N_2 being N_2O an intermediate product (Zumft, 1997). Complementarily, the diversity of potential N_2O reducers goes beyond *sensu stricto* denitrifiers (Jones *et al.*, 2013), which makes it difficult to identify the main players in N_2O production and consumption. Identifying these players will improve our understanding of these processes in the field and promote more sustainable fertilizer use.

While experiments on micro and mesocosm scales have been performed to identify microbial drivers of the N cycle and specifically linked to N_2O fluxes from soils (e.g. Ishii *et al.*, 2011; Jung *et al.*, 2014), field conditions cannot be reproduced. Surely such micro and mesocosm studies give important insights into the microbes involved in N cycle and specific processes. However,

the dynamic system of soil/plants in addition to weather instability results in a complex microbial community that goes beyond cultivable microorganisms. Thus, in this study by combining intensive *in situ* N_2O monitoring, high-throughput sequencing of soil microbial 16S rRNA gene and powerful statistical methods we have identified potential microbial taxa that might be explored to improve the sustainability of sugarcane as a bioenergy crop as a result of their biotechnological potential and/or their role in nutrient cycling.

Materials and methods

Site description and experimental design

The experimental site is located in Piracicaba, São Paulo state, Brazil (22°41′19.34″S; 47°38′41.97″W). More than 50% of the sugarcane of Brazil is grown in this state (Conab, 2013). The site's altitude is 575 m, the mean annual temperature is 21°C, and the mean annual precipitation is 1390 mm. The climate is defined as humid tropical with a rainy summer and a dry winter and is considered to be a Cwa type according to the Köppen classification system (Critchfield, 1960). The soil is classified as Haplic Ferralsol (FAO, 2006); fertility parameters are listed in Table S1. Prior to the experiment, the sugarcane crop of the plant cycle was harvested without burning; thus, the field was in the first ratoon stage. After harvesting, the site was subdivided into two subsites. The straw was left onto soil of subsite 1 (10 t ha^{-1}) and removed mechanically of subsite 2. The removal of straw simulated the scenario that occurs when the sugarcane straw is used as raw material for biofuel production.

Ten different treatments based on combination of organic, mineral fertilizers and straw were applied to the experimental site. The plots ($n = 4$; 8 × 9 m) were randomly distributed and separated from each other by 2 m borders. All treatments received single superphosphate as the phosphorus source (17 kg ha^{-1} of P). The source of potassium was potassium chloride (100 kg ha^{-1} of K) or vinasse (normal or concentrated). The nitrogen source was ammonium nitrate applied at a rate of 100 kg N ha^{-1}. The treatments without straw are described as the following: Ctrl (without N) – mineral fertilizer containing P and K; MN (mineral N) – mineral fertilizer containing N, P, and K; MN_V (mineral N plus vinasse) – vinasse as K source (1.10^5 l ha^{-1}) and mineral fertilizer containing N and P; V (vinasse) – vinasse as K source (1.10^5 l ha^{-1}) and mineral fertilizer containing P; CV (concentrated vinasse) – concentrated vinasse as K source (1.10^4 l ha^{-1}) and mineral fertilizer containing P. The same treatments were used with straw (Ctrl_S; MN_S; MN_V_S; V_S; and CV_S). Fertilization was carried out on 19 November, 2012. Vinasse was applied to the total area, while the mineral fertilizers and concentrated vinasse were applied in bands parallel to the crop line as is usually performed in commercial areas. There was no treatment with CV and N because they could be applied to different sides of the plant; thus, the combination of C and N that results in high N_2O

emissions was avoided. The composition of the vinasse and concentrated vinasse is presented in Table S2.

Greenhouse gas flux measurement and emission estimation

Soil GHG fluxes were measured using the chamber-based method (Varner et al., 2003). In each plot, a PVC chamber (30 cm in diameter) was placed on the soil in which it was partially inserted (3 cm deep). After closing the chambers, 60 ml samples were collected at time points 1, 10, 20, and 30 min using syringes and stored under pressure in 20-ml-evacuated penicillin flasks sealed with gas-impermeable butyl-rubber septa (Bellco Glass 2048). All samplings were performed between 8:00 and 12:00 a.m. The first 31 samplings were collected within 60 days after fertilization, followed by 4 weekly samplings and finally fortnightly intervals until the crop was harvested. Each gas chamber flux was calculated from slope regression between the gas concentration and collection time according to Carmo et al. (2013). Measurements of atmospheric pressure, chamber height, and air temperature were taken during gas sampling to determine the air chamber volume and to calculate GHGs emissions.

The gas samples were analyzed in a GC-2014 model gas chromatograph with electron capture for N_2O (Shimadzu, Kyoto, Japan). Soil temperature was measured at a depth of 10 cm to assist in the interpretation of the results. All of the chambers were installed at the fertilized sugarcane line position. GHG emissions were calculated by linear interpolation of fluxes between sampling events (Allen et al., 2010). We estimated that the fertilized bands accounted for 20% of the total experimental area; the remaining 80% of the area consisted of the spaces between the rows. Fertilizer emission factors (EF) were calculated according to Allen et al. (2010) and Carmo et al. (2013).

Soil sampling and soil chemical analysis

Soil sampling and gas collection initiated 1 day after fertilization. After collecting the gases, a composite soil sample 0–10 cm deep from three points was collected from each plot at the fertilizer band position. The samples were stored at −20°C. Soil moisture was determined by gravimetry, and mineral N content was determined by colorimetry of the soil extracts (KCl 2 mol l^{-1}) using flow injection analysis (FIAlab 2500) based on the methods proposed by Kamphake et al. (1967) for NO_3^- and Krom (1980) for NH_4^+.

DNA extraction, 16S rRNA gene amplification and sequencing

DNA was extracted from 250 mg of soil using the PowerSoil DNA Isolation Kit (Mobio Laboratories, Carlsbad, CA, USA) according to the manufacturer's protocol. We performed DNA extraction in samples from eight time points of each treatment corresponding to 240 samples (10 treatments × 8 time points × 3 replicates). Archaeal and bacterial community struc-tures were assessed by sequencing the V4 region of the 16S rRNA gene using primers 515F (GTGCCAGCMGCCGCGG-TAA) and 806R (TAATCTWTGGGVHCATCAGG) (Caporaso et al., 2010) and the LIB-L kit for unidirectional sequencing. Duplicate 50 μl PCRs were performed as follows: 5 μl Roche 10 × PCR buffer with $MgCl_2$, 2 U of Roche Taq DNA polymerase, 2 μl of dNTP (10 mM), 1 μl each primer (5 pmol $μl^{-1}$), and 1 μl of DNA template. Thermocycling conditions consisted of an initial denaturation at 94 °C for 5 min, 30 cycles at 95 °C for 30 s, 53 °C for 1 min, 72 °C for 1 min, and an final extension at 72 °C for 10 min. Technical replicates were pooled and purified using the Qiagen PCR purification kit. The 240 samples were equimolar mixed and then sequenced (Macrogen Inc. Company, South Korea) on a Roche 454 automated sequencer using the GS-FLX Titanium system (454 Life Sciences, Brandford, CT, USA).

Sequence data analysis

Sequence data were processed using the workflow of the NG6 system (Mariette et al., 2012), installed on a local server, which system depends on Mothur version 1.31.1 (Schloss et al., 2009). The flowgrams were demultiplexed and sizes between 350 and 750 flows were selected. The flowgrams were corrected using the shhh.flows command, which is the Mothur implementation of the original PyroNoise algorithm (Quince et al., 2011). Next, the results of the different sff files were combined for further analysis. The merged sequences were aligned to the bacterial and archaeal reference alignment provided on the Mothur website (http://www.mothur.org/wiki/Silva_reference_alignment) based on the SILVA 102 release of the SSURef database (Quast et al., 2013). Only reads that fell into region 13 862–22 580 of the reference alignment were kept. The pre.cluster command was used to reduce typical sequence errors due to sequencing PCR products from high-diversity DNA samples. This command assigns sequences that are within two mismatches to the most abundant sequence (Huse et al., 2010). Chimeric sequences were identified and removed using the chimera.uchime command (Edgar et al., 2011). Operational taxonomic units (OTUs) were formed at a maximum distance of 0.03 using the dist.seqs command and average neighbor clustering. All sequences were taxonomically classified using the Mothur implementation of the RDP classifier (Wang et al., 2007) using the training set (version 9) provided on the Mothur website (http://www.mothur.org/wiki/RDP_reference_files) and a bootstrap cutoff of 80%. OTUs were defined at an identity cutoff of 97%. We are aware that the Mothur developers state this training set as of poor quality and it is small in size. However, RDP taxonomy is one of the only sets that take subgroups of Acidobacteria into account. For each OTU, consensus taxonomy was determined using the classify.otu command. Representative sequences for each OTU were re-aligned to the Silva reference alignment and a neighbor joining tree was created using the clear-cut program (Sheneman et al., 2006). Taxonomic classification and OTU clustering data were combined into the BIOM format (McDonald et al., 2012) for further downstream statistical analysis with the Phyloseq (McMurdie & Holmes, 2013) R package.

The raw 454 pyrosequencing data of the 16S rRNA are available at the European Nucleotide Archive (ENA) (https://www.ebi.ac.uk/ena/) under the study Accession Number PRJEB8973.

Statistical analyses

Differential analyses of the count data (Anders & Huber, 2010) were applied to verify the effect of different management practices on the abundance of operational taxonomic units (OTU). Time points were not considered for these analyses. The control (Ctrl) treatment was used as the reference, and the negative binomial generalized linear model was applied using the DESeq2 package (Love et al., 2014) in R. OTU fold changes were considered statically significant using $P < 0.01$ as the criterion.

Generalized linear mixed models (Bolker et al., 2009) were used to test the effect of different management practices, soil parameters, and soil taxa abundance on N_2O fluxes. The sampling event was used as the random factor. First, we fitted a global model using the 'lme4' R package version 1.1–7 (Bates et al., 2014). Model selection was performed using the 'MuMIn' R package version 1.10.5 (Barton, 2014) with the Akaike information criterion (AIC) for ranking. To verify the effect of OTUs on N_2O fluxes, we performed screening steps to reduce the number of parameters before applying the 'dredge' function, which imposes limited number of parameters. From each global model, we consecutively fitted nested models selecting the OTUs with P values <0.05, <0.1, and <0.2, until we obtained approximately 15 OTUs. Thus, the 'dredge' function was applied to select the best fitted model based on all parameter combinations. The AIC was used as a criterion for model selection. In general, P values in generalized linear mixed models should be only used as guides to test nested models (Bolker et al., 2009). Thus, once the best model for each management practice was selected, and all OTUs were considered relevant for explaining N_2O fluxes from soil even at $P > 0.05$.

Results

GHG fluxes and emissions

Taking into account management practices and soil physicochemical factors, we found by the best fitted model that organic C addition ($P < 0.001$), N fertilization ($P < 0.001$), straw ($P = 0.016$), and moisture ($P = 0.004$) increased the fluxes of N_2O from the soil. In contrast, NO_3^- reduced ($P = 0.065$) the fluxes of N_2O from the soil. Ammonium (Fig. S1) and soil temperature did not explain N_2O fluxes according to the best fitted model. The intercept of the model highly correlated with soil moisture ($r^2 = 0.949$). The relationship of nitrate with N_2O production is well known. In our study, the models showed an association but not a cause–effect relationship because straw and organic fertilization increased the N_2O fluxes but also reduced soil NO_3^- (e.g., by immobilization; Fig. S2). Thus, N fertilization

by itself (i.e., available N and not the total N amount) better explained the N_2O fluxes than soil inorganic N. The parameter estimation for fixed effects in global and best fitted models used to explain the N_2O fluxes from soil considering physicochemical parameters and management practices are presented in Table S3. Despite the high N_2O peaks verified in treatments without straw (Fig. 1), annual N_2O emissions were greater in the plots receiving treatment with straw (Fig. 2). We assumed that the effect of high moisture in the treatments with straw resulted in longer slopes with high N_2O fluxes compared to treatments without straw (illustrated in Fig. 1).

During the first 60 days of sampling, the cumulative N_2O emissions reached more than 70% of the total emissions observed during the sugarcane cycle for all treatments (Fig. S3). At the 89th day, the cumulative emissions reached at least 85%. After 60 days, the soil inorganic N content reached a similar content to that before fertilization (Figs S1 and S2), which was expected as sugarcane plants were actively taking up nutrients from the soil. Therefore, we selected soil samples from the first 60 days to assess the microbial community related to N redox in soil.

In accordance with the modeling showing that organic fertilization, mineral fertilization, and straw addition increased N_2O fluxes (Fig. 1), the calculated annual emissions were also affected by these management practices (Fig. 2). The emission factors (EF) showed that the amount of N from fertilizer released as N_2O was equivalent to the following: (i) 210 g ha^{-1} when only mineral nitrogen was applied to the soil without straw; (ii) 2157 g ha^{-1} when mineral nitrogen was combined with vinasse; (iii) 1141 g ha^{-1} only with vinasse; and (iv) 503 g ha^{-1} when concentrated vinasse was used as fertilizer. In the treatments with straw, the N-N_2O released from the fertilizers increased to 1060 g ha^{-1} in the treatment containing mineral nitrogen; 3380 g ha^{-1} in the treatment with mineral nitrogen combined with vinasse; 1677 g ha^{-1} in the treatment with vinasse; and 688 g ha^{-1} when concentrated vinasse was applied.

Archaeal and Bacterial microbial community

Overall sequencing of 16S rRNA gene marker yielded >1.3 million reads with an average length of 283 bp. After quality trimming, a total of 836 750 sequences were obtained from 240 samples. These sequences covered 21 657 OTUs; a total of 7037 of these OTUs were represented by more than five reads. Differential analyses showed the effects of the management practices on 423 OTUs. Time points were not discriminate for this purpose. From these, 32 belonged to Acidobacteria (Fig. 3c), 46 to Actinobacteria (Fig. 3f), 29 to Bacterioide-

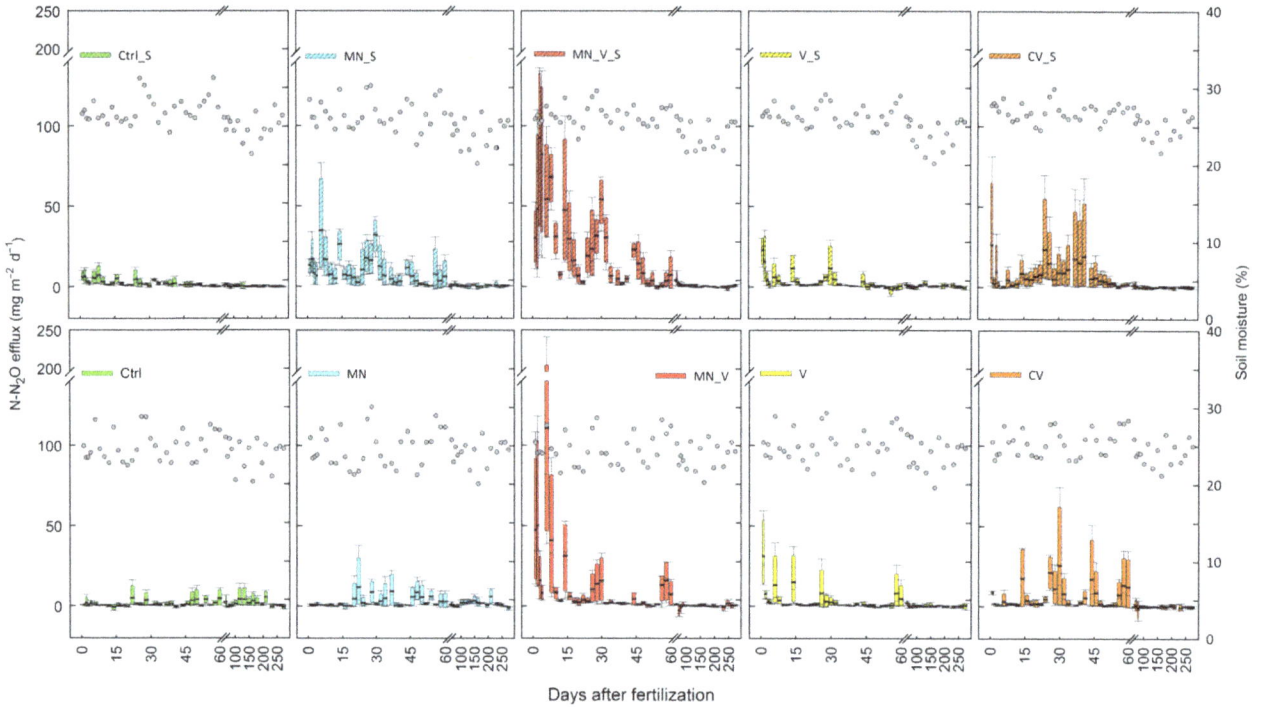

Fig. 1 Box-plot representation of nitrous oxide fluxes from soil with sugarcane crop after soil fertilization. Ctrl: no nitrogen addition; MN: mineral nitrogen fertilization; V: vinasse; CV: concentrated vinasse; S: straw left on the soil. *Daily measurements for samplings 1–4; 3–4 times/week measurements for samplings 5–31; once a week measurements for samplings 32–35; fortnightly measurements for samplings 36–47 (just before the crop harvesting). The first sampling was 1 day after fertilization.

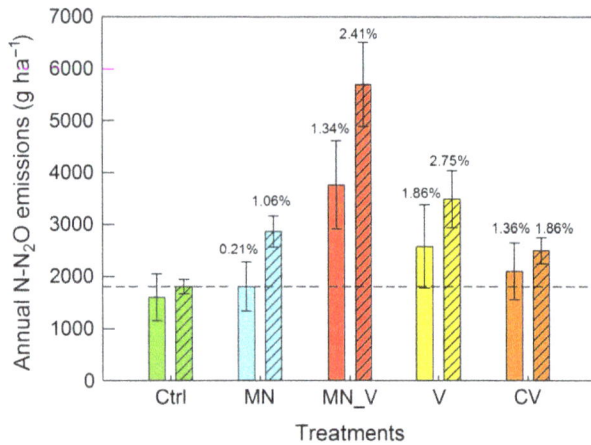

Fig. 2 Annual nitrous oxide emissions from soils with sugarcane and respective emission factors (%). Bars are the standard errors. Reference lines are the calculated emissions from the control (Ctrl) treatments without straw (dotted line) and with straw (dashed line). Ctrl: no nitrogen addition; MN: mineral nitrogen fertilization; MN_V: mineral nitrogen plus vinasse fertilization; V: vinasse; CV: concentrated vinasse. Transversal lines: treatments with straw left on soil.

tes (Fig. 3b), 24 to Chloroflexi (Fig. 3a), 58 to Firmicutes (Fig. 3d), 13 to Gemmatimonadetes (Fig. 3b), 3 to Nitrospirae (Fig. 3b), 8 to Planctomycetes (Fig. 3b), 124

to Proteobacteria (Fig. 3e), 2 to Thaumarcheota (Fig. 3b), 3 to Thermotogae (Fig. 3b), 5 to Verrucomicrobia (Fig. 3b), and 76 are unclassified bacteria or belong to other phyla.

Interestingly, taxa classified as *Chloroflexales* were increased in abundance only in the treatments in which the straw was left on the soil. All OTUs in this order belonged to the genus *Roseiflexus*. In contrast, the differential analysis showed that the members of *Ktedobacterales* were underrepresented in soils with straw. There was a significant effect of straw in reducing the abundance of 12 OTUs from the *Ktedobacterales* order, while in the plots without straw, only 4 OTUs were underrepresented. These OTUs were underrepresented 23 times with straw and only five times without straw (Fig. 3a), suggesting an inhibitory effect of straw on *Ktedobacterales* members.

The taxa recognized as ammonia oxidizers and affected by the treatments with straw and mineral N were from Thaumarcheota genus *Nitrososphaera* (OTU-123, OTU-333, Fig. 3b) and bacterial genus *Nitrosospira* (OTU-487, Fig. 3e, Fig. S4) (*Nitrosomonadales* order). *Nitrososphaera* OTUs were overrepresented only in treatments with straw, while *Nitrosospira* sp. OTU increased abundance in treatments with the addition of mineral N. The thousand most archaeal OTUs belong to the

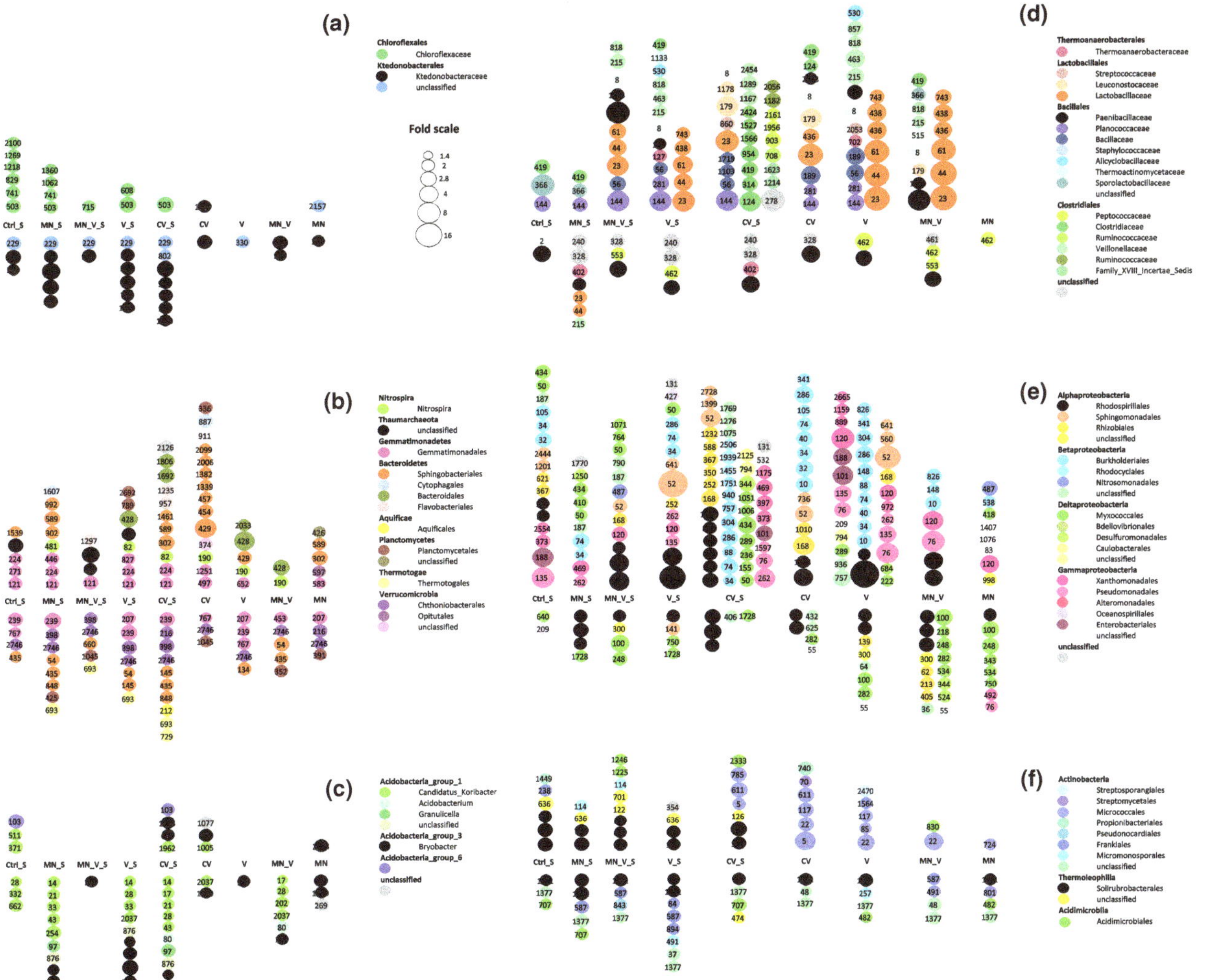

Fig. 3 Effect of different treatments on OTUs belonging to Chloroflexi (a), Acidobacteria (c), Firmicutes (d), Proteobacteria (e), Actinobateria (f), or other phyla (b). Numbers represent the OTU identification. Color groups at Family (a, d), Order (b, e, f), or Genus (c) level. Circle size indicates the fold change of the respective OTU when compared to the control treatment (Ctrl). Overrepresented OTUs are in the upper part of the plot and underrepresented OTU are in the below part of the plot. Ctrl_S: no N with straw; MN_S: mineral nitrogen with straw; MN_V_S: mineral nitrogen plus vinasse with straw; V_S: vinasse with straw; CV_S: concentrated vinasse with straw; CV: concentrated vinasse; V: vinasse; MN_V: mineral nitrogen plus vinasse; MN: mineral nitrogen.

genus *Nitrososphaera*. Within the entire 16S rRNA dataset, there was a clear dominance of the OTU 487 classified as ammonia oxidizer bacteria, which representative sequence presents 99% of identity with *N. multiformis*. The other member of the *Nitrosomonadales* order (OTU 2796, *Gallionellaceae*) was low abundant. Among the nitrite oxidizers, only *Nitrospira* spp. (OTU-82, OTU-190, OTU-481, Fig. 3b) showed overrepresentation in treatments with straw and vinasse. The OTU-136 (Fig. 3d) classified as *Ammoniphilus* sp. also related to nitrogen cycle increased abundance in treatment with the addition of vinasse. *Ammoniphilus* is obligatory oxalotrophic,

haloalkalitolerant bacteria and requires high ammonium concentration for growth (Zaitsev *et al.*, 1998).

Vinasse organic fertilization resulted in an increase in 37 Firmicutes taxa; however, the *Lactobacillales* order was the most responsive in terms of fold change (Fig. 3d). The *Lactobacillales* OTUs were also highly abundant based on the model intercept analysis (Fig. S5). Many OTUs (24) within *Clostridiales* were stimulated by the different treatments. However, 18 of these OTUs had low-positive fold changes with treatments including concentrated vinasse and straw left on the soil (Fig. 3d). Shifts in members of family *Leuconostocaceae* only occurred with concentrated vinasse treatment, while shifts in members of family *Alicyclobacillaceae* were observed with normal vinasse treatment. OTU-215 and OTU-818 belonging to genera *Megasphaera* and *Dialister*, respectively, were shifted in all treatments with normal vinasse (Fig. 3d).

Regarding the phylum Actinobacteria, members of class *Thermoleophilia* were clearly overrepresented in treatments including the presence of straw. Order *Micrococales* from the class *Actinobacteria* was overrepresented in soils with the absence of straw, while the OTU belonging to the order *Micronomosporales* was overrepresented in soils with straw combined with mineral nitrogen fertilization (Fig. 3f).

In general, straw on the soil and the input of nutrients (organic or mineral) reduced the abundance of OTUs within phylum Acidobacteria. However, no pattern among specific Acidobacteria OTUs and fertilization practices was observed. The presence of straw resulted in 36 negative fold changes within 19 different OTUs; in treatment without straw, there were only 12 negative fold changes within 10 OTUs. Only 9 OTUs belonging to phylum Acidobacteria slightly increased their abundance in soil with straw (Fig. 3c).

Overall, Proteobacteria was favored by organic fertilization and inhibited by the absence of straw (Fig. 3e). The treatments without straw resulted in 38 negative fold changes within 25 OTUs, while treatment with straw resulted in 19 negative fold changes within 14 OTUs (Fig. 3e). In treatments with organic fertilization, *Alphaproteobacteria* OTU-52, OTU-71, and OTU-317 that were classified as *Novosphingobium*, *Gluconacetobacter* and *Acetobacter*, respectively, were highly overrepresented (Fig. 3e). In addition to organic fertilization with concentrated vinasse, members of order *Rhodocyclales* (*Betaproteobacteria*) needed soil covered with straw to be overrepresented. The ability of members of order *Rhodocyclales* to use complex organic sources and perform important steps of the nitrogen cycle, such as denitrification during waste treatment, has been previously described (Hesselsoe *et al.*, 2009). The class *Deltaproteobacteria* is another group that was overrepresented in soil with straw. Additionally, the members of this group were underrepresented in soil without straw (Fig. 3e), indicating a condition of dependency on straw. Interestingly, there were different effects at the OTU level within the *Anaeromyxobacter* spp. (*Deltaproteobacteria*) in the different treatments. From this taxon, OTU-50 and OTU-434 were overrepresented in soil with straw, OTU-100 and OTU-248 were inhibited with the input of mineral nitrogen, and OTU-289 had a positive fold change with the addition of concentrated vinasse. Among the members of *Gammaproteobacteria*, OTU-76 and OTU-120 belonging to genera *Acetinobacter* and *Dyella*, respectively, were highly abundant in the treatments without straw when compared with the treatments with straw. Moreover, OTU-120, assigned to *Dyella* sp., was increased in soil with mineral nitrogen treatment without straw.

OTU-121 and OTU-224 of the phylum Gemmatimonadetes were overrepresented in treatments with straw (Fig. 3b). Within the phylum Verrucomicrobia, OTUs belonging to the order *Chthoniobacterales* presented negative fold changes in all treatments (Fig. 3b). One of these OTUs (OTU-398, classified as *Candidatus Xiphinematobacter*) was underrepresented only in the treatments with straw. Members of this genus are obligate endosymbionts of nematodes of commercial importance (Vandekerckhove *et al.*, 2000), suggesting that straw might reduce the amount of this parasite in soil. Therefore, this taxon might be considered as a bioindicator of soil quality.

Microbial taxa to explain N$_2$O fluxes

After identifying the management practices (i.e., straw, mineral nitrogen fertilization, and organic fertilization) that increased N$_2$O fluxes, we statistically checked what OTUs explain gas fluxes under the respective conditions. As a threshold, only OTUs that were represented at least twice in treatments with the same management practice were used to compose the respective model. These included were 64 OTUs related to straw, 24 OTUs related to mineral nitrogen, and 64 OTUs related to organic fertilization. In contrast to the differential analysis, we used frequency-normalized data to troubleshoot convergence failures. The global models that explain N$_2$O fluxes for the different management practices using microbial taxa narrowed the field ($P < 0.05$) to 19 OTUs for straw, six for mineral nitrogen, and 14 for organic fertilization (Table 1). The best and global models fitted with microbial OTUs under different managements to explain N$_2$O are presented in Table 1. The most represented orders in these models were *Burkholderiales* (6 OTUs), *Bacillales* (5 OTUs), *Lactobacillales* (4 OTUs), *Myxococcales* (3 OTUs), and *Xanthomonadales* (3 OTUs). All nested models generated to obtain approximately 15 OTUs prior to application of the 'dredge' function (see Materials and methods for details) as well as the global and best models are presented in a supplemental file. The relative abundances of OTUs presented in Table 1 are illustrated for Firmicutes (Fig. S6), Proteobacteria (Fig. S7) and other phyla (Fig. S8). There was an effect of vinasse on *Lactobacillus*, concentrated vinasse on *Leuconostoc* and straw on *Rummeliibacillus* OTUs, significantly explaining the N$_2$O fluxes from soil (Fig. S6). Furthermore, the effect of straw and vinasse on relative OTU abundance within *Anaeromyxobacter* spp. was prominent (Fig. S7). The effects of straw on the OTUs

Table 1 OTUs that explain N_2O fluxes in soils under different sugarcane management practices (S = straw; MN = mineral nitrogen fertilization; V = vinasse fertilization) by the global models and by the best models

| OTU | Global model§ | | | Best model | | | Taxonomy | | | |
	S	MN	V	S	MN	V	Phylum	Order	Family	Genus
103				†			Acidobacteria			Acidobacteria Gp6
5	***			*			Actinobacteria	Actinomycetales	Micrococcaceae	Arthrobacter
231	*						Actinobacteria	Solirubrobacterales	Unclassified	
429	*			**			Bacteroidetes	Sphingobacteriales	Sphingobacteriaceae	Sphingobacterium
503						**	Chloroflexi	Chloroflexales	Chloroflexaceae	Roseiflexus
23		*	***		*	***	Firmicures	Lactobacillales	Lactobacillaceae	Lactobacillus
61		**	*		**	***	Firmicures	Lactobacillales	Lactobacillaceae	Lactobacillus
436	*						Firmicures	Lactobacillales	Lactobacillaceae	Lactobacillus
179			*			*	Firmicutes	Lactobacillales	Leuconostocaceae	Leuconostoc
56	*						Firmicutes	Bacillales	Bacilaceae	Bacillus
136	***	**	*	***	**	**	Firmicutes	Bacillales	Paenibacillaceae	Ammoniphilus
144	*						Firmicutes	Bacillales	Paenibacillaceae	Rummeliibacillus
189			*			***	Firmicutes	Bacillales	Bacilaceae	Unclassified
366	***			***			Firmicutes	Bacillales	Sporolactobacillaceae	Sporolactobacillus
124		*				*	Firmicutes	Clostridiales	Clostridiaceae	Clostridium
215					**		Firmicutes	Selenomonadales	Veillonellaceae	Megasphaera
121			***			***	Gemmatimonadetes	Gemmatimonadales	Gemmatimonadaceae	Gemmatimonas
224	**						Gemmatimonadetes	Gemmatimonadales	Gemmatimonadaceae	Gemmatimonas
34	**		**				Proteobacteria	Burkholderiales	Comamonadaceae	Unclassified
40		*					Proteobacteria	Burkholderiales	Burkholderiaceae	Burkholderia
74		*				†	Proteobacteria	Burkholderiales	Oxalobacteraceae	Unclassified
88	*		*	†			Proteobacteria	Burkholderiales	Burkholderiaceae	Burkholderia
341	*						Proteobacteria	Burkholderiales	Burkholderiaceae	Ralstonia
826						***	Proteobacteria	Burkholderiales	Unclassified	
101						*	Proteobacteria	Enterobacteriales	Enterobacteriaceae	Unclassified
50		*			**		Proteobacteria	Myxococcales	Cystobacteraceae	Anaeromyxobacter
289	***			*			Proteobacteria	Myxococcales	Cystobacteraceae	Anaeromyxobacter
434				*			Proteobacteria	Myxococcales	Cystobacteraceae	Anaeromyxobacter
317	*						Proteobacteria	Rhodospirillales	Acetobacteraceae	Acetobacter
641			***			***	Proteobacteria	Sphingomonadales	Sphingomonadaceae	Sphingomonas
131	*			**			Proteobacteria	Unclassified		
120		*				**	Proteobacteria	Xanthomonadales	Xanthomonadaceae	Dyella
373	**			*			Proteobacteria	Xanthomonadales	Xanthomonadaceae	Arenimonas
469	*						Proteobacteria	Xanthomonadales	Sinobacteraceae	JTB255
333	*						Thaumarchaeota	Nitrososphaerales	Nitrososphaeraceae	Nitrososphaera
172				‡			Unclassified			
206		*		**			Unclassified			
242	**	*	*	†		‡	Unclassified			

§Only OTUs that presented $P < 0.05$ are shown. Complete list of OTUs are presented in a supplemental file.
***$P < 0.001$; **$P < 0.01$; *$P < 0.05$; †$P < 0.1$; ‡not significant but is part of the best model selected by AIC.

that explain N_2O fluxes within the archaeal genus *Nitrososphaera* and bacterial genera *Roseiflexus* and *Gemmatimonas* are shown in Fig. S8.

Discussion

Based on our field experiment, we expect to guide further efforts for mining specific members in the microbial community for sustainable biofuel production. A promising example for mining is our finding of the dominance of archaeal *Nitrososphaera* and bacterial *Nitrosospira* as ammonia oxidizers instead of *Nitrosomonas* commonly found in soils and used as model for ammonia oxidizing bacteria. One of the practices to reduce N_2O emissions in sugarcane is the use of nitrification inhibitors. However, the available inhibitors are expensive and some are applied in high amounts (up to 5% of the N is applied as nitrification inhibitor compounds) into soil (Soares *et al.*, 2015). Furthermore, the available nitrification inhibitors were initially addressed to *Nitrosomonas* (Slangen & Kerkhoff, 1984) and some of them have different effect on *Nitrososphaera* and *Nitrosospira* (Shen *et al.*, 2013). Therefore, based on our results, the use and development of specific products for these taxa might be addressed.

The genus *Roseiflexus* were present in all treatments, however, overrepresented in treatments with straw (Fig. 3a). Of the six *Chloroflexi* classes, *Chloroflexia* are able to perform photosynthesis (Gupta *et al.*, 2013). Despite the fact that mixotrophy metabolism have been reported for *Roseiflexus* spp. (Klatt *et al.*, 2013), it is difficult to suggest that photosynthesis occurred below the straw layer due to the absence/low incidence of light. Until recently, no other morphological, physiological, biochemical, or molecular traits were known to be shared by different groups from phylum Chloroflexi (Gupta *et al.*, 2013). Nevertheless, Dellas *et al.* (2013) have described, using *Roseiflexus* sp. as reference organism, that the members of class *Chloroflexia* present a singular mevalonate pathway. In other words, they present a singular pathway to synthesize the precursor of all isoprenoids, which are compounds of biotechnological interest for a broad range of industrial areas (Maury *et al.*, 2005). Additionally, some results have shown that members of phylum Chloroflexi are promising cellulose degraders (Stott *et al.*, 2009) and the utilization of plant polymers may be widespread across the Chloroflexi phylum (Hug *et al.*, 2013). Many efforts have being addressed to obtain cellulases by screening and sequencing new organisms as they have potential to become the largest group of industrially used enzymes worldwide (Wilson, 2009). These results highlight *Roseiflexus* as an important target for optimization of the C yield per harvested area from crops used for biofuel production.

Large amounts of N are exported from the sugarcane crop as compared to the N applied as fertilizer in field, what makes this crop one of the most efficient crops for bioenergy. One of the explanations is that sugarcane benefits from N_2-fixing processes (Boddey *et al.*, 2003; Urquiaga *et al.*, 2012). Many bacterial isolates and bacteria consortia have been used to test this hypothesis without consistent results (Boddey *et al.*, 2003; Martinez De Oliveira *et al.*, 2006; Schultz *et al.*, 2014). The reason might be that a great number of N_2-fixing organisms exist in the field, both in soils and in plants, non-endophytic microorganisms contribute to N fixation (Fischer *et al.*, 2012). Although nitrogenase activity has not yet been confirmed in *Roseiflexus* spp. strains, transcriptomic profiles similar to *Oscillochloris trichoides* have shown that *Roseiflexus* spp. fix atmospheric N and this activity is more intense during the night (Klatt *et al.*, 2013).

Thus, deeper studies concerning soil living *Roseiflexus* spp. and their functional traits would contribute to better understanding of N cycle in soils.

Here, we showed that organic fertilization was the management practice that most affected the abundance of Firmicutes in soils (Fig. 3d). As proposed by Sharmin *et al.* (2013), Firmicutes-type bacteria play a dominant role in the sugarcane industry due to their physiological ability to ferment sugars. Firmicutes, in particular lactic acids producers, are historically recognized as contaminants by the ethanol industry because their metabolites effectively inhibit yeast ethanol production (Muthaiyan *et al.*, 2011). Thus, because most of the OTUs of *Lactobacillus* and *Leuconostoc* were not observed in treatments without vinasse application and *Lactobacillus* is the dominant taxa in vinasse (Assis Costa *et al.*, 2015), vinasse residues might be the main source of these taxa to the soil (Fig. S6). As verified in this work and by Carmo *et al.* (2013), vinasse increases N_2O emissions from sugarcane fields. The main reason might be because vinasse residues are rich in labile C and have high biological oxygen demand (BOD) values (Carmo *et al.*, 2013; Christofoletti *et al.*, 2013); thus, carbon and moisture are the key factors related with N_2O release. General fermenters and *Lactobacillus* are present in vinasse and when vinasse is applied to soil those microorganisms might be additional important factor for the N_2O balance. In accordance with Philippot *et al.* (2011), we also propose that the increase in the proportion of these organisms in soils here studied might result in increases in N_2O release as truncated denitrification pathways are found in *Lactobacillus* spp. (Shapleigh, 2013). Four of the taxa that explained N_2O fluxes from soil in our study were probably added to soil by vinasse organic fertilization. These taxa are from the genera *Lactobacillus* and *Leuconostoc* and were not detected in treatments without vinasse application (Fig. S6).

Another important observation is that nitric oxide (NO) is a natural product from amino acids reduction during fermentation processes as well from amino acids oxidation (Medinets *et al.*, 2015). NO can also be used as an electron acceptor by microorganisms such as the nondenitrifier *Anaeromyxobacter* spp., which have the ability to reduce NO and N_2O (Sanford *et al.*, 2012). Thus, the question arises of whether the proportion of nitrous oxide reducers can improve the N_2O balance. Nishizawa *et al.* (2014) showed that the inoculation into the soil of N_2-generating denitrifiers (*Azoarcus* sp., *Niastella* sp. and *Burkholderia* sp.) reduced the release of N_2O. Nevertheless, it is not adequate in fields with mineral fertilization because denitrifiers also tend to reduce the efficiency of the applied N. One promising approach is the overexpression of the *nosZ* gene by denitrifiers to improve the N_2O balance, for instance, when N is not dependent on artificially fixed sources (Itakura *et al.*, 2013). In theory, microorganisms lacking genes related to NxO generation (NO and N_2O) or those with a greater affinity for NxO than other N species could also be applied to fertilized fields (Miyahara *et al.*, 2010). Sanford *et al.* (2012) proposed that the proportion of nitrous oxide reducers would improve the N_2O balance between the soil and the atmosphere, with special attention paid to the nondenitrifier *Anaeromixobacter* spp. This genus was among the most abundant taxa identified in this study (OTUs-50, -100, -133, -150, -236, -289, -434, -764, -1051, -1435, and -1593; Figs S5 and S7). Microorganisms that use trace gas as nutrient source or electron acceptors require efficient enzymatic systems because the concentration of these gases is very low (Conrad, 1996). By consensus, the more specialized the function is, the more diversity is required as the rare (micro) organisms exert these functions as has been shown for methane consumption (Pester *et al.*, 2010). Thus, it is not clear whether trace gas consumption is a specialized niche or whether diversity is not required to maintain active specialized niches. Because *Anaeromyxobacter* spp. can use sources other than NxO as an electron acceptor (i.e., iron oxides; Hori *et al.*, 2010), we assume that the reduction of trace gases is not a specialty of *Anaeromyxobacter*. According to the thermodynamic theory, oxygen>nitrogen>manganese>iron>sulfur oxides are reduced sequentially (Ponnamperuma, 1972; Thauer *et al.*, 1977). Therefore, in theory when NxO are available in the environment, *Anaeromyxobacter* spp. tend to use these gases rather than other oxides. Thus, NxO reduction does not represent the niche of these microorganisms and the reduction of NxO does not require specialized microorganisms. OTUs classified as *Anaeromyxobacter* spp. were identified both by the global models and by the best fitted models explaining N_2O fluxes from soil with straw and soil with mineral N fertilization (Table 1). To date, there are four strains of *Anaeromyxobacter* spp. with available genome sequences that are able to reduce NO and N_2O; however, they do not possess the capacity to produce NO from NO_2^- (Sanford *et al.*, 2012). Our findings linking this taxon to N_2O fluxes reinforce the hypothesis that *Anaeromyxobacter* spp. are important contributors to N_2O consumption. However, in this present study, some OTUs of *Anaeromyxobacter* spp. were overrepresented, while others were underrepresented under the management practices that increase N_2O emissions. These results indicate that the selection of groups related to N_2O balance at different taxonomical levels should be considered. For instance, our results suggest that different OTUs might share the same ability to

perform a specific function, but they might respond differently under conditions of stress (i.e., fertilizers).

Most OTUs identified in this study seem to be related to processes involved in N_2O release. For example, one OTU of *Nitrososphaera* significantly explained the fluxes of N_2O on the global model in straw management (Table 1, Fig. 3b, Fig. S8). This archaeal genus is well recognized as an ammonia oxidizer (Spang *et al.* 2010) with the capacity for N_2O production from ammonia oxidation (Jung *et al.*, 2014; Stieglmeier *et al.*, 2014).

Jones *et al.* (2014) showed that 33% of the *nosZ* gene (encoding nitrous oxide reductase) sequences in diverse environmental samples were most closely related to *Gemmatinoma* ssp., while 18% of all clones grouped with *nosZ* sequences were from phyla Bacteroidetes and Chloroflexi. These groups are members of the poorly characterized Clade II of the nitrous oxide-reducing microbial community (Jones *et al.*, 2014), and they were represented in our fitted models explaining N_2O fluxes from sugarcane soil. However, due to their poor level of characterization and lack of experimental data tracing N_2O production or consumption such as for *Anaeromyxobacter* spp., it is not possible to support a cause–effect relationship between these taxa and N_2O fluxes. It is also not possible to infer if they might contribute to N_2O production or consumption. In contrast, the well-characterized denitrifying Proteobacteria were identified within the OTUs explaining N_2O fluxes. For instance, strains of *Burkholderia* and *Dyella* have been tested to reduce the amount of N_2O released into the atmosphere after organic fertilization (Nishizawa *et al.*, 2014). However, they might increase the N losses from the soil to atmosphere through denitrification and consequently reduce synthetic fertilizer efficiency. The OTU classified as *Ammoniphilus* sp. significantly explained N_2O fluxes in all fitted models (Table 1). Despite the fact that this genus is known to require high concentrations of ammonium and use oxalates exclusively as a C source, to date no metabolic pathways that could result in N_2O have been described for members of this genus. The available *Ammoniphilus* sp. type strains are not able to reduce NO_3^- and are strictly aerobic (Zaitsev *et al.*, 1998).

In an ideal context, a nitrogen fixer such as *Azoarcus* sp. that lacks the steps of denitrification prior to the production of N_2O (Krause *et al.*, 2006) might be used to improve the N balance in bioenergy crops. *Azoarcus* strains have been tested for potential reduction of N_2O emissions from rice fields (Ishii *et al.*, 2011; Nishizawa *et al.*, 2014). For sugarcane, an analogous situation could arise with members of the family *Acetobacteraceae*. All of the known N-fixing organisms that are members of this family belong to the genera *Acetobacter* and *Gluconacetobacter* (Dutta & Gachhui, 2006). The association between these two genera and sugarcane was reported by James *et al.* (1994). These organisms are of special interest for sugarcane crop as they have the ability to fix N in aerobic conditions like well-drained soils. Additionally, *Acetobacter* and *Gluconacetobacter* can use a broad range of organic sources, including alcohols present in the vinasse (James *et al.*, 1994; Parnaudeau *et al.*, 2008). In our study, OTUs (OTU-71 and OTU-317, Fig. 3e) belonging to these two genera were strongly favored by vinasse organic fertilization and as showed by Assis Costa *et al.* (2015) they are also present in the ethanol production process. The same *Acetobacter* OTU favored by vinasse organic fertilization significantly explained N_2O fluxes from soil with straw by the global fitted model. Thus, *Acetobacter* and *Gluconacetobacter* should be highlighted as the closest representatives to an ideal situation for the improvement of N-cycling in sugarcane, both by acting as N-fixing organisms and potential drivers of processes related to N_2O release. Furthermore, *Acetobacter* and *Gluconacetobacter* could be added to vinasse used as organic fertilizer in sugarcane fields, and thus, contributing to redefine vinasse from a source to a sink of N_2 and NxO. However, it must be acknowledged that the survival of microorganisms inoculated in the soil system is a challenge.

To our knowledge, this is the first study relating specific microbial taxa to the respective functional response (N_2O release) along the time and under different management practices in tropical soil. Here, we show that the limitation of linking microbial taxa with functional traits is particularly challenging because the microbial responses are at different taxonomical levels such as at OTU and genus levels, as observed for *Anaeromyxobacter* and *Roseiflexus*, respectively.

The effects of fertilization and straw on N_2O emissions have been previously reported (Carmo *et al.*, 2013), however, not taken into account the microbial community. Besides the common Proteobacteria model organisms for denitrification, we have found taxa recently described as potential drivers of N_2O production and consumption (Jones *et al.*, 2013). Additionally, we have identified taxa with potential biotechnological properties that might improve the sustainability of bioethanol by increasing C yields and improving N efficiency in sugarcane fields.

Acknowledgements

We thank AC Vitti, AM Yasuda, KS Lourenço, L Martinelli for technical assistance and W van der Werf, L Hemerik for data analyses instructions. This study was supported by grants from CAPES/NUFFIC (037/2012; BEX 0258/13-0); São Paulo Research Foundation-FAPESP (2012/50694-6) and Joint Research Projects in Biobased Economy (FAPESP/NWO 13/50365-5); The National Council of Technological and Scientific Development (CNPq) (471886/2012-2). Publication 5868 of the Netherlands Institute of Ecology (NIOO-KNAW).

References

Allen DE, Kingston G, Rennenberg H, Dalal RC, Schmidt S (2010) Effect of nitrogen fertilizer management and waterlogging on nitrous oxide emission from subtropical sugarcane soils. *Agriculture Ecosystems & Environment*, **136**, 209–217.

Anders S, Huber W (2010) Differential expression analysis for sequence count data. *Genome Biology*, **11**, R106.

Assis Costa OY, Souto BM, Tupinamba DD *et al.* (2015) Microbial diversity in sugarcane ethanol production in a Brazilian distillery using a culture-independent method. *Journal of Industrial Microbiology & Biotechnology*, **42**, 73–84.

Barton K (2014) MuMIn: Multi-model inference. [Computer software manual]. Available at: http://cran.rproject.org/web/packages/MuMIn/MuMIn.pdf (accessed 04 February 2014).

Bates D, Maechler M, Bolker B, Walker S (2014) lme4: Linear mixed-effects models using Eigen and S4. [Computer software manual]. Available at: http://cran.r-project.org/web/packages/lme4/lme4.pdf (accessed 03 February 2014).

Boddey RM, Urquiaga S, Alves BJR, Reis V (2003) Endophytic nitrogen fixation in sugarcane: present knowledge and future applications. *Plant and Soil*, **252**, 139–149.

Bolker BM, Brooks ME, Clark CJ, Geange SW, Poulsen JR, Stevens MHH, White J-SS (2009) Generalized linear mixed models: a practical guide for ecology and evolution. *Trends in Ecology & Evolution*, **24**, 127–135.

Bouwman AF, Beusen AHW, Griffioen J *et al.* (2013) Global trends and uncertainties in terrestrial denitrification and N_2O emissions. *Philisophical Transactions of the Royal Society of London, B Biological Sciences*, **386**, 20130112.

Caporaso JG, Lauber CL, Walters WA *et al.* (2010) Global patterns of 16S rRNA diversity at a depth of millions of sequences per sample. *Proceedings of the National Academy of Sciences of the United States of America*, **108**, 4516–4522.

Carmo JB, Filoso S, Zotelli LC *et al.* (2013) Infield greenhouse gas emissions from sugarcane soils in Brazil: effects from synthetic and organic fertilizer application and crop trash accumulation. *Global Change Biology Bioenergy*, **5**, 267–280.

Christofoletti CA, Escher JP, Correia JE, Urbano Marinho JF, Fontanetti CS (2013) Sugarcane vinasse: environmental implications of its use. *Waste Management*, **33**, 2752–2761.

Conab (2013) *Acompanhamento da Safra Brasileira: Cana-de-açúcar, safra 2013/2014* Companhia Nacional de Abastecimento, Brasília, 19 p.

Conrad R (1996) Soil microorganisms as controllers of atmospheric trace gases (H_2, CO, CH_4, OCS, N_2O, and NO). *Microbiological Reviews*, **60**, 609–640.

Critchfield HJ (1960) *General climatology*. Prentice-Hall, Englewood Cliffs. 465 p.

Dellas N, Thomas ST, Manning G, Noel JP (2013) Discovery of a metabolic alternative to the classical mevalonate pathway. *eLife*, **2**, e0067.

Dutta D, Gachhui R (2006) Novel nitrogen-fixing *Acetobacter nitrogenifigens* sp nov., isolated from Kombucha tea. *International Journal of Systematic and Evolutionary Microbiology*, **56**, 1899–1903.

Edgar RC, Haas BJ, Clemente JC, Quince C, Knight R (2011) UCHIME improves sensitivity and speed of chimera detection. *Bioinformatics*, **27**, 2194–2200.

FAO (2006) *World Reference Base for Soil Resources, 2006: A Framework for International Classification, Correlation and Communication*. Food and Agriculture Organization of the United Nations, Rome.

Fischer D, Pfitzner B, Schmid M *et al.* (2012) Molecular characterisation of the diazotrophic bacterial community in uninoculated and inoculated field-grown sugarcane (Saccharum sp.). *Plant and Soil*, **356**, 83–99.

Galloway JN, Dentener FJ, Capone DG *et al.* (2004) Nitrogen cycles: past, present and future. *Biogeochemistry*, **70**, 153–226.

Gupta RS, Chander P, George S (2013) Phylogenetic framework and molecular signatures for the class Chloroflexi and its different clades; proposal for division of the class Chloroflexi class. nov into the suborder *Chloroflexineae* subord. nov., consisting of the emended family *Oscillochloridaceae* and the family *Chloroflexaceae* fam. nov., and the suborder *Roseiflexineae* subord. nov., containing the family *Roseiflexaceae* fam. nov. *Antonie Van Leeuwenhoek International Journal of General and Molecular Microbiology*, **103**, 99–119.

Hayatsu M, Tago K, Saito M (2008) Various players in the nitrogen cycle: diversity and functions of the microorganisms involved in nitrification and denitrification. *Soil Science and Plant Nutrition*, **54**, 33–45.

Hesselsoe M, Fuereder S, Schloter M *et al.* (2009) Isotope array analysis of Rhodocyclales uncovers functional redundancy and versatility in an activated sludge. *ISME Journal*, **3**, 1349–1364.

Hori T, Mueller A, Igarashi Y, Conrad R, Friedrich MW (2010) Identification of iron-reducing microorganisms in anoxic rice paddy soil by C-13-acetate probing. *ISME Journal*, **4**, 267–278.

Hug LA, Castelle CJ, Wrighton KC et al. (2013) Community genomic analyses constrain the distribution of metabolic traits across the Chloroflexi phylum and indicate roles in sediment carbon cycling. *Microbiome*, **1**, 22–22.

Huse SM, Welch DM, Morrison HG, Sogin ML (2010) Ironing out the wrinkles in the rare biosphere through improved OTU clustering. *Environmental Microbiology*, **12**, 1889–1898.

Ishii S, Ohno H, Tsuboi M, Otsuka S, Senoo K (2011) Identification and isolation of active N$_2$O reducers in rice paddy soil. *ISME Journal*, **5**, 1936–1945.

Itakura M, Uchida Y, Akiyama H et al. (2013) Mitigation of nitrous oxide emissions from soils by *Bradyrhizobium japonicum* inoculation. *Nature Climate Change*, **3**, 208–212.

James EK, Reis VM, Olivares FL, Baldani JI, Dobereiner J (1994) Infection of sugarcane by the nitrogen-fixing bacterium *Acetobacter diazotrophicus*. *Journal of Experimental Botany*, **45**, 757–766.

Jones CM, Graf DRH, Bru D, Philippot L, Hallin S (2013) The unaccounted yet abundant nitrous oxide-reducing microbial community: a potential nitrous oxide sink. *ISME Journal*, **7**, 417–426.

Jones CM, Spor A, Brennan FP et al. (2014) Recently identified microbial guild mediates soil N2O sink capacity. *Nature Climate Change*, **4**, 801–805.

Jung M-Y, Well R, Min D et al. (2014) Isotopic signatures of N2O produced by ammonia-oxidizing archaea from soils. *ISME Journal*, **8**, 1115–1125.

Kamphake LJ, Hannah SA, Cohen JM (1967) Automated analysis for nitrate by hydrazine reduction. *Water Research*, **1**, 205–216.

Kao MCJ, Gesmann M, Gheri F (2014) FAOSTAT: A complementary package to the FAOSTAT database and the Statistical Yearbook of the Food and Agricultural Organization of the United Nations [Computer software manual]. Available at: http://cran.r-project.org/web/packages/FAOSTAT/index.html (accessed 02 June 2014).

Khalil K, Mary B, Renault P (2004) Nitrous oxide production by nitrification and denitrification in soil aggregates as affected by O-2 concentration. *Soil Biology & Biochemistry*, **36**, 687–699.

Klatt CG, Liu ZF, Ludwig M, Kuhl M, Jensen SI, Bryant DA, Ward DM (2013) Temporal metatranscriptomic patterning in phototrophic Chloroflexi inhabiting a microbial mat in a geothermal spring. *ISME Journal*, **7**, 1775–1789.

Krause A, Ramakumar A, Bartels D et al. (2006) Complete genome of the mutualistic, N-2-fixing grass endophyte *Azoarcus* sp strain BH72. *Nature Biotechnology*, **24**, 1385–1391.

Krom MD (1980) Spectrophotometric determination of ammonia – a study of a modified Berthelot reaction using salicylate and dichloroisocyanurate. *Analyst*, **105**, 305–316.

Love MI, Huber W, Anders S (2014) Moderated estimation of fold change and dispersion for RNA-Seq data with DESeq2. *Genome Biology*, **15**, 550.

Macedo IC, Seabra JEA, Silva JEAR (2008) Green house gases emissions in the production and use of ethanol from sugarcane in Brazil: the 2005/2006 averages and a prediction for 2020. *Biomass and Bioenergy*, **32**, 582–595.

Mariette J, Escudie F, Allias N, Salin G, Noirot C, Thomas S, Klopp C (2012) NG6: integrated next generation sequencing storage and processing environment. *BMC Genomics*, **13**, 462.

Martinez De Oliveira AL, Canuto EDL, Urquiaga S, Reis VM, Baldani JI (2006) Yield of micropropagated sugarcane varieties in different soil types following inoculation with diazotrophic bacteria. *Plant and Soil*, **284**, 23–32.

Maury J, Asadollahi MA, Moller K, Clark A, Nielsen J (2005) Microbial isoprenoid production: an example of green chemistry through metabolic engineering. *Biotechnology for the Future*, **100**, 19–51.

McDonald D, Clemente J, Kuczynski J et al. (2012) The Biological Observation Matrix (BIOM) format or: how I learned to stop worrying and love the ome-ome. *GigaScience*, **1**, 7.

McMurdie PJ, Holmes S (2013) phyloseq: an R package for reproducible interactive analysis and graphics of microbiome census data. *PLoS ONE*, **8**, e61217.

Medinets S, Skiba U, Rennenberg H, Butterbach-Bahl K (2015) A review of soil NO transformation: associated processes and possible physiological significance on organisms. *Soil Biology and Biochemistry*, **80**, 92–117.

Miyahara M, Kim S-W, Fushinobu S et al. (2010) Potential of aerobic denitrification by *Pseudomonas stutzeri* TR2 to reduce nitrous oxide emissions from wastewater treatment plants. *Applied and Environmental Microbiology*, **76**, 4619–4625.

Muthaiyan A, Limayem A, Ricke SC (2011) Antimicrobial strategies for limiting bacterial contaminants in fuel bioethanol fermentations. *Progress in Energy and Combustion Science*, **37**, 351–370.

Mutton MA, Rossetto R, Mutton MJR (2010) Agricultural use of stillage. In: *Sugarcane Bioethanol - R&D for Productivity and Sustainability* (ed. Cortez LAB), pp. 423–440. Editora Edgard Blücher, São Paulo.

Nishizawa T, Quan A, Kai A et al. (2014) Inoculation with N-2-generating denitrifier strains mitigates N2O emission from agricultural soil fertilized with poultry manure. *Biology and Fertility of Soils*, **50**, 1001–1007.

Parnaudeau V, Condom N, Oliver R, Cazevieille P, Recous S (2008) Vinasse organic matter quality and mineralization potential, as influenced by raw material, fermentation and concentration processes. *Bioresource Technology*, **99**, 1553–1562.

Pester M, Bittner N, Deevong P, Wagner M, Loy A (2010) A 'rare biosphere' microorganism contributes to sulfate reduction in a peatland. *ISME Journal*, **4**, 1591–1602.

Philippot L, Hallin S, Schloter M (2007) Ecology of denitrifying prokaryotes in agricultural soil. *Advances in Agronomy*, **96**, 249–305.

Philippot L, Andert J, Jones CM, Bru D, Hallin S (2011) Importance of denitrifiers lacking the genes encoding the nitrous oxide reductase for N2O emissions from soil. *Global Change Biology*, **17**, 1497–1504.

Ponnamperuma FN (1972) The Chemistry of Submerged Soils. *Advances in Agronomy*, **24**, 29–96.

Quast C, Pruesse E, Yilmaz P et al. (2013) The SILVA ribosomal RNA gene database project: improved data processing and web-based tools. *Nucleic Acids Research*, **41**, 590–596.

Quince C, Lanzen A, Davenport RJ, Turnbaugh PJ (2011) Removing noise from pyrosequenced amplicons. *BMC Bioinformatics*, **12**, 38.

Sanford RA, Wagner DD, Wu Q et al. (2012) Unexpected nondenitrifier nitrous oxide reductase-gene diversity and abundance in soils. *Proceedings of the National Academy of Sciences of the United States of America*, **109**, 19709–19714.

Schloss PD, Westcott SL, Ryabin T et al. (2009) Introducing mothur: open-source, platform-independent, community-supported software for describing and comparing microbial communities. *Applied and Environmental Microbiology*, **75**, 7537–7541.

Schultz N, JA Silva, Sousa JS et al. (2014) Inoculation of sugarcane with diazotrophic bacteria. *Revista Brasileira De Ciencia Do Solo*, **38**, 407–414.

Shapleigh J (2013) Denitrifying Prokaryotes. In: *The Prokaryotes* (eds Rosenberg E, Delong E, Lory S, Stackebrandt E, Thompson F), pp. 405–425. Springer, Berlin Heidelberg.

Sharmin F, Wakelin S, Huygens F, Hargreaves M (2013) Firmicutes dominate the bacterial taxa within sugar-cane processing plants. *Scientific Reports*, **3**, 3107.

Shen T, Stieglmeier M, Dai J, Urich T, Schleper C (2013) Responses of the terrestrial ammonia-oxidizing archaeon Ca. Nitrososphaera viennensis and the ammonia-oxidizing bacterium Nitrosospira multiformis to nitrification inhibitors. *FEMS Microbiology Letters*, **344**, 121–129.

Sheneman L, Evans J, Foster JA (2006) Clearcut: a fast implementation of relaxed neighbor joining. *Bioinformatics*, **22**, 2823–2824.

Slangen JHG, Kerkhoff P (1984) Nitrification inhibitors in agriculture and horticulture: a literature review. *Fertilizer Research*, **5**, 1–76.

Soares JR, Cantarella H, Vargas VP, Carmo JB, Martins AA, Sousa RM, Andrade CA (2015) Enhanced-efficiency fertilizers in nitrous oxide emissions from urea applied to sugarcane. *Journal of Environmental Quality*, **44**, 423–430.

Spang A, Hatzenpichler R, Brochier-Armanet C et al. (2010) Distinct gene set in two different lineages of ammonia-oxidizing archaea supports the phylum Thaumarchaeota. *Trends in Microbiology*, **18**, 331–340.

Stieglmeier M, Mooshammer M, Kitzler B, Wanek W, Zechmeister-Boltenstern S, Richter A, Schleper C (2014) Aerobic nitrous oxide production through N-nitrosating hybrid formation in ammonia-oxidizing archaea. *ISME Journal*, **8**, 1135–1146.

Stott MB, Dunfield PF, Crowe MA (2009) Class of Chloroflexi-like thermophilic cellulose degrading bacteria. Google Patents No. CA, 2716728, A1.

Thauer RK, Jungermann K, Decker K (1977) Energy-conservation in chemotropic anaerobic bacteria. *Bacteriological Reviews*, **41**, 100–180.

Toyoda S, Yano M, Nishimura S-I et al. (2011) Characterization and production and consumption processes of N$_2$O emitted from temperate agricultural soils determined via isotopomer ratio analysis. *Global Biogeochemical Cycles*, **25**, GB2008.

Urquiaga S, Xavier RP, De Morais RF et al. (2012) Evidence from field nitrogen balance and N-15 natural abundance data for the contribution of biological N-2 fixation to Brazilian sugarcane varieties. *Plant and Soil*, **356**, 5–21.

Vandekerckhove TTM, Willems A, Gillis M, Coomans A (2000) Occurrence of novel verrucomicrobial species, endosymbiotic and associated with parthenogenesis in Xiphinema americanum-group species (Nematoda, Longidoridae). *International Journal of Systematic and Evolutionary Microbiology*, **50**, 2197–2205.

Varner RK, Keller M, Robertson JR et al. (2003) Experimentally induced root mortality increased nitrous oxide emission from tropical forest soils. *Geophysical Research Letters*, **30**, 1144.

Wang Q, Garrity GM, Tiedje JM, Cole JR (2007) Naive Bayesian classifier for rapid assignment of rRNA sequences into the new bacterial taxonomy. *Applied and Environmental Microbiology*, **73**, 5261–5267.

Wilson DB (2009) Cellulases and biofuels. *Current Opinion in Biotechnology*, **20**, 295–299.

Zaitsev GM, Tsitko IV, Rainey FA, Trotsenko YA, Uotila JS, Stackebrandt E, Salkinoja-Salonen MS (1998) New aerobic ammonium-dependent obligately oxalotrophic bacteria: description of Ammoniphilus oxalaticus gen. nov., sp. nov. and Ammoniphilus oxalivorans gen. nov., sp. nov. *International Journal of Systematic Bacteriology*, **48**, 151–163.

Zhu X, Burger M, Doane TA, Horwath WR (2013) Ammonia oxidation pathways and nitrifier denitrification are significant sources of N$_2$O and NO under low oxygen availability. *Proceedings of the National Academy of Sciences of the United States of America*, **110**, 6328–6333.

Zumft WG (1997) Cell biology and molecular basis of denitrification. *Microbiology and Molecular Biology Reviews*, **61**, 533–616.

6

Winter oilseed production for biofuel in the US Corn Belt: opportunities and limitations

AARON J. SINDELAR[1], MARTY R. SCHMER[1], RUSSELL W. GESCH[2], FRANK FORCELLA[2], CARRIE A. EBERLE[2], MATTHEW D. THOM[2] and DAVID W. ARCHER[3]

[1]*Agroecosystem Management Research Unit, USDA-ARS, 251 Food Industry Complex, UNL-East Campus, Lincoln, NE 68583, USA,* [2]*North Central Soil Conservation Research Lab, USDA-ARS, 803 Iowa Ave, Morris, MN 56267, USA,* [3]*Northern Great Plains Research Laboratory, USDA-ARS, 1701 10th Ave SW, Mandan, ND 58554, USA*

Abstract

Interest from the US commercial aviation industry and commitments established by the US Navy and Air Force to use renewable fuels has spurred interest in identifying and developing crops for renewable aviation fuel. Concern regarding greenhouse gas emissions associated with land-use change and shifting land grown for food to feedstock production for fuel has encouraged the concept of intensifying current prominent cropping systems through various double cropping strategies. Camelina (*Camelina sativa* L.) and field pennycress (*Thlaspi arvense* L.) are two winter oilseed crops that could potentially be integrated into the corn (*Zea mays* L.)–soybean [(*Glycine max* (L.) Merr.] cropping system, which is the prominent cropping system in the US Corn Belt. In addition to providing a feedstock for renewable aviation fuel production, integrating these crops into corn–soybean cropping systems could also potentially provide a range of ecosystem services. Some of these include soil protection from wind and water erosion, soil organic C (SOC) sequestration, water quality improvement through nitrate reduction, and a food source for pollinators. However, integration of these crops into corn–soybean cropping systems also carries possible limitations, such as potential yield reductions of the subsequent soybean crop. This review identifies and discusses some of the key benefits and constraints of integrating camelina or field pennycress into corn–soybean cropping systems and identifies generalized areas for potential adoption in the US Corn Belt.

Keywords: Camelina, cropping system intensification, double cropping, pennycress, renewable jet fuel, second-generation biofuels, winter oilseeds

Introduction

Producing sustainable renewable fuels with little to no competition with land used for food production while simultaneously reducing society's carbon (C) footprint is an important goal of the second-generation biofuel crop industry. Land-use changes and potential subsequent greenhouse gas emission increases are main concerns associated with the production of crops for bioenergy (e.g., Fargione *et al.*, 2008; Searchinger *et al.*, 2008; Kim *et al.*, 2009; Dale *et al.*, 2011). The intensification of current cropping systems with a bioenergy crop through double or relay cropping is one strategy that may alleviate these concerns regarding land-use change (Heaton *et al.*, 2013; Moore & Karlen, 2013). In the Corn Belt Region of the United States (Fig. 1), corn-based

cropping systems dominate the agricultural landscape, and cropland is often fallow during the winter months (Fig. 2). Because of this, winter annual oilseed crops offer potential for biofuel production while coexisting with other crops grown for food [e.g., corn, soybean, wheat (*Triticum aestivum* L.)]. While the inclusion of winter oilseeds into cropping systems where corn is grown continuously may also be an option, this discussion is centered on corn–soybean cropping systems because it is the dominant cropping system in the US Corn Belt, with an estimated area of 27 million hectares in the region (Table 1). In this cropping system, the oilseed crop would ideally be interseeded into corn before or immediately after corn harvest (Fig. 3) and would be harvested prior to soybean planting in a sequential double cropping system or while soybean is growing in a relay double cropping system. Benefits and limitations exist with winter oilseed crop integration into corn–soybean systems. However, the added value of winter

Correspondence: Aaron J. Sindelar
e-mail: aaron.sindelar@ars.usda.gov

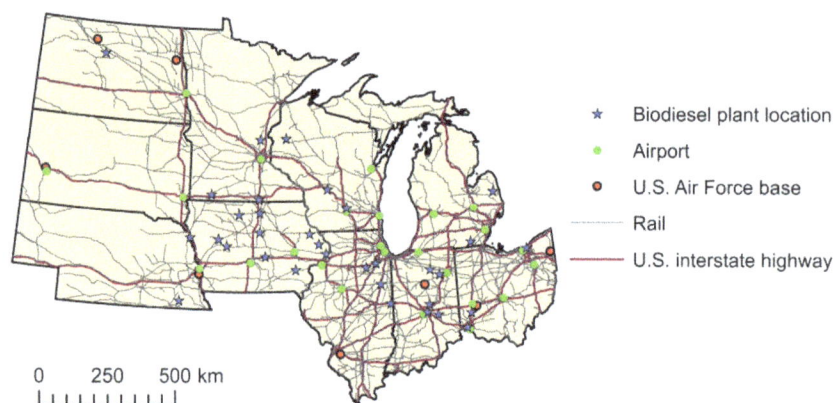

Fig. 1 Select infrastructure for biodiesel production in the US Corn Belt. Airports were those with enplanements of ≥250 000 people.

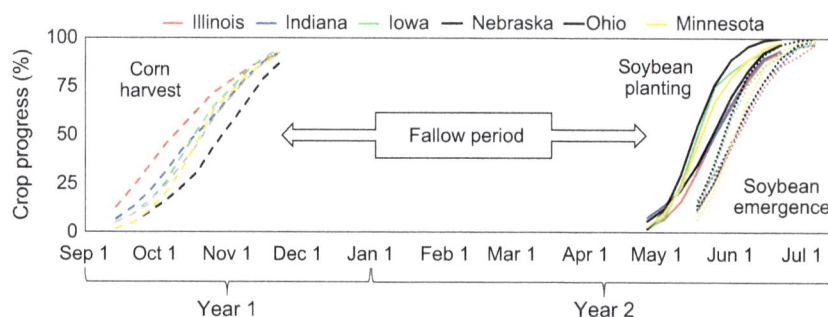

Fig. 2 Average (2009–2013) corn harvest progress (long dash line), soybean planting progress (solid line), and soybean emergence progress (short dash line) for selected states in the US Corn Belt (www.nass.usda.gov).

cropping diversity and winter cash crops may also offer relief for some of the previously reported limitations. This review identifies and discusses agronomic, environmental, and socioeconomic benefits and challenges associated with the possible implementation of this system.

Renewable fuel demands

The 2013 adjustment of the US Renewable Fuel Standard (RFS2) requires 7.3 and 10.4 billion L of biomass-based diesel and advanced biofuels, respectively, in an effort to increase domestic fuel production and decrease reliance on foreign sources (United States Environmental Protection Agency, 2013, 2015). The aviation industry also has significant interest in adopting renewable fuels that reduce carbon emissions while simultaneously being economically competitive with petroleum-based fuels. At the 2008 Aviation and Environment Summit, commitments were set to maintain current carbon emission levels by 2020 and halve emissions based on 2005 levels by 2050 (Air Transportation Action Group, 2015). In attempt to shift its reliance to sustainable fuel, the US

Federal Aviation Administration established the goal of using approximately 4 billion L yr^{-1} of renewable jet fuel by 2018 (FAA, 2011). The US Air Force and US Navy have also set renewable fuel consumption goals, where the Air Force goal is that 50% of noncontingency operations will use an alternative drop-in fuel blend by 2025 (U.S. Air Force, 2013). The Navy has set a target that 50% of total energy consumed will be from renewable sources by 2020 (U.S. Navy, 2010). Therefore, renewable fuels are planned to serve as an integral fuel source for ground- and air-based transportation in the near future and beyond. Additionally, a recent report by the US Environmental Protection Agency has concluded that emissions from aircraft endangers human health because of their contribution to global warming (United States Environmental Protection Agency, 2015), which will likely result in emission regulations imposed on the commercial aviation industry (Mouawad & Davenport, 2015).

Traditional oilseed feedstocks used for biofuel production include soybean, rapeseed/canola (*Brassica napus*), palm (*Elaeis guineensis*), and sunflower (*Helianthus annus*) (Moser, 2012). However, as noted by

Table 1 Acreage estimates for 2- and 3-year corn-soybean and alfalfa-based cropping systems in the U.S Corn Belt from 2009–13[*]

State	Corn-soybean		Corn-corn-soybean		Alfalfa-based	
	Area, ha	Distribution, %[†]	Area, ha	Distribution, %	Area, ha	Distribution, %
Illinois	4 281 810	19.8	1 153 150	21.1	31 550	1.4
Iowa	5 417 130	25.1	1 164 080	21.3	140 410	6
Indiana	2 353 080	10.9	466 910	8.5	29 450	1.3
Michigan	426 590	2	151 780	2.8	384 950	16.6
Minnesota	2 923 730	13.5	661 830	12.1	363 020	15.6
Nebraska	2 161 620	10	1 024 380	18.7	258 800	11.1
North Dakota	517 060	2.4	87 070	1.6	124 830	5.4
Ohio	1 337 550	6.2	153 190	2.8	60 650	2.6
South Dakota	1 732 970	8	311 320	5.7	220 290	9.5
Wisconsin	428 490	2	301 570	5.5	707 850	30.5
Total	21 580 030		5 475 280		2 321 800	

*Classification accuracy of corn, soybean, and alfalfa area by USDA-NASS Cropland Data Layers is available in a Table S1.
†Distribution is the proportion of a cropping system for a given state compared to the U.S. Corn Belt region.

Moser (2012), the sole use of these traditional oilseed feedstocks will likely not meet production demands mandated by the RFS2. For example, Hill *et al.* (2006) estimated that dedicating all US soybean production to biofuel would only satisfy 6% of the US diesel demand. Also, the use of prominent crops like soybean for biofuel production diminishes its opportunity to be used for food (Cassman & Liska, 2007; Tilman *et al.*, 2009; Karp & Richter, 2011), which is of increasing importance as world population is estimated to surpass nine billion by 2044 (U.S. Census Bureau, 2014). Finally, soybean production has been reported to have adverse effects on the soil environment, specifically through greater wind and water erosion susceptibility and less soil organic carbon (SOC) when compared to corn-based cropping systems (Laflen & Moldenhauer, 1979; Varvel & Wilhelm, 2008, 2011).

Alternative crops and cover crops

The aforementioned concerns have resulted in expansion of research on alternative oilseed crops for bio-based diesel and aviation fuel. Examples of second-generation oilseed crops include camelina, field pennycress, crambe (*Crambe abyssinica*), cuphea, (*Cuphea* spp.) oilseed radish (*Raphanus sativus* L.), Ethiopian mustard (*Brassica carinata*), white mustard (*Sinapis alba* L.), and flax (*Linum usitatissimum* L.), to name a few (Johnson *et al.*, 2007; Moser, 2012). These crops have several desirable characteristics including greater seed oil concentration than soybean, reported lower agricultural input requirements (Putnam *et al.*, 1993; Zubr, 1997), and potential for little or no disruption when integrated into prominent cropping systems. They can also provide environmental benefits like protection from soil erosion,

increased C inputs for potential SOC sequestration, and providing food resources and habitat for pollinators and other beneficial insects. Camelina and field pennycress are particularly viewed as attractive options for biodiesel or aviation fuel due to their desirable fatty acid profiles in addition to high oil contents (Moser *et al.*, 2009; Boateng *et al.*, 2010; Moser, 2012; Fan *et al.*, 2013). Also, significant greenhouse gas emissions reductions are expected from second-generation oilseed crops either for ground transportation or aviation fuels, especially when used in a double cropping system (Shonnard *et al.*, 2010; Krohn & Fripp, 2012; Fan *et al.*, 2013).

Double cropping strategies

Simply defined, double cropping is the agricultural practice of growing two crops on the same unit of land during the same growing season. In traditional double cropping systems, the second crop is planted following the harvest of the first. Double cropping of biomass crops has been viewed to be a potential option in northern areas of the US Corn Belt (Iowa, Minnesota, and Wisconsin), where the traditional growing season is often regarded as short (Moore & Karlen, 2013). For example, Feyereisen *et al.* (2013) concluded that winter rye (*Secale cereale* L.) could be successfully grown in a corn–soybean cropping system across the US Corn Belt, thus potentially alleviating the issue of displacing cereal crops with energy crops and avoiding significant land-use change. Additionally, land area suitable for double cropping in the US Corn Belt has increased in the last 30 years and is projected to further increase as a result of future climate change (Seifert & Lobell, 2015). A particular double cropping strategy that may help to mitigate risk, particularly in northern environments, is relay

Fig. 3 Pennycress growth following corn in a double cropping system near Ithaca, NE (top), and simultaneous camelina and soybean growth in a relay cropping system near Morris, MN (bottom).

cropping. In this type of system, the second crop (summer annual) is interplanted into the first crop (winter annual) prior to maturity, which extends the growing season for the second crop (Fig. 3). Research in the North Central USA has shown promising results with this system. For example, Gesch *et al.* (2014) reported that soybean relay-cropped with winter camelina produced greater seed and more total gross income compared to a sequential soybean–camelina double cropping system and was economically competitive with full-season monocropped soybean. Moreover, relay cropping soybean with winter camelina was shown to produce at least 50% more seed oil per hectare than growing a sole full-season soybean crop. Research similar to that performed by Gesch *et al.* (2014) will need to be conducted in other environments and on other oilseed crops to identify double cropping strategies that maximize productivity and profitability.

Potential ecosystem value

While the intensification of corn–soybean cropping systems with a winter oilseed crop would be to primarily increase opportunities for renewable diesel or hydroprocessed renewable jet fuel, its inclusion may also indirectly affect cellulosic biofuel (ethanol) from agricultural residues. Corn stover, the aboveground

biomass remaining after grain is harvested, is forecasted to be a primary feedstock for lignocellulosic biofuel production with estimated stover harvests ranging from 54 to 123 Tg yr^{-1} for the USA (Graham *et al.*, 2007; Muth *et al.*, 2013). However, harvesting corn stover, or other crop residues, without other organic amendments can have adverse effects on soil erosion susceptibility (Mann *et al.*, 2002; Jin *et al.*, 2015; Kenney *et al.*, 2015), SOC (Wilhelm *et al.*, 2004; Stetson *et al.*, 2012; Schmer *et al.*, 2014), and general soil fertility (Karlen *et al.*, 1984; Blanco-Canqui & Lal, 2009; Sindelar *et al.*, 2013). Our hypothesis is that inclusion of winter oilseed crops into corn–soybean cropping systems where corn stover is harvested may reduce, or even potentially offset in some cases, organic C losses associated with corn stover removal. For example, Blanco-Canqui *et al.* (2014) found that the inclusion of a winter rye cover crop helped offset SOC losses when stover was removed in a continuous corn system in Nebraska. Osborne *et al.* (2014) reported that a winter lentil (*Lens culinaris* Medik.)/slender wheatgrass [*Elymus trachycaulus* (Link) Gould ex Shinners] cover crop combination reduced the impact of corn stover removal on SOC in a corn–soybean cropping system in South Dakota. Research is needed to determine whether the response of oilseed cover crops would result in a deviation from this response. Corn stover removal has also been shown to increase residual nitrate-N in the soil profile following corn, particularly at N fertilizer rates that economically optimize grain yield as stover, that can often immobilize N, is absent from the soil surface (Sindelar *et al.*, 2015). Strock *et al.* (2004) found that a corn–soybean cropping system with a winter rye cover crop reduced subsurface tile water drainage discharge and flow-weighted mean nitrate concentration. Drury *et al.* (2014) reported that the inclusion of a winter wheat cover crop in a corn–soybean cropping system reduced nitrate-N loss by 14 to 16%. A recent study by Kladivko *et al.* (2014) used the Root Zone Water Quality Model to simulate residual nitrate-N dynamics in the Mississippi River Basin and concluded that residual nitrate-N loadings would be reduced by approximately 20% if cover crops were integrated into corn–soybean cropping systems. In western Minnesota, R.W. Gesch (unpublished data) recently found that winter camelina seeded in late summer as a cover crop produced as much as 1544 kg ha^{-1} (DW) of aboveground biomass containing 50 kg ha^{-1} of N by the onset of winter. Therefore, there may be opportunity to mitigate the risk of N loss if a winter oilseed crop is integrated into a corn–soybean cropping system where corn stover is removed.

Winter oilseed crops may provide other ecosystem services beyond soil and water quality improvements, including weed and pest control, and insect and wildlife habitat (Heaton *et al.*, 2013). Johnson *et al.* (2015) reported that pennycress reduced weed biomass by >80%. Although not explicitly discussed by the authors, this response may have been related to the presence of glucosinolates, a class of secondary metabolites found in mustards (Moser, 2012), which can inhibit growth of other plants (Gimsing & Kirkegaard, 2009). This might be of practical interest in organic production systems and may also serve as a use for coproducts that remain following oil extraction. For example, Vaughn *et al.* (2005) found that defatted pennycress seed meal inhibited seedling germination and emergence of wheat and arugula [*Eruca vesicaria* (L.) Cav. Susp. *sativa* (Mill.) Thell.]. Control of pests by oilseed residue has also been reported. Mojtahedi *et al.* (1991) found that soils amended with rapeseed tissues had lower populations of root-knot nematode (*Meloidogyne* spp.).

Oilseed crops can also enhance beneficial insect populations and diversity, as several studies on the ecology of pollinators and mass-flowering crops have demonstrated. Bumble bee (*Bombus* spp.) density was found to be related positively to rapeseed availability on the landscape (Westphal *et al.*, 2003), with a higher abundance in rapeseed fields compared to wheat and Miscanthus (*Miscanthus* × *giganteus*) (Stanley & Stout, 2013). This effect can extend even to crops maturing later in the season; the density of bumble bees in late flowering sunflower has been shown to be enhanced by early flowering rapeseed, an effect known as temporal spillover (Riedinger *et al.*, 2014). Recently, Eberle *et al.* (2015) evaluated pollinator visitation and nectar production of camelina and pennycress grown as cover crops in Minnesota and found that visitation rates for these two oilseeds reached 67 and 22 insects min^{-1}, respectively (112 m^2-area). They also reported that, during the period of anthesis (Table 2), camelina and pennycress flowers produced 100 and 13 kg of sugar ha^{-1} in their nectar, respectively, the former of which was even greater than that of winter canola (84 kg ha^{-1}), a widely recognized pollinator-friendly plant. Both camelina and

pennycress specifically were cited by the presidentially mandated Pollinator Health Task Force (2015) as being useful early season forage resources for pollinators, which may assist in improving health of both domesticated and wild pollinators.

Solitary bee biodiversity is also enhanced by mass flowering in agricultural habitats, including increased abundance (Holzschuh *et al.*, 2013) and species richness (Diekotter *et al.*, 2014) in nearby seminatural areas, as well as increased species richness within rapeseed fields (Stanley & Stout, 2013). Pollination services rendered by bumble bees and solitary bees are increasingly important as they are recognized as more efficient pollinators in many crops and native plants compared to honey bees (*Apis* spp.), and in some cases, the only pollinator (Klein *et al.*, 2007, and sources therein). Furthermore, the declines in managed honey bee colonies as a result of a number of interacting factors (pests, disease, pesticides, stress, inadequate diet) places an increased burden on native pollinators for agricultural production and ecosystem function.

Mass-flowering crops like rapeseed are likewise attractive to pollinating syrphid flies (*Syrphidae* spp.), whose ecosystem services also include biological control resulting from predaceous larval stages that consume crop pests such as aphids (*Aphis* spp.) (Schmidt *et al.*, 2003; Haenke *et al.*, 2014). Syrphids are major visitors of pennycress and camelina, with both plants serving as a foraging resource during times when few other crops or native plants are flowering (Groeneveld & Klein, 2014). In a relay or double cropping system including a winter oilseed followed by soybean, there is potential for a temporal spillover of biological control of soybean aphid (*Aphis glycines*), analogous to the temporal spillover of bumble bees mentioned above. There is need for further study on this topic, as enhancement of natural enemies such as syrphid flies could provide a direct value by reducing pest populations and outbreaks, which in turn would reduce the economic and environmental costs of pesticide use. If oilseeds are to provide pollinator bene-

Table 2 Anthesis interval of winter oilseeds in Morris, Minnesota, and Brookings, South Dakota, 2011–2014*

Location	Year	Camelina		Pennycress		Canola	
		Begin	End	Begin	End	Begin	End
Morris, MN	2011	28 May	3 July	–	–	–	–
	2012	23 April	29 May	7 May	29 May	7 May	19 June
	2013	7 June	25 June	7 June	25 June	23 May	7 July
	2014	23 May	4 June	23 May	9 June	27 May	9 July
Brookings, SD	2013	19 June	3 July	19 June	3 July	–	–
	2014	23 May	3 June	–	–	–	–

*Ground coverage of flowers ≥1%.

fits, it will be important that they are incorporated into cropping systems in ways that enhance these benefits and do not negate them (e.g., by increasing need for insecticide applications).

Corn–Soybean cropping system demographics

A geographic analysis (2010–13) estimated that 21.6 and 5.5 million hectares are classified as 2-year (corn–soybean) and 3-year (corn–corn–soybean) cropping systems, respectively, in the US Corn Belt (Table 1). The analysis also found that crop sequences in the 2-year system were quite consistent (51 and 49%). This indicates that there are no discernable temporal acreage fluctuations as a result of crop phase changes with this cropping system. When individual states were compared, Illinois and Iowa combined, accounted for 45% of the total 2-year corn–soybean system acreage in the US Corn Belt, while Minnesota, Indiana, and Nebraska individually accounted for an additional 14, 11, and 10%, respectively. Conversely, North Dakota, Michigan, and Wisconsin (<2.5%, individually) accounted for low proportions of the 2-year cropping system acreage. Further county-level extrapolation of the acreage data displayed dense areas of 2-year corn–soybean cropping systems in central Illinois, western Iowa, and southwestern Minnesota (Fig. 4a), with a high volume of counties with 2-year corn–soybean acreage ranging from 50 000 to 150 000 hectares. Additionally, at least ten counties in Nebraska and South Dakota had acreage ranging from 50 000 to 100 000 hectares. The prevalence of 2-year corn–soybean cropping systems was <1000 hectares in western areas of North Dakota, South Dakota, and Nebraska, northern areas of Minnesota, Wisconsin, and Michigan, and southeastern Ohio. This is a function of agricultural row crop production being limited in these areas with greater production of small grains and other shorter-season and less water-intensive crops, and as a result of the prevalence of grasslands (North Dakota, South Dakota, and Nebraska) and forests (Minnesota, Wisconsin, Michigan, and Ohio) in those areas.

Total acreage of the 3-year cropping system is approximately one-fourth of that observed for the 2-year system (Table 1). However, the spatial layout of dense areas of 3-year cropping systems was generally similar to those observed for the 2-year cropping system, with some slight differences (Fig. 4b). In this case, clusters of counties with the greatest 3-year cropping system acreage (>30 000 hectares) were located in south central Nebraska and northern Illinois. The analysis also indicated that all states except Michigan and Ohio had multiple counties that had at least 10 000 hectares of the 3-year cropping system.

Adoption considerations and potential challenges

Several environmental, agronomic, and socioeconomic issues may exist as a result of intensifying a corn–soybean cropping system with a winter oilseed crop in the US Corn Belt. Additionally, these constraints may be spatially dependent. Moore & Karlen (2013) concluded that double cropping may be a feasible cropping strategy in the north central USA, pending further research. However, the system they discussed would be primarily biomass-driven, thus allowing both crops to be harvested prior to physiological maturity. The cropping system of interest in this review would require both crops (winter oilseed and soybean) to reach physiological maturity within the same calendar year, as seed/grain harvest is the primary goal of both crops. However, recent work in the north central USA has shown that double cropping winter oilseeds with soybean is attainable if agronomic management decisions (e.g., soybean cultivar selection) are modified appropriately (Gesch & Archer, 2013; Johnson et al., 2015). This will be discussed in detail later.

Environmental

The inclusion of a cover crop into a cropping system is formally the concept of double cropping, where two crops are grown during the same season on the same unit of land (Gesch & Archer, 2013; Moore & Karlen, 2013). The concept of a double cropping system is to better utilize the resources available during the active growing season, particularly growing degree unit (GDU) availability. Both Helsel & Wedin (1981) and Moore & Karlen (2013) concluded that areas in the north central region, where the growing season is often regarded as short when compared with other areas of the USA, can support double cropping systems if at least one, if not both, of the crops is harvested for forage. The biomass harvest of the first crop would theoretically result in the planting of the second crop at a date that would not significantly deviate from a traditional planting date, barring unfavorable weather conditions (e.g., wet soils) that interfere with timely planting. Winter oilseed growth could help reduce the potential for planting delays in wet conditions by utilizing water during late fall and early spring periods and by improving soil structure to increase internal drainage and support field equipment.

In double cropping systems where the first crop is harvested for seed, planting of the second crop may often occur after the 'traditional' planting window (Fig. 2). For example, Johnson et al. (2015) planted soybean following pennycress maturation in early- to mid-June in a 2-year study in Minnesota, while Gesch &

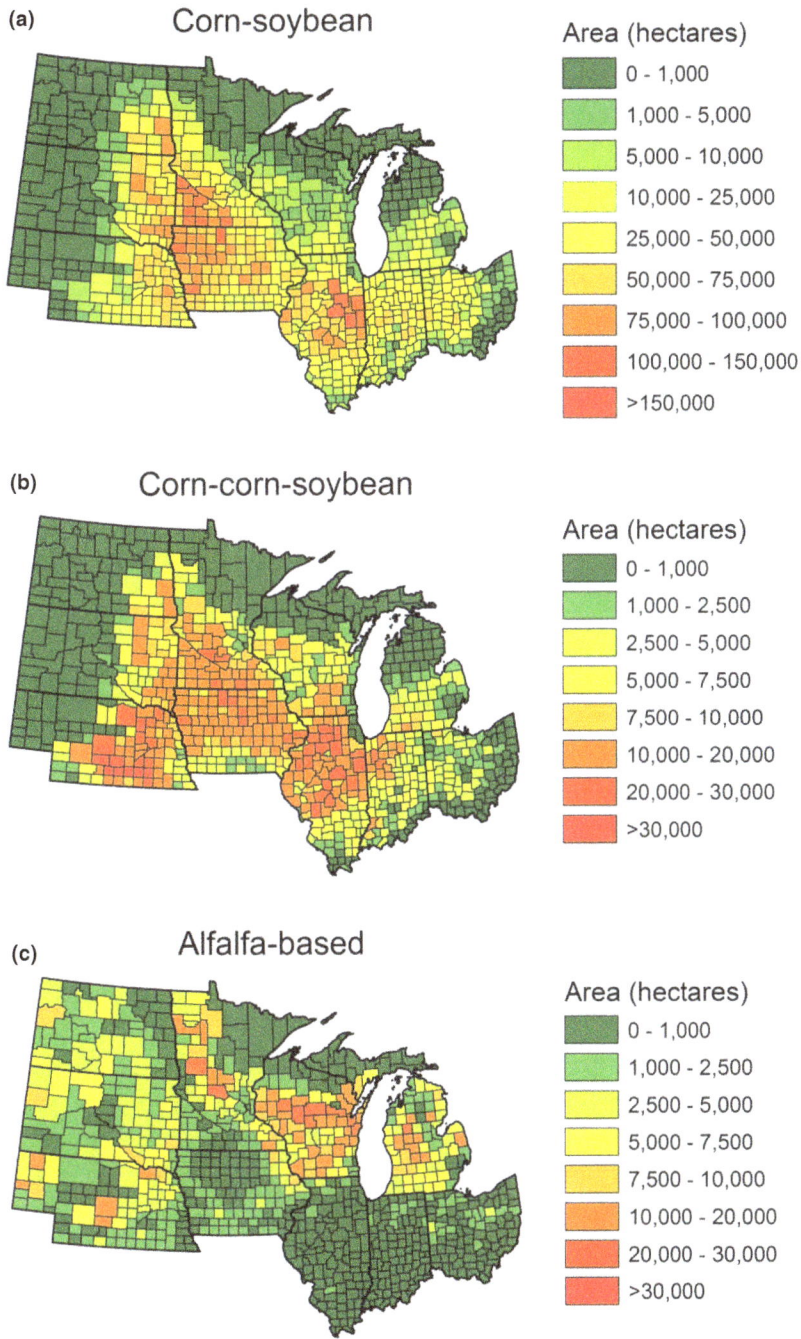

Fig. 4 Acreage estimates (2010–2013) by county in the US Corn Belt for (a) 2-year corn–soybean, (b) 3-year corn–soybean, and (c) alfalfa-based cropping systems.

Archer (2013) planted soybeans following camelina in late June to early July in a 2-year study in Minnesota. These dates were at least 2 to 3 weeks later than recommended for optimum soybean production in that state. In a research trial in Nebraska, soybeans were planted after pennycress harvest in early- to mid-June (M.R. Schmer & A.J. Sindelar, unpublished data). As GDU availability will often be one of the primary limiting factors to grain production in double cropping systems, particularly in the north central USA, its availability was estimated for the second crop based on 30-year normal temperature data from 15 June to the average first frost (≤ 0 °C) date (Fig. 5a). This analysis found that GDU availability over this time period ranged from 956 to 3070 GDU across the US Corn Belt region and 1396 to 3070 in counties with at least 1000 hectares of 2-year corn–soybean cropping systems. The limited amount of GDUs over this time period in Wisconsin, Minnesota, and North Dakota may result in either a forage crop or soybean being the preferred second crop in a double cropping system. The selection of indeterminate soybean cultivars (those that continue vegetative growth during reproductive development) may be better suited for compressed growing seasons than other major crops with a determinate growth process (e.g., corn, grain sorghum). Foley *et al.* (1986) evaluated yield performance of determinate and indeterminate soybean lines in Minnesota and found that yields of indeterminate lines were often greater. While quantification of GDU requirements of soybean is difficult, Kandel & Akyüz (2012) estimated that maturity group (MG) 00 soybean required at least 1679 GDU to reach physiological maturity. Another report by Kandel & Akyüz (2011) estimated that the GDU requirement of a MG 0 soybean ranged from 1542 to 1837 GDU. While the lower requirement of these ranges should not be considered to be the absolute minimum GDU requirement, it does at least suggest that there is some degree of risk that soybean growing in a double cropping system may not often reach physiological maturity in far northern areas of the US Corn Belt. However, the risk of the growing season ending prior to the secondary crop reaching physiological maturity should decrease if a proper soybean cultivar is selected to account for the condensed growing season. This issue may become less problematic if relay cropping strategy is implemented. Gesch *et al.* (2014) demonstrated that soybean could be produced successfully by interplanting it into winter camelina in the spring at a normal or near-normal time for western Minnesota. Furthermore, this technique allowed the use of a standard Maturity Group soybean for the region (MG I), whereas sequential double cropping required the use of a shorter-maturity soybean variety (MG 00).

Water availability through precipitation is the other primary environmental variable that may influence the success of a double cropping system. However, the overall influence of cover crops on subsequent crop yields is conflicted. For example, Olson *et al.* (2014) found no negative effects of hairy vetch (*Vicia villosa* Roth) and rye cover crops on corn and soybean production, respectively, in Illinois. In Nebraska, a winter rye cover crop reduced subsequent corn grain yield in one of 3 years, but had no effect in the other two (Kessavalou & Walters, 1997). Acuña & Villamil (2014) evaluated the effects of several cover crop species and mixtures on subsequent soybean production in Illinois and found that there was no yield reduction when compared with the conventional treatment. In South Dakota, seeding a cover crop mixture following wheat did not

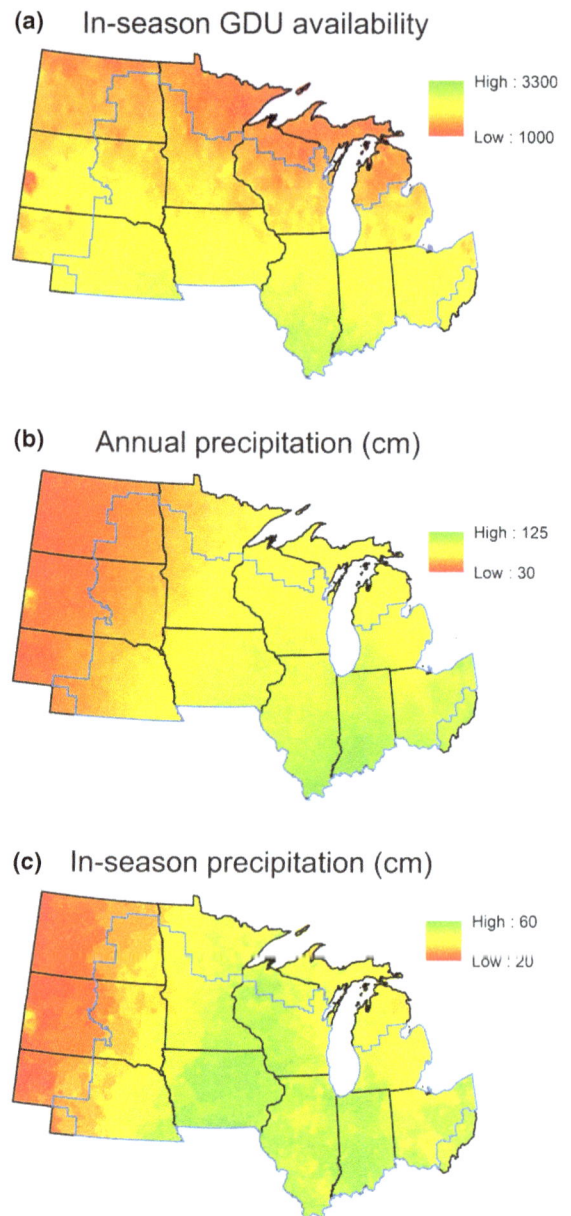

(a) In-season GDU availability

High : 3300
Low : 1000

(b) Annual precipitation (cm)

High : 125
Low : 30

(c) In-season precipitation (cm)

High : 60
Low : 20

Fig. 5 Long-term (30 year) (a) in-season growing degree unit (GDU) availability, (b) annual precipitation, and (c) in-season precipitation in the US Corn Belt. Area within the blue outline identifies the generalized geography of ≥1000 hectares county^{-1} of 2-year corn–soybean cropping systems.

affect corn grain yield the following year in a low or high water stress environment, but did reduce yield at a site with moderate water stress (Reese *et al.*, 2014). At this site, cumulative soil water in the 30- to 60-cm depth was reduced in treatments with the cover crop mixture by 15%. However, no soil water changes occurred in the low- or high-stress environments with the use of the cover crop mixture. Lastly, winter rye cover crops in Minnesota reduced soybean yield by one-third in

1 year, but had no influence in another for both short- and full-season cultivars (Forcella, 2013). With regard to winter oilseeds, in a camelina–soybean double cropping study, Gesch & Johnson (2015) found that water use for sequentially and relay-cropped soybean in Minnesota was greater than continuous soybean, but absolute differences were not large. These authors concluded that these double cropping systems can be used feasibly in most rain-fed agricultural areas in the US Corn Belt region.

The region of interest in this review exhibits a considerable range in average annual precipitation, with as little as 33 cm received in western Nebraska to 124 cm in Illinois (or 41 to 124 cm within the region that supports 1000 ha county^{-1} corn–soybean cropping system) (Fig. 5b). When in-season precipitation was calculated according to the same parameters explained for GDU availability, a spatial precipitation gradient from west to east that was similar to annual precipitation was observed (Fig. 5c). Over the hypothetical growing season for soybean, in-season precipitation ranged from 20 to 60 cm across the region (24 to 60 cm within 1000-hectare county^{-1} corn–soybean geography). While in-season and annual precipitation can serve as an initial gauge, the overall suitability will ultimately be determined by water demand by the combined needs of crop (T) and evaporative demand (E), which is defined as evapotranspiration (ET). As water demand can fluctuate on both macro- (Robinson & Nielsen, 2015) and micro-scales (Sadler et al., 2002), to explicitly identify or rule out suitable areas at this point would be difficult, if not inappropriate. However, as evaporative demand generally decreases as the distance from the southern US Plains increases, a logical hypothesis is that the core area of the US Corn Belt would be under less restriction. Furthermore, we speculate that locations restricted by water availability will primarily be located in central and western areas of Nebraska, South Dakota, and North Dakota. However, even exclusion of areas in these states is not necessarily certain nor appropriate at this point because of the confounding influence of supplemental irrigation. According to the USDA-National Agricultural Statistics Service (2008), approximately 3.5 million hectares of agricultural land is irrigated in Nebraska, accounting for 15% of total irrigated land in the USA In comparison, 1.25 million hectares are irrigated in the remaining nine states in the USA Corn Belt, combined. Therefore, opportunities for cropping system intensification with winter oilseed cover crops will likely exist in many areas in the western Corn Belt, particularly in Nebraska, as a result of supplemental irrigation. In these areas, the potential increase in water demand would need to be weighed against any limitations in irrigation water supply. Further identification of irrigated land in these regions would be necessary to identify acreage where cropping system intensification would not be practical.

Finally, landscape attributes, specifically slope, could also affect land availability for a corn–soybean cropping system with stover removal and winter oilseed crop inclusion. In several areas in the US Corn Belt, scenarios exist where at least 1000 hectares of corn–soybean cropping systems reside in areas with landscape slopes of 6% or greater (Fig. 6). A point of emphasis by many has been that corn stover removal rates need to be at levels that simultaneously maintain SOC and control wind and water erosion, and that areas exist where corn stover removal may not be appropriate (e.g., Mann et al., 2002; Nelson, 2002; Wilhelm et al., 2007; Johnson et al., 2010). As water erosion is affected by landscape slope, this becomes a critical attribute for determining landscape suitability for such a cropping system. Graham et al. (2007) reported that the area appropriate for corn stover removal decreases significantly as erosion susceptibility increases.

Coincidentally, however, there may be situations where the addition of a winter oilseed crop may allow for stover removal when previously declared unacceptable as a result of landscape erosion constraints. In a 2-year study at a sloped site in Iowa, Kaspar et al. (2001) found that a winter rye cover crop interseeded with soybean reduced interrill erosion and runoff in both years, while an oat (Avena sativa L.) cover crop reduced erosion and runoff in 1 year. Bonner et al. (2014) used an integrated modeling framework to evaluate the effect of cover crops on corn stover supplies and found that their inclusion would increase sustainable stover removal rates across a range of slope gradients, although it became less impactful at slopes >4%. Therefore, based on initial results from this study, it seems that the amount of harvestable stover in this scenario may still be rather small.

Agronomic

From a production standpoint, soybean yield reduction associated with delaying the planting of the crop in a double cropping system is probable, regardless of geographic location. The review of 17 studies where soybean planting was delayed revealed a clear yield response when planting was delayed by at least 19 days, although the magnitude varied among sites (Table 3). In this review, all but four site-years exhibited a yield loss of 4 to 28% when planting was delayed. Two cases existed where no yield losses occurred with delayed planting in Minnesota and Wisconsin, although planting was delayed into late May by 19 and 20 days, respectively (Leuschen et al., 1992; Pedersen & Lauer,

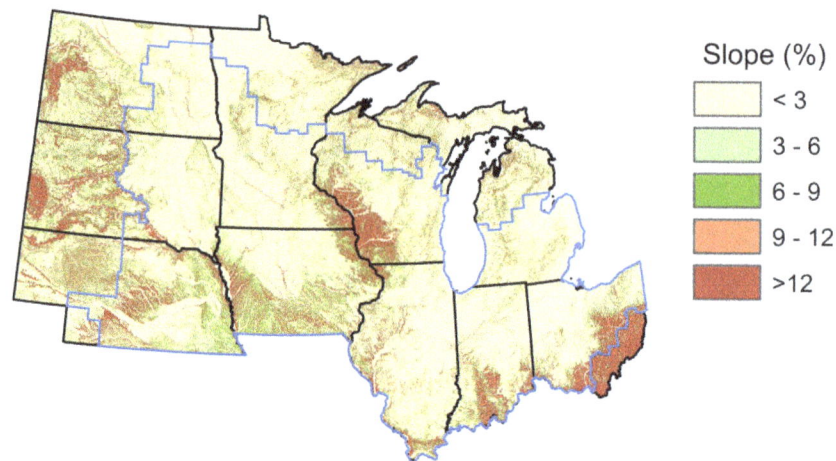

Fig. 6 Landscape slope in the US Corn Belt. Area within the blue outline identifies the generalized geography of ≥1000 hectares of 2-year corn–soybean cropping systems.

2003). Two studies used in this review evaluated soybean production in double cropping systems (LeMahieu & Brinkman, 1990; Gesch & Archer, 2013). In Wisconsin, LeMahieu & Brinkman (1990) reported that planting in late June (delay of 41 days) as a result of growing a winter rye crop for biomass reduced subsequent soybean yield by 26%. In comparison, delaying soybean planting over the same period without a double crop reduced soybean yield by 27%. In Minnesota, Gesch & Archer (2013) reported that soybean yield decreased by 20% when it was planted following camelina harvest when compared with continuous soybean sown in early May (52-day difference between planting dates). Based on these results, producers must be aware of and amendable to probable soybean yield reductions associated with grain-based double cropping systems, regardless of location in the US Corn Belt. However, it is important to consider is that the reduction in soybean yield may not necessarily equate to less net profit for the cropping system when the economic value of the oilseed crop is considered.

Several authors (e.g., Carter & Boerma, 1979; Wilcox & Frankenberger, 1987; Specht et al., 2012) have stated that reductions in soybean yield as a function of delayed planting is a result of the reduction of main stem nodes. Specht et al. (2012) also noted that a new node is produced on the main stem every 3.7 days, regardless of planting date. Therefore, soybean plants planted after 'traditional' planting dates have less opportunity for nodal development. An observed reduction in main stem node number reported by Carter & Boerma (1979) also coincided with a reduction in plant height by 14 to 39% when planting was delayed. Because of this physiological response and the aforementioned planting date studies, there are likely limited scenarios where delaying soybean planting to mid- to

late June would not result in some degree of yield reduction. However, additional research is clearly needed to identify crop management practices that reduce this yield gap as a result of delayed planting in double cropping systems.

An agronomic-associated landscape constraint for winter oilseed adoption may be temporal and spatial proximity to alfalfa (*Medicago sativa* L.)-based cropping systems. Cropping systems where alfalfa precedes at least 1 year of corn can be quite attractive as a result of the legume N credit supplied to the subsequent corn crop (Fox & Piekielek, 1988; Bundy & Andraski, 1993), and scenarios can exist where second-year corn following alfalfa does not require additional N fertilizer (Yost et al., 2014). Alfalfa with high protein concentration is of particular value to the dairy industry, as it has been shown to increase milk production (Wu & Satter, 2000; Olmos Colmenero & Broderick, 2006), among other reasons. Therefore, areas or states with a prominent dairy industry and subsequent large areas of alfalfa production (Table 1, Fig. 4c) will require thoughtful planning and placement of cropping systems utilizing some types of winter oilseed crops (e.g., pennycress and camelina). This is because their temporal or spatial 'escape' into an alfalfa-based cropping system may negatively influence quality and/or palatability. Although most research on the reduction of alfalfa forage quality is dedicated to other common weeds, a general conclusion is that polycultures of alfalfa and nontarget plant species often produce less dry matter, have lower forage quality, and are less palatable than monoculture alfalfa (Temme et al., 1979; Marten et al., 1987; Fisher et al., 1988). Moyer & Hironaka (1993) reported that the proportion of digestible crude protein by sheep and subsequent digestible energy of pennycress was lower than that with alfalfa, despite crude protein concentration being similar

Table 3 Reported responses of soybean grain yield to delayed planting across the US Corn Belt

State	Site(s)	Study length, years	Average planting date		Yield change (D$_1$–D$_2$), kg ha^{-1}	Reference
			Date 1 (D$_1$)	Date 2 (D$_2$)		
Illinois	Urbana	3	13 May	18 June	−458 (13%)	Beaver & Johnson (1981)
	Brownston	2	20 May	14 June	+478 (28%)	Beaver & Johnson (1981)
Illinois	Urbana	2	12 May	14 June	−715 (22%)	Anderson & Vasilas (1985)
Indiana	West Lafayette	3	12 May	22 June	−531 (21%)	Wilcox & Frankenberger (1987)
Indiana	West Lafayette	2	10 May	10 June	−620 (15%)	Robinson et al. (2009)
Iowa	Ames	3	14 May	8 June	−205 (8%)	Krell et al. (2005)
Iowa	Newton	3	11 May	7 June	−200 (7%)	Perez-Bidegain et al. (2007)
Iowa	Ames, Crawfordsville, De Witt, Nashua, Nevada, Whiting	4	29 April	8 June	−1055 (25%)	De Bruin & Pedersen (2008)
Minnesota	Morris	4	3 May	22 May	−215 (7%)‡	Leuschen et al. (1992)
	Lamberton	3	4 May	25 May	+38 (2%)	Leuschen et al. (1992)
	Waseca	4	8 May	28 May	−303 (10%)	Leuschen et al. (1992)
Minnesota	Waseca, Lamberton, Madison, Rosemount	3	15 May	14 June	−590 (23%)	Coulter et al. (2011)
Minnesota*	Morris	2	8 May	29 June	−463 (20%)	Gesch & Archer (2013)
Nebraska	Clay Center	3	7 May	15 June	−390 (13%)	Elmore (1990)
North Dakota	Fargo, Casselton	2	21 May	17 June	−820 (26%)	Helms et al. (1990)
Ohio	Hoytville, South Charlston	3	1 May	10 June	−834 (25%)	Beuerlein, (2013)
Wisconsin†	Arlington	2	12 May	22 June	−1004 (27%)	LeMahieu & Brinkman (1990)
		2	12 May	22 June	−941 (26%)	LeMahieu & Brinkman (1990)
Wisconsin	Arlington	4	15 May	13 June	−572 (17%)	Oplinger & Philbrook (1992)
Wisconsin	Hancock	2	1 May	30 May	−565 (21%)	Grau et al. (1994)
Wisconsin	Arlington	4	5 May	25 May	−170 (4%)	Pedersen & Lauer (2003)
	Hancock	4	10 May	30 May	−600 (14%)*	Pedersen & Lauer (2003)

*Soybeans were double cropped, with camelina being the preceding crop.
†Soybeans were double cropped, with barley being the preceding crop.
‡Authors reported no statistical change.

between the two plant species. Pennycress can also have adverse effects on livestock and their corresponding milk and meat products if excess quantities are consumed due to high erucic acid levels in its seed oil and glucosinolates, specifically sinigrin, in its seed and vegetative tissues (Best & MCIntyre, 1975; Warwick et al., 2002; Sedbrook et al., 2014). For example, consumption of large quantities of plants containing glucosinolates by dairy cattle can result in off-flavored milk products (Fenwick et al., 1982). Also, beef cattle carcasses have been shown to have slight to strong odor when fed high proportions of pennycress seed immediately before processing, yet no discernable odor existed when the time period between feeding and processing was 24 h or greater (Whiting et al., 1958) Despite this effect on animals and corresponding products, myrosinase, the enzyme responsible for the degradation of sinigrin into the toxic compounds isothiocynate and allyl isothiocynate, can be successfully deactivated through a heat treatment prior to seed crushing (Sedbrook et al., 2014). Furthermore, Hojilla-Evangelista et al. (2014) reported

that two protein recovery methods, saline extraction and acid precipitation, performed on defatted pennycress seed also can effectively recover protein extracts that are sinigrin-free. This finding is significant, as protein content in pennycress seed meal can range from 20 to 27% (Hojilla-Evangelista et al., 2013; Selling et al., 2013), and thus could be a serviceable livestock feed. Conversely, camelina should be less problematic than pennycress in this regard. Research has shown that camelina seed meal is low in glucosinolates (Schuster & Friedt, 1998), can serve as a potential livestock feed (Korsrud et al., 1978; Moriel et al., 2011; Szumacher-Strabel et al., 2011; Colombini et al., 2013), and is approved by the US Food and Drug Administration for use in cattle and chicken feeds (up to 10% of total ration). Long-term management plans in addition to short-term strategies should also be considered in areas with dense dairy production or fields that may be seeded to alfalfa in the near future. However, a counterpoint to this issue is the recent availability of glyphosate-tolerant alfalfa cultivars that can aid in control of nontarget plants. Therefore, any issues regarding winter oilseed production in or near alfalfa-based cropping systems may be alleviated if widespread adoption of glyphosate-tolerant alfalfa cultivars occurs.

Both pennycress and camelina are considered minor weeds to select crops. For pennycress, most yield losses associated with its presence have been reported with canola, wheat, and safflower (*Carthamus tinctorius* L.) in Canada (Warwick *et al.*, 2002). Camelina is a minor weed in flax (Putnam *et al.*, 1993), resulting in its alternate name as *false flax*. Despite both crops being historically regarded as a weed and a recent risk assessment by the USDA-Animal and Plant Health Inspection Service (APHIS) classifying pennycress as a high risk potential for spread and impact (USDA-APHIS, 2015), we hypothesize their impact as a volunteer weed on corn and soybean production to be low. This is because chemical control of both plants is historically quite successful. Francis & Warwick (2009) concluded that agronomic impact of camelina in corn–soybean cropping systems should not be an issue when proper herbicide chemistries are used. Pennycress is also quite susceptible to multiple herbicide chemistries (Warwick *et al.*, 2002), including glyphosate and glufosinate (Sedbrook *et al.*, 2014). Therefore, a pre-emergence herbicide application at corn or soybean planting should provide adequate control of most pennycress or camelina seeds as they germinate, and any escapes can be managed with postemergence chemistries. Additionally, the majority of soybean currently grown in the US Corn Belt has resistance to glyphosate. In other situations, herbicide chemistries such as 2,4-dichlorophenoxyacetic acid, imazethapyr, and acifluorfen should also successfully control escapes.

From a field management standpoint, a critical practice associated with oilseed and cover crop productivity is proper stand establishment. Studies that have evaluated a winter oilseed–soybean double cropping system have typically followed wheat (Phippen & Phippen, 2012; Gesch & Archer, 2013). Depending on type, wheat harvest occurred no later than mid-August, thus allowing adequate time for fall growth of the winter oilseed when direct-seeded into the soil. Gesch & Cermak (2011) demonstrated that although stand establishment of winter camelina can differ by year and tillage system (e.g., conventional vs. no-till) in the northern US Corn Belt, they found that final plant population density varied little by planting date when seeded from early September to mid-October. However, the greatest seed yields and oil content were generally obtained for plants seeded in late September to early October. Similar results have been obtained for pennycress (R.W. Gesch, unpublished data). Winter camelina has also been successfully established in North Dakota in September and

even as a dormant seeding in late October (D.W. Archer, unpublished data). Moreover, Gesch *et al.* (2014) have shown that winter camelina seeded in mid- to late October in west central Minnesota, just weeks before permanent soil freezing occurred, exhibited adequate winter survival and spring growth. This indicates that, in some of the southern regions of the US Corn Belt (e.g., Iowa, Illinois, and Nebraska), it may be possible to successfully establish winter camelina and pennycress following harvest of corn for grain. Nevertheless, this type of management schedule may not be applicable to many corn–soybean cropping systems unless (1) the entire corn plant is harvested for silage prior to reaching physiological maturity or (2) an early maturing hybrid is used. Typically, corn hybrids are selected based on maturities that best exploit the growing season when grain yield optimization is the primary goal. This practice, in turn, results in little growing season remaining after the corn plants reach physiological maturity, grain is allowed to dry to an appropriate moisture content, and harvest occurs (Fig. 2). Therefore, aerial seeding of the winter oilseed crop may be necessary in most areas, particularly for crops/cultivars that require vernalization. Baker & Griffis (2009) concluded that aerial seeding of winter rye in northern areas of the US Corn Belt may help mitigate growth limitations imposed by a short growing season. A survey by Singer (2008) found that 8% of cover crops were aerially seeded in a portion of the Corn Belt (Illinois, Indiana, Iowa, and Minnesota), compared to 68 and 21% for direct drilling and broadcast spreading, respectively. However, 62% of cover crops in this area were not harvested. Although not explicitly discussed by Singer (2008), we speculate that the proportion of cover crops aerially seeded in this area is a function of cost, and that seed from few, if any, of these cover crops would have been harvested. Additionally, we hypothesize this proportion would increase if cash-positive winter crops were used or if incentives for cover crops are provided.

Benefits do exist for aerially seeding winter cover crops when compared with other traditional methods. These include offering a greater temporal window for seeding, elimination of machinery traffic, which may help to reduce soil compaction, and less time requirement as aerial seeding can often be performed in a shorter amount of time than other traditional methods (Wilson *et al.*, 2013). To some, however, the disadvantages outweigh the benefits. These include a short-time period for fall growth, cost of aerial application, seed predation, and proper stand establishment (Wilson *et al.*, 2014). Barnett & Comeau (1980) evaluated aerial seeding effects on the germination of wheat, oat, and barley (*Hordeum vulgare* L.) and found that the number of plants germinated was substantially less than direct-seeded or broadcast treatments. Therefore, to compensate for poor germination of aerially seeded crops, seeding rates may need to be increased, which would subsequently increase input costs. However, both camelina and pennycress have very small seeds compared to, for example, winter rye (about 0.8, 0.8, and 33 g 1000^{-1} seeds, respectively), which would serve to lower aerial seeding costs on a unit-area basis (Robinson, 1987; Phippen et al., 2010; Miedaner et al., 2014).

Other considerations include information and techniques needed in managing these crops, and the need for continued breeding efforts to improve characteristics of these crops. For example, risk of seed loss due to shatter can be a serious concern for camelina and pennycress, with complete yield loss potentially occurring in a single untimely rain and wind event (D.W. Archer, unpublished data). Fortunately, genetic similarities of camelina and pennycress with *Arabidopsis* have resulted in expedited progress toward the genomic understanding of both crops (Gehringer et al., 2006; Dorn et al., 2013, 2015; Nguyen et al., 2013). For pennycress, this can potentially accelerate its domestication through improvement of several traits including seed oil quality and quantity improvement, reduction of glucosinolates, seed dormancy manipulation, flowering and senescence synchronization, and pod shatter reduction (Sedbrook et al., 2014). Also, availability of herbicides labeled for use in these crops is limited, presenting a challenge for in-season weed management, particularly in no-till production systems. One possible solution to this issue is herbicide-resistant cultivars, such as camelina with resistance to acetolactate synthase inhibitor herbicides (Walsh et al., 2012). While initial research has demonstrated that management practices can be adjusted to overcome some of these constraints (e.g., Gesch et al., 2014), additional crop development and agronomic management research is clearly needed.

Socioeconomic

Despite most cover crop research generally occurring in the past quarter century, its benefits have long been recognized (e.g., Odland & Knoblauch, 1938; Beale et al., 1955; Moschler et al., 1967; Mitchell & Teel, 1977). Some of these include erosion control (Kaspar et al., 2001), soil organic matter additions (Ding et al., 2006), nitrate-N leaching reductions (Kladivko et al., 2014), and weed suppression (Teasdale, 1996). Furthermore, the addition of a cover crop into a corn-based cropping system with stover removal may help to offset some of the adverse effects on soil properties (Fronning et al., 2008; Blanco-Canqui et al., 2014) and may even increase profitability of these cropping systems (Pratt et al., 2014). However, despite these advantages, adoption by

farmers is low to date. A survey by Singer et al. (2007) found that 18% of farmers in Illinois, Indiana, Iowa, and Minnesota had used a cover crop, and only 6% of the land area for the average-sized farm was seeded to cover crops in 2005. When farmers do indeed integrate cover crops into their cropping systems, it is because of the benefits associated with cropping system diversification, erosion control, organic matter addition, weed suppression, and N addition by leguminous cover crops (Mallory et al., 1998; Stivers-Young and Tucker, 1999; Snapp et al., 2005; Singer et al., 2007). However, as at least partially reflected in low adoption rates, many farmers feel the disadvantages outweigh the benefits. A farmer survey by Mallory et al. (1998) identified cost, potential influence on the subsequent crop, and weed problems as main concerns. Surveys by Stivers-Young & Tucker (1999) and Snapp et al. (2005) also identified cost as a concern. Despite the cash crop potential of winter oilseed crops, the limited market for their seed still makes these cover crops a cost concern. Additionally, prospective winter oilseed crops like pennycress and camelina are regarded as weeds in some cropping systems. This may further dissuade farmers from adopting these crops, particularly as pennycress seed requires a maturation period, which can increase its persistence in the seedbank (Warwick et al., 2002). However, use of nondormant pennycress cultivars (Isbell et al., 2015) may decrease this persistence, and both camelina and pennycress are easily controlled by commercially available herbicides.

From a direct-cost standpoint, cropping systems integrating winter cover or oilseed crops may be less profitable, especially if they have adverse effects on subsequent crops. Bollero & Bullock (1994) estimated a net loss of $11 hectare^{-1} when a hairy vetch cover crop preceded corn and grain sorghum [*Sorghum bicolor* (L.) Moench] in Illinois. Mallory et al. (1998) reported that when only the sole N value of a legume cover crop was considered, it was not economically advantageous when compared with fertilizer N. However, recent research has shown that corn-based cropping systems with stover removal maintain or increase profitability when an annual ryegrass (*Lolium multiflorum* Lam.) or crimson clover (*Trifolium incarnatum* L.) cover crop is added to the cropping system (Pratt et al., 2014). Gesch et al. (2014) reported lower net return for a sequential double crop camelina–soybean system when compared with monoculture soybean. However, in that same study, they also reported similar net returns between monoculture soybean and a relay-cropped camelina–soybean system when the camelina was treated with glyphosate seven to 10 days before harvest. Therefore, market values of winter oilseeds likely will need to be established before an increase in adoption of winter oilseed crops occurs. These initial findings are promising and certainly warrant further research on the topic, particularly with winter oilseed crops that may further provide monetary value.

Conclusion

Inclusion of winter oilseed crops into corn–soybean cropping systems in the US Corn Belt may simultaneously provide a cash-positive commodity crop used for renewable diesel or hydroprocessed aviation fuel in addition to the wide range of ecosystem services often observed with cover crops. The geographic analysis conducted for this discussion identified 21.5 and 5.6 million hectares that are grown in corn–soybean and corn–corn–soybean cropping systems in the US Corn Belt, with the densest areas generally located in Illinois, Iowa, and southern Minnesota. However, what remains unknown is the proportion of these cropping systems that ultimately would be used as adoption of biomass cover crops in the US Corn Belt is low to date. A survey by Singer *et al.* (2007) reported 56% of farmers would adopt cover crops if financial cost sharing for the cover crop was available. Consequently, the additional value of the seed from a winter oilseed crop quite possibly may incentivize farmers to adopt these crops.

Environmental, agronomic, and socioeconomic obstacles resulting from integrating winter oilseed crops into Midwestern corn–soybean cropping systems do exist, and further research is needed in all of these areas in an attempt to reduce or minimize them. Initial research has already demonstrated that management practices can be adjusted to overcome some of these constraints (e.g., Gesch *et al.*, 2014). Management practices for a winter oilseed crop plausibly may vary slightly among areas within the US Corn Belt, much like management practices for corn and soybean can often vary slightly across the region. Regardless of these constraints, a corn–soybean cropping system where corn stover is removed and a winter oilseed crop is grown theoretically offers great potential to produce multiple independent sources for food, fuel, and fiber. However, more research is needed regarding the environmental and economic viability and sustainability of this intensified cropping system over a larger geographical area throughout the Corn Belt region and other areas in the USA.

Acknowledgements

This review contributes to CenUSA Bioenergy, which is supported by Agriculture and Food Research Initiative Competitive Grant No. 2011-68005-30411 from the USDA National Institute of Food and Agriculture. The USDA is an equal opportunity employer and provider.

References

Acuña JCM, Villamil MB (2014) Short-term effects of cover crops and compaction on soil properties and soybean production in Illinois. *Agronomy Journal*, **106**, 860–870.

Air Transport Action Group (2015) *Our climate plan.* Available at: http://aviation-benefits.org/environmental-efficiency/aviation-and-climate-change/our-climate-plan/ (accessed 21 March 2015).

Anderson LR, Vasilas BL (1985) Effects of planting date on two soybean cultivars: seasonal dry matter accumulation and seed yield. *Crop Science*, **25**, 999–1004.

Baker JM, Griffis TJ (2009) Evaluating the potential use of cover crops in corn-soybean systems for sustainable co-production of food and fuel. *Agricultural and Forest Meteorology*, **149**, 2120–2132.

Barnett GM, Comeau JE (1980) Seeding cereals by air and ground. *Canadian Journal of Plant Science*, **60**, 1147–1155.

Beale OW, Nutt GB, Peele TC (1955) The effects of mulch tillage on runoff, erosion, soil properties, and crop yields. *Soil Science Society of America Journal*, **19**, 244–247.

Beaver JS, Johnson RR (1981) Response of determinate and indeterminate soybeans to varying cultural practices in the northern USA. *Agronomy Journal*, **73**, 833–838.

Best KF, McIntyre GI (1975) The biology of Canadian weeds. 9. *Thlaspi arvense* L.. *Canadian Journal of Plant Science*, **55**, 279–292.

Beuerlein JE (2013) Yield of indeterminate and determinate semidwarf soybean for several planting dates, row spacings, and seeding rates. *Journal of Production Agriculture*, **4**, 300–303.

Blanco-Canqui H, Lal R (2009) Crop residue removal impacts on soil productivity and environmental quality. *Critical Reviews in Plant Sciences*, **28**, 139–163.

Blanco-Canqui H, Ferguson RB, Jin VL, Schmer MR, Wienhold BJ, Tatarko J (2014) Can cover crop and manure maintain soil properties after stover removal from irrigated no-till corn? *Soil Science Society of America Journal*, **78**, 1368–1377.

Boateng AA, Mullen CA, Goldberg NM (2010) Producing stable pyrolysis liquids from the oil-seed presscakes of mustard family plants: pennycress (*Thlaspi arvense* L.) and camelina (*Camelina sativa*). *Energy & Fuels*, **24**, 6624–6632.

Bollero GA, Bullock DG (1994) Cover cropping systems for the central Corn Belt. *Journal of Production Agriculture*, **7**, 55–58.

Bonner IJ, Muth DJ, Koch JB, Karlen DL (2014) Modeled impacts of cover crops and vegetative barriers on corn stover availability and soil quality. *BioEnergy Research*, **7**, 576–589.

Bundy LG, Andraski TW (1993) Soil and plant nitrogen availability tests for corn following alfalfa. *Journal of Production Agriculture*, **6**, 200–206.

Carter TE, Boerma HR (1979) Implications of genotype × planting date and row spacing interactions in double-cropped soybean cultivar development. *Crop Science*, **19**, 607–610.

Cassman KG, Liska AJ (2007) Food and fuel for all: realistic or foolish? *Biofuels, Bioproducts, & Biorefining*, **1**, 18–23.

Colombini S, Broderick GA, Galasso I, Martinelli T, Rapetti L, Russo R, Reggiani R (2013) Evaluation of *Camelina sativa* (L.) Crantz meal as an alternative protein source in ruminant rations. *Journal of the Science of Food and Agriculture*, **94**, 736–743.

Coulter JA, Sheaffer CC, Haar MJ, Wyse DL, Orf JH (2011) Soybean cultivar response to planting date and seeding rate under organic management. *Agronomy Journal*, **103**, 1223–1229.

Dale VH, Kline KL, Wright LL, Perlack RD, Downing M, Graham RL (2011) Interactions among bioenergy feedstock choices, landscape dynamics, and land use. *Ecological Applications*, **21**, 1039–1054.

De Bruin JL, Pedersen P (2008) Soybean seed yield response to planting date and seeding rate in the Upper Midwest. *Agronomy Journal*, **100**, 696–703.

Diekotter T, Peter F, Jauker B, Wolters V, Jauker F (2014) Mass-flowering crops increase richness of cavity-nesting bees and wasps in modern agro-ecosystems. *Global Change Biology Bioenergy*, **6**, 219–226.

Ding G, Liu X, Herbert S, Novak J, Amarasiriwardena D, Xing B (2006) Effect of cover crop management on soil organic matter. *Geoderma*, **130**, 229–239.

Dorn KM, Fankhauser JD, Wyse DL, Marks MD (2013) *De novo* assembly of the pennycress (*Thlaspi arvense*) transcriptome provides tools for the development of a winter cover crop and biodiesel feedstock. *The Plant Journal*, **75**, 1028–1038.

Dorn KM, Fankhauser JD, Wyse DL, Marks MD (2015) A draft genome of field pennycress (*Thlaspi arvense*) provides tools for the domestication of a new winter biofuel crop. *DNA Research*, **22**, 121–131.

Drury CF, Tan CS, Welacky TW *et al.* (2014) Reducing nitrate loss in tile drainage water with cover crops and water-table management systems. *Journal of Environmental Quality*, **43**, 587–598.

Eberle CA, Thom MD, Nemec KT *et al.* (2015) Using cash cover crops: pennycress, camelina, and canola to provision pollinators. *Industrial Crops and Products*, doi:10.1016/j.indcrop.2015.06.026.

Elmore RW (1990) Soybean cultivar response to tillage systems and planting date. *Agronomy Journal*, **82**, 69–73.

Fan J, Shonnard DR, Kalnes TM, Johnsen PB, Rao S (2013) A life cycle assessment of pennycress (*Thlaspi arvense* L.)-derived jet fuel and diesel. *Biomass and Bioenergy*, **55**, 87–100.

Fargione J, Hill J, Tilman D, Polasky S, Hawthorne P (2008) Land clearing and the biofuel carbon debt. *Science*, **319**, 1235–1238.

Federal Aviation Administration (2011). *FAA Desination 2025*. Available at: http://www.faa.gov/about/plans_reports/media/Destination2025.pdf (accessed 19 March 2015).

Fenwick GR, Heaney RK, Mullin WJ, VanEtten CH (1982) Glucosinolates and their breakdon products in food and food plants. *Critical Reviews in Food Science and Nutrition*, **18**, 123–201.

Feyereisen GW, Camargo GGT, Baxter RE, Baker JM, Richard TL (2013) Cellulosic biofuel potential of a winter rye double crop across the U.S. corn-soybean belt. *Agronomy Journal*, **103**, 631–642.

Fisher AJ, Dawson JH, Appleby AP (1988) Interference of annual weeds in seedling alfalfa (*Medicago sativa*). *Weed Science*, **36**, 583–588.

Foley TC, Orf JH, Lambert JW (1986) Performance of related determinate and indeterminate soybean lines. *Crop Science*, **26**, 5–8.

Forcella F (2013) Short- and full-season soybean in stale seedbeds versus rolled-crimped winter rye mulch. *Renewable Agriculture and Food Systems*, **29**, 92–99.

Fox RH, Piekielek WP (1988) Fertilizer N equivalence of alfalfa, birdsfoot trefoil, and red clover for succeeding corn crops. *Journal of Production Agriculture*, **1**, 313–317.

Francis A, Warwick SI (2009) The biology of Canadian weeds. 142. *Camelina alyssum* (Mill.) Thell.; *C microcarpa* Andrz. Ex DC.; *C. sativa* (L.) Crantz. *Canadian Journal of Plant Science*, **142**, 791–810.

Fronning BE, Thelen KD, Min D-H (2008) Use of manure, compost, and cover crops to supplant crop residue carbon in corn stover removed cropping systems. *Agronomy Journal*, **100**, 1703–1710.

Gehringer A, Friedt W, Lühs W, Snowdon RJ (2006) Genetic mapping of agronomic traits in false flax (*Camelina sativa* subsp. *sativa*). *Genome*, **49**, 1555–1563.

Gesch RW, Cermak SC (2011) Sowing date and tillage effects on fall-seeded camelina in the northern Corn Belt. *Agronomy Journal*, **103**, 980–987.

Gesch RW, Archer DW (2013) Double-cropping with winter camelina in the northern Corn Belt to produce fuel and food. *Industrial Crops and Products*, **44**, 718–725.

Gesch RW, Johnson JMF (2015) Water use in a winter camelina-soybean dual cropping systems. *Agronomy Journal*, **107**, 1098–1104.

Gesch RW, Allen BL, Archer DW *et al.* (2014) Agronomic comparisons of several Brassica species in the Corn Belt as feedstock for hydrotreated jet fuel. Association for the Advancement of Industrial Crops, 26th Annual Meeting, Athens, Greece.

Gimsing AL, Kirkegaard JA (2009) Glucosinolates and biofumigation: fate of glucosinolates and their hydrolysis products in soil. *Phytochemistry Review*, **8**, 299–310.

Graham RL, Nelson R, Sheehan J, Perlack RD, Wright LL (2007) Current and potential U.S. corn stover supplies. *Agronomy Journal*, **99**, 1–11.

Grau CR, Oplinger ES, Adee EA, Hinkens EA, Martinka MJ (1994) Planting date and row width effect on severity of brown stem rot and soybean productivity. *Journal of Production Agriculture*, **7**, 347–351.

Groeneveld JH, Klein A-M (2014) Pollination of two oil-producing plant species: Camelina (*Camelina sativa* L. Crantz) and pennycress (*Thlaspi arvense* L.) double-cropping in Germany. *GCB Bioenergy*, **6**, 242–251.

Haenke S, Kovacs-Hostyanszki A, Frund J, Batary P, Jauker B, Tscharntke T, Holzschuh A (2014) Landscape configuration of crops and hedgerows drives local syrphid fly abundance. *Journal of Applied Ecology*, **51**, 505–513.

Heaton EA, Schulte LA, Berti M, Langeveld H, Zegada-Lizarazu W, Parrish D, Monti A (2013) Managing a second-generation crop portfolio through sustainable intensification: examples from the USA and the EU. *Biofuels, Bioproducts, & Biorefining*, **7**, 702–714.

Helms TC, Hurburgh CR, Lussenden RL, Whited DA (1990) Economic analysis of increased protein and decreased yield due to delayed planting of soybean. *Journal of Production Agriculture*, **3**, 367–371.

Helsel ZR, Wedin WF (1981) Harvested dry matter from single and double-cropping systems. *Agronomy Journal*, **73**, 895–900.

Hill J, Nelson E, Tilman D, Polasky S, Tiffany D (2006) Environmental, economic, and energetic costs and benefits of biodiesel and ethanol biofuels. *Proceedings of the National Academy of Sciences*, **103**, 11206–11210.

Hojilla-Evangelista MP, Evangelista RL, Isbell TA, Selling GW (2013) Effects of cold-pressing and seed cooking on functional properties of protein in pennycress (*Thlaspi arvense* L.) seed and press cakes. *Industrial Crops and Products*, **45**, 223–229.

Hojilla-Evangelista MP, Selling GW, Berhow MA, Evangelista RL (2014) Preparation, composition, and functional properties of pennycress (*Thlaspi arvense* L.) seed protein isolates. *Industrial Crops and Products*, **55**, 173–179.

Holzschuh A, Dormann CF, Tscharntke T, Steffan-Dewenter I (2013) Mass-flowering crops enhance wild bee abundance. *Oecologia*, **172**, 477–484.

Isbell TA, Cermak SC, Dierig DA, Eller FJ, Marek LF (2015) Registration of Katelyn *Thlaspi arvense* L. (pennycress) with improved nondormant traits. *Journal of Plant Registrations*, **9**, 212–215.

Jin VA, Schmer MR, Wienhold BJ, *et al.* (2015) Twelve years of stover removal increases soil erosion potential without impacting yield. *Soil Science Society of America Journal*, **79**, 1169–1178.

Johnson JMF, Coleman MD, Gesch R, Jaradat A, Mitchell RD, Wilhelm WW (2007) Biomass-bioenergy crops in the United States: a changing paradigm. *The Americas Journal of Plant Science and Biotechnology*, **1**, 1–28.

Johnson JMF, Karlen DL, Andrews SS (2010) Conservation considerations for sustainable bioenergy feedstock production: If, what, where, and how much? *Journal of Soil and Water Conservation*, **65**, 88A–91A.

Johnson GA, Kantar MB, Betts KJ, Wyse DL (2015) Field pennycress production and weed control in a double crop system with soybean in Minnesota. *Agronomy Journal*, **107**, 532–540.

Kandel H, Akyüz FA (2011) *Heat units and development stages of soybean*. North Dakota State University Extension Service. Available at: http://www.ag.ndsu.edu/cpr/plant-science/heat-units-and-development-stages-of-soybean-7-7-11. (accessed 13 March 2015).

Kandel H, Akyüz FA (2012) *Growing degree model for North Dakota soybean*. North Dakota State University Extension Service. Available at: http://www.ag.ndsu.edu/cpr/plant-science/growing-degree-day-model-for-north-dakota-soybean-6-28-12 (accessed 13 March 2015).

Karlen DL, Hung PG, Campbell RB (1984) Crop residue removal effects on corn yield and fertility of a Norfolk sandy loam. *Soil Science Society of America Journal*, **48**, 868–872.

Karp A, Richter GM (2011) Meeting the challenge of food and energy security. *Journal of Experimental Botany*, **62**, 3263–3271.

Kaspar TC, Radke JK, Laflen JM (2001) Small grain cover crops and wheel traffic effects on infiltration, runoff, and erosion. *Journal of Soil and Water Conservation*, **56**, 160–164.

Kenney I, Blanco-Canqui H, Presley DR, Rice CW, Janssen K, Olson B (2015) Soil and crop response to stover removal from rainfed and irrigated corn. *GCB Bioenergy*, **7**, 219–230.

Kessavalou A, Walters DT (1997) Winter rye as a cover crop following soybean under conventional tillage. *Agronomy Journal*, **89**, 68–74.

Kim H, Kim S, Dale BE (2009) Biofuels land use change, and greenhouse gas emissions: some unexplored variables. *Environmental Science & Technology*, **43**, 961–967.

Kladivko EJ, Kaspar TC, Jaynes DB, Malone RW, Singer J, Morin XK, Searchinger T (2014) Cover crops in the upper midwestern United States: potential adoption and reduction of nitrate leaching in the Mississippi River Basin. *Journal of Soil and Water Conservation*, **69**, 279–291.

Klein A-M, Vaissière BE, Cane JH, Steffan-Dewenter I, Cunningham SA, Kremen C, Tscharntke T (2007) Importance of pollinators in changing landscapes for world crops. *Proceedings of the Royal Society B: Biological Sciences*, **274**, 303–313.

Korsrud GO, Keith MO, Bell JM (1978) A comparison of the nutritional value of crambe and camelina seed meals with egg and casein. *Canadian Journal of Animal Science*, **58**, 493–499.

Krell RK, Pedigo LP, Rice ME, Westgate ME, Hill JH (2005) Using planting date to manage bean pod mottle virus in soybean. *Crop Protection.*, **24**, 909–914.

Krohn BJ, Fripp M (2012) A life cycle assessment of biodiesel derived from the "niche filling" energy crop camelina in the USA. *Applied Energy*, **92**, 92–98.

Laflen JM, Moldenhauer WC (1979) Soil and water losses from corn-soybean rotations. *Soil Science Society of America Journal*, **43**, 1213–1215.

LeMahieu PJ, Brinkman MA (1990) Double-cropping soybean after harvesting small grains as forage in the north central USA. *Journal of Production Agriculture*, **3**, 385–389.

Leuschen WE, Ford JH, Evans SD *et al.* (1992) Tillage, row spacing, and planting date effects on soybean following corn or wheat. *Journal of Production Agriculture*, **5**, 254–260.

Mallory EB, Posner JL, Baldock JO (1998) Performance, economics, and adoption of cover crops in Wisconsin cash grain rotations: on-farm trials. *American Journal of Alternative Agriculture*, **13**, 2–11.

Mann L, Tolbert V, Cushman J (2002) Potential environmental effects of corn (*Zea mays* L.) stover removal with emphasis on soil organic matter and erosion. *Agriculture, Ecosystems & Environment*, **89**, 149–166.

Marten GC, Sheaffer CC, Wyse DL (1987) Forage nutritive value and palatability of perennial weeds. *Agronomy Journal*, **79**, 980–986.

Miedaner T, Schwegler DD, Wilde P, Reif JC (2014) Association between line per se and testcross performance for eight agronomic and quality traits in winter rye. *Theoretical and Applied Genetics*, **127**, 33–41.

Mitchell WH, Teel MR (1977) Winter-annual cover crops for no-tillage corn production. *Agronomy Journal*, **69**, 569–573.

Mojtahedi H, Santo GS, Hang AN, Wilson JH (1991) Suppression of root-knot nematode populations with selected rapeseed cultivars as green manure. *Journal of Nematology*, **23**, 170–174.

Moore KJ, Karlen DL (2013) Double cropping opportunities for biomass crops in the north central USA. *Biofuels*, **4**, 605–615.

Moriel P, Nayigihugu V, Cappellozza B. *et al.* (2011) Camelina meal and crude glycerin as feed supplements for developing replacement beef heifers. *Journal of Animal Science*, **89**, 4314–4324.

Moschler WW, Shear GM, Hallock DL, Sears RD, Jones GD (1967) Winter cover crops for sod-planted corn: their selection and management. *Agronomy Journal*, **59**, 547–551.

Moser BR (2012) Biodiesel from alternative oil feedstocks: camelina and field pennycress. *Biofuels*, **3**, 193–209.

Moser BR, Knothe G, Vaughn SF, Isbell TA (2009) Production and evaluation of biodiesel from field pennycress (*Thlaspi arvense* L.) oil. *Energy & Fuels*, **23**, 4149–4155.

Mouawad J, Davenport C (2015). E.P.A. takes step to cut emissions from planes. *The New York Times*. 10 June, 2015. Available at: http://www.nytimes.com/2015/06/11/business/energy-environment/epa-says-it-will-set-rules-for-airplane-emissions.html (accessed 14 June 2015).

Moyer JR, Hironaka R (1993) Digestible energy and protein content of some annual weeds, alfalfa, bromegrass, and tame oats. *Canadian Journal of Plant Science*, **73**, 1305–1308.

Muth DJ Jr, Bryden KM, Nelson RG (2013) Sustainable agricultural residue removal for bioenergy: a spatially comprehensive US national assessment. *Applied Energy*, **102**, 403–417.

Nelson RG (2002) Resource assessment and removal analysis for corn stover and wheat straw in the Eastern and Midwestern United States – rainfall and wind-induced soil erosion methodology. *Biomass and Bioenergy*, **22**, 349–363.

Nguyen HT, Silva JE, Podicheti R, Macrander J (2013) Camelina seed transcriptome: a tool for meal and oil improvement and translational research. *Plant Biotechnology Journal*, **11**, 759–769.

Odland TE, Knoblauch HC (1938) The value of cover crops in continuous corn culture. *Journal of the American Society of Agronomy*, **30**, 22–29.

Olmos Colmenero JJ, Broderick GA (2006) Effect of dietary crude protein concentration on milk production and nitrogen utilization in lactating dairy cows. *Journal of Dairy Science*, **89**, 1704–1712.

Olson K, Ebelhar SA, Lang JM (2014) Long-term effects of cover crops on crop yields, soil organic stocks and sequestration. *Open Journal of Soil Science*, **4**, 284–292.

Oplinger ES, Philbrook BD (1992) Soybean planting date, row width, and seeding rate response in three tillage systems. *Journal of Production Agriculture*, **5**, 94–99.

Osborne SL, Johnson JMF, Jin VL, Hammerbeck AL, Varvel GE, Schumacher TE (2014) The impact of corn residue removal on soil aggregates and particulate organic matter. *Bioenergy Research*, **7**, 559–567.

Pedersen P, Lauer JG (2003) Soybean agronomic response to management systems in the Upper Midwest. *Agronomy Journal*, **95**, 1146–1151.

Perez-Bidegain M, Cruse RM, Ciha A (2007) Tillage system by planting date interaction effects on corn and soybean yield. *Agronomy Journal*, **99**, 630–636.

Phippen WB, Phippen ME (2012) Soybean seed yield and quality as a response to field pennycress residue. *Crop Science*, **52**, 2767–2773.

Phippen WB, John B, Phippen ME, Isbell T (2010) Planting date, herbicide, and soybean rotation studies with field pennycress (*Thlaspi arvense* L.). Association for the Advancement of Industrial Crops, 22nd Annual Meeting, Ft Collins, Colorado, USA.

Pollinator Health Task Force (2015) *Pollinator Research Action Plan*. Available at: https://www.whitehouse.gov/sites/default/files/microsites/ostp/Pollinator%20Research%20Action%20Plan%202015.pdf (accessed 17 June 2015).

Pratt MR, Tyner WE, Muth DJ, Kladivko EJ (2014) Synergies between cover crops and corn stover removal. *Agricultural Systems*, **130**, 67–76.

Putnam DH, Budin JT, Field LA, Breene WM (1993) Camelina: A promising low-input oilseed. In: *New Crops* (eds Janick J, Simon JE), pp. 314–322. Wiley, New York, USA.

Reese CL, Clay DE, Clay SA, Bich AD, Kennedy AC, Hansoen SA, Moriles J (2014) Winter cover crops impact on corn production in semiarid regions. *Agronomy Journal*, **106**, 1479–1488.

Riedinger V, Renner M, Rundlöf M, Steffan-Dewenter I, Holzschuh A (2014) Early mass-flowering crops mitigate pollinator dilution in late-flowering crops. *Landscape Ecology*, **29**, 425–435.

Robinson RG (1987) Camelina: a useful research crop and a potential oilseed crop. *University of Minnesota Agricultural Experiment Station Bulletin*, **579–1987**, 1–12.

Robinson C, Nielsen D (2015) The water conundrum of planting cover crops in the Great Plains: when is an inch not an inch? *Crops & Soils*, **48**, 25–31.

Robinson AP, Conley SP, Volenec JJ, Santini JB (2009) Analysis of high yielding, early-planted soybean in Indiana. *Agronomy Journal*, **101**, 131–139.

Sadler EJ, Camp CR, Evans DE, Millen JA (2002) Spatial variation of corn response to irrigation. *Transactions of the American Society of Agricultural Engineers*, **45**, 1869–1881.

Schmer MR, Jin VL, Wienhold BJ, Varvel GE, Follett RF (2014) Tillage and residue management effects on soil carbon and nitrogen under irrigated continuous corn. *Soil Science Society of America Journal*, **78**, 1987–1996.

Schmidt MH, Lauer A, Purtauf T, Thies C, Schaefer M, Tscharntke T (2003) Relative importance of predators and parasitoids for cereal aphid control. *Proceedings of the Royal Society B-Biological Sciences*, **270**, 1905–1909.

Schuster A, Friedt W (1998) Glucosinolate content and composition as parameters of quality of *Camelina* seed. *Industrial Crops and Products*, **7**, 297–302.

Searchinger T, Heimlich R, Houghton RA *et al.* (2008) Use of U.S. croplands for biofuels increases greenhouse gases through emissions from land-use change. *Science*, **319**, 1238–1240.

Sedbrook JC, Phippen WB, Marks MD (2014) New approaches to facilitate rapid domestication of a wild plant to an oilseed crop: example pennycress (*Thlaspi arvense* L.). *Plant Science*, **227**, 122–132.

Seifert CA, Lobell DB (2015) Response of double cropping suitability to climate change in the United States. *Environmental Research Letters*, **10**, doi:10.1088/1748-9326/10/2/024002.

Selling GW, Hojilla-Evangelista MP, Evangelista RL, Isbell T, Price N, Doll KN (2013) Extraction of proteins from pennycress seeds and press cake. *Industrial Crops and Products*, **41**, 113–119.

Shonnard DR, Williams L, Kalnes TN (2010) Camelina-derived jet fuel and diesel: sustainable advanced biofuels. *Sustainable Energy*, **29**, 382–392.

Sindelar AJ, Lamb JA, Sheaffer CC, Rosen CJ, Jung HG (2013) Fertilizer nitrogen rate effects on nutrient removal by corn stover and cobs. *Agronomy Journal*, **105**, 437–445.

Sindelar AJ, Coulter JA, Lamb JA, Vetsch JA (2015) Nitrogen, stover, and tillage management affect nitrogen use efficiency in continuous corn. *Agronomy Journal*, **107**, 843–850.

Singer JW (2008) Corn Belt assessment of cover crop management and preferences. *Agronomy Journal*, **100**, 1670–1672.

Singer JW, Nusser SM, Alf CJ (2007) Are cover crops being used in the US Corn Belt? *Journal of Soil and Water Conservation*, **62**, 353–358.

Snapp SS, Swinton SM, Labarta R *et al.* (2005) Evaluating cover crops for benefits, costs and performance within cropping system niches. *Agronomy Journal*, **97**, 322–332.

Specht JE, Rees JM, Zoubek GL *et al.* (2012) *Soybean planting date-When and why*. University of Nebraska Extension Bulletin, **EC145**, 1–7.

Stanley DA, Stout JC (2013) Quantifying the impacts of bioenergy crops on pollinating insect abundance and diversity: a field-scale evaluation reveals taxon-specific responses. *Journal of Applied Ecology*, **50**, 335–344.

Stetson SJ, Osborne SL, Schumacher TE *et al.* (2012) Corn residue removal impact on topsoil organic carbon in a corn-soybean rotation. *Soil Science Society of America Journal*, **76**, 1399–1406.

Stivers-Young LJ, Tucker FA (1999) Cover-cropping practices of vegetable producers in western New York. *Horttechnology*, **9**, 459–465.

Strock JS, Porter PM, Russelle MP (2004) Cover cropping to reduce nitrate loss through subsurface drainage in the northern U.S. corn belt. *Journal of Environmental Quality*, **33**, 1010–1016.

Szumacher-Strabel M, Cieślak A, Zmora P, Pers-Kamczyc E, Bielińska S, Stanisz M, Wójtowski J (2011) Camelina sativa cake improved unsaturated fatty acids in ewe's milk. *Journal of the Science of Food and Agriculture*, **91**, 2031–2037.

Teasdale JR (1996) Contribution of cover crops to weed management in sustainable agricultural systems. *Journal of Production Agriculture*, **9**, 475–479.

Temme DG, Harvey RG, Fawcett RS, Young AW (1979) Effects of annual weed control on alfalfa forage quality. *Agronomy Journal*, **71**, 51–54.

Tilman D, Socolow R, Foley JA *et al.* (2009) Beneficial biofuels- the food, energy, and environment trilemma. *Science*, **325**, 270–271.

United States Air Force (2013) *Air Force energy strategic plan*. Available at: http://www.safie.hq.af.mil/shared/media/document/AFD-091208-027.pdf (accessed 21 April 2015).

United States Census Bureau (2014) *International database, world population: 1950–2050*. Available at: http://www.census.gov/population/international/data/idb/worldpopgraph.php (accessed 12 March 2015).

United States Environmental Protection Agency (2013) *Regulatory announcement: EPA finalizes 2013 renewable fuel standards*. EPA-420-F-13-048. Available at: http://www.epa.gov/otaq/fuels/renewablefuels/documents/420f13048.pdf (accessed 20 March 2015).

United States Environmental Protection Agency (2015) *Proposed finding that greenhouse gas emissions from aircraft cause or contribute to air pollution that may reasonability be anticipated to endanger public health and welfare and advance notice of proposed rulemaking*. EPA-HQ-OAR-2014-0828. Available at: http://epa.gov/otaq/documents/aviation/aircraft-ghg-pr-anprm-2015-06-10.pdf (accessed 11 June 2015).

United States Navy (2010) *A Navy energy vision for the 21st century*. Available at: http://dtic.mil/ndia/2010navy/Cullom_Revised.pdf (accessed 21 April 2015)

USDA-APHIS (2015) *Weed risk assessment for Thlaspi arvense L. (Brassicacae) – Field pennycress*. Available at: http://www.aphis.usda.gov/plant_health/plant_pest_info/weeds/downloads/wra/Thlaspi-arvense.pdf (accessed 21 April 2015).

USDA-NASS (2008) *2007 Farm and ranch irrigation survey*. Available at: http://www.agcensus.usda.gov/Publications/2007/Online_Highlights/Farm_and_Ranch_Irrigation_Survey/index.php (accessed 11 March 2015).

Varvel GE, Wilhelm WW (2008) Soil carbon levels in irrigated western Corn Belt rotations. *Agronomy Journal*, **100**, 1180–1184.

Varvel GE, Wilhelm WW (2011) No-tillage increases soil profile carbon and nitrogen under long-term rainfed cropping systems. *Soil & Tillage Research*, **114**, 28–36.

Vaughn SF, Isbell TA, Weisleder D, Berhow MA (2005) Biofumigant compounds released by field pennycress (*Thlaspi arvense*) seedmeal. *Journal of Chemical Ecology*, **31**, 167–177.

Walsh DT, Babiker EM, Burke IC, Hulbert SH (2012) Camelina mutants resistant to acetolactate synthase inhibitor herbicides. *Molecular Breeding*, **30**, 1053–1063.

Warwick SI, Francis A, Susko DJ (2002) The biology of Canadian weeds. 9. *Thlaspi arvense* L. (updated). *Canadian Journal of Plant Science*, **82**, 803–823.

Westphal C, Steffan-Dewenter I, Tscharntke T (2003) Mass flowering crops enhance pollinator densities at a landscape scale. *Ecology Letters*, **6**, 961–965.

Whiting F, Young D, Phillips AH, Munro WB, Steves HL (1958) The effects on the odour and flavour of the meat of feeding refuse screenings with a high stinkweed (*Thlaspi Arvense* L.) content to fattening cattle. *Canadian Journal of Animal Science*, **38**, 48–52.

Wilcox JR, Frankenberger EM (1987) Indeterminate and determinate soybean responses to planting date. *Agronomy Journal*, **79**, 1074–1078.

Wilhelm WW, Johnson JMF, Hatfield JL, Voorhees WB, Linden DR (2004) Crop and soil productivity response to corn residue removal. *Agronomy Journal*, **96**, 1–17.

Wilhelm WW, Johnson JMF, Karlen DL, Lightle DT (2007) Corn stover to sustain soil organic carbon further constrains biomass supply. *Agronomy Journal*, **99**, 1665–1667.

Wilson ML, Baker JM, Allan DL (2013) Factors affecting successful establishment of aerially seeded winter rye. *Agronomy Journal*, **105**, 1868–1877.

Wilson ML, Allan DL, Baker JM (2014) Aerially seeding cover crops in the northern US Corn Belt: limitations, future research needs, and alternative practices. *Journal of Soil and Water Conservation*, **69**, 67A–72A.

Wu Z, Satter LD (2000) Milk production during the complete lactation of dairy cows fed diets containing different amounts of protein. *Journal of Dairy Science*, **83**, 1042–1051.

Yost MA, Morris TF, Russelle MP, Coulter JA (2014) Second-year corn after alfalfa often requires no fertilizer nitrogen. *Agronomy Journal*, **106**, 659–669.

Zubr J (1997) Oil-seed crop: *Camelina sativa*. *Industrial Crops and Products*, **6**, 113–119.

7

Bioethanol from maize cell walls: genes, molecular tools, and breeding prospects

ANDRES F. TORRES[1,2], RICHARD G. F. VISSER[1] and LUISA M. TRINDADE[1]

[1]*Wageningen UR Plant Breeding, Wageningen University and Research Centre, P.O. Box 386, Wageningen 6700 AJ, the Netherlands, [2]Graduate School Experimental Plant Sciences, Wageningen University, Droevendaalsesteeg 1, 6708 PB Wageningen, the Netherlands*

Abstract

In the last decade, cellulosic ethanol has caught the growing interest of governments and private investors worldwide as it brings the promise of responsible renewable-energy and an opportunity to depart from an oil-reliant economy. Alongside advances in bioprocessing technologies, the development of specialized bioenergy crops is seen as a pressing industrial necessity, and while C4 perennials (e.g., Miscanthus, switchgrass, and sugarcane) have been coined the most promising candidates for the production of lignocellulosic biomass, maize should not be overlooked. In this review, we have addressed the benefits of advancing maize as a second-generation bioenergy feedstock. We have also analyzed current knowledge on the maize cell wall and promising genetic strategies for its modification, given that lignocellulose recalcitrance represents the most crucial breeding target in bioenergy crop research programs. In addition to lignin, a focus on the underlying genetic basis of cellulose, hemicellulose, and ferulate cross-linking patterns, as well as their regulation, has been warranted. A comprehensive overview of the state-of-art of genomic and phenotyping strategies available for bioenergy crop research is also provided. Overall, maize represents an outstanding model organism for understanding complex cell wall characteristics and defining the path for breeders looking to improve this and other promising bioenergy grasses. With an extensive array of dedicated agronomic and genomic resources at hand, we believe that breeding maize with improved processing amenability is a likely prospect but would like to remind readers that advances in high-biomass yielding properties, improved agronomic hardiness, and enhanced processing efficiency will also be necessary.

Keywords: bioethanol, breeding, cell wall, genes, lignocellulose, maize, molecular tools

Introduction

As we enter the third millennium, it seems difficult to ignore the societal and environmental consequences of our incommensurate reliance on finite fossil fuels. Alongside the guarantee for energetic security, climate change and its detrimental effects on the environment and agriculture have instigated a global pursuit for sustainable energy alternatives (Vermerris *et al.*, 2007; Wyman, 2007; Huber & Dale, 2009). In particular, substitutes for fossil-based transportation fuels have become a pressing necessity, as our mobility sector currently accounts for over one-third of global greenhouse gas (GHG) emissions (Wyman, 2007). And so as society departs from its oil-reliant economy, researchers, governments, and private investors worldwide have grown increasingly resolute on 'cellulosic ethanol' – a viable near-term alternative to petrol (Schubert, 2006; Waltz, 2008).

Cellulosic ethanol essentially derives from lignocellulose, arguably the most abundant renewable carbon substrate on earth (Schubert, 2006; Dale, 2007; Vermerris *et al.*, 2007). As lignocellulose production requires less agricultural and energetic inputs, cellulosic ethanol could outperform gasoline- and starch-based ethanol as the transportation fuel with lowest GHG emissions and greatest net-energetic outputs (Farrell *et al.*, 2006; Dale, 2007; Wang *et al.*, 2011, 2012). This condition can only be met, however, when processing technologies have matured, as these are currently too energy intensive. Despite extensive revamps in funding and unrelenting governmental support, cellulosic ethanol is yet to transcend the demonstration plant and achieve widescale commercialization (Bacovsky, 2010; Larsen *et al.*, 2012; Saddler

Correspondence: Luisa M. Trindade
e-mail: luisa.trindade@wur.nl

et al., 2012). With the first commercial endeavors underway (Bacovsky, 2010; Larsen *et al.*, 2012; Brown & Brown, 2013), progress in the commercialization of cellulosic ethanol could be conditioned by the instability of oil prices, market incentives, and governmental policies (Sorda *et al.*, 2010; Brown & Brown, 2013). To survive this uncertain scenario, cellulosic ethanol will need to overcome a series of technical and economic hurdles to compete neck-to-neck with fossil-based transportation fuels.

The feedstock problem

The conversion of biomass into transportation fuels can be effectively achieved through a variety of technological routes, including advanced thermochemical technologies (e.g., Fischer–Tropsch synthesis, gasification or catalytic pyrolysis; Vermerris *et al.*, 2007; Huber & Dale, 2009; Brown & Brown, 2013). Nevertheless, cellulosic ethanol production via biochemical pathways is currently the most commercially represented technology in the sector (Wyman, 2007; Brown & Brown, 2013) and therefore constitutes the referential focus of this review. By 2014, five commercial-scale cellulosic ethanol projects are expected to start operations, and their performance will crucially influence the future of cellulosic fuel policy and economic incentives (Brown & Brown, 2013).

During the production of cellulosic ethanol, the polysaccharide fraction (cellulose and hemicellulose) of plant lignocellulose is enzymatically depolymerized and much like in starch-based ethanol platforms, the resulting sugars are fermented into hydrous ethanol. Lignocellulose, however, has evolved to resist enzymatic degradation, and its efficient depolymerization into fermentable sugars is the predominant technical bottleneck in the system (Wyman, 2007). To circumvent this problem, thermochemical pretreatments are typically employed to increase the accessibility of biomass polysaccharides to hydrolytic enzymes (Mosier *et al.*, 2005). This accessory procedure greatly improves fermentable sugar yields, but it also significantly increases production costs and reduces the energetic and environmental performance of the conversion system (Mosier *et al.*, 2005; Wyman, 2007).

Experts ultimately coincide that the commercial future of cellulosic ethanol is pending on innovations that can reduce the use of costly pretreatments, while simultaneously improving fermentable sugar yields (Wyman, 2007; Huber & Dale, 2009). At its core, research in the field has prioritized advances in the techno-economic efficiency of thermochemical pretreatments, as well as biotechnological endeavors aimed at increasing the yields of enzymatic hydrolysis and fermentation processes. Alongside these advances, however, the choice of feedstock used in the industry will also play a determinant role in the efficiency and profitability of the industry (Vermerris *et al.*, 2007; Wyman, 2007; Carroll & Somerville, 2009; Torres *et al.*, 2013).

Based on the constraints faced by current conversion technologies (including thermochemical routes), cellulosic ethanol will need to be produced from an inexpensive, readily abundant and sustainable substrate (Farrell *et al.*, 2006; Vermerris *et al.*, 2007; Huber & Dale, 2009). In addition, because lignocellulose recalcitrance is a critical barrier to the efficient production of cellulosic fuels, improving the ease with which lignocellulosic materials are consumed in processing facilities would lead to higher energetic yields and greater economic gains. Crops that entirely meet this criterion are not yet available, but genetic improvement programs are underway and optimistic prospects exist for the creation of lignocellulosic feedstocks that can effectively accommodate the needs of the fast-growing cellulosic ethanol industry.

Maize makes sense

Fast-growing C4 perennials, like Miscanthus, switchgrass, and sugarcane, have been coined the most promising candidates for the industrial production of lignocellulosic biomass. These species are principally coveted for their high biomass yields (Fig. 1), broad geographic adaptation, superior carbon sequestration, and efficient nutrient utilization (Carroll & Somerville, 2009; Weijde *et al.*, 2013). Additionally, when used for the production of bio-based fuels, C4 perennials will expectedly offer the greatest net-energetic outputs in relation to other bioenergy feedstocks (Farrell *et al.*, 2006; Von Blottnitz & Curran, 2007; Yuan *et al.*, 2008). The commercial success of upcoming perennials, however, will rely on the availability of superior cultivars that increase the competitiveness of the industry, while sustainably meeting projected market volumes (Weijde *et al.*, 2013). Breeding objectives include increasing biomass yields and yield stability under low-input agricultural systems, enhancing pest and disease resistance, and modifying biomass composition for improved industrial processing (Carroll & Somerville, 2009; Weijde *et al.*, 2013).

With the first cellulosic ethanol commercial plants on the way (Brown & Brown, 2013), a reliable and abundant feedstock is a pressing necessity (Zegada-Lizarazu *et al.*, 2013). Because C4 perennials cannot be readily implemented on a wide-commercial scale, maize will prove instrumental to the development and commercial success of the cellulosic ethanol industry (Schubert, 2006; Vermerris *et al.*, 2007; Carpita & Mccann, 2008;

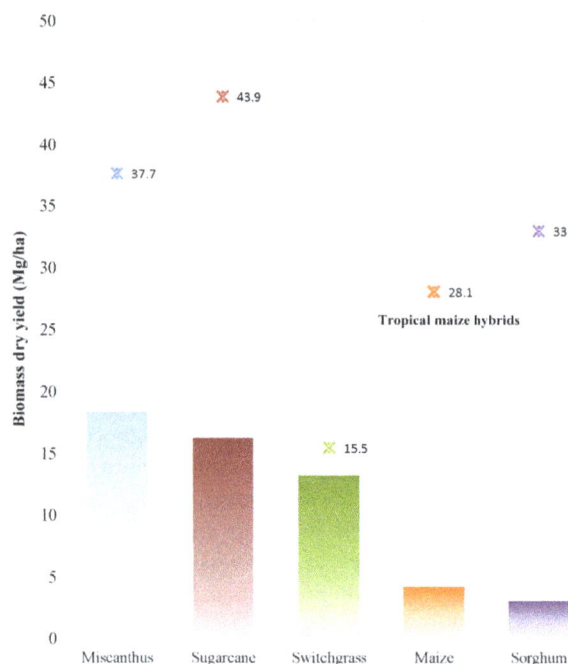

Fig. 1 Mean and potential annual dry biomass yields (Mg ha^{-1}) for relevant C4 energy grasses. For Miscanthus and switchgrass, delayed harvest yields (after winter) are reported. Colored asterisks correspond to highest reported yields in literature, except for sugarcane, for which highest reported yields were extracted from Faostat (2011). Mean and maximum yield values were calculated or extracted from (Clifton-Brown *et al.*, 2004; Heaton *et al.*, 2008) for Miscanthus (Faostat, 2011) for Sugarcane (Heaton *et al.*, 2008; Dweikat *et al.*, 2012) for switchgrass (Faostat, 2011; White *et al.*, 2012) for maize and (Faostat, 2011; Dweikat *et al.*, 2012) for Sorghum. Reported yields do not correspond to comparative trials using standardized conditions (e.g., soil, temperature, season, etc.) and should be regarded as potentiality indicators.

Dweikat *et al.*, 2012; Torres *et al.*, 2013; Weijde *et al.*, 2013). Currently, around 1300 million tons of dry maize stover are produced worldwide; and after factoring the effects of stover removal on soil erosion and nutrient depletion, experts believe that between 40 and 65% of all maize agricultural residues can be sustainably harvested for advanced fuel production (Kim & Dale, 2004; Graham *et al.*, 2007; Youngs & Somerville, 2012). Combined with agricultural crop residues from Sorghum (another promising annual bioenergy grass), this much biomass can contribute significantly to the industry's present and future feedstock needs (Vermerris *et al.*, 2007; Weijde *et al.*, 2013). Furthermore, implementing the technology required for cellulosic fuel production entails significant capital investments and financial risks (Schubert, 2006; Waltz, 2008; Huber & Dale, 2009). Experts have also envisioned that the first commercial cellulosic fuel plants should operate in the vicinity of

starch-based ethanol facilities and use maize stover as their lignocellulosic substrate (Schubert, 2006; Vermerris *et al.*, 2007; Carpita & Mccann, 2008). By doing so, nascent enterprises will reduce financial burdens by benefitting from the commercially effective maize-farming, processing and transportation infrastructure (Carpita & Mccann, 2008; Dweikat *et al.*, 2012).

In the future, grower's acceptance of bioenergy perennials will also impact the prevalence of maize as a lignocellulosic feedstock (Carpita & Mccann, 2008). This perspective takes into consideration the high costs and financial risks associated with the setup of new plantations, the amount of years needed before these reach maximum productivity, the loss of growing flexibility that only comes with the choice of annual bioenergy crops like maize and sorghum and the subjective preferences/prejudices of farmers (Carpita & Mccann, 2008; Dweikat *et al.*, 2012; Zegada-Lizarazu *et al.*, 2013). Ultimately, this factor can open unexplored avenues for the production of 'energy-dedicated' maize varieties that could potentially compete with other promising C4 species. With a wealth of agronomic and genomic resources, advancing maize with high-biomass yielding properties and improved nutrient use efficiency is a likely prospect (Carpita & Mccann, 2008; White, 2010; Dweikat *et al.*, 2012; White *et al.*, 2012). Photoperiod-sensitive hybrids derived from crosses between temperate and tropical varieties, for instance, are a proof-of-concept example for the derivation of maize into an energy-dedicated species (Fig. 1). These temperate × tropical maize (TTM) hybrids typically allocate the majority of their biomass into the stover and can yield up to 28.1 Mg ha^{-1} annual dry biomass in cropping systems supplemented with nitrogen (N) fertilizer (White *et al.*, 2012), and up to 21.3 Mg ha^{-1} annual dry biomass without supplemental N fertilization (White, 2010). Because TTM hybrids can also accumulate high amounts of soluble sugars in their stems (~50% more when compared to commercial hybrids), these can expectedly yield comparable amounts of ethanol (~8000 L ha^{-1}) per hectare under no supplemental N fertilization as commercial grain hybrids supplemented with N (~10 500 L ha^{-1}; White *et al.*, 2012). Although preliminary in nature, these results demonstrate the potential behind breeding endeavors looking to advance maize outside its classical framework. Understandably, before TTM hybrids can be considered for dedicated lignocellulose production, major advances in nutrient use efficiency, climatic hardiness, biotic resistance, and seed productivity will need to be achieved through genetic improvement and crop management (Dweikat *et al.*, 2012; White *et al.*, 2012). In particular, claims regarding enhanced biomass productivity in the absence of fertilization should be evaluated carefully, given that such cropping systems would rapidly deplete nutrient soil reserves when the crop fails to return nutrients back to the soil. As of today, however, the extensive genetic diversity of maize remains largely unexploited (Box 1) and opportunities exist for the introgression of useful exotic traits that can expedite the advance of dual-purpose and energy-dedicated maize cultivars for the cellulosic ethanol industry (Lewis *et al.*, 2010; Lorenz *et al.*, 2010; Cairns *et al.*, 2012; Muttoni *et al.*, 2012).

Box 1.
The unexploited diversity of maize
Progress in the development of maize for cellulosic fuel production should not be confined to the exclusive utilization of commercially available germplasm. Breeding endeavors in maize have predominantly focused on advancing grain yield and yield stability, and only a minority have specialized on exploiting useful biomass characteristics (Lauer *et al.*, 2001). A natural outcrosser, maize is remarkably diverse, with most of its desirable traits yet to be utilized (Yan *et al.*, 2009; Hallauer *et al.*, 2010; Muttoni *et al.*, 2012). This unexploited diversity has been preserved in gene bank collections at numerous international research centers and is publically available upon request. In addition, public and private efforts like the Latin American Maize Project (Salhuana *et al.*, 1991), the Germplasm Enhancement of Maize project (Pollak, 2003), and on-going work at the International Maize and What Improvement Centre (CIMMYT) are making immense contributions towards the evaluation and classification of exotic germplasm, as well as its adaptation into elite material. The success of these and similar projects will prove indispensable to the incorporation of novel characteristics; all of which offer possibilities to improve the biomass potential and agronomic sustainability of this crop species.

Building upon the maize cell wall: from gene to phenotype

A comprehensive recount of the state-of-art of maize breeding for the cellulosic ethanol industry would encompass a broad range of subjects spanning over the allotted length of this article. We have focused on the maize cell wall, nonetheless, as we firmly believe that advancing biomass feedstocks that best match the processing conditions used in the industry can improve the commercial and environmental performance of cellulosic ethanol production (Torres *et al.*, 2013). This sec-

tion describes current knowledge on the genes involved in the synthesis of the main components of maize cell walls (cellulose, hemicellulose and lignin), their regulation and strategies to optimize biomass composition for bioethanol production.

Cellulose

Improving the relative content and industrial quality of cellulose is a pivotal strategy toward the development of advanced lignocellulosic feedstocks. On the one hand, a higher abundance of cell wall polysaccharides per unit of biomass will conceivably increase the amount of harvestable energy per unit of land. Alterations in cellulose ultrastructure, which simplify its enzymatic depolymerization, on the other hand, are expected to improve the processing efficiency and economics of biomass-to-ethanol conversion technologies. Cellulose is a highly recalcitrant substrate and properties presumed to limit its enzymatic degradability include its high degree of polymerization and high crystallinity index (Mansfield *et al.*, 1999; Park *et al.*, 2010).

Modifying cellulose assembly and deposition patterns in maize, however, is a challenging undertaking that will require a thorough understanding of its complex biosynthetic machinery. At present, 12 members of the maize cellulose synthase (*CesA*) gene family have been annotated and characterized (Holland *et al.*, 2000; Appenzeller *et al.*, 2004). Based on sequence orthology, these genes presumably encode the catalytic subunits of the maize cellulose synthase complex (CSC; Holland *et al.*, 2000; Appenzeller *et al.*, 2004; Penning *et al.*, 2009). In accordance with the functional specialization of *CesA* isoforms in Arabidopsis (Taylor *et al.*, 2003; Desprez *et al.*, 2007; Persson *et al.*, 2007), rice (Tanaka *et al.*, 2003) and barley(Burton *et al.*, 2004), expression studies reveal that at least three specific maize *CesAs* (namely *ZmCesA10*, 11 and 12) are required during secondary cell wall formation, while the rest are assumed to be involved for primary cell wall deposition (Appenzeller *et al.*, 2004; Penning *et al.*, 2009).

The CSC also appears to interact with a wide array of plasma membrane-associated proteins; most suspected necessary for normal cellulose microfibril assembly, crystallization, orientation, and patterning (Somerville, 2006). In maize, a gene orthologous to the Arabidopsis *Cobra-Like4* isoform was cloned from the *brittle stalk-2* (*bk-2*) mutant (Ching *et al.*, 2006; Sindhu *et al.*, 2007); a naturally occurring phenotype characterized by stalks which break easily under mechanical pressure. This finding and other recent breakthroughs in fundamental cell wall research would suggest that cellulose content and ultrastructure are targets of multiple regulatory mechanisms extending further than the CSC and its associated proteins. Targeted alterations in cellulose content or molecular quality are yet to be reported for maize. Genetic engineering approaches will need to be carefully evaluated, however, as perturbations to the cellulose synthesis machinery could lead to phenotypes with decreased vigor or other undesirable biomass characteristics. Alternatively, allelic variants of crucial cellulose biosynthesis genes could be characterized and used directly in classical breeding schemes. Harris et al. (2009) uncovered an Arabidopsis mutant (irx 1-2) exhibiting a point-mutation at the C-terminal transmembrane region of the CesA3. The resulting phenotype displayed lower cellulose crystallinity (<~30%) and improved cell wall digestibility relative to wild type, but no profound perturbation on growth and fitness (Harris et al., 2009, 2012).

Hemicellulose

Research efforts looking to improve the yields and industrial quality of lignocellulosic crops have paid less recognition to the benefits that could arise from modifying the hemicellulosic fraction of plant cell walls. Current advances in the development of novel xylanases and C5-fermenting microorganisms, however, have opened the possibility to exploit this polysaccharide for the production of cellulosic ethanol and other side-stream bio-commodities (Becker & Boles, 2003; Bera et al., 2011; Zhang et al., 2011). In addition, because hemicellulose binds to cellulose microfibrils and threads them via cross-links with lignin (Carpita, 1996; Grabber et al., 2000), hemicellulose plays a crucial role in the structural integrity and recalcitrant nature of the cell wall. By elucidating the genetic mechanisms controlling hemicellulose biosynthesis, it should be possible to identify genetic variants that improve cell wall digestibility.

Although it is well recognized that plant hemicelluloses are synthesized in the ER/Golgi and mobilized to the growing cell wall via secreted vesicles (Carpita, 1996; Carpita & Mccann, 2010), limited information exists with respect to the enzymatic complexes directing their biosynthesis. Hemicellulosic cell wall polysaccharides appear to be synthesized by members of the *Cellulose Synthase Like* (*Csl*) gene family; a multigene complex highly homologous to the *CesA* family. Richmond and Somerville (2000, 2001) originally ascribed *Csl* gene products a processive glycosyltransferase (GT) function after observing that all *Csl* proteins possess a conserved domain defining their ability to catalyze the characteristic β-linkage common to cell wall polysaccharides. Thus far, expression studies suggest that primary wall xyloglucans (Cocuron et al., 2007), (gluco)mannans (Dhugga et al., 2004; Liepman et al., 2005) and grass-specific mixed linkage glucans (Doblin et al., 2009) are all synthesized by *Csl* encoded enzymes. By contrast, extensive

evidence indicates that the xylan backbones of secondary wall glucuronoxylan (GX) in dicots and (glucurono) arabinoxylan (GAX) in grasses are synthesized by non-processive GTs (Table 1; Aspeborg et al., 2005; Brown et al., 2005; Persson et al., 2005; Mitchell et al., 2007; Bosch et al., 2011). Advances in our understanding of the synthesis of GX in model dicots (e.g., Arabidopsis) will prove fundamental to the development of bioenergy grasses with tailored hemicellulose composition, as GAX represents the major noncellulosic polysaccharide in monocots. In maize, Bosch et al. (2011) have identified two GT47 sequences (*GRMZM2G100143* and *GRMZM2G059825*) displaying preferential expression in internodes undergoing secondary cell wall deposition. Both genes are homologous to the reduced-xylan deposition *IRX10* and *IRX10L* mutants of Arabidopsis and are likely candidates for the biosynthesis of GAX.

Efforts have also been devoted toward characterizing enzymes mediating GX and GAX branching reactions. Recent breakthroughs include the identification of the reduced wall acetylation (Lee et al., 2011) and Glucuronic Acid Substitution of Xylan (*GUX*; Mortimer et al., 2010) genes from Arabidopsis, as well as the Xylan Arabinosyltransferase (*XAT*) genes from rice and wheat (Anders et al., 2012) and the Xylosyl Arabinosyl Substitution of Xylan (*XAX*) gene from rice (Chiniquy et al., 2012). Exciting new evidence would also suggest that homologous sidegroup transferases differ in their enzymatic affinities and produce unique branching patterns. The functionally divergent *GUX1* and *GUX2* homologues from Arabidopsis appear to decorate distinct domains of the same xylan molecule either in evenly spaced long intervals or in tightly clustered patterns (Bromley et al., 2013). It is yet to be determined, however, whether GXs differ in the proportion, length, and distribution of substitution patterns (Bromley et al., 2013), or whether heterogeneous GXs have different affinities and functions in the plant cell wall.

Research on the genetic mechanisms controlling hemicellulosic branching is increasingly appealing for the production of advanced lignocellulosic feedstocks. Presently, a prominent view suggests that reducing the complexity of hemicelluloses would enhance their extractability and improve the overall degradability of lignocellulosic biomass (Appeldoorn et al., 2010; Mortimer et al., 2010; Van Eylen et al., 2011). In maize, the high substitution frequency of GAX has proven detrimental to the enzymatic conversion of cell wall polysaccharides following dilute acid pretreatment (Appeldoorn et al., 2010; Van Eylen et al., 2011). Based on the work of Van Eylen et al. (2011) and Appeldoorn et al. (2010), reductions in the frequency of acetic acid, uronic acid and arabinose side groups in GAX would concomitantly lead to a reduction in the use of costly enzymatic cocktails

Table 1 Genes involved in cell wall xylan biosynthesis

Gene	Species	GT sub-class	Presumed function	References
IRX8, FRA8, FH8, PARVUS	Arabidopsis	GT47 (Fra-8, FH8) GT8 (IRX-8, Parvus)	Synthesis of a unique β-d-Xyl-(1 → 3)-α-l-Rha-(1 → 2)-α-d-GalA-(1 → 4)-d-Xyl tetrasaccharide commonly found at the reducing end of glucuronoxylan (GX). This tetrasaccharide may act as a primer sequence for the initiation of short xylan chains which presumably splice together to form longer xylan polymers	(Brown et al., 2007; Peña et al., 2007; York & O'neill, 2008; Lee et al., 2009)
IRX9, IRX9-L, IRX-10, IRX-10L, IRX14, IRX14-L	Arabidopsis	GT43 (IRX9, IRX9-L, IRX14, IRX14-L) GT47 (IRX-10, IRX-10L)	Elongation of GX oligomeric backbones	(Brown et al., 2009; Wu et al., 2009; Keppler & Showalter, 2010; Wu et al., 2010)
GUX-1, GUX-2	Arabidopsis	GT8	Addition of glucuronic acid and 4-O-methylglucuronic acid side branches on GX backbone	(Mortimer et al., 2010; Bromley et al., 2013)
TaGT43-4, TaGT47-13, TaGT75-4	Wheat	GT43 (TaGT43-4) GT47 (TaGT47-13) GT75 (TaGT75-4)	Presumably involved in glucuronoarabinoxylan (GAX) biosynthesis, although specific functions are yet undefined. TaGT43-4 and TaGT47-13 are, respectively, orthologous to Arabidopsis IRX14 AND IRX10	(Zeng et al., 2010)
TaXAT2; OsXAT2, OsXAT3	Wheat, Rice	GT61	Arabinosylation of the xylan backbones of GAX	(Anders et al., 2012)
OsIRX9, OsIRX9L, OsIRX14	Rice	GT8	Presumably involved in the synthesis and elongation of the xylan backbone of GAX	(Chiniquy et al., 2013)
OsXAX1	Rice	GT61	Putatively involved in the β-(1,2) xylosyl substitution of α-(1,3) arabinosyl residues of GAX	(Chiniquy et al., 2012)
GRMZM2G100143, GRMZM2G059825	Maize	GT47	Presumably involved in the synthesis of GAX in secondary cell walls. GRMZM2G100143 and GRMZM2G059825 are homologues to IRX10 and IRX10-L	(Bosch et al., 2011)

and the formation of acetyl-based fermentation inhibitors during maize cellulosic ethanol conversion. Contradictorily, an alternative strategy to enhance the enzymatic accessibility of cell wall polysaccharides would entail increasing the abundance of 'favorable' side-chain substitutions in the backbones of hemicelluloses. This approach is grounded on the assumption that highly branched xylan polymers have a reduced adsorption-affinity to cellulose and improved water solubility (Kabel et al., 2007). More recently, Torres et al. (2013) demonstrated that the compounded effect of reduced cell wall lignin and high GAX arabinose-to-xylose ratio significantly improved the enzymatic conversion efficiency of mildly pretreated maize stem materials. Clearly, insights as to how GAX substitution patterns affect the strength and recalcitrance of the cell wall under different processing conditions are still necessary. However, it appears that maize harbors extensive genetic variation in the degree and (presumably) distribution of GAX substitution patterns (Torres et al., 2013), thus opening the possi-

bility to tailor maize cell wall hemicelluloses to the dynamic demands of the industry.

Lignin

The genetic and metabolic mechanisms that lead to the formation of lignin have been extensively studied and are well understood. For recent compendiums describing the structure, biosynthesis and biological function of this complex biopolymer, readers should refer to the work of Zhong & Ye (2009), Bonawitz and Chapple (2010), Vanholme et al. (2010) and Courtial et al. (2013b).

In the cell wall, lignin and other phenolic aromatics chemically cross-link to each other and to hemicellulose to produce an unyielding matrix that cohesively links and masks cell wall polysaccharides (Carpita, 1996). Evidence suggests that lignin reduces the effectiveness of enzymatic saccharification processes by adsorbing and nonproductively binding to hydrolytic enzymes (Berlin et al., 2006; Nakagame et al., 2010) and by physically shielding cellulose microfibrils from enzymatic attack (Selig *et al.*, 2007). As a consequence, efforts looking to reduce the inherent recalcitrance of bioenergy feedstocks have focused on understanding how variations in lignin content, composition and structure can improve cell wall degradability.

Currently, the most accepted notion is that reductions in cell wall lignin concentration can contribute positively to the ease with which cell walls are deconstructed. Supporting this claim are studies on the conversion efficiency of the *brown midrib* mutants of maize (*bm*) and sorghum (*bmr*; Barrière *et al.*, 2004; Vermerris *et al.*, 2007; Saballos *et al.*, 2008; Barrière *et al.*, 2013) and other species exhibiting genetically engineered reductions in lignin content (Chen & Dixon, 2007; Fu *et al.*, 2011; Jung *et al.*, 2012). Modifying lignin composition with respect to its monomeric constituents has also been coined a promising approach for enhancing biomass degradability. In maize and other bioenergy grasses, perturbations in monolignol metabolism favoring lower syringyl/guaiacyl ratios have been associated to reductions in biomass recalcitrance (Saballos *et al.*, 2008; Fu *et al.*, 2011; Fornalé *et al.*, 2012; Jung *et al.*, 2012). However, as favorable changes in monolignol ratios are often accompanied by reductions in lignin content (Chen & Dixon, 2007; Saballos *et al.*, 2008; Fornalé *et al.*, 2012; Jung *et al.*, 2012), it is still difficult to ascertain whether monolignol balance truly affects degradability properties (Grabber *et al.*, 2009). More recently, the concept of redesigning lignin *in planta* has gained momentum (Eudes *et al.*, 2012; Vanholme *et al.*, 2012; Zhang *et al.*, 2012). Fundamentally, this novel strategy exploits the combinatorial plasticity of the lignin polymerization process, which allows for the incorporation of 'un-conventional' monolignols into the lignin polymer (Vanholme *et al.*, 2010, 2012). This strategy ultimately allows for the creation of crops with customized lignin polymers displaying enhanced solubility, extractability, and chemical valorization. As an example, Eudes *et al.* (2012) induced the expression of a hydroxycinnamoyl-CoA hydratase-lyase from *Pseudomonas fluorescens* in Arabidopsis, to divert the metabolism of regular C_6C_3 monolignols in favor of atypical C_6C_1 aromatics, naturally present in lignin in trace amounts. Compared with wild-type controls, engineered lines showed a higher incorporation of the atypical aromatic in lignin, and a concomitant reduction in the degree of lignin polymerization. The engineered lines also displayed improved enzymatic saccharification efficiency following thermochemical pretreatment. (Eudes *et al.*, 2012).

In maize, classical breeding approaches have proven successful in the targeted modification of lignin for improved cell wall degradability properties. Extensive surveys of experimental populations and mutant panels have revealed the vast extent of lignin variation and enzymatic digestibility properties available in forage maize (Argillier *et al.*, 1996; Fontaine *et al.*, 2003a; Marita *et al.*, 2003; Méchin *et al.*, 2005; Barrière *et al.*, 2009, 2010, 2013) and have served as platforms for the identification of quantitative trait loci (QTL) underlying maize lignification characteristics relevant to cellulosic ethanol production (Méchin *et al.*, 2001; Cardinal *et al.*, 2003; Barrière *et al.*, 2008, 2009; Thomas *et al.*, 2010; Courtial *et al.*, 2013a). More recently, Lorenzana *et al.* (2010) and Torres *et al.* (2013) have demonstrated the strong negative correlation ($r > -0.65$) that exists between maize cell wall lignin content and enzymatic conversion efficiency after dilute-acid pretreatment. From these studies, it has become apparent that variation in lignin content leading to improved bioconversion efficiency is highly heritable, making it possible to select and advance dedicated maize feedstocks that can improve the efficiency and economics of biomass-to-ethanol conversion technologies.

Genetic engineering has also been explored as a viable strategy for the modification of lignin content and composition in maize. Piquemal *et al.* (2002) and He *et al.* (2003) used an antisense/gene approach to independently produce transgenic lines with reduced Caffeic-acid O-methyltransferase activity, thereby mimicking the naturally occurring *bm3* phenotype. In both studies, the resulting transgenics displayed significant reductions in whole-plant lignin content as well as improved *in vitro* enzymatic digestibility. More recently, Fornalé *et al.* (2012) used RNA interference to produce engineered lines with reduced cinnamyl/alcohol/dehydrogenase activity; and one was selected for extensive characterization. Although the selected transgenic displayed a slight reduction in lignin content and improved cell wall digestibility in leaf midribs, its stems showed no change in lignin accumulation or improved enzymatic digestibility relative to the wild-type control (Fornalé *et al.*, 2012). These results ultimately strengthen the notion that a systematic understanding of lignin biosynthesis is elemental if we seek to maximize the beneficial effects, and avoid the detrimental consequences, of engineered perturbations in lignin metabolic fluxes. Extensive evidence suggests that targeted alterations in lignin properties are often accompanied by compensatory mechanisms that can either restore the original phenotype or reduce the phenotypic effect of a profound metabolic alteration (Barrière *et al.*, 2003; Fontaine *et al.*, 2003a; Shi *et al.*, 2006; Guillaumie *et al.*, 2007a; Vermerris *et al.*, 2007; Barrière *et al.*, 2009). To illustrate this, when the *bm3* gene was introgressed into different genetic backgrounds, the resulting lines exhibited clear differences in lignin content and overall digestibility (Gentinetta *et al.*, 1990; Barrière *et al.*, 2003; Shi *et al.*,

2006). Accordingly, effective lignin engineering strategies need to consider the effects of pathway cross-talk mechanisms, spatial expression, and allelic redundancy to achieve desired results.

Deconstructing the matrix: the role of ferulate cross-links

In grasses, GAX molecules cross-link to each other via esterified diferulic bridges and to lignin via ferulic/diferulic ether bonds (Grabber *et al.*, 2000, 2004), thereby forming a matrix that encases the cellulosic skeleton of the plant cell wall. It is commonly understood that both diferulate cross-linking between xylans and ferulate cross-linking of xylans to lignin occur at the plant cell wall via oxidative coupling reactions, essentially devoid of enzymatic control (Grabber *et al.*, 1995; Ralph *et al.*, 1995; Grabber *et al.*, 2000). By contrast, ferulates are expectedly esterified to the arabinosyl residues of GAX through an enzymatically driven process occurring at the Golgi (Grabber *et al.*, 2004). To date, however, none of the genes involved in this process have been identified.

Unambiguous evidence from cell wall mimetic studies has demonstrated that both, xylan-to-xylan ferulate bridging (Grabber *et al.*, 1998a) and ferulate-to-lignin cross-links (Grabber *et al.*, 1998b, 2009) limit the enzymatic depolymerization of cell wall polysaccharides. Understandably, strategies that could reduce the incidence of ferulate cross-links in the cell wall have the potential to improve cell wall degradability properties relevant to cellulosic ethanol production. For instance, numerous studies in maize have revealed the extent of genetic variation potentially available in cell wall ferulate content, as well as its negative relationship with cell wall digestibility properties (Argillier *et al.*, 1996; Fontaine *et al.*, 2003b; Barrière *et al.*, 2008; Barros-Rios *et al.*, 2012). Similarly, Jung & Phillips (2010) have identified a putative maize mutation – seedling ferulate ester – which has been shown to reduce the content of etherified and esterified ferulates in the cell wall and increase cell wall digestibility, without affecting plant growth and yield. And while highly promising, the influence of ferulate cross-linking on degradability properties needs to be analyzed within the context of cellulosic ethanol production systems, considering that the physical, thermochemical, and enzymatic mechanisms underlying cell wall degradation processes in animal rumen and biomass-to-ethanol conversion platforms are not strictly similar (Jung & Bernardo, 2012; Torres *et al.*, 2013).

Transcription factors

Transcription factors regulate the quantitative, spatial and temporal expression of gene networks and define the differentiation of plant tissues, organs and other architectural features. Within the same organism, plant cell walls can vary greatly in their compositional and structural constitution among functionally divergent cell types (Zhong & Ye, 2007). The elucidation of the regulatory mechanisms that control cell wall differentiation will facilitate the tailoring of biomass yield and quality traits in a more coordinated and targeted fashion (Petersen *et al.*, 2012; Yang *et al.*, 2013).

In the last decade, numerous studies in Arabidopsis (as well as other species) have uncovered a vast array of key transcriptional regulators involved in secondary cell wall biosynthesis and differentiation. From these studies, it has become apparent that members of the *NAC* (e.g., *NST1*, *NST2*, *VND6*, and *VND7*) protein family act as master regulators of secondary cell wall deposition (Mitsuda *et al.*, 2007; Zhong & Ye, 2007) with different members displaying cell type specific expression patterns (Kubo *et al.*, 2005; Zhong *et al.*, 2006, 2008; Yamaguchi *et al.*, 2010). These master regulators appear to control downstream transcriptional cascades, which in turn activate cell wall lignin and carbohydrate biosynthetic pathways (Zhong *et al.*, 2008). In fact, *MYB* transcription factors have been highlighted as targets of *NAC* master regulators and have been shown to directly or indirectly activate cell wall deposition processes (Zhong *et al.*, 2008, 2010). For instance, while Arabidopsis *MYB46* and *MYB83* appear to globally regulate secondary cell wall deposition (Zhong *et al.*, 2007; Mccarthy *et al.*, 2009; Zhong & Ye, 2012), *MYB58*, *MYB63*, and *MYB85* have been shown to specifically regulate lignin biosynthesis (Zhong *et al.*, 2008; Zhou *et al.*, 2009). Much work is needed, however, before we entirely comprehend the complex transcriptional network governing cell wall deposition processes. In particular, the identification of novel modulators and downstream targets of *NAC* master regulators and a better understanding of their spatial regulation in specific cell types will prove beneficial to the development of effective cell wall engineering strategies. In maize, advances in functional genomics are rapidly unraveling the identity of NAC and MYB transcription factors presumably involved in cell wall biosynthesis and differentiation (Fornalé *et al.*, 2006, 2010; Bosch *et al.*, 2011; Zhong *et al.*, 2011).

Interestingly, despite gaps in our understanding of cell wall regulatory processes, several studies have demonstrated successful approaches to alter cell wall biosynthesis through controlled modulations of transcription factors (Eudes *et al.*, 2012; Petersen *et al.*, 2012; Yang *et al.*, 2013). Noteworthy, Yang *et al.* (2013) 'rewired' the secondary cell wall deposition network of Arabidopsis using a sophisticated approach that enabled for simultaneous cell wall alterations in specific tissues. Firstly, *c4h* lignin-deficient mutants were transformed with a functional *C4H* variant containing the vessel-specific promoter of *VND6* (pVND6), which allowed for the targeted recovery of cell wall lignification in stem vessels. The resulting pVND6::C4H lines were then transformed with an engineered construct of *NST1* coupled to the promoter sequence of *IRX8* (pIRX8); itself known to be a downstream target of *NST1*. By doing so, an artificial positive feedback loop was created whereby the expression of *NST1*, a master regulator of secondary wall formation in stem fibers, was specifically enhanced in tissues undergoing secondary wall deposition. Ultimately, pVND6::*CH4*-pIRX8::*NST1* lines showed wild-type vigor and growth, functional stem vessels, and increased cell wall deposition in fiber cells without over lignification. Also, the double-transgenics displayed higher fermentable sugar release relative to wild-type following pretreatment and enzymatic saccharification. Clearly, advances in our understanding of cell wall regulatory mechanisms have expanded our potential to precisely engineer biomass yield and quality characteristics, while circumventing the detrimental effects on yield and vigor commonly associated with transgenic approaches targeting cell wall metabolic fluxes.

Advancing energy maize: tools and concepts

Phenotyping tools

The greatest challenge in 'bioenergy crop' research and breeding programs is the screening of thousands of genetic variants to evaluate, map, and select traits that enhance the conversion potential of biomass into liquid fuels. Currently, numerous analytical platforms are in place for the exhaustive analysis of cell wall compositional and conversion efficiency parameters (Table 2). This comprehensive toolkit ranges from simple enzymatic assays to evaluate the saccharification efficiency of lignocellulosic substrates, to state-of-art chromatographic tools used to pinpoint the compositional diversity and ultrastructure of cell wall polymers. With the advent of highly precise weighing and liquid-handling robotic workstations, standard compositional quantification methods and bioconversion assays have been successfully down-scaled and automated to accommodate high-throughput analyses (Chundawat *et al.*, 2008; Gomez *et al.*, 2010; Santoro *et al.*, 2010; Selig *et al.*, 2010; Studer *et al.*, 2010). Notwithstanding, phenotyping tools that provide additional layers of information, like imaging techniques used to study the effects of pretreatments on biomass substrates or methods, which allow for the quantitative partitioning of biomass fibers (i.e., ratio of rind to pith in maize internodes), are yet to be adapted into automated systems.

More efficient and economical alternatives to robotic platforms have also been proposed; the most promising of which are based on spectroscopic methods, such as near-infrared (NIR), Fourier-transformed infrared (FT-IR) and pyrolysis molecular beam mass spectroscopy (Py-MBMS; Labbe *et al.*, 2005; Vermerris *et al.*, 2007; Philip Ye *et al.*, 2008). In these systems, a core set of samples is exhaustively analyzed using conventional chemical assays to build calibration models that can link compositional information to specific spectral variants. Once the model has been established, the biochemical properties of unknown samples can be predicted based on their spectral fingerprint. Although these screening tools convey considerable capital investments, their principle advantage is that spectral acquisition is fast, simple and does not require chemical consumables. For maize, NIRS is routinely employed in a commercial setting for the assessment of complex forage quality traits including the analysis of cell wall digestibility properties (Lauer *et al.*, 2001; Vermerris *et al.*, 2007; Lewis *et al.*, 2010). Within the scope of biomass research for cellulosic ethanol, several reports have demonstrated the successful application of NIRS for the prediction of polysaccharide, neutral sugar, lignin and ferulate content, as well as bioconversion efficiency (Barrière *et al.*, 2008; Lorenz *et al.*, 2009; Wolfrum & Lorenz, 2009; Jung & Phillips, 2010; Lorenzana *et al.*, 2010; Chavigneau *et al.*, 2012).

Genomic and molecular tools

Because of its global relevance as an agricultural and industrial staple, maize remains at the forefront of fundamental developments in molecular and genomic technologies. Currently, maize geneticists and breeders worldwide benefit from an extensive infrastructure of genotyping platforms, expression analyses repositories and powerful experimental populations. In addition, a draft sequence of the maize genome is now available (Schnable *et al.*, 2009), and numerous resequencing projects have updated our knowledge on the evolution, diversity, and complex heterotic nature of this crop species (Gore *et al.*, 2009; Lu *et al.*, 2009; Lai *et al.*, 2010; Chia *et al.*, 2012; Wallace *et al.*, 2014). Complemented by powerful data-mining resources (e.g., POPcorn, MaizeGDB, Panzea), marker discovery, and gene annotation in maize are advancing rapidly.

Classical linkage analysis will prove integral to the identification of QTL influencing complex biomass accumulation and cell wall architectural traits. Lorenzana *et al.* (2010), for instance, surveyed the testcrosses of 223 recombinant inbred lines from the IBM population (Lee *et al.*, 2002) for variation in different biomass characteristics, including conversion efficiency after dilute acid pretreatment. Despite the appreciably lim-

Table 2 High-throughput techniques available for the analysis of cell wall traits relevant to cellulosic ethanol production

Analysis	Technique	Used for determining	Description	References
Cell wall composition	Chromatography	Polysaccharide monomeric composition, lignin content	Cell wall polysaccharides are extracted and digested through a two-stage acid hydrolysis. Released monosaccharides are quantified via HPLC. Klasson lignin estimates are obtained gravimetrically	(Demartini *et al.*, 2011)
	Analytical Pyrolysis	Polysaccharide content, lignin content, lignin monomeric composition	Lignocellulosic samples are pyrolyzed and the resulting fragments are analyzed via GC/MS or MB/MS	(Fontaine *et al.*, 2003a; Sykes *et al.*, 2009; Studer *et al.*, 2011)
	Spectroscopy	Polysaccharide content, lignin content, lignin monomeric composition	Chemical composition is predicted through an array of spectroscopic platforms (NIR, FT-IR, Raman, NMR) based on calibration models linking compositional information to specific spectral patterns	(Lupoi *et al.*, 2013)
Cell wall ultrastructure	Chromatography, Electrophoresis	Hemicellulose and pectin degree of polymerization and degree of substitution	Cell wall polysaccharides are chemically extracted and digested using selective hydrolytic enzymes. The resulting oligosaccharides are separated and quantified via chromatographic (HPLC) or electrophoretic (PACE, CE) techniques, and/or identified through mass spectrometry (MALDI, NMR)	(Lerouxel *et al.*, 2002; Obel *et al.*, 2009; Persson *et al.*, 2010)
	Spectroscopy	Lignin content, monomeric composition and linkage analysis	Lignin is isolated from the cell wall and analyzed using one- and two-dimensional NMR spectroscopy	(Foston *et al.*, 2012; Lupoi *et al.*, 2013)
	Immuno-profiling	Hemicellulose, pectin and glycoprotein degree of polymerization and degree of substitution	Cell wall polysaccharides are chemically extracted and/or digested using selective hydrolytic enzymes. The resulting fractions are fixed onto microarrays (CoMPP) or ELISA microplates (Glycome Profiling) and probed using mAbs and CBMs with specificity for polysaccharide epitopes	(Moller *et al.*, 2007; Knox, 2008; Duceppe *et al.*, 2012; Pattathil *et al.*, 2012; Demartini *et al.*, 2013)
Cell wall recalcitrance	Enzymatic	Bioconversion efficiency of lignocellulosic substrates	The NREL LAP-009 bioconversion assay has been automated and downscaled to via robotic platforms. To accommodate an accessory pretreatment step, the most sophisticated systems rely on stackable 96-well metallic reactor plates that can withstand the chemical loads, pressure, and high temperatures used in industry	(Chundawat *et al.*, 2008; Gomez *et al.*, 2010; Santoro *et al.*, 2010; Selig *et al.*, 2010; Studer *et al.*, 2010)

ited degree of variation available in the population (e.g., lignin content on cell wall basis ranged from 20.3 to 21.9% across the experimental panel), the authors uncovered 152 small effect QTLs for a variety of cell wall and cellulosic ethanol-relevant characters. Knowledge obtained from linkage studies should also be complemented with findings from a wealth of forage maize studies elucidating crucial QTLs for cell-wall digestibility, lignin content, and lignin composition. Also, the advent of high-throughput single nucleotide

polymorphism genotyping platforms, sophisticated biometric models and high-resolution mapping panels (including the powerful Nested Association Mapping Panel of maize) will expectedly expedite genomewide association studies for biomass yield and quality characterics (Riedelsheimer *et al.*, 2012; Windhausen *et al.*, 2012; Massman *et al.*, 2013; Wallace *et al.*, 2014).

Functional genomics will also contribute immensely to our understanding of the genetic and biochemical mechanisms governing the construction of the plant cell wall. In maize, expression studies using diverse developmental models have led to the identification, annotation, and functional classification of numerous genes involved in cell wall biosynthesis (Shi *et al.*, 2006; Guillaumie *et al.*, 2007a,b; Penning *et al.*, 2009; Bosch *et al.*, 2011). Expression analyses of the elongating maize internode have proven particularly appealing, as these have provided a developmental snapshot for the deposition of the highly recalcitrant secondary cell wall. Forward- and reverse-genetic assessments of mutagenized maize populations are also powerful tools for identifying and underpinning the function of cell wall genes. In particular, gene tagging through transposon insertional mutagenesis, in combination with high-throughput genomic/phenomic platforms, has simplified the generation, discovery and cloning of cell wall mutants. Within the framework of the Cell Wall Genomics project (http://cellwall.genomics.purdue.edu), Vermerris *et al.* (2007) have conceptualized the use of NIR and Py-BMS platforms to identify novel cell wall mutants from the UniformMu population. Using the same mutant collection, Penning *et al.* (2009) have shown the versatility of

next-generation sequencing for the identification of mutants in specific cell wall genes, with the goal of better understanding their role in cell wall metabolic processes. Without doubt, the wealth of dedicated genomic resources currently available for maize make it an outstanding model organism for understanding complex biomass characteristics and defining the path for breeders looking to improve this crop for a bio-based economy.

Transgenic approaches

Conventional bioengineering strategies have been extensively used for the production of novel phenotypes with improved biomass characteristics. Knock-out, antisense construct and RNA-interference technologies have been the de facto routes for studying the effects of targeted alterations in cell wall metabolic fluxes and regulatory networks.

More recently, protein engineering and heterologous expression systems have broadened the horizons of energy crop bioengineering (Table 3). Heterologous

Table 3 Examples of studies using heterologous expression systems for the improvement of bioenergy crops

Species	Target polymer	Approach	Results	References
Arabidopsis	Lignin	Expression of a *Pseudomonas fluorescens* hydroxycinnamoyl-CoA hydratase-lyase	The overproduction of atypical C_6C_1 monolignols leads to the formation of sidechain truncated lignin with a lower degree of polymerization. Transformed lines displayed improved lignin extractability and enzymatic conversion after mild pretreatment	(Eudes *et al.*, 2012)
Arabidopsis	Lignin	Expression of *Clarkia breweri* monolignol 4-O-methyltransferase (MOMT) engineered via iterative saturation mutagenesis	In engineered lines, MOMT etherifies the para-hydroxyl group of lignin monomeric precursors, necessary for oxidative cross-coupling. Engineered lines showed a marked reduction in lignin content, the accumulation of novel phenolic esters and improved digestibility	(Zhang *et al.*, 2012)
Poplar	Lignin	Expression of a *Petroselinum crispum* tyrosine-rich glyco-peptide and targeted accumulation in the cell wall	Engineered lines displayed normal levels of lignification and increased saccharification after pretreatment with proteases. The incorporation of the glyco-peptide into the lignin polymer is yet to be confirmed	(Liang *et al.*, 2008)
Maize	Cellulose	Expression of a thermostable endocellulase E1 (Cel5A) from *Acidothermus cellulolyticus*	Relative to wild-type, engineered lines displayed improved digestibility after mild pretreatment. The exact mechanism through which the enzyme affects cell wall recalcitrance is yet to be uncovered	(Brunecky *et al.*, 2011)
Maize	Hemicellulose	Expression of an engineered cell wall degrading xylanase containing a thermoregulated intein sequence	Transformed lines were able to produce their own xylanase and release up to 60% cell wall glucose after enzymatic hydrolysis following mild thermochemical pretreatment	(Shen *et al.*, 2012)

gene transfer has been pursued as a means to redesign cell wall polymers *in planta*; proving particularly successful in the creation of de novo lignin configurations exhibiting higher solubility and extractability (Liang *et al.*, 2008; Eudes *et al.*, 2012; Zhang *et al.*, 2012). Also, the expression of microbial cellulases and other exogenous cell wall modifying enzymes has proven a viable strategy for the production of lignocellulosic crops with the ability to guide their own 'self-digestion'. Noteworthy, Shen *et al.* (2012) engineered a cell wall degrading xylanase containing a thermoregulated intein sequence that could self-splice and restore the catalytic activity of the enzyme at high temperatures. When subjected to mild thermochemical pretreatment (55 °C), maize lines transformed with the engineered enzyme were able to produce their own xylanase and release up to 60% cell wall glucose after enzymatic hydrolysis. Moreover, because the xylanase only becomes active after thermochemical treatment, the transgenics showed normal seed development, fertility, and biomass accumulation. Along other exemplary works, Shen *et al.* (2012) demonstrate that it is fundamentally possible to control the accumulation and timely expression of exogenous CWD enzymes *in planta* and circumvent the repercussions on plant health commonly associated with heterologous gene expression.

The way forward

Genetic engineering has an immense appeal for the production of efficient bioenergy crops, especially when considering that promising perennial species either have complex genomes, difficult reproductive patterns or limited genetic variation for relevant cell wall characteristics. Notwithstanding, we are just beginning to learn about the intricate regulation of cell wall biosynthetic processes and we are still far from fully comprehending how targeted perturbations in cell wall metabolic fluxes will affect plant vigor and agronomic fitness. Also worthy of consideration, while public acceptance of genetically modified (GM) crops for bioenergy purposes might be higher than for GM food and feed commodities, unyielding governmental regulations (particularly in Europe) can stall, delay or discourage the deployment of GM energy grasses.

To circumvent the technical challenges and political issues related to GM technologies, we believe that advancing maize for the cellulosic ethanol industry can be effectively achieved by harnessing the standing variation available in commercial germplasm through modern selection tools. The convergence of classical selection schemes with inexpensive genotyping, advanced biometric models and double-haploid production technologies, has led to the conceptualization of 'next-generation'

breeding platforms with the potential to accelerate maize cultivar development and commercial release (Riedelsheimer *et al.*, 2012; Wallace *et al.*, 2014). In addition, the advent of high-throughput bioconversion assays and cell wall phenotyping technologies can expedite selection for complex biomass and cell wall characters without the need for an in-depth understanding of cell wall genetic mechanisms. Cell wall functional genomic and classical linkage studies should by no means be underestimated, however, as they will still constitute the fundamental base upon which to guide biomass breeding programs. We should also remember that the unexploited variation concealed within exotic germplasm offers great opportunities for the transformation of maize into a biomass- or energy-dedicated feedstock and modern selection tools are opening avenues for the rapid incorporation of rare alleles into elite material.

Conclusions

The economic impact of maize cell wall modifications

Over the last decade, diverse studies have demonstrated that bioenergy crops diverging in cell wall constitution respond differentially to the combined operations of pretreatment and enzymatic hydrolysis. These findings have invariably led to the recognition that the processing efficiency and environmental performance of biomass-to-ethanol conversion systems can be greatly improved through the adequate selection of biomass substrates. Remarkably, most technoeconomic assessments of the feasibility of cellulosic ethanol refineries appear to disregard this evidence, and only a handful of projective studies support the notion that the economics of the industry could be improved through the utilization of biomass feedstocks with enhanced processing amenability.

Notwithstanding, analyses of genetic variants in maize, switchgrass, poplar and sugarcane, have indicated that reductions in the chemical, enzymatic, and energetic stringency of biomass-to-ethanol conversions systems can be achieved through the utilization of genotypes displaying highly degradable cell walls. In fact, Torres *et al.* (2013) has even demonstrated that industrially competitive saccharification yields at milder processing conditions are accompanied by a 95% reduction in the production of toxic inhibitors that can affect fermentation efficiencies and down-stream process economics. And while the extent of these beneficial effects are yet to be confirmed on large-scale trials, it becomes clear that most comparative analyses of the economic and environmental performance of ethanol refineries are underestimating the impact of biomass composition on the overall efficiency of the industry.

Beyond cellulosic ethanol and the plant cell wall

In addition to biochemical pathways, thermochemical routes are also regarded frontrunners for the production of cellulosic biofuels. Based on comparative life-cycle and technoeconomic analyses, however, neither technology has a clear competitive environmental or commercial advantage in the industry (Wright & Brown, 2007a, b; Anex et al., 2010). Irrespective of the uncertainty over which conversion route(s) will ultimately prevail, the successful deployment of maize as a lignocellulosic substrate will adhere to the same incontrovertible principles.

To begin with, the plant cell wall will indubitably remain a central focus of bio-based maize breeding endeavors. Extensive evidence has demonstrated the influence biomass composition exerts on the economic, environmental, and technical efficiency of biomass-to-fuel conversion systems. And while cell wall 'ideotypes' will be largely determined by the conversion route (e.g., higher lignin content is favored by fast-pyrolysis conversion routes), all knowledge pertaining the maize cell wall (i.e., biosynthesis, phenotyping tools, and genomic approaches for modification) can be universally extrapolated toward the selection of specific cell wall compositional profiles that can best match the conversion system.

Notwithstanding, maize breeding for improved agronomic and environmental efficiency will also have great implications for the industry and cannot be disregarded. Being a central pillar to global food security, maize demand for human and animal nutrition will greatly expand by 2050 (Long & Ort, 2010; Tilman et al., 2011). Understandably, bio-based maize will ideally encompass dual-purpose hybrids combining both, optimal grain yield and high stover productivity (Vermerris et al., 2007; Weijde et al., 2013). Simultaneously improving grain and stover yields is a feasible undertaking (Lauer et al., 2001; Lorenz et al., 2009; Lewis et al., 2010; Lorenz et al., 2010), but maize production will also be constrained by the urgencies of modern agriculture (Long & Ort, 2010; Tilman et al., 2011). In this regard, ongoing endeavors have achieved major accomplishments in uncovering and exploiting novel genetic diversity for climate-related stresses and sustainable production under lower agricultural inputs (Cairns et al., 2012; Edmeades, 2013). Ultimately, the incorporation of agronomic 'hardiness' in dual-purpose hybrids will improve the economics and environmental performance of the industry (regardless of the conversion route) by lowering the GHG footprint of maize production, offsetting the conversion of virgin agricultural soils and reducing farm-to-plant transportation distances (Wright & Brown, 2007a,b; Wang et al., 2011;

Cairns et al., 2012; Wang et al., 2012; Edmeades, 2013). The diversification of maize into an energy-dedicated species should be examined with caution; however, as socioeconomic and environmental concerns are likely to arise if energy-dedicated maize is to replace grain maize production. To avoid a food-over-fuel debacle, biomass-dedicated maize will only make sense if it can be produced on marginal soils and compete with the high yields, agronomic hardiness and soil-recovery properties displayed by other promising bioenergy grasses.

Acknowledgements

We gratefully acknowledge Tim van der Weijde for his help with the production of Fig. 1, as well as his insightful comments on the manuscript. Within the framework of the Carbohydrate Competence Centre, this research has been financially supported by the European Union, the European Regional Development Fund, and the Northern Netherlands Provinces (Samenwerkingsverband Noord-Nederland), KOERS NOORD.

References

Anders N, Wilkinson MD, Lovegrove A et al. (2012) Glycosyl transferases in family 61 mediate arabinofuranosyl transfer onto xylan in grasses. *Proceedings of the National Academy of Sciences*, **109**, 989–993.

Anex RP, Aden A, Kazi FK et al. (2010) Techno-economic comparison of biomass-to-transportation fuels via pyrolysis, gasification, and biochemical pathways. *Fuel*, **89**, S29–S35.

Appeldoorn MM, Kabel MA, Van Eylen D, Gruppen H, Schols HA (2010) Characterization of oligomeric xylan structures from corn fiber resistant to pretreatment and simultaneous saccharification and fermentation. *Journal of Agricultural and Food Chemistry*, **58**, 11294–11301.

Appenzeller L, Doblin M, Barreiro R et al. (2004) Cellulose synthesis in maize: isolation and expression analysis of the cellulose synthase (CesA) gene family. *Cellulose*, **11**, 287–299.

Argillier O, Barrière Y, Lila M, Jeanneteau F, Gélinet K, Ménanteau V (1996) Genotypic variation in phenolic components of cell-walls in relation to the digestibility of maize stalks. *Agronomie*, **16**, 123–130.

Aspeborg H, Schrader J, Coutinho PM et al. (2005) Carbohydrate-active enzymes involved in the secondary cell wall biogenesis in hybrid aspen. *Plant Physiology*, **137**, 983–997.

Bacovsky D (2010) How close are second-generation biofuels? *Biofuels, Bioproducts and Biorefining*, **4**, 249–252.

Barrière Y, Guillet C, Goffner D, Pichon M (2003) Genetic variation and breeding strategies for improved cell wall digestibility in annual forage crops. A review. *Animal Research*, **52**, 193–228.

Barrière Y, Ralph J, Méchin V et al. (2004) Genetic and molecular basis of grass cell wall biosynthesis and degradability. II. Lessons from brown-midrib mutants. *Comptes Rendus Biologies*, **327**, 847–860.

Barrière Y, Thomas J, Denoue D (2008) QTL mapping for lignin content, lignin monomeric composition, p hydroxycinnamate content, and cell wall digestibility in the maize recombinant inbred line progeny F838× F286. *Plant Science*, **175**, 585–595.

Barrière Y, Méchin V, Lafarguette F et al. (2009) Toward the discovery of maize cell wall genes involved in silage quality and capacity to biofuel production. *Maydica*, **54**, 161.

Barrière Y, Charcosset A, Denoue D, Madur D, Bauland C, Laborde J (2010) Genetic variation for lignin content and cell wall digestibility in early maize lines derived from ancient landraces. *Maydica*, **55**, 65.

Barrière Y, Chavigneau H, Delaunay S et al. (2013) Different mutations in the ZmCAD2 gene underlie the maize brown-midrib1 (bm1) phenotype with similar effects on lignin characteristics and have potential interest for bioenergy production. *Maydica*, **58**, 6–20.

Barros-Rios J, Malvar RA, Jung H-JG, Bunzel M, Santiago R (2012) Divergent selection for ester-linked diferulates in maize pith stalk tissues. Effects on cell wall composition and degradability. *Phytochemistry*, **83**, 43–50.

Becker J, Boles E (2003) A modified saccharomyces cerevisiae strain that consumes l-arabinose and produces ethanol. *Applied and Environmental Microbiology*, **69**, 4144–4150.

Bera A, Ho NY, Khan A, Sedlak M (2011) A genetic overhaul of Saccharomyces cerevisiae 424A(LNH-ST) to improve xylose fermentation. *Journal of Industrial Microbiology & Biotechnology*, **38**, 617–626.

Berlin A, Balakshin M, Gilkes N, Kadla J, Maximenko V, Kubo S, Saddler J (2006) Inhibition of cellulase, xylanase and β-glucosidase activities by softwood lignin preparations. *Journal of Biotechnology*, **125**, 198–209.

Bonawitz ND, Chapple C (2010) The genetics of lignin biosynthesis: connecting genotype to phenotype. *Annual Review of Genetics*, **44**, 337–363.

Bosch M, Mayer C-D, Cookson A, Donnison IS (2011) Identification of genes involved in cell wall biogenesis in grasses by differential gene expression profiling of elongating and non-elongating maize internodes. *Journal of Experimental Botany*, **62**, 3545–3561.

Bromley JR, Busse-Wicher M, Tryfona T, Mortimer JC, Zhang Z, Brown D, Dupree P (2013) GUX1 and GUX2 glucuronyltransferases decorate distinct domains of glucuronoxylan with different substitution patterns. *The Plant Journal*, **74**, 423–434.

Brown TR, Brown RC (2013) A review of cellulosic biofuel commercial-scale projects in the United States. *Biofuels, Bioproducts and Biorefining*, **7**, 235–245.

Brown DM, Zeef LA, Ellis J, Goodacre R, Turner SR (2005) Identification of novel genes in Arabidopsis involved in secondary cell wall formation using expression profiling and reverse genetics. *The Plant Cell Online*, **17**, 2281–2295.

Brown DM, Goubet F, Wong VW, Goodacre R, Stephens E, Dupree P, Turner SR (2007) Comparison of five xylan synthesis mutants reveals new insight into the mechanisms of xylan synthesis. *The Plant Journal*, **52**, 1154–1168.

Brown DM, Zhang Z, Stephens E, Dupree P, Turner SR (2009) Characterization of IRX10 and IRX10-like reveals an essential role in glucuronoxylan biosynthesis in Arabidopsis. *The Plant Journal*, **57**, 732–746.

Brunecky R, Selig M, Vinzant T, Himmel M, Lee D, Blaylock M, Decker S (2011) In planta expression of *A. cellulolyticus* Cel5A endocellulase reduces cell wall recalcitrance in tobacco and maize. *Biotechnology for Biofuels*, **4**, 1.

Burton RA, Shirley NJ, King BJ, Harvey AJ, Fincher GB (2004) The CesA gene family of barley. Quantitative analysis of transcripts reveals two groups of co-expressed genes. *Plant Physiology*, **134**, 224–236.

Cairns J, Sonder K, Zaidi P et al. (2012) 1 maize production in a changing climate: impacts, adaptation, and mitigation strategies. *Advances in Agronomy*, **114**, 1.

Cardinal A, Lee M, Moore K (2003) Genetic mapping and analysis of quantitative trait loci affecting fiber and lignin content in maize. *Theoretical and Applied Genetics*, **106**, 866–874.

Carpita NC (1996) Structure and biogenesis of the cell walls of grasses. *Annual Review of Plant Biology*, **47**, 445–476.

Carpita NC, Mccann MC (2008) Maize and sorghum: genetic resources for bioenergy grasses. *Trends in Plant Science*, **13**, 415–420.

Carpita NC, Mccann MC (2010) The maize mixed-linkage (1→ 3), (1→ 4)-β-d-glucan polysaccharide is synthesized at the Golgi membrane. *Plant Physiology*, **153**, 1362–1371.

Carroll A, Somerville C (2009) Cellulosic biofuels. *Annual Review of Plant Biology*, **60**, 165–182.

Chavigneau H, Goué N, Delaunay S et al. (2012) QTL for floral stem lignin content and degradability in three recombinant inbred line (RIL) progenies of Arabidopsis thaliana and search for candidate genes involved in cell wall biosynthesis and degradability. *OJGen*, **2**, 7–20.

Chen F, Dixon RA (2007) Lignin modification improves fermentable sugar yields for biofuel production. *Nature Biotechnology*, **25**, 759–761.

Chia J-M, Song C, Bradbury PJ et al. (2012) Maize HapMap2 identifies extant variation from a genome in flux. *Nature Genetics*, **44**, 803–807.

Ching A, Dhugga KS, Appenzeller L, Meeley R, Bourett TM, Howard RJ, Rafalski A (2006) Brittle stalk 2 encodes a putative glycosylphosphatidylinositol-anchored protein that affects mechanical strength of maize tissues by altering the composition and structure of secondary cell walls. *Planta*, **224**, 1174–1184.

Chiniquy D, Sharma V, Schultink A et al. (2012) XAX1 from glycosyltransferase family 61 mediates xylosyltransfer to rice xylan. *Proceedings of the National Academy of Sciences*, **109**, 17117–17122.

Chiniquy D, Varanasi P, Oh T et al. (2013) Three novel rice genes closely related to the Arabidopsis IRX9, IRX9L, and IRX14 genes and their roles in xylan biosynthesis. *Frontiers in Plant Science*, **4**, doi: 10.3389/fpls.2013.00083.

Chundawat SP, Balan V, Dale BE (2008) High-throughput microplate technique for enzymatic hydrolysis of lignocellulosic biomass. *Biotechnology and Bioengineering*, **99**, 1281–1294.

Clifton-Brown JC, Stampfl PF, Jones MB (2004) Miscanthus biomass production for energy in Europe and its potential contribution to decreasing fossil fuel carbon emissions. *Global Change Biology*, **10**, 509–518.

Cocuron J-C, Lerouxel O, Drakakaki G et al. (2007) A gene from the cellulose synthase-like C family encodes a β-1, 4 glucan synthase. *Proceedings of the National Academy of Sciences*, **104**, 8550–8555.

Courtial A, Méchin V, Reymond M, Grima-Pettenati J, Barrière Y (2013a) Colocalizations between several QTLs for cell wall degradability and composition in the F288× F271 early maize RIL progeny raise the question of the nature of the possible underlying determinants and breeding targets for biofuel capacity. *BioEnergy Research*, doi: 10.1007/s12155-013-9358-8.

Courtial A, Soler M, Chateigner-Boutin A-L et al. (2013b) Breeding grasses for capacity to biofuel production or silage feeding value: an updated list of genes involved in maize secondary cell wall biosynthesis and assembly. *Maydica*, **58**, 67–102.

Dale BE (2007) Thinking clearly about biofuels: ending the irrelevant 'net energy' debate and developing better performance metrics for alternative fuels. *Biofuels, Bioproducts and Biorefining*, **1**, 14–17.

Demartini JD, Studer MH, Wyman CE (2011) Small-scale and automatable high-throughput compositional analysis of biomass. *Biotechnology and Bioengineering*, **108**, 306–312.

Demartini JD, Pattathil S, Miller JS, Li H, Hahn MG, Wyman CE (2013) Investigating plant cell wall components that affect biomass recalcitrance in poplar and switchgrass. *Energy & Environmental Science*, **6**, 898–909.

Desprez T, Juraniec M, Crowell EF et al. (2007) Organization of cellulose synthase complexes involved in primary cell wall synthesis in Arabidopsis thaliana. *Proceedings of the National Academy of Sciences*, **104**, 15572–15577.

Dhugga KS, Barreiro R, Whitten B et al. (2004) Guar seed ß-mannan synthase is a member of the cellulose synthase super gene family. *Science*, **303**, 363–366.

Doblin MS, Pettolino FA, Wilson SM et al. (2009) A barley cellulose synthase-like CSLH gene mediates (1, 3; 1, 4)-β-D-glucan synthesis in transgenic Arabidopsis. *Proceedings of the National Academy of Sciences*, **106**, 5996–6001.

Duceppe M-O, Bertrand A, Pattathil S et al. (2012) Assessment of genetic variability of cell wall degradability for the selection of alfalfa with improved saccharification efficiency. *BioEnergy Research*, **5**, 904–914.

Dweikat I, Weil C, Moose S et al. (2012) Envisioning the transition to a next-generation biofuels industry in the US Midwest. *Biofuels, Bioproducts and Biorefining*, **6**, 376–386.

Edmeades G (2013) *Progress in Achieving and Delivering Drought Tolerance in Maize-An Update*. ISAAA, Ithaca, NY.

Eudes A, George A, Mukerjee P et al. (2012) Biosynthesis and incorporation of side-chain-truncated lignin monomers to reduce lignin polymerization and enhance saccharification. *Plant Biotechnology Journal*, **10**, 609–620.

FAOSTAT (2011) United Nations Food and Agriculture Organization Statistical Database. Available at: http://faostat.fao.org (accessed 3 April 2013).

Farrell AE, Plevin RJ, Turner BT, Jones AD, O'hare M, Kammen DM (2006) Ethanol can contribute to energy and environmental goals. *Science*, **311**, 506–508.

Fontaine A-S, Bout S, Barrière Y, Vermerris W (2003a) Variation in cell wall composition among forage maize (Zea mays L.) inbred lines and its impact on digestibility: analysis of neutral detergent fiber composition by pyrolysis-gas chromatography-mass spectrometry. *Journal of Agricultural and Food Chemistry*, **51**, 8080–8087.

Fontaine A, Briand M, Barriere Y (2003b) Genetic variation and QTL mapping of para-coumaric and ferulic acid. *Maydica*, **48**, 75–84.

Fornalé S, Sonbol F-M, Maes T, Capellades M, Puigdomenech P, Rigau J, Caparros-Ruiz D (2006) Down-regulation of the maize and Arabidopsis caffeic acid O-methyl-transferase genes by two new maize R2R3-MYB transcription factors. *Plant Molecular Biology*, **62**, 809–823.

Fornalé S, Shi X, Chai C et al. (2010) ZmMYB31 directly represses maize lignin genes and redirects the phenylpropanoid metabolic flux. *The Plant Journal*, **64**, 633–644.

Fornalé S, Capellades M, Encina A et al. (2012) Altered lignin biosynthesis improves cellulosic bioethanol production in transgenic maize plants down-regulated for cinnamyl alcohol dehydrogenase. *Molecular Plant*, **5**, 817–830.

Foston M, Samuel R, Ragauskas AJ (2012) 13C cell wall enrichment and ionic liquid NMR analysis: progress towards a high-throughput detailed chemical analysis of the whole plant cell wall. *Analyst*, **137**, 3904–3909.

Fu C, Mielenz JR, Xiao X et al. (2011) Genetic manipulation of lignin reduces recalcitrance and improves ethanol production from switchgrass. *Proceedings of the National Academy of Sciences*, **108**, 3803–3808.

Gentinetta E, Bertolini M, Rossi I, Lorenzoni C, Motto M (1990) Effect of brown midrib-3 mutant on forage and yield in maize. *Journal of Genetics & Breeding*, **44**, 21–26.

Gomez LD, Whitehead C, Barakate A, Halpin C, Mcqueen-Mason SJ (2010) Automated saccharification assay for determination of digestibility in plant materials. *Biotechnology for Biofuels*, **3**, 23–23.

Gore MA, Chia J-M, Elshire RJ et al. (2009) A first-generation haplotype map of maize. *Science*, **326**, 1115–1117.

Grabber JH, Hatfield RD, Ralph J, Zoń J, Amrhein N (1995) Ferulate cross-linking in cell walls isolated from maize cell suspensions. *Phytochemistry*, **40**, 1077–1082.

Grabber JH, Hatfield RD, Ralph J (1998a) Diferulate cross-links impede the enzymatic degradation of non-lignified maize walls. *Journal of the Science of Food and Agriculture*, **77**, 193–200.

Grabber JH, Ralph J, Hatfield RD (1998b) Ferulate cross-links limit the enzymatic degradation of synthetically lignified primary walls of maize. *Journal of Agricultural and Food Chemistry*, **46**, 2609–2614.

Grabber J, Ralph J, Hatfield R (2000) Cross-linking of maize walls by ferulate dimerization and incorporation into lignin. *Journal of Agriculture and Food Chemistry*, **48**, 6106–6113.

Grabber JH, Ralph J, Lapierre C, Barrière Y (2004) Genetic and molecular basis of grass cell-wall degradability. I. Lignin–cell wall matrix interactions. *Comptes Rendus Biologies*, **327**, 455–465.

Grabber JH, Mertens DR, Kim H, Funk C, Lu F, Ralph J (2009) Cell wall fermentation kinetics are impacted more by lignin content and ferulate cross-linking than by lignin composition. *Journal of the Science of Food and Agriculture*, **89**, 122–129.

Graham RL, Nelson R, Sheehan J, Perlack R, Wright LL (2007) Current and potential US corn stover supplies. *Agronomy Journal*, **99**, 1–11.

Guillaumie S, Pichon M, Martinant J-P, Bosio M, Goffner D, Barrière Y (2007a) Differential expression of phenylpropanoid and related genes in brown-midrib bm1, bm2, bm3, and bm4 young near-isogenic maize plants. *Planta*, **226**, 235–250.

Guillaumie S, San-Clemente H, Deswarte C et al. (2007b) MAIZEWALL. Database and developmental gene expression profiling of cell wall biosynthesis and assembly in maize. *Plant Physiology*, **143**, 339–363.

Hallauer AR, Miranda Filho J, Carena MJ (2010) Germplasm. In: *Quantitative Genetics in Maize Breeding* (eds Hallauer AR, Miranda Filho J, Carena MJ), pp. 531–576. Springer, New York.

Harris D, Stork J, Debolt S (2009) Genetic modification in cellulose-synthase reduces crystallinity and improves biochemical conversion to fermentable sugar. *GCB Bioenergy*, **1**, 51–61.

Harris DM, Corbin K, Wang T et al. (2012) Cellulose microfibril crystallinity is reduced by mutating C-terminal transmembrane region residues CESA1A903V and CESA3T942I of cellulose synthase. *Proceedings of the National Academy of Sciences*, **109**, 4098–4103.

He X, Hall MB, Gallo-Meagher M, Smith RL (2003) Improvement of forage quality by downregulation of maize-methyltransferase. *Crop Science*, **43**, 2240–2251.

Heaton EA, Dohleman FG, Long SP (2008) Meeting US biofuel goals with less land: the potential of Miscanthus. *Global Change Biology*, **14**, 2000–2014.

Holland N, Holland D, Helentjaris T, Dhugga KS, Xoconostle-Cazares B, Delmer DP (2000) A comparative analysis of the plant cellulose synthase (CesA) gene family. *Plant Physiology*, **123**, 1313–1324.

Huber GW, Dale BE (2009) Grassoline at the pump. *Scientific American*, **301**, 52–59.

Jung H-JG, Bernardo R (2012) Comparison of cell wall polysaccharide hydrolysis by a dilute acid/enzymatic saccharification process and rumen microorganisms. *BioEnergy Research*, **5**, 319–329.

Jung H, Phillips R (2010) Putative seedling ferulate ester *(sfe)* maize mutant: morphology, biomass yield, and stover cell wall composition and rumen degradability. *Crop Science*, **50**, 403–418.

Jung JH, Fouad WM, Vermerris W, Gallo M, Altpeter F (2012) RNAi suppression of lignin biosynthesis in sugarcane reduces recalcitrance for biofuel production from lignocellulosic biomass. *Plant Biotechnology Journal*, **10**, 1067–1076.

Kabel MA, Van Den Borne H, Vincken J-P, Voragen AG, Schols HA (2007) Structural differences of xylans affect their interaction with cellulose. *Carbohydrate Polymers*, **69**, 94–105.

Keppler BD, Showalter AM (2010) IRX14 and IRX14-LIKE, two glycosyl transferases involved in glucuronoxylan biosynthesis and drought tolerance in Arabidopsis. *Molecular Plant*, **3**, 834–841.

Kim S, Dale BE (2004) Global potential bioethanol production from wasted crops and crop residues. *Biomass and Bioenergy*, **26**, 361–375.

Knox JP (2008) Revealing the structural and functional diversity of plant cell walls. *Current Opinion in Plant Biology*, **11**, 308–313.

Kubo M, Udagawa M, Nishikubo N et al. (2005) Transcription switches for protoxylem and metaxylem vessel formation. *Genes & Development*, **19**, 1855–1860.

Labbe N, Rials TG, Kelley SS, Cheng Z-M, Kim J-Y, Li Y (2005) FT-IR imaging and pyrolysis-molecular beam mass spectrometry: new tools to investigate wood tissues. *Wood Science and Technology*, **39**, 61–76.

Lai J, Li R, Xu X et al. (2010) Genome-wide patterns of genetic variation among elite maize inbred lines. *Nature Genetics*, **42**, 1027–1030.

Larsen J, Haven MØ, Thirup L (2012) Inbicon makes lignocellulosic ethanol a commercial reality. *Biomass and Bioenergy*, **46**, 36–45.

Lauer J, Coors J, Flannery P (2001) Forage yield and quality of corn cultivars developed in different eras. *Crop Science*, **41**, 1449–1455.

Lee M, Sharopova N, Beavis WD, Grant D, Katt M, Blair D, Hallauer A (2002) Expanding the genetic map of maize with the intermated B73× Mo17 (IBM) population. *Plant Molecular Biology*, **48**, 453–461.

Lee C, Teng Q, Huang W, Zhong R, Ye Z-H (2009) The F8H glycosyltransferase is a functional paralog of FRA8 involved in glucuronoxylan biosynthesis in Arabidopsis. *Plant and Cell Physiology*, **50**, 812–827.

Lee C, Teng Q, Zhong R, Ye Z-H (2011) The four Arabidopsis reduced wall acetylation genes are expressed in secondary wall-containing cells and required for the acetylation of xylan. *Plant and Cell Physiology*, **52**, 1289–1301.

Lerouxel O, Choo TS, Séveno M, Usadel B, Faye LC, Lerouge P, Pauly M (2002) Rapid structural phenotyping of plant cell wall mutants by enzymatic oligosaccharide fingerprinting. *Plant Physiology*, **130**, 1754–1763.

Lewis MF, Lorenzana RE, Jung H-JG, Bernardo R (2010) Potential for simultaneous improvement of corn grain yield and stover quality for cellulosic ethanol. *Crop Science*, **50**, 516–523.

Liang H, Frost CJ, Wei X, Brown NR, Carlson JE, Tien M (2008) Improved sugar release from lignocellulosic material by introducing a tyrosine-rich cell wall peptide gene in poplar. *CLEAN–Soil, Air, Water*, **36**, 662–668.

Liepman AH, Wilkerson CG, Keegstra K (2005) Expression of cellulose synthase-like (Csl) genes in insect cells reveals that CslA family members encode mannan synthases. *Proceedings of the National Academy of Sciences of the United States of America*, **102**, 2221–2226.

Long SP, Ort DR (2010) More than taking the heat: crops and global change. *Current Opinion in Plant Biology*, **13**, 240–247.

Lorenz A, Coors J, De Leon N, Wolfrum E, Hames B, Sluiter A, Weimer P (2009) Characterization, genetic variation, and combining ability of maize traits relevant to the production of cellulosic ethanol. *Crop Science*, **49**, 85–98.

Lorenz A, Gustafson T, Coors J, Leon ND (2010) Breeding maize for a bioeconomy: a literature survey examining harvest index and stover yield and their relationship to grain yield. *Crop Science*, **50**, 1–12.

Lorenzana RE, Lewis MF, Jung H-JG, Bernardo R (2010) Quantitative trait loci and trait correlations for maize stover cell wall composition and glucose release for cellulosic ethanol. *Crop Science*, **50**, 541–555.

Lu Y, Yan J, Guimaraes CT et al. (2009) Molecular characterization of global maize breeding germplasm based on genome-wide single nucleotide polymorphisms. *Theoretical and Applied Genetics*, **120**, 93–115.

Lupoi JS, Singh S, Simmons BA, Henry RJ (2013) Assessment of lignocellulosic biomass using analytical spectroscopy: an evolution to high-throughput techniques. *BioEnergy Research*, doi: 10.1007/s12155-013-9352-1.

Mansfield SD, Mooney C, Saddler JN (1999) Substrate and enzyme characteristics that limit cellulose hydrolysis. *Biotechnology Progress*, **15**, 804–816.

Marita JM, Vermerris W, Ralph J, Hatfield RD (2003) Variations in the cell wall composition of maize brown midrib mutants. *Journal of Agricultural and Food Chemistry*, **51**, 1313–1321.

Massman JM, Jung H-JG, Bernardo R (2013) Genomewide selection versus marker-assisted recurrent selection to improve grain yield and stover-quality traits for cellulosic ethanol in maize. *Crop Science*, **53**, 58–66.

Mccarthy RL, Zhong R, Ye Z-H (2009) MYB83 is a direct target of SND1 and acts redundantly with MYB46 in the regulation of secondary cell wall biosynthesis in Arabidopsis. *Plant and Cell Physiology*, **50**, 1950–1964.

Méchin V, Argillier O, Hébert Y, Guingo E, Moreau L, Charcosset A, Barrière Y (2001) Genetic analysis and QTL mapping of cell wall digestibility and lignification in silage maize. *Crop Science*, **41**, 690–697.

Méchin V, Argillier O, Rocher F et al. (2005) In search of a maize ideotype for cell wall enzymatic degradability using histological and biochemical lignin characterization. *Journal of Agricultural and Food Chemistry*, **53**, 5872–5881.

Mitchell RA, Dupree P, Shewry PR (2007) A novel bioinformatics approach identifies candidate genes for the synthesis and feruloylation of arabinoxylan. *Plant Physiology*, **144**, 43–53.

Mitsuda N, Iwase A, Yamamoto H, Yoshida M, Seki M, Shinozaki K, Ohme-Takagi M (2007) NAC transcription factors, NST1 and NST3, are key regulators of the formation of secondary walls in woody tissues of Arabidopsis. *The Plant Cell Online*, **19**, 270–280.

Moller I, Sørensen I, Bernal AJ et al. (2007) High-throughput mapping of cell-wall polymers within and between plants using novel microarrays. *The Plant Journal*, **50**, 1118–1128.

Mortimer JC, Miles GP, Brown DM et al. (2010) Absence of branches from xylan in Arabidopsis gux mutants reveals potential for simplification of lignocellulosic biomass. *Proceedings of the National Academy of Sciences*, **107**, 17409–17414.

Mosier N, Wyman C, Dale B, Elander R, Lee Y, Holtzapple M, Ladisch M (2005) Features of promising technologies for pretreatment of lignocellulosic biomass. *Bioresource Technology*, **96**, 673–686.

Muttoni G, Palacios-Rojas N, Galicia L, Rosales A, Pixley KV, De Leon N (2012) Cell wall composition and biomass digestibility diversity in Mexican maize (Zea mays L) landraces and CIMMYT inbred lines. *Maydica*, **58**, 21–33.

Nakagame S, Chandra RP, Saddler JN (2010) The effect of isolated lignins, obtained from a range of pretreated lignocellulosic substrates, on enzymatic hydrolysis. *Biotechnology and Bioengineering*, **105**, 871–879.

Obel N, Erben V, Schwarz T, Kühnel S, Fodor A, Pauly M (2009) Microanalysis of plant cell wall polysaccharides. *Molecular Plant*, **2**, 922–932.

Park S, Baker JO, Himmel ME, Parilla PA, Johnson DK (2010) Research Cellulose crystallinity index: measurement techniques and their impact on interpreting cellulase performance. *Biotechnology for Biofuels*, **3**, 1–10.

Pattathil S, Avci U, Miller JS, Hahn MG (2012) Immunological approaches to plant cell wall and biomass characterization: glycome profiling. In: *Biomass Conversion* (ed. Himmel ME), pp. 61–71. Humana Press, New York.

Peña MJ, Zhong R, Zhou G-K et al. (2007) Arabidopsis irregular xylem8 and irregular xylem9: implications for the Complexity of Glucuronoxylan Biosynthesis. *The Plant Cell Online*, **19**, 549–563.

Penning BW, Hunter Iii CT, Tayengwa R et al. (2009) Genetic resources for maize cell wall biology. *Plant Physiology*, **151**, 1703–1728.

Persson S, Wei H, Milne J, Page GP, Somerville CR (2005) Identification of genes required for cellulose synthesis by regression analysis of public microarray data sets. *Proceedings of the National Academy of Sciences of the United States of America*, **102**, 8633–8638.

Persson S, Paredez A, Carroll A et al. (2007) Genetic evidence for three unique components in primary cell-wall cellulose synthase complexes in Arabidopsis. *Proceedings of the National Academy of Sciences*, **104**, 15566–15571.

Persson S, Sørensen I, Moller I, Willats W, Pauly M (2010) Dissection of plant cell walls by high-throughput methods. *Annual Plant Reviews: Plant Polysaccharides, Biosynthesis and Bioengineering*, **41**, 43–64.

Petersen PD, Lau J, Ebert B et al. (2012) Engineering of plants with improved properties as biofuels feedstocks by vessel-specific complementation of xylan biosynthesis mutants. *Biotechnology for Biofuels*, **5**, 1–19.

Philip Ye X, Liu L, Hayes D, Womac A, Hong K, Sokhansanj S (2008) Fast classification and compositional analysis of cornstover fractions using Fourier transform near-infrared techniques. *Bioresource Technology*, **99**, 7323–7332.

Piquemal J, Chamayou S, Nadaud I et al. (2002) Down-regulation of caffeic acid O-methyltransferase in maize revisited using a transgenic approach. *Plant Physiology*, **130**, 1675–1685.

Pollak LM (2003) The History and Success of the public–private project on germplasm enhancement of maize (GEM). In: *Advances in Agronomy* (ed. Sparks DL), pp. 45–87. Academic Press, Waltham, MA.

Ralph J, Grabber JH, Hatfield RD (1995) Lignin-ferulate cross-links in grasses: active incorporation of ferulate polysaccharide esters into ryegrass lignins. *Carbohydrate Research*, **275**, 167–178.

Richmond TA, Somerville CR (2000) The cellulose synthase superfamily. *Plant Physiology*, **124**, 495–498.

Richmond TA, Somerville CR (2001) Integrative approaches to determining Csl function. *Plant Molecular Biology*, **47**, 131–143.

Riedelsheimer C, Czedik-Eysenberg A, Grieder C et al. (2012) Genomic and metabolic prediction of complex heterotic traits in hybrid maize. *Nature Genetics*, **44**, 217–220.

Saballos A, Vermerris W, Rivera L, Ejeta G (2008) Allelic association, chemical characterization and saccharification properties of brown midrib mutants of sorghum (Sorghum bicolor (L.) Moench). *BioEnergy Research*, **1**, 193–204.

Saddler JN, Mabee WE, Simms R, Taylor M (2012) The biorefining story: progress in the commercialization of biomass-to-ethanol. In: *Forests in Development: A Vital Balance* (eds Schlichter T, Montes L), pp. 39–52. Springer, New York.

Salhuana W, Jones Q, Sevilla R (1991) The Latin American Maize Project: model for rescue and use of irreplaceable germplasm. *Diversity*, **7**, 40–42.

Santoro N, Cantu S, Tornqvist C-E et al. (2010) A high-throughput platform for screening milligram quantities of plant biomass for lignocellulose digestibility. *Bioenergy Research*, **3**, 93–102.

Schnable PS, Ware D, Fulton RS et al. (2009) The B73 maize genome: complexity, diversity, and dynamics. *Science*, **326**, 1112–1115.

Schubert C (2006) Can biofuels finally take center stage? *Nature Biotechnology*, **24**, 777–784.

Selig MJ, Viamajala S, Decker SR, Tucker MP, Himmel ME, Vinzant TB (2007) Deposition of lignin droplets produced during dilute acid pretreatment of maize stems retards enzymatic hydrolysis of cellulose. *Biotechnology Progress*, **23**, 1333–1339.

Selig MJ, Tucker MP, Sykes RW et al. (2010) ORIGINAL RESEARCH: lignocellulose recalcitrance screening by integrated high-throughput hydrothermal pretreatment and enzymatic saccharification. *Industrial Biotechnology*, **6**, 104–111.

Shen B, Sun X, Zuo X et al. (2012) Engineering a thermoregulated intein-modified xylanase into maize for consolidated lignocellulosic biomass processing. *Nature Biotechnology*, **30**, 1131–1136.

Shi C, Koch G, Ouzunova M, Wenzel G, Zein I, Lübberstedt T (2006) Comparison of maize brown-midrib isogenic lines by cellular UV-microspectrophotometry and comparative transcript profiling. *Plant Molecular Biology*, **62**, 697–714.

Sindhu A, Langewisch T, Olek A et al. (2007) Maize Brittle stalk2 encodes a COBRA-like protein expressed in early organ development but required for tissue flexibility at maturity. *Plant Physiology*, **145**, 1444–1459.

Somerville C (2006) Cellulose synthesis in higher plants. *Annual Review of Cell and Developmental Biology*, **22**, 53–78.

Sorda G, Banse M, Kemfert C (2010) An overview of biofuel policies across the world. *Energy Policy*, **38**, 6977–6988.

Studer MH, Demartini JD, Brethauer S, Mckenzie HL, Wyman CE (2010) Engineering of a high-throughput screening system to identify cellulosic biomass, pretreatments, and enzyme formulations that enhance sugar release. *Biotechnology and Bioengineering*, **105**, 231–238.

Studer MH, Demartini JD, Davis MF et al. (2011) Lignin content in natural Populus variants affects sugar release. *Proceedings of the National Academy of Sciences*, **108**, 6300–6305.

Sykes R, Yung M, Novaes E, Kirst M, Peter G, Davis M (2009) High-throughput screening of plant cell-wall composition using pyrolysis molecular beam mass spectroscopy. In: *Biofuels* (ed. Mielenz JR), pp. 169–184. Springer, New York.

Tanaka K, Murata K, Yamazaki M, Onosato K, Miyao A, Hirochika H (2003) Three distinct rice cellulose synthase catalytic subunit genes required for cellulose synthesis in the secondary wall. *Plant Physiology*, **133**, 73–83.

Taylor NG, Howells RM, Huttly AK, Vickers K, Turner SR (2003) Interactions among three distinct CesA proteins essential for cellulose synthesis. *Proceedings of the National Academy of Sciences*, **100**, 1450–1455.

Thomas J, Guillaumie S, Verdu C, Denoue D, Pichon M, Barriere Y (2010) Cell wall phenylpropanoid-related gene expression in early maize recombinant inbred lines differing in parental alleles at a major lignin QTL position. *Molecular Breeding*, **25**, 105–124.

Tilman D, Balzer C, Hill J, Befort BL (2011) Global food demand and the sustainable intensification of agriculture. *Proceedings of the National Academy of Sciences*, **108**, 20260–20264.

Torres AF, Van Der Weijde T, Dolstra O, Visser RG, Trindade LM (2013) Effect of maize biomass composition on the optimization of dilute-acid pretreatments and enzymatic saccharification. *BioEnergy Research*, **6**, 1038–1051.

Van Eylen D, Van Dongen F, Kabel M, De Bont J (2011) Corn fiber, cobs and stover: enzyme-aided saccharification and co-fermentation after dilute acid pretreatment. *Bioresource Technology*, **102**, 5995–6004.

Vanholme R, Demedts B, Morreel K, Ralph J, Boerjan W (2010) Lignin biosynthesis and structure. *Plant Physiology*, **153**, 895–905.

Vanholme R, Morreel K, Darrah C, Oyarce P, Grabber JH, Ralph J, Boerjan W (2012) Metabolic engineering of novel lignin in biomass crops. *New Phytologist*, **196**, 978–1000.

Vermerris W, Saballos A, Ejeta G, Mosier NS, Ladisch MR, Carpita NC (2007) Molecular breeding to enhance ethanol production from corn and sorghum stover. *Crop Science*, **47**, S-142–S-153.

Von Blottnitz H, Curran MA (2007) A review of assessments conducted on bio-ethanol as a transportation fuel from a net energy, greenhouse gas, and environmental life cycle perspective. *Journal of Cleaner Production*, **15**, 607–619.

Wallace J, Larsson S, Buckler E (2014) Entering the second century of maize quantitative genetics. *Heredity*, **112**, 30–38.

Waltz E (2008) Cellulosic ethanol booms despite unproven business models. *Nature Biotechnology*, **26**, 8–9.

Wang MQ, Han J, Haq Z, Tyner WE, Wu M, Elgowainy A (2011) Energy and greenhouse gas emission effects of corn and cellulosic ethanol with technology improvements and land use changes. *Biomass and Bioenergy*, **35**, 1885–1896.

Wang M, Han J, Dunn JB, Cai H, Elgowainy A (2012) Well-to-wheels energy use and greenhouse gas emissions of ethanol from corn, sugarcane and cellulosic biomass for US use. *Environmental Research Letters*, **7**, 045905.

Weijde T, Alvim Kamei CL, Torres AF, Vermerris W, Dolstra O, Visser RGF, Trindade LM (2013) The potential of C4 grasses for cellulosic biofuel production. *Frontiers in Plant Science*, **4**, 1–18.

White WG (2010) *Sugar, Biomass and Biofuel Potential of Temperate by Tropical Maize Crosses*. University of Illinois, Urbana-Champaign, IL.

White WG, Vincent ML, Moose SP, Below FE (2012) The sugar, biomass and biofuel potential of temperate by tropical maize hybrids. *GCB Bioenergy*, **4**, 496–508.

Windhausen VS, Wagener S, Magorokosho C et al. (2012) Strategies to subdivide a target population of environments: results from the CIMMYT-led maize hybrid testing programs in Africa. *Crop Science*, **52**, 2143–2152.

Wolfrum EJ, Lorenz AJ (2009) Correlating detergent fiber analysis and dietary fiber analysis data for corn stover collected by NIRS. *Cellulose*, **16**, 577–585.

Wright M, Brown RC (2007a) Establishing the optimal sizes of different kinds of biorefineries. *Biofuels, Bioproducts and Biorefining*, **1**, 191–200.

Wright MM, Brown RC (2007b) Comparative economics of biorefineries based on the biochemical and thermochemical platforms. *Biofuels, Bioproducts and Biorefining*, **1**, 49–56.

Wu A-M, Rihouey C, Seveno M et al. (2009) The Arabidopsis IRX10 and IRX10-LIKE glycosyltransferases are critical for glucuronoxylan biosynthesis during secondary cell wall formation. *The Plant Journal*, **57**, 718–731.

Wu A-M, Hörnblad E, Voxeur A, Gerber L, Rihouey C, Lerouge P, Marchant A (2010) Analysis of the Arabidopsis IRX9/IRX9-L and IRX14/IRX14-L pairs of glycosyltransferase genes reveals critical contributions to biosynthesis of the hemicellulose glucuronoxylan. *Plant Physiology*, **153**, 542–554.

Wyman CE (2007) What is (and is not) vital to advancing cellulosic ethanol. *TRENDS in Biotechnology*, **25**, 153–157.

Yamaguchi M, Goué N, Igarashi H et al. (2010) VASCULAR-RELATED NAC-DOMAIN6 and VASCULAR-RELATED NAC-DOMAIN7 effectively induce transdifferentiation into xylem vessel elements under control of an induction system. *Plant Physiology*, **153**, 906–914.

Yan J, Shah T, Warburton ML, Buckler ES, Mcmullen MD, Crouch J (2009) Genetic characterization and linkage disequilibrium estimation of a global maize collection using SNP markers. *PLoS ONE*, **4**, e8451.

Yang F, Mitra P, Zhang L et al. (2013) Engineering secondary cell wall deposition in plants. *Plant Biotechnology Journal*, **11**, 325–335.

York WS, O'neill MA (2008) Biochemical control of xylan biosynthesis—which end is up? *Current Opinion in Plant Biology*, **11**, 258–265.

Youngs H, Somerville C (2012) Development of feedstocks for cellulosic biofuels. *F1000 Biology Reports*, **4**, 1–11.

Yuan JS, Tiller KH, Al-Ahmad H, Stewart NR, Stewart CN Jr (2008) Plants to power: bioenergy to fuel the future. *Trends in Plant Science*, **13**, 421–429.

Zegada-Lizarazu W, Parrish D, Berti M, Monti A (2013) Dedicated crops for advanced biofuels: consistent and diverging agronomic points of view between the USA and the EU-27. *Biofuels, Bioproducts and Biorefining*, **7**, 715–731.

Zeng W, Jiang N, Nadella R, Killen TL, Nadella V, Faik A (2010) A glucurono (arabino)xylan synthase complex from wheat contains members of the GT43, GT47, and GT75 families and functions cooperatively. *Plant Physiology*, **154**, 78–97.

Zhang D, Vanfossen A, Pagano R et al. (2011) Consolidated pretreatment and hydrolysis of plant biomass expressing cell wall degrading enzymes. *Bioenergy Research*, **4**, 276–286.

Zhang K, Bhuiya M-W, Pazo JR, Miao Y, Kim H, Ralph J, Liu C-J (2012) An engineered monolignol 4-O-methyltransferase depresses lignin biosynthesis and confers novel metabolic capability in Arabidopsis. *The Plant Cell Online*, **24**, 3135–3152.

Zhong R, Ye Z-H (2007) Regulation of cell wall biosynthesis. *Current Opinion in Plant Biology*, **10**, 564–572.

Zhong R, Ye Z-H (2009) Transcriptional regulation of lignin biosynthesis. *Plant Signaling & Behavior*, **4**, 1028–1034.

Zhong R, Ye Z-H (2012) MYB46 and MYB83 bind to the SMRE sites and directly activate a suite of transcription factors and secondary wall biosynthetic genes. *Plant and Cell Physiology*, **53**, 368–380.

Zhong R, Demura T, Ye Z-H (2006) SND1, a NAC domain transcription factor, is a key regulator of secondary wall synthesis in fibers of Arabidopsis. *The Plant Cell Online*, **18**, 3158–3170.

Zhong R, Richardson EA, Ye Z-H (2007) The MYB46 transcription factor is a direct target of SND1 and regulates secondary wall biosynthesis in Arabidopsis. *The Plant Cell Online*, **19**, 2776–2792.

Zhong R, Lee C, Zhou J, Mccarthy RL, Ye Z-H (2008) A battery of transcription factors involved in the regulation of secondary cell wall biosynthesis in Arabidopsis. *The Plant Cell Online*, **20**, 2763–2782.

Zhong R, Lee C, Ye Z-H (2010) Evolutionary conservation of the transcriptional network regulating secondary cell wall biosynthesis. *Trends in Plant Science*, **15**, 625–632.

Zhong R, Lee C, Mccarthy RL, Reeves CK, Jones EG, Ye Z-H (2011) Transcriptional activation of secondary wall biosynthesis by rice and maize NAC and MYB transcription factors. *Plant and Cell Physiology*, **52**, 1856–1871.

Zhou J, Lee C, Zhong R, Ye Z-H (2009) MYB58 and MYB63 are transcriptional activators of the lignin biosynthetic pathway during secondary cell wall formation in Arabidopsis. *The Plant Cell Online*, **21**, 248–266.

8

Relevance of environmental impact categories for perennial biomass production

MORITZ WAGNER and IRIS LEWANDOWSKI

Department of Biobased Products and Energy Crops, Institute of Crop Science, University of Hohenheim (340 b), 70593 Stuttgart, Germany

Abstract

The decarbonization of the economy will require large quantities of biomass for energy and biomaterials. This biomass should be produced in sufficient quantities and in a sustainable way. Perennial crops in particular are often cited in this context as having low environmental impacts. One example of such crops is miscanthus, a tall perennial rhizomatous C4 grass with high yield potential. There are many studies which have assessed the global warming potential (GWP) of miscanthus cultivation. This is an important impact category which can be used to quantify the environmental benefit of perennial crops. However, the GWP only describes one impact of many. Therefore, the hypothesis of this study was that a holistic assessment also needs to include other impact categories. A life cycle assessment (LCA) with a normalization step was conducted for perennial crops to identify relevant impact categories. This assessed the environmental impact of both miscanthus and willow cultivation and the subsequent combustion for heat production in eighteen categories using a system expansion approach. This approach enables the inclusion of fossil reference system hot spots and thus the evaluation of the net benefits and impacts of perennial crops. The normalized results clearly show the benefits of the substitution of fossil fuels by miscanthus or willow biomass in several impact categories (e.g. for miscanthus: climate change -303.47 kg CO_2 eq./MWh$_{th}$; terrestrial acidification: -0.22 kg SO_2 eq./MWh$_{th}$). Negative impacts however occur, for example, in the impact categories marine ecotoxicity and human toxicity (e.g. for miscanthus: $+1.20$ kg 1.4-DB eq./MWh$_{th}$ and $+68.00$ kg 1.4-DB eq./MWh$_{th,}$ respectively). The results of this study clearly demonstrate the necessity of including more impact categories than the GWP in order to be able to assess the net benefits and impacts of the cultivation and utilization of perennial plants holistically.

Keywords: combustion, environmental performance, global warming potential, life cycle assessment, miscanthus, normalization, perennial crop, willow

Introduction

In 2009, the European Commission set mandatory targets for the production and promotion of energy from renewable resources. The EU renewable energy directive stipulates that, by the year 2020, 20% of total EU energy consumption should come from renewable sources and at least 10% of petrol and diesel consumption for transport should be supplied through biofuels (European Commission, 2009). The European Commission expects the use of renewable energy to increase considerably over the next decades and its proportion of gross final energy consumption to reach values of up to 55% by the year 2050. In the energy roadmap 2050, the European Commission also emphasizes the need for large quantities of biomass for heat, electricity and transport to achieve the goal of the decarbonization of

the economy (European Commission, 2011). There is a wide range of biomass resources available, which can be potentially exploited for bioenergy production. Of these, dedicated energy crops such as miscanthus have emerged as a promising future feedstock for biomass-based energy production. For this reason, miscanthus was chosen as the main representative perennial crop for this study. In addition to miscanthus, willow short rotation coppice was included in this study to examine whether there are any differences between woody perennials and perennial grasses.

Miscanthus is a tall perennial rhizomatous C4 grass, which can yield up to 25 Mg ha^{-1} yr^{-1} (dry matter) in Central Europe after a two-year establishment period and can be harvested annually over a twenty-year cultivation period (Lewandowski *et al.*, 2000; Christian *et al.*, 2008; Felten *et al.*, 2013; Iqbal *et al.*, 2015). It is a low-input crop with a high nitrogen, land-use and energy efficiency (Lewandowski & Schmidt, 2006) and has the potential to remove CO_2 from the atmosphere through

Correspondence: Moritz Wagner
e-mail: mowagner@uni-hohenheim.de

carbon sequestration (Clifton-Brown *et al.*, 2007). In an editorial regarding the environmental benefits of miscanthus, Voigt (2015) recommends *Miscanthus x giganteus* 'as the energy crop of choice'. In the context of sustainability requirements, it is important to assess the performance of each crop in economic, social and ecological terms. This study focuses on evaluating the ecological performance of the utilization of perennial energy crops. One option available for such an evaluation is life cycle assessment (LCA). Life cycle assessment is a method which is standardized by two ISO norms – 14040 and 14044 (ISO, 2006a,b). In the last ten years, several papers have been published which use LCA to assess the potential environmental impacts and benefits of miscanthus. The impact categories examined in these are presented in Table 1.

As shown in Table 1, most of the studies carried out to evaluate the environmental performance of miscanthus focus on one impact category – the global warming potential (GWP). In the EU, political support for bioenergy aims at reducing greenhouse gas (GHG) emissions from fossil fuels. Therefore, any LCA study on bioenergy includes an assessment of the GWP.

Agriculture contributes significantly to GWP. The agricultural sector is responsible for about 10–12% of total anthropogenic emissions of greenhouse gases globally (Smith *et al.*, 2007). However, agriculture also has an influence on other impact categories, such as eutrophication potential (EP) and acidification potential

(AP) (EEA, 2005; Rice & Herman, 2012). In the European Union, the agricultural sector is responsible for 93.3% of ammonia emissions (Eurostat, 2015), which are a main driver of AP. In addition, the use of mineral and organic fertilizers on agricultural land leads to a gross nitrogen surplus of 51 kg nitrogen ha^{-1} yr^{-1} and a gross phosphorus surplus of 2 kg P ha^{-1} yr^{-1} (Eurostat, 2012, 2013). These nutrients can enter groundwater, for example, through nitrate leaching and lead to marine eutrophication. High concentrations of nutrients in water can pose health risks for humans (Di & Cameron, 2002). Nitrate leaching is only one example of the manifold emissions released by the entire agricultural value chain and the subsequent biomass utilization. From this, it can be concluded that, when trying to assess the environmental performance of perennial crops, the estimation of GWP alone is too simplistic. An analysis of the studies on the environmental performance of perennial crops listed in Table 1 confirms this conclusion. Jeswani *et al.* (2015) found that, when considering the GWP of second-generation biofuels, the production of the feedstock for the ethanol plant – the cultivation of the biomass – is the most important hot spot. However, the influence of the feedstock on other impact categories is relatively small. For the impact categories abiotic resource depletion (ADP, elements), AP, EP and freshwater aquatic ecotoxicity potential (FAETP), the main driver is the subsequent conversion of the biomass. Considering GWP alone grossly underestimates the

Table 1 Impact categories used in LCA studies on miscanthus

Authors	Method	GWP	EP	AP	ADP	POCP	ODP	TET	FET	MET	HT
Jeswani *et al.* (2015)	CML	x	x	x	x	x	x	x	x	x	x
Monti *et al.* (2009)	CML	x	x	x	x	–	x	x	x	x	x
Godard *et al.* (2013)	CML; USES-LCA 2.0; CED	x	x	x	x	x	x	x	–	–	–
Styles *et al.* (2015)	CML	x	x	x	x	–	–	–	–	–	–
Nguyen & Hermansen (2015)	EDIP 97; IPCC; Impact 2002+	x	x	x	–	–	–	–	–	–	–
Murphy *et al.* (2013)	CML	x	x	x	–	–	–	–	–	–	–
Brandão *et al.* (2011)	CML	x	x	x	–	–	–	–	–	–	–
Tonini *et al.* (2012)	EDIP 2003	x	x	–	–	–	–	–	–	–	–
Sanscartier *et al.* (2014)	IPCC	x	–	–	–	–	–	–	–	–	–
Styles & Jones (2008)	IPCC	x	–	–	–	–	–	–	–	–	–
Felten *et al.* (2013)	IPCC	x	–	–	–	–	–	–	–	–	–
Dwivedi *et al.* (2015)	n.a.	x	–	–	–	–	–	–	–	–	–
Brandão *et al.* (2010)	IPCC	x	–	–	–	–	–	–	–	–	–
Scown *et al.* (2012)	n.a.	x	–	–	–	–	–	–	–	–	–
Roy *et al.* (2015)	n.a.	x	–	–	–	–	–	–	–	–	–
Wang *et al.* (2012)	n.a.	x	–	–	–	–	–	–	–	–	–
Iqbal *et al.* (2015)	IPCC	x	–	–	–	–	–	–	–	–	–
Parajuli *et al.* (2015)	Stepwise 2006	x	–	–	–	–	–	–	–	–	–
Smeets *et al.* (2009)	IPCC	x	–	–	–	–	–	–	–	–	–

GWP, global warming potential; EP, eutrophication potential; AP, acidification potential; ADP, abiotic depletion potential; POCP, photochemical ozone creation potential; ODP, ozone depletion potential; TET, terrestrial ecotoxicity; FET, freshwater ecotoxicity; MET, marine water ecotoxicity; HT, human toxicity.

influence of the conversion stage on the environmental performance of the second-generation biofuels. Godard *et al.* analysed the environmental performance of heat produced from different feedstocks (flax shives, miscanthus, cereal straw, linseed straw and triticale as whole plant). Using economic allocation, heat produced from miscanthus has the lowest GWP, but scores worse in all the other impact categories in comparison with flax shives as feedstock. If an allocation based on mass is used, heat produced from miscanthus has the best environmental performance in all selected impact categories (Godard *et al.*, 2013). If GWP alone is analysed, it is not only impossible with economic allocation to select the feedstock with the best environmental performance, but it is also impossible to thoroughly analyse the impact of different allocation procedures. The comparison of different perennial crops revealed that marine water ecotoxicity is the most affected impact category after normalization. It is 20–30 times higher than the other categories. Switchgrass, for example, achieved very low values in this important impact category. For this reason, it is a very suitable crop for sites near rivers or coastlines (Monti *et al.*, 2009). In order to select the biomass crop best adapted to specific conditions, it is essential to have a complete picture of the environmental performance of each crop. Various studies on short rotation coppice (poplar and willow), which analysed the environmental performance of the cultivation and utilization in several impact categories, confirm the hypothesis that a number of categories need to be assessed. González-García *et al.* (2012a) showed that, for poplar plantations, in addition to GWP, the impact categories ADP, AP, EP, FE and ME were the most significant after a normalization step. Further results showed that the selection of the most environmentally friendly energy conversion pathway for willow chips largely depends on which impact categories analysed (González-García *et al.*, 2012b). The same applies to the management practice of willow plantations (González-García *et al.*, 2012c).

In this study, an LCA was conducted according to the ISO standards 14040 and 14044 to analyse the environmental performance of miscanthus cultivation and utilization in eighteen different impact categories in comparison with a fossil reference (ISO, 2006a,b). This was done employing the widely used ecoinvent database (version 3.1) and openLCA, an open source LCA software. One objective was to identify those impact categories that need to be included in a holistic assessment of environmental impacts and benefits of the production and utilization of perennial crops, such as miscanthus. To compare the importance of the different impact categories analysed, a normalization step was carried out. According to ISO, normalization is defined as

'calculation of the magnitude of category indicator results relative to reference information' (ISO, 2006a). Normalization factors were taken from the ReCiPe methodology. The result for each impact category is divided by the respective emissions caused by an average European citizen in the year 2000. This results in values without units, which show the calculated emissions as a proportion of the emissions of an average European citizen. Through this additional calculation, it is possible to compare the importance of different impact categories (Goedkoop *et al.*, 2008). A hot spot analysis was also conducted. It reveals which processes are responsible for the largest share of emissions in each impact category.

Through the normalization and the hot spot analysis, it is possible to determine not only the relevant impact categories in the cultivation and utilization of perennial crops, but also which processes or emission sources are most important for each category.

This study aimed to provide guidelines for future research on the environmental performance of perennial crops, with regard to both the choice of relevant impact categories and the focus on data for the most important processes and emission sources.

Material and methods

Scope and boundaries

The scope of this study is a cradle-to-grave analysis of the environmental performance of the cultivation of miscanthus (*Miscanthus x giganteus*) and willow (*Salix viminalis*) short rotation coppice (variety 'Tora') and subsequent combustion in a biomass-fuelled boiler. In order to compare this performance with a fossil reference (heat produced through combustion of light fuel oil), a system expansion approach was applied. This approach enables the inclusion of fossil reference system hot spots. The outcome of this analysis shows the net benefits and impacts through the substitution of fossil fuel by the energetic utilization of miscanthus and willow chips. One megawatt hour of heat (MWh$_{th}$) was chosen as the functional unit. These systems are described in Fig. 1. The system boundaries include the production of the mineral fertilizers and the pesticides used, the production of the propagation material (miscanthus rhizomes and willow cuttings) and the land management (soil preparation, planting, mulching, fertilizing, spraying of pesticides, harvesting, recultivation) over a twenty-year cultivation period. The miscanthus was mulched in the first year and harvested from the second year onwards; the willow plantation was harvested from the fourth year on and then in three-year cycles. Both crops were harvested with a self-propelled forage harvester. The biomass is then transported to a biomass heater where it is combusted to produce heat. The coarse ash is rich in potassium and phosphorus and is used as fertilizer. The fly ash is disposed of in landfill.

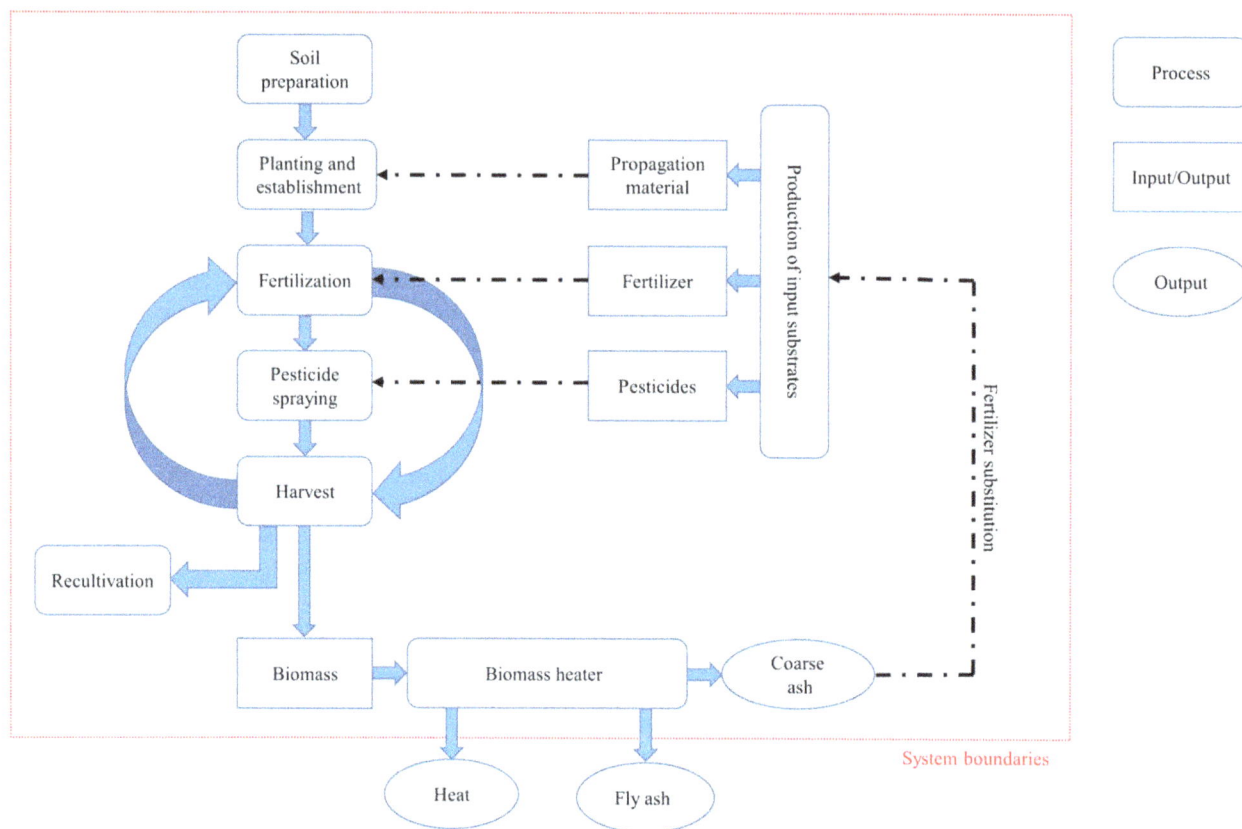

Fig. 1 System description and boundaries for miscanthus and willow biomass production and subsequent utilization in a biomass heater.

Life cycle inventory

The data for the cultivation process used in this LCA study was obtained from a multiannual field trial at Ihinger Hof, a research station of the University of Hohenheim. The Ihinger Hof is located in southwest Germany (48.75°N and 8.92°E). The soil belongs to the soil class Haplic Luvisol. The mean annual temperature for the measurement period was 9.2 °C, and the average annual rainfall was 707.5 mm. The experimental design of the trial is described in Iqbal *et al.* (2015). Data on cultivation practices, fertilizer and pesticide inputs as well as the yields was available for a 10-year period from 2002 to 2012. For both perennial crops, three different fertilizer regimes were applied: N1 with 0 kg of nitrogen, N2 with 40 kg nitrogen and N3 with 80 kg nitrogen per year and hectare in the form of calcium ammonium nitrate. Potassium and phosphate fertilizer levels were the same in all three application regimes. For miscanthus, herbicides only were applied (described in Iqbal *et al.*, 2015). For willow, one insecticide (Karate Zeon, Syngenta, active ingredient 100 g l^{-1} lambda-cyhalothrin) was applied in 2004 at a rate of 0.075 l ha^{-1}. Three herbicides (3 l ha^{-1} Durano, Monsanto, active ingredient 360 g l^{-1} glyphosate; 5 l ha^{-1} U-46 M-Fluid, Nufarm, active ingredient 500 g l^{-1} MCPA; and 2 l ha^{-1} Starane 180, Syngenta, active ingredient 180 g l^{-1} fluroxypyr) were applied in 2006. In the following years, no pesticides were applied. The principle data for the cultivation

of miscanthus and willow used in this analysis is summarized in Table 2. As yield data was only available for the first ten years, it was predicted for the rest of the 20-year cultivation period. For willow, the average of the three measured harvests (years 4, 7 and 10) was taken to estimate the yield for years 11 to 20. For miscanthus, the average of year four to ten was taken for this prediction. The yields of the first three years were excluded in the estimation because the crop was still in its establishment period and has lower yield than after full establishment. However, the yield data inputted into the LCA is the average yield over the whole cultivation period including the establishment phase. Background data for the environmental impacts associated with the production of the input substrates and the cultivation processes (soil preparation, harvesting) was taken from the ecoinvent database version 3.1 (Weidema *et al.*, 2013).

Direct N_2O and NO emissions from mineral fertilizers were estimated according to Bouwman *et al.* (2002). Indirect N_2O emissions from mineral fertilizers and N_2O emissions from harvest residues were calculated according to IPCC (2006). Ammonia emissions were estimated using emission factors from the Joint EMEP/CORINAIR Atmospheric Emission Inventory Guidebook (EMEP/CORINAIR, 2001). Nitrate leaching to groundwater was calculated according to the SQCB – NO_3 model described in Faist Emmenegger *et al.* (2009). Phosphate and phosphorus emissions to surface water

Table 2 Summary of in- and outputs of each perennial crop

Values in kg yr^{-1} ha^{-1}	Miscanthus			Willow		
	N1	N2	N3	N1	N2	N3
N	0	40	80	0	40	80
K$_2$O	128	128	128	64	64	64
P$_2$O$_5$	32	32	32	32	32	32
Pesticides	1.375	1.375	1.375	0.504	0.504	0.504
Dry Matter Yield	16404	19684	20333	16013	17755	20583

and groundwater as well as heavy metal emissions to agricultural soils were calculated according to Nemecek & Kägi (2007). The nitrogen, phosphorus and heavy metal emissions are summarized for the respective crops and fertilizer levels in Table S1.

As no data for the transport of the input substrates (fertilizer, pesticides and propagation material) to the farmer and the biomass to the biomass heater were available, a transport distance of 150 km for the input material and 50 km for the biomass, both by truck, was assumed. The average field-to-farm distance was assumed to be 2 km. The emission stage for the truck used was EUR5. The process data for the transportation of the input material and the biomass was taken from the ecoinvent database (Weidema *et al.*, 2013).

The biomass heater used in this LCA study is a furnace of 300-kW capacity for heat production. The background data for the emissions associated with the combustion of the different biomasses is taken from the ecoinvent database. The data set is based on a Froling Turbomat 320-kW woodchip boiler. The thermal efficiency is assumed to be 75%. As stipulated in the process description of the ecoinvent database, this thermal efficiency is lower than in the technical specification, because it represents the average annual operation, including start and stop phases (Weidema *et al.*, 2013). As there is not enough specific information available regarding the emissions from the combustion of miscanthus, a straw combustion process was used as a worst-case assumption. Where miscanthus-specific emissions factors were available, the straw combustion process was adapted accordingly. This was the case for carbon monoxide, sulphur dioxide, hydrogen chloride, nitrogen oxides and particulates. The emission factors are based on Dahl & Obernberger (2004). A scenario analysis with an improved emission setting was performed to analyse the impact of this assumption.

Miscanthus has a water content of around 15% at the time of harvest, so a further drying process was not necessary. This corresponds to a calorific value of 4.3 kWh kg^{-1} fresh biomass. The wood chips have a water content of 50% at the time of harvest. The chips are then stored on the farm where natural drying is employed. This process results in a water content of around 20%, which corresponds to a calorific value of 3.86 kWh kg^{-1} fresh biomass.

For all willow fertilization levels and the N2 and N3 miscanthus variants, the use of the coarse ash as fertilizer allows the crops to be cultivated without additional input of mineral phosphate or potassium fertilizers. Therefore, the boundaries

applied in this study only include nitrogen fertilizer. For the N1 miscanthus variant, an additional input of 4 kg P$_2$O$_5$ and 14 kg K$_2$O was necessary. Information on ash content, amount of fly and coarse ash and the nutrient as well as the heavy metal content of the coarse ash can be found in Table S2. The fly ash is disposed to landfill.

Choice of impact categories

This LCA study used the life cycle impact assessment method ReCiPe, which consists of eighteen different impact categories (Goedkoop *et al.*, 2008). All mid-point indicators described in the ReCiPe methodology were included. The following impact categories were considered: climate change (CC), which corresponds to global warming potential (GWP); ozone depletion (OD); terrestrial acidification (TA); freshwater eutrophication (FE); marine eutrophication (ME); human toxicity (HT); photochemical oxidant formation (POF); particulate matter formation (PMF); terrestrial ecotoxicity (TET); freshwater ecotoxicity (FET); marine ecotoxicity (MET); ionizing radiation (IR); agricultural land occupation (ALO); urban land occupation (ULO); natural land transformation (NLT); mineral resource depletion (MRD); fossil fuel depletion (FD); and water depletion (WD). Characterization and normalization factors were taken from Goedkoop *et al.* (2008). A normalization factor for the impact category water depletion is not available in the ReCiPe methodology. For this reason, only absolute values are given for this impact category.

Results

Life cycle impact assessment (LCIA)

Table 3 presents the environmental impact in the different impact categories per MWh$_{th}$ of the miscanthus and willow cultivation and subsequent combustion of the biomass. The results are shown for the N2 fertilization level. With these data, it is possible to compare the environmental performance of cultivation and combustion for the two perennial crops in different impact categories. However, due to the different reference units, it is not possible to compare the significance of the different impact categories themselves. For that, a normalization step is necessary.

Table 3 LCIA of the combustion of miscanthus and willow (fertilization level N2) per MWh_{th}

Impact category	Miscanthus	Willow	Reference unit
Fossil fuel depletion	7.3780	7.2425	kg oil eq.
Agricultural land occupation	135.7814	168.6545	m^2*a
Photochemical oxidant formation	0.5002	0.7597	kg NMVOC
Particulate matter formation	0.2008	0.6375	kg PM_{10} eq.
Marine ecotoxicity	1.7175	1.4394	kg 1,4-DB eq.
Natural land transformation	0.0072	0.0074	m^2
Ozone depletion	2.91E-06	0.0829	kg CFC-11 eq.
Terrestrial ecotoxicity	0.1781	0.0030	kg 1,4-DB eq.
Freshwater eutrophication	0.0220	0.0258	kg P eq.
Freshwater ecotoxicity	0.7532	1.5774	kg 1,4-DB eq.
Mineral resource depletion	2.2788	2.8192	kg Fe eq.
Urban land occupation	0.5211	0.5168	m^2*a
Human toxicity	84.3484	6.8041	kg 1,4-DB eq.
Water depletion	97.6863	103.8735	m^3
Marine eutrophication	0.1626	0.2581	kg N eq.
Ionising radiation	4.3592	4.5147	kg U235 eq.
Climate Change	37.4125	40.1806	kg CO_2 eq.
Terrestrial acidification	0.5058	0.5015	kg SO_2 eq.

Comparison of the environmental performance of the cultivation and combustion of miscanthus and willow

Figure 2 presents a comparison between the environmental performance of the cultivation and utilization of miscanthus and willow. For each crop, the LCIA results of the fertilizer level N2 (40 kg nitrogen) are shown. In fifteen of the eighteen impact categories analysed, there are no differences in the rankings of the impact categories between the two perennial crops. The exceptions are human toxicity, terrestrial ecotoxicity and particulate matter formation. In the case of human toxicity potential, the values for willow are significantly lower than for miscanthus. This is in part due to the higher uptake of heavy metals by willow than by miscanthus. These are partially removed from the system through the disposal of the fly ash (which is rich in heavy metals) to landfill. Another reason is the fact that the combustion process of miscanthus produces higher emissions. These are also the main cause of the significantly higher terrestrial ecotoxicity. Differences in the emissions from the combustion of the biomasses are also responsible for the lower particulate matter formation with miscanthus than with willow. Heat produced

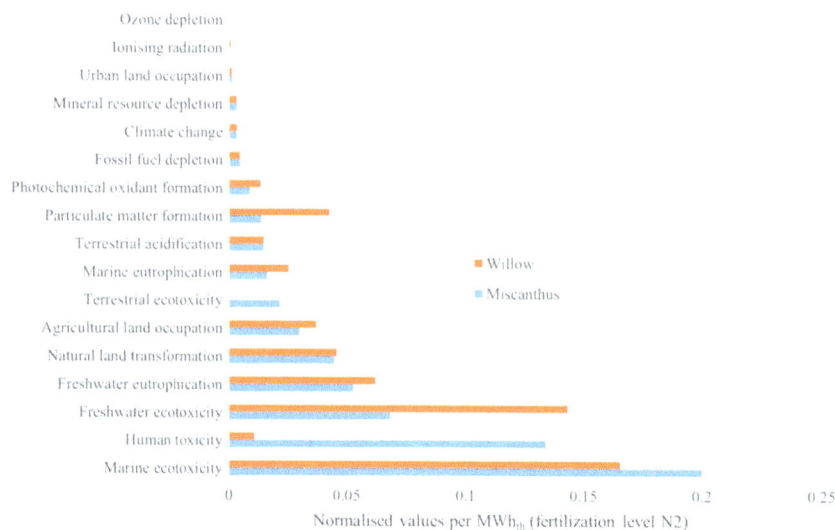

Fig. 2 Assessment of the environmental performance of the cultivation and utilization of miscanthus and willow.

by the combustion of willow chips also has a higher marine eutrophication caused by slightly higher nitrate leaching rates during cultivation. Nitrate is a main source of the marine eutrophication potential. A further difference in the environmental performance of the utilization of the two crops is the freshwater ecotoxicity. Here, the differences in emissions from the combustion process are the main driver.

Life cycle emissions of miscanthus cultivation and combustion using a different combustion scenario

The LCIA results reveal that most of the main differences seen in the comparison of the cultivation and utilization of the two biomasses stem from the combustion process rather than the cultivation phase. The results for the comparison between miscanthus and willow cultivation and combustion using a second combustion scenario for the miscanthus biomass are presented in Fig. 3. A scenario analysis was performed with a boiler with emission characteristics comparable to wood combustion. Under this setting, the differences between willow and miscanthus are much less pronounced.

Life cycle emissions of miscanthus cultivation and combustion using a system expansion approach

In Fig. 4, the results of the LCIA of the miscanthus cultivation and utilization are shown for the three different fertilization regimes using a system expansion approach. The values include the emissions avoided through the substitution of heat produced from a conventional furnace using light fuel oil by heat produced

from a biomass heater. In this approach, negative values represent burdens avoided by the substitution of fossil by renewable fuels, while positive values represent an additional impact due to the use of the biomass heater. The substitution of fossil fuels by miscanthus biomass leads to burdens avoided especially in the impact categories fossil fuel depletion, climate change and terrestrial acidification. However, it causes additional impacts in the categories marine ecotoxicity, human toxicity, agricultural land occupation, freshwater eutrophication and terrestrial ecotoxicity. It should be noted that these results are strongly depending on the fossil reference used. In the case of the substitution of heat produced from hard coal instead of light fuel oil, the use of miscanthus biomass would lead to an avoided burden in the impact category human toxicity instead of an additional impact (data not shown).

In most impact categories analysed, there are only small differences between the three fertilization levels. These are mainly caused by differences in yield and the amount of fertilizer used. Marine eutrophication, for which nitrate leaching is an important driver, increases significantly from N1 to N3 due to the higher nitrate leaching through the additional input of nitrogen fertilizer.

Life cycle emissions of willow cultivation and combustion using a system expansion approach

The substitution of heat produced from the combustion of light heating fuel by heat produced from willow chips leads to burdens avoided especially in the impact categories fossil fuel depletion, climate change and

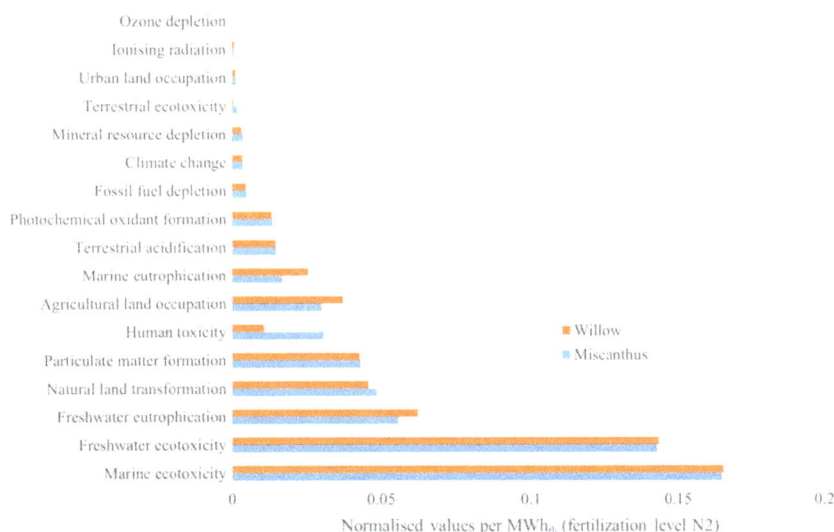

Fig. 3 Scenario analysis of the environmental performance of combustion of miscanthus and willow.

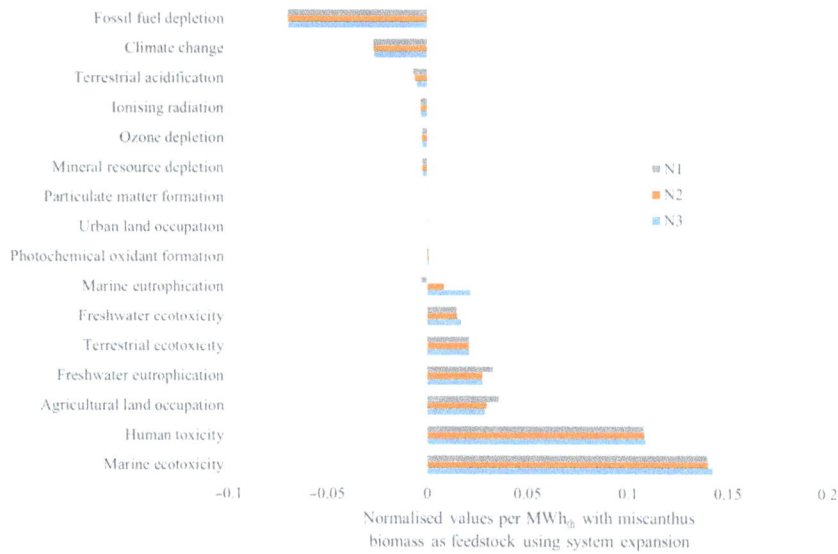

Fig. 4 Assessment of the environmental performance of the cultivation and utilization of miscanthus using a system expansion approach.

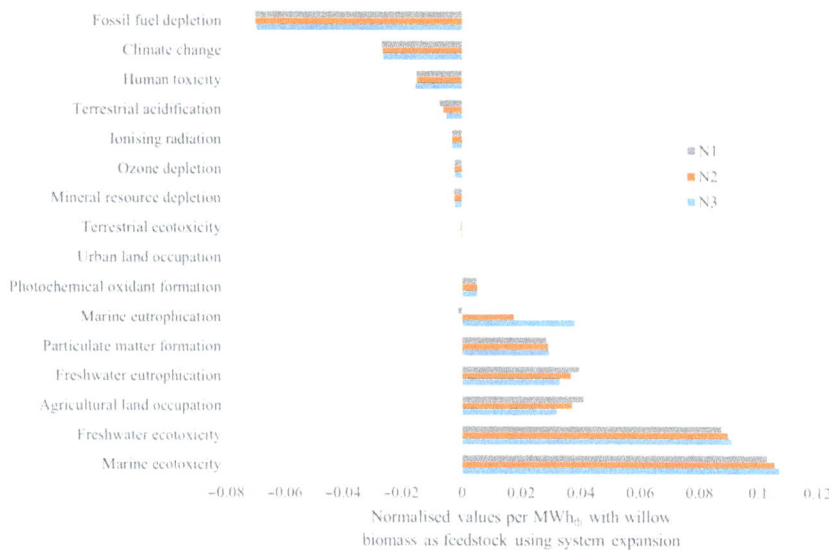

Fig. 5 Assessment of the environmental performance of the cultivation and utilization of willow using a system expansion approach.

human toxicity (see Fig. 5). However, it also causes additional impacts in the categories marine ecotoxicity, freshwater ecotoxicity, agricultural land occupation, freshwater eutrophication and particulate matter formation.

The differences between the results of the three fertilization levels (e.g. in marine eutrophication) can be explained in the same way as for miscanthus. The results for the impact category urban land occupation are mapped in Fig. 5; however, the normalized values are too small to be visible (−9.6 E-05).

Comparison of the environmental performance of the cultivation and combustion of miscanthus and willow using a system expansion approach

Figure 6 shows a comparison of heat produced from miscanthus and willow biomass at fertilization level N2 using a system expansion approach. Because the fossil reference is identical for both crops, the reasons for the differences in environmental performance between heat produced from miscanthus and willow are the same as for the normalized values without system expansion.

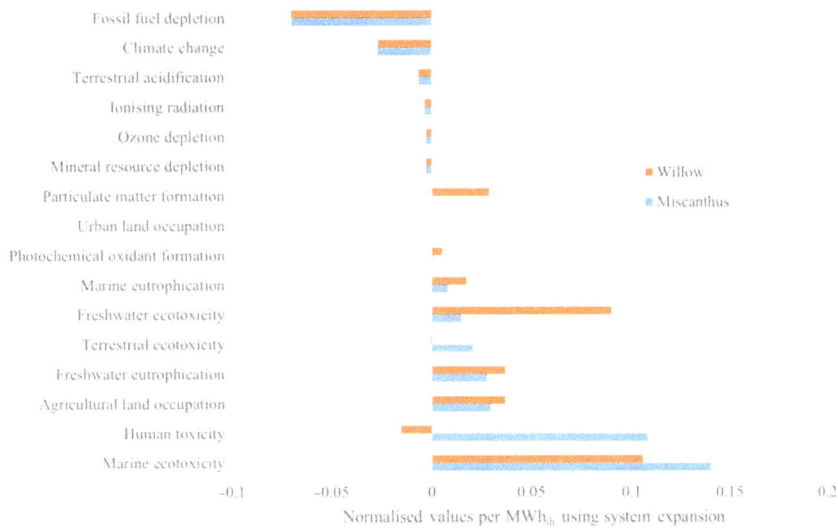

Fig. 6 Comparison of the environmental performance of the cultivation and utilization of miscanthus and willow (fertilization level N2) using a system expansion approach.

While the results for both crops are very similar in impact categories such as climate change, fossil fuel depletion and ionizing radiation, there are substantial differences in other impact categories, especially freshwater ecotoxicity and human toxicity.

Results for natural land transformation (data not shown) using a system expansion approach are lower than −0.7 (in normalized values) and are therefore much lower than the results of all other impact categories.

Hot spot analysis

The hot spot analysis reveals which processes are responsible for the largest share of emissions in each impact category. The grouping 'cultivation' summarizes all cultivation steps up to and including harvest, the production and transport of input substrates (e.g. mineral fertilizers), and the transport of the coarse ash from the biomass heater back to the field. The transport of the input substrates and ash each accounts for <1% of the total emission of the respective impact category. For this reason, they are not represented individually in the hot spot analysis (Figs 7–9). The grouping 'biomass transport' represents the environmental impacts of the transport of the biomass from the field to the biomass heater. The grouping 'combustion' indicates the proportion of total emissions associated with the combustion process. Overall, the results of the hot spot analysis show no large differences between the utilization of miscanthus and willow for most of the impact categories.

In each of the impact categories shown in Fig. 7, the combustion of the biomass has a share of over 90% of total emissions. The negative values seen for the

cultivation of willow can be explained by the higher uptake of heavy metals. The emissions associated with biomass transport have a substantial impact especially on the impact categories natural land transformation, urban land occupation, fossil fuel depletion and mineral resource depletion (see Fig. 8). The combustion process has an impact of over 50% on almost all impact categories shown in Fig. 8, in particular terrestrial acidification and photochemical oxidant formation. The differences in impact of the cultivation stage of miscanthus and willow on the photochemical oxidant formation are much smaller in absolute than in percentage terms.

The cultivation stage has a substantial impact on the categories shown in Fig. 9. In the case marine eutrophication, nitrate leaching is the main cause. Nitrous oxide emissions from the use of mineral nitrogen fertilizer are also an important driver of climate change. The impact of cultivation on freshwater eutrophication is mainly due to phosphor emissions to ground water and surface water. Its impact on ozone depletion stems from mineral fertilizer production and agricultural management. The cultivation stage is also responsible for over 99% of agricultural land occupation.

Discussion

Normalization and system expansion

The normalization of the results is a useful way of assessing the importance of different impact categories. It shows the impact of perennial crop production and utilization in each category and thus helps in the selection of the relevant ones for an assessment of their

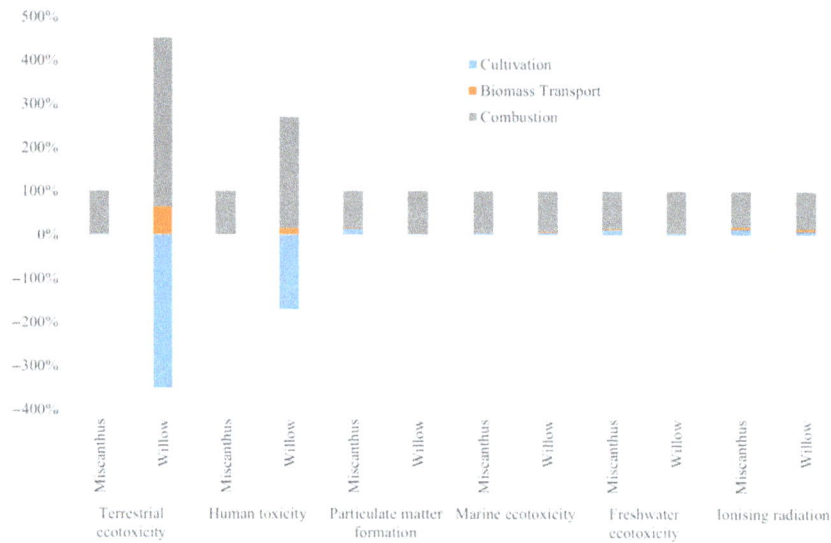

Fig. 7 Hot spot analysis of environmental impacts of cultivation and combustion of two perennial crops. Bars show the contribution of the processes to the overall emissions in each impact category.

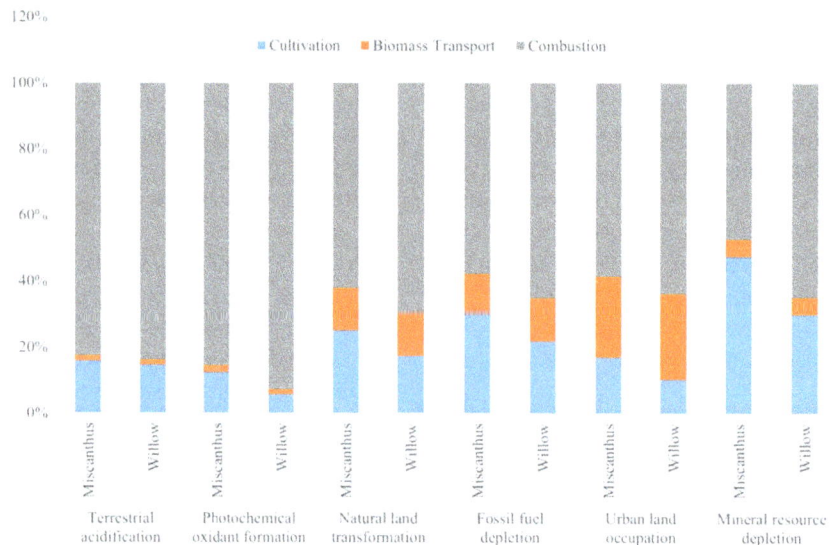

Fig. 8 Hot spot analysis of environmental impacts of cultivation and combustion of two perennial crops. Bars show the contribution of the processes to the overall emissions in each impact category.

environmental performance. However, a system expansion approach is necessary to reveal the net benefits and impacts of biomass utilization. Impact categories with a low ranking before a system expansion approach was applied, for example climate change and terrestrial acidification, show substantial benefits in avoided burdens after system expansion. In impact categories such as marine ecotoxicity and freshwater ecotoxicity, which have relatively high normalized values, the substitution of fossil fuels leads to additional impacts on the environment. If the normalized values alone are analysed, without a comparison to a fossil reference, the benefits of the utilization of perennial crops are substantially undervalued.

Choice of relevant impact categories

The hypothesis of this study was that a holistic assessment of the environmental performance needs to include other impact categories than just global warming potential (climate change). As presented in Fig. 6, the cultivation and utilization of the two analysed crops show no significant differences in the impact category climate change. In order to choose a biomass or utilization pathway on the basis of its environmental performance, it is also necessary to compare other impact categories. As shown in this study, the substitution of fossil fuel by miscanthus or willow chips leads to net

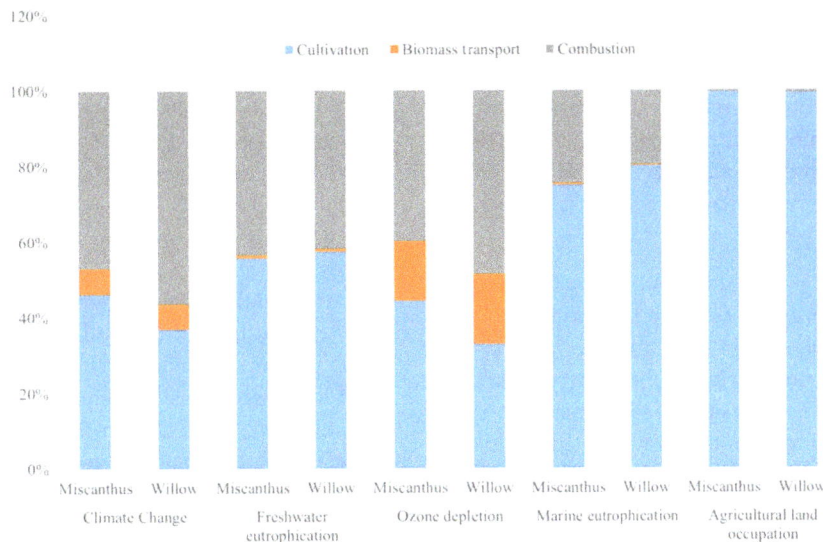

Fig. 9 Hot spot analysis of environmental impacts of cultivation and combustion of two perennial crops. Bars show the contribution of the processes to overall emissions in each impact category.

benefits in the impact category climate change. However, this substitution also leads to additional impacts on the environment in other categories. If climate change alone is assessed, other substantial environmental burdens are ignored. This again emphasizes the need to include more impact categories.

In order to assess the environmental performance of the cultivation and utilization holistically, the impact categories that show substantial benefits and those that show strong negative impacts on the environment need to be included. While there are only small differences between the two crops in the impact categories with net benefits (fossil fuel depletion, climate change, terrestrial acidification), there are substantial differences in the categories with the strongest impact (e.g. human toxicity, freshwater ecotoxicity). This is mainly due to differences in the heavy metal uptake (both in the amount and in the kind of heavy metal) of the crops and differences in emissions associated with the combustion process. This emphasizes the difficulties in preselecting impact categories and the need to analyse several impact categories when assessing the environmental performance of perennial crop-based value chains.

The assessment of the environmental impact of the cultivation and utilization of perennial crops carries a risk of double-counting emissions. For example, particulate matter formation has a strong impact on human toxicity and there is an overlap between mineral resource depletion and fossil fuel depletion. Nevertheless, as shown in Fig. 6, the normalized results for human toxicity and particulate formation can differ substantially. For this reason, both impact categories should be included, despite the double counting. However, the

correlation between them should be clearly stated and integrated in the evaluation of their respective relevance.

The normalized results, however, are not necessarily the sole indicator in the assessment of impact categories' relevance. The normalization does not, for example, include social preferences or specific perspectives of the company commissioning the study. Another important point not included in the normalization is the preload of the specific environment. For example, at a site where the initial acidification is low and the buffer capacity of the soils is high, terrestrial acidification might not be the most urgent issue. The selection of impact categories is thus always dependent on the specific conditions and the questions to be answered by the study. Nevertheless, normalization of the life cycle impact assessment results is a crucial step in the assessment and comparison of the magnitude of different impact categories for biomass production and utilization.

Uncertainties in assessing the environmental performance of the cultivation and utilization perennial crops

1. Yield. One important influence on the environmental performance of perennial crops is the yield. With increasing yields, the environmental impact per tonne biomass is decreasing if the input of fertilizers and pesticides remains the same. There are only few field data for miscanthus yield performance over a ten-year or longer period. Those reports on long-term yields which are available indicate an entire range of developments: from stable yields over long periods;

through year-to-year variations; to yield decreases after an early peak (Gauder *et al.*, 2012; Iqbal *et al.*, 2015). More long-term field trials at different locations are necessary to get a reliable data basis. In the present study, the uncertainty caused by the yield was reduced through the availability of yield data for both crops under very similar conditions from field trials over a 10-year period.

An important aspect in environmental impact assessment of perennial energy crops is the accounting for yield variations over plantation time and the length of the productive period. Low yields in the first years and the fact that woody perennials only are harvested every third year are also an important reason why the whole cultivation period of perennial crops should be considered instead of only one year. The first harvest of miscanthus is in the second year and for willow only feasible from the fourth year onwards. Therefore, the environmental impact of the establishment period is broken down on the subsequent years. If the cultivation period is shorter, the total impact of the establishment on the environmental performance of the harvested biomass is increasing.

Besides that, there are uncertainties regarding the influence of nitrogen fertilizer in the yield development of miscanthus. In the field trial, which provided the underlying data for this study, there were significant differences in the yield of miscanthus between the three nitrogen fertilizer levels (Iqbal *et al.*, 2015). Similar results were found in field trials with miscanthus in the US Midwest, where the yield increased significantly with nitrogen fertilization (Arundale *et al.*, 2014). In contrast, multiannual trials in England showed no significant yield differences under different nitrogen fertilization levels (Christian *et al.*, 2008). If it would be possible to maintain high yields while decreasing the inputs of mineral fertilizers, it would improve the environmental performance significantly. On a location with good soils with a high nitrogen content and a high nitrogen deposition rate, it is reasonable to assume that no nitrogen fertilizer is applied. On the other hand, on poor sites the nitrogen fertilizer use should be included. Therefore, a recommended approach for LCA in perennial energy crops is to calculate with fertilization levels that are equivalent to the withdrawal of nutrients by the biomass.

2. Emission factors and calculation models. The hot spot analysis in this study revealed that the emissions associated with the use of mineral fertilizer – especially nitrogen fertilizers – have a huge impact on the environmental performance of the cultivation stage. The nitrate emissions, for example, are a main driver for the marine eutrophication potential.

However, recent studies show that the nitrate leaching under perennial plants is much lower than under annual crops (Lesur *et al.*, 2014; Pugesgaard *et al.*, 2015). These results suggest that the data for nitrate leaching used in this study – which were calculated with a common agricultural model for nitrate leaching – are probably higher than actually experienced in perennial energy crop production. This emphasizes the need for emission models, which are adapted to the distinctive features of perennial crops.

3. CO_2 sequestration. In the last years, there were several papers published which highlighted the potential of miscanthus to sequester carbon in the soil (Kahle *et al.*, 2001; Clifton-Brown *et al.*, 2007; Brandão *et al.*, 2011; Felten & Emmerling, 2012). However, there are still huge uncertainties regarding the amount of CO_2, which will be sequestered, and the time frame of the sequestration (Harris *et al.*, 2015). Due to these uncertainties, the sequestration of carbon in the soil through the cultivation of perennial plants was not included in this study. Other LCA studies, which included the sequestration, showed that the cultivation of miscanthus could, under certain conditions, act as a real carbon sink (Brandão *et al.*, 2011; Godard *et al.*, 2013). Even if the carbon is only sequestered for the cultivation process and released again after the recultivation of the site, there still is a positive environmental impact in perennial crop production, which should be accounted for through the GWP based on a 20-year horizon (20-year GWP).

4. Missing impact categories. Miscanthus cultivation has a positive impact on the biodiversity with more weed vegetation and open-ground bird species (Semere & Slater, 2007a) and on the abundance of invertebrate populations (Semere & Slater, 2007b). The prolonged fallow period of perennial crops improves the soil quality, and the soil cover over winter reduces the erosion. However, it is not possible yet to include these positive effects of perennial crops in a LCA study. Therefore, approaches should be further developed for including these equally important environmental impacts into a holistic impact assessment. There are already some approaches to include these impacts. Oberholzer *et al.* (2012), for example, developed a method to include the impact of agricultural practice on soil quality in LCA. However, to date, this has only occasionally been used in LCA studies due to its complexity and huge data requirements. There are also approaches to integrate biodiversity aspects in LCA. Finnan *et al.* (2012) included a biodiversity indicator for miscanthus in their assessment of the environmental impacts of bioenergy plans. However, there

are still several shortcomings in the biodiversity indicators presently available for LCA (Souza *et al.*, 2015). Therefore, further research is necessary to allow a realistic assessment of the impact of agriculture or land use in general on biodiversity.

5. Indirect land-use change. While there are many positive effects of perennial biomass crops on the environment, the expansion of their cultivation area still bears a risk. If their cultivation is not restricted to marginal or unused land, an increase in their production area can lead to food production displacement. These food crops then need to be produced elsewhere. This indirect land-use change can lead to substantial negative impacts on the environment.

6. Utilization. As shown in the hot spot analysis, the emissions associated with the combustion process have a substantial impact on the different categories. While the data basis for the combustion of willow chips is adequate, there is insufficient information available on emissions from the combustion of miscanthus. In this study, a straw combustion process was taken as a worst-case assumption. In practice, the combustion of miscanthus would produce less emissions than shown here and have a lower impact on the environment. A reason for that is the higher chloride and sulphur content of straw in comparison with miscanthus biomass. These elements lead to harmful emissions in the combustion process (Spliethoff & Hein, 1998; Iqbal & Lewandowski, 2014)

A holistic environmental impact assessment of perennial biomass crops requires additional impact categories other than climate change (GWP). The GWP is a basic impact category due to the importance of climate change and the fact that the reduction in GHG emissions is one of the positive environmental contributions of perennial crop production. Based on the results of this study, an assessment of the environmental performance of the cultivation and combustion of perennial crops should also include the impact categories fossil fuel depletion, terrestrial acidification, freshwater ecotoxicity and human toxicity. However, the results of the study also show that the relevance of impact categories can differ depending on the crop and the utilization pathway. In order to resolve this issue, the choice of the relevant impact categories should be an iterative process. The first step is to analyse the relevance of the different impacts categories for the respective study goals and boundaries using initial data, a normalization step and a system expansion approach. After the determination of the relevant impact categories, the quality of the data important for these categories can be improved and the goal and scope adapted if necessary.

The choice of impact categories, emission models and data basis should also consider the special features of perennial crops in order to be able to assess the environmental benefits of their production. These include, for example, reduction in nitrate leaching, soil carbon sequestration and maintenance of biodiversity. Therefore, it is recommended that impacts on soil quality and biodiversity are included in future environmental impact studies and methodologies for integrating them into LCA are developed. In addition, nitrogen emission models currently available need to be adapted to actual data of perennial crop performance.

Acknowledgements

This work was supported by a grant from the Ministry of Science, Research and the Arts of Baden-Württemberg within the framework of the Bioeconomy Research Program Baden-Württemberg. The authors are grateful to the staff of the research station Ihinger Hof, especially Thomas Truckses, for managing the field trial and providing the basic data for this study. The authors would also like to thank to Nicole Gaudet and Yasir Iqbal for editing the manuscript.

References

Arundale RA, Dohleman FG, Voigt TB, Long SP (2014) Nitrogen Fertilization Does Significantly Increase Yields of Stands of *Miscanthus x giganteus* and *Panicum virgatum* in Multiyear Trials in Illinois. *BioEnergy Research*, **7**, 408–416.

Bouwman AF, Boumans LJM, Batjes NH (2002) Modeling global annual N_2O and NO emissions from fertilized fields. *Global Biogeochemical Cycles*, **16**, 1–9.

Brandão M, Clift R, Milà i Canals L, Basson L (2010) A Life-Cycle Approach to Characterising Environmental and Economic Impacts of Multifunctional Land-Use Systems: An Integrated Assessment in the UK. *Sustainability*, **2**, 3747–3776.

Brandão M, Milà i Canals L, Clift R (2011) Soil organic carbon changes in the cultivation of energy crops: Implications for GHG balances and soil quality for use in LCA. *Biomass and Bioenergy*, **35**, 2323–2336.

Christian DG, Riche AB, Yates NE (2008) Growth, yield and mineral content of *Miscanthus x giganteus* grown as a biofuel for 14 successive harvests. *Industrial Crops and Products*, **28**, 320–327.

Clifton-Brown JC, Breuer J, Jones MB (2007) Carbon mitigation by the energy crop, Miscanthus. *Global Change Biology*, **13**, 2296–2307.

Dahl J, Obernberger I (2004) Evaluation of the combustion characteristics of four perennial energy crops (*Arundo donax, Cynara cardunculus, Miscanthus x giganteus* and *Panicum virgatum*). 2nd World Conference on Biomass for Energy, Industry and Climate Protection, Rome, Italy.

Di HJ, Cameron KC (2002) Nitrate leaching in temperate agroecosystems: sources, factors and mitigating strategies. *Nutrient Cycling in Agroecosystems*, **46**, 237–256.

Dwivedi P, Wang W, Hudiburg T *et al.* (2015) Cost of abating greenhouse gas emissions with cellulosic ethanol. *Environmental Science and Technology*, **49**, 2512–2522.

EEA (2005) Source apportionment of nitrogen and phosphorus inputs into the aquatic environment. In: EEA Report No. 7. pp. 48. European Environment Agency, Copenhagen.

EMEP/CORINAIR (2001) *Joint EMEP/CORINAIR Atmospheric Emission Inventory Guidebook*, 3rd edn. European Environment Agency, Copenhagen.

European Commission (2009) Directive 2009/28/EC of the European parliament and of the council of 23 April 2009 on the promotion of the use of energy from renewable sources. Official Journal of the European Union. Brussels.

European Commission (2011) *Energy Roadmap 2050*. Communication from the Commission to the European Parliament, the Council, the European Economic and Social Committee and the Committee of the Regions, Brussels.

Eurostat (2012) *Agri-Environmental Indicators – Gross Nitrogen Balance*. Eurostat, Luxembourg City.

Eurostat (2013) *Agri-Environmental Indicators – Risk of Pollution by Phosphorus*. Eurostat, Luxembourg City.

Eurostat (2015) *Agriculture – Ammonia Emission Statistics*. Eurostat, Luxembourg City.

Faist Emmenegger M, Reinhard J, Zah R (2009) Sustainability Quick Check for Biofuels – intermediate background report. Agroscope Reckenholz-Tänikon Research Station ART, Dübendorf, Switzerland.

Felten D, Emmerling C (2012) Accumulation of *Miscanthus*-derived carbon in soils in relation to soil depth and duration of land use under commercial farming conditions. *Journal of Plant Nutrition and Soil Science*, **175**, 661–670.

Felten D, Fröba N, Fries J, Emmerling C (2013) Energy balances and greenhouse gas-mitigation potentials of bioenergy cropping systems (Miscanthus, rapeseed, and maize) based on farming conditions in Western Germany. *Renewable Energy*, **55**, 160–174.

Finnan J, Styles D, Fitzgerald J, Connolly J, Donnelly A (2012) Using a strategic environmental assessment framework to quantify the environmental impact of bioenergy plans. *GCB Bioenergy*, **4**, 311–329.

Gauder M, Graeff-Hönninger S, Lewandowski I, Claupein W (2012) Long-term yield and performance of 15 different *Miscanthus* genotypes in southwest Germany. *Annals of Applied Biology*, **160**, 126–136.

Godard C, Boissy J, Gabrielle B (2013) Life-cycle assessment of local feedstock supply scenarios to compare candidate biomass sources. *GCB Bioenergy*, **5**, 16–29.

Goedkoop M, Heijungs R, Huijbregts M, De SA, Struijs J, Van ZR (2008) ReCiPe 2008. A life cycle impact assessment method which comprises harmonised category indicators at the midpoint and the endpoint level; First edition Report I. Characterisation. VROM, Den Haag, The Netherlands.

González-García S, Bacenetti J, Murphy R, Fiala M (2012a) Present and future environmental impact of poplar cultivation in Po valley (Italy) under different crop management systems. *Journal of Cleaner Production*, **26**, 56–66.

González-García S, Iribarren D, Susmozas A, Dufour J, Murphy RJ (2012b) Life cycle assessment of two alternative bioenergy systems involving Salix spp. biomass: Bioethanol production and power generation. *Applied Energy*, **95**, 111–122.

González-García S, Mola-Yudego B, Dimitrou I, Aronsson P, Murphy R (2012c) Environmental assessment of energy production based on long term commercial willow plantations in Sweden. *Science of the Total Environment*, **421–422**, 201–219.

Harris ZM, Spake R, Taylor G (2015) Land use change to bioenergy: A meta-analysis of soil carbon and GHG emissions. *Biomass and Bioenergy*, **82**, 27–39.

IPCC (2006) Guidelines for National Greenhouse Gas Inventories. In: *Prepared by the National Greenhouse Gas Inventories Programme* (eds Eggleston HS, Buendia L, Miwa K, Ngara T, Tanabe K), IGES, Hayama, Japan.

Iqbal Y, Lewandowski I (2014) Inter-annual variation in biomass combustion quality traits over five years in fifteen Miscanthus genotypes in south Germany. *Fuel Processing Technology*, **121**, 47–55.

Iqbal Y, Gauder M, Claupein W, Graeff-Hönninger S, Lewandowski I (2015) Yield and quality development comparison between miscanthus and switchgrass over a period of 10 years. *Energy*, **89**, 268–276.

ISO (2006a) *ISO 14040: Environmental Management — Life Cycle Assessment — Principles and Framework*, 2nd edn. ISO, Geneva.

ISO (2006b) *ISO 14044: Environmental Management — Life Cycle Assessment — Requirements and Guidelines*. ISO, Geneva.

Jeswani HK, Falano T, Azapagic A (2015) Life cycle environmental sustainability of lignocellulosic ethanol produced in integrated thermo-chemical biorefineries. *Biofuels, Bioproducts and Biorefining*, **9**, 661–676.

Kahle P, Beuch S, Boelcke B, Leinweber P, Schulten HR (2001) Cropping of Miscanthus in Central Europe: biomass production and influence on nutrients and soil organic matter. *European Journal of Agronomy*, **15**, 171–184.

Lesur C, Bazot M, Bio-Berif F, Mary B, Jeuffroy MH, Loyce C (2014) Assessing nitrate leaching during the three-first years of *Miscanthus x giganteus* from on-farm measurements and modeling. *GCB Bioenergy*, **6**, 439–449.

Lewandowski I, Schmidt U (2006) Nitrogen, energy and land use efficiencies of miscanthus, reed canary grass and triticale as determined by the boundary line approach. *Agriculture, Ecosystems and Environment*, **112**, 335–346.

Lewandowski I, Clifton-Brown JC, Scurlock JMO, Huismand W (2000) Miscanthus: European experience with a novel energy crop. *Biomass and Bioenergy*, **19**, 209–227.

Monti A, Fazio S, Venturi G (2009) Cradle-to-farm gate life cycle assessment in perennial energy crops. *European Journal of Agronomy*, **31**, 77–84.

Murphy F, Devlin G, McDonnell K (2013) Miscanthus production and processing in Ireland: An analysis of energy requirements and environmental impacts. *Renewable and Sustainable Energy Reviews*, **23**, 412–420.

Nemecek T, Kägi T (2007) Life cycle inventories of agricultural production systems - Data v2.0. In: EcoInvent Report No.15. pp. 360. Agroscope Reckenholz-Tänikon Research Station ART, Zürich and Dübendorf, Amersfoort.

Nguyen TLT, Hermansen JE (2015) Life cycle environmental performance of miscanthus gasification versus other technologies for electricity production. *Sustainable Energy Technologies and Assessments*, **9**, 81–94.

Oberholzer HR, Freiermuth Knuchel R, Weisskopf P, Gaillard G (2012) A novel method for soil quality in life cycle assessment using several soil indicators. *Agronomy for Sustainable Development*, **32**, 639–649.

Parajuli R, Sperling K, Dalgaard T (2015) Environmental performance of Miscanthus as a fuel alternative for district heat production. *Biomass and Bioenergy*, **72**, 104–116.

Pugesgaard S, Schelde K, Larsen SU, Lærke PE, Jørgensen U (2015) Comparing annual and perennial crops for bioenergy production – influence on nitrate leaching and energy balance. *GCB Bioenergy*, **7**, 1136–1149.

Rice KC, Herman JS (2012) Acidification of Earth: An assessment across mechanisms and scales. *Applied Geochemistry*, **27**, 1–14.

Roy P, Dutta A, Deen B (2015) Greenhouse gas emissions and production cost of ethanol produced from biosyngas fermentation process. *Bioresource Technology*, **192**, 185–191.

Sanscartier D, Deen B, Dias G, Maclean HL, Dadfar H, Mcdonald I, Kludze H (2014) Implications of land class and environmental factors on life cycle GHG emissions of Miscanthus as a bioenergy feedstock. *GCB Bioenergy*, **6**, 401–413.

Scown CD, Nazaroff WW, Mishra U et al. (2012) Corrigendum: Lifecycle greenhouse gas implications of US national scenarios for cellulosic ethanol production. *Environmental Research Letters*, **7**, 1–9.

Semere T, Slater FM (2007a) Ground flora, small mammal and bird species diversity in miscanthus (*Miscanthus x giganteus*) and reed canary-grass (*Phalaris arundinacea*) fields. *Biomass and Bioenergy*, **31**, 20–29.

Semere T, Slater FM (2007b) Invertebrate populations in miscanthus (*Miscanthus x giganteus*) and reed canary-grass (*Phalaris arundinacea*) fields. *Biomass and Bioenergy*, **31**, 30–39.

Smeets EMW, Lewandowski I, Faaij APC (2009) The economical and environmental performance of miscanthus and switchgrass production and supply chains in a European setting. *Renewable and Sustainable Energy Reviews*, **13**, 1230–1245.

Smith P, Martino D, Cai Z et al. (2007) Agriculture. In: *Climate Change 2007: Mitigation. Contribution of Working Group III to the Fourth Assessment Report of the Intergovernmental Panel on Climate Change* (eds Metz B, Davidson OR, Bosch PR, Dave R, Meyer LA), pp. 497–540. Cambridge University Press, Cambridge, UK and New York, NY, USA.

Souza DM, Teixeira RFM, Ostermann OP (2015) Assessing biodiversity loss due to Land use with Life Cycle Assessment: are we there yet? *Global Change Biology*, **21**, 32–47.

Spliethoff H, Hein KRG (1998) Effect of co-combustion of biomass on emissions in pulverized fuel furnaces. *Fuel Processing Technology*, **54**, 189–205.

Styles D, Jones MB (2008) Miscanthus and willow heat production - An effective land-use strategy for greenhouse gas emission avoidance in Ireland? *Energy Policy*, **36**, 97–107.

Styles D, Gibbons J, Williams AP et al. (2015) Consequential life cycle assessment of biogas, biofuel and biomass energy options within an arable crop rotation. *GCB Bioenergy*, **7**, 1305–1320.

Tonini D, Hamelin L, Wenzel H, Astrup T (2012) Bioenergy Production from Perennial Energy Crops: A Consequential LCA of 12 Bioenergy Scenarios including Land Use Changes. *Environmental Science and Technology*, **46**, 13521–13530.

Voigt TB (2015) Are the environmental benefits of *Miscanthus x giganteus* suggested by early studies of this crop supported by the broader and longer-term contemporary studies? *GCB Bioenergy*, **7**, 567–569.

Wang S, Wang S, Hastings A, Pogson M, Smith P (2012) Economic and greenhouse gas costs of Miscanthus supply chains in the United Kingdom. *GCB Bioenergy*, **4**, 358–363.

Weidema BP, Bauer C, Hischier R et al. (2013) The ecoinvent database: Overview and methodology. Data quality guideline for the ecoinvent database version 3.

9

Soil fungal and bacterial responses to conversion of open land to short-rotation woody biomass crops

CHAO XUE[1,2], CHRISTOPHER RYAN PENTON[2,3], BANGZHOU ZHANG[2,4], MENGXIN ZHAO[2,5], DAVID E. ROTHSTEIN[6], DAVID J. MLADENOFF[7], JODI A. FORRESTER[7], QIRONG SHEN[1] and JAMES M. TIEDJE[2]

[1]Jiangsu Collaborative Innovation Center for Solid Organic Waste Utilization and National Engineering Research Center for Organic-based Fertilizers, Department of Plant Nutrition, Nanjing Agricultural University, Nanjing, Jiangsu 210095, China, [2]Center of Microbial Ecology, Michigan State University, East Lansing, MI 48824, USA, [3]School of Letters and Sciences, Faculty of Science and Mathematics, Arizona State University, Mesa, AZ 85212, USA, [4]State Key Laboratory of Marine Environmental Science and Key Laboratory of the Ministry of Education for Coast and Wetland Ecosystems, School of Life Sciences, Xiamen University, Xiamen, Fujian 360102, China, [5]State Key Joint Laboratory of Environment Simulation and Pollution Control, School of Environment, Tsinghua University, Beijing 100084, China, [6]Department of Forestry, Michigan State University, East Lansing, MI 48824, USA, [7]Department of Forest and Wildlife Ecology, University of Wisconsin, Madison, WI 53706, USA

Abstract

Short-rotation woody biomass crops (SRWCs) have been proposed as an alternative feedstock for biofuel production in the northeastern US that leads to the conversion of current open land to woody plantations, potentially altering the soil microbial community structures and hence functions. We used pyrosequencing of 16S and 28S rRNA genes in soil to assess bacterial and fungal populations when 'marginal' grasslands were converted into willow (Salix spp.) and hybrid poplar (Populus spp.) plantations at two sites with similar soils and climate history in northern Michigan (Escanaba; ES) and Wisconsin (Rhinelander; RH). In only three growing seasons, the conversion significantly altered both the bacterial and fungal communities, which were most influenced by site and then vegetation. The fungal community showed greater change than the bacterial community in response to land conversion at both sites with substantial enrichment of putative pathogenic, ectomycorrhizal, and endophytic fungi associated with poplar and willow. Conversely, the bacterial community structures shifted, but to a lesser degree, with the new communities dissimilar at the two sites and most correlated with soil nutrient status. The bacterial phylum Nitrospirae increased after conversion and was negatively correlated to total soil nitrogen, but positively correlated to soil nitrate, and may be responsible for nitrate accumulation and the increased N_2O emissions previously reported following conversion at these sites. The legacy effect of a much longer grassland history and a second dry summer at the ES site may have influenced the grassland (control) microbial community to remain stable while it varied at the RH site.

Keywords: grassland, poplar, short-rotation woody biomass crop, soil bacterial community, soil fungal community, willow

Introduction

Biofuel production, including plant-derived bioethanol, biodiesel, and biogas, has the potential to significantly reduce global dependence on fossil fuels and provide a sustainable strategy to partially meet the growing energy demands (Demirbas, 2008; Ghatak, 2011). In 2005, 14.3% of the US corn harvest produced 1.48×10^{10} l of ethanol, and soybean oil extracted from 1.5% of the US soybean

Correspondence: Qirong Shen
e-mail: shenqirong@njau.edu.cn and James M. Tiedje
e-mail: tiedjej@msu.edu

harvest produced 2.56×10^8 l of biodiesel (Hill et al., 2006). The US federal government adopted a goal to annually produce 36 billion gallons of alternative fuels by 2022 (EISA, 2007). To meet just half of this target, the production of corn-based bioethanol would require 65.8% of the 2005 corn harvest. The most likely outcome of increased production of corn-based ethanol would be higher food prices and a potential for food-related security crises. Short-rotation woody biomass crops (SRWCs) such as hybrid poplar (Populus spp.) and willow (Salix spp.) have been proposed as alternative options for biofuel feedstocks due to their fast growth and short harvest cycles (Labrecque & Teodorescu, 2005; Aylott et al., 2008).

The availability of limited arable area for food crop production is a major constraint for many countries, hence the proposal to use marginal lands for biofuel crops (Hill *et al.*, 2006; Gopalakrishnan *et al.*, 2009). Studies have indicated that marginal agricultural land can contribute to biomass production without impacting food production on prime cropland (Tang *et al.*, 2010; Cai *et al.*, 2011; Zhuang *et al.*, 2011). In the northern Great Lake States (MI, WI and MN), widespread planting of hybrid poplar and willow on marginal open lands has been viewed as the best option for sustaining a regional bioenergy production system (Davis *et al.*, 2012; Zalesny *et al.*, 2012), and there is a long history of hybrid poplar breeding efforts in the region dating to the first energy crisis of the 1970s (Hansen, 1991). Open lands targeted for conversion consist of a mixture of marginal agricultural land, abandoned farms, hayfields, and pasturelands and make up a significant portion of the landscape matrix of this region (Mladenoff *et al.*, 2015).

Soil microbial communities, particularly bacteria and fungi, have the potential to influence biofuel crop establishment on marginal lands through plant–microbe interactions that include nutrient acquisition, growth promotion, the alleviation of environmental stress, and disease bio-control (Glick, 1995; Lugtenberg & Kamilova, 2009; Compant *et al.*, 2010). Plants also affect the soil microbial community (Loon, 2007; Bever *et al.*, 2012) by shaping the diversity and composition through niche establishment, secretion of root exudates, symbiosis, and litter chemistry and directly altering soil structure, nutrient availability, and cycling (Yang & Crowley, 2000; Garbeva *et al.*, 2004; Schweitzer *et al.*, 2008; Thoms *et al.*, 2010). Furthermore, microbial community structure is also largely driven by land use, vegetation, soil pH, and drainage (aeration) (Lauber *et al.*, 2009; Rousk *et al.*, 2010; Nacke *et al.*, 2011; Hanson *et al.*, 2012).

The establishment of SRWC plantations on marginal lands often entails conversion of existing grasslands to tree plantations. Previous studies at Escanaba (ES), Rhinelander, and other sites document that this conversion can have substantial impacts on soil nitrogen cycling and greenhouse gas emissions (Nikièma *et al.*, 2012; Palmer *et al.*, 2013). However, there is a knowledge gap concerning how such conversions, including new crop and tillage, affect soil microbial communities on marginal lands in the Great Lakes region, and the role that microbial community shifts may play in driving the observed increases in greenhouse gas emissions and other system functions.

To assess the impact of grassland to SRWC conversion, we examined the effects of converting hayfields to willow and hybrid poplar plantations on soil bacterial and fungal communities after three growing seasons at two sites in northern Michigan and Wisconsin through pyrosequencing of 16S and 28S rRNA genes in soil DNA. Our objectives were to determine the following: (i) the degree to which microbial communities shifted in response to land conversion, (ii) the degree to which community shifts were consistent between the two sites, and (iii) which taxa changed in this shift and the attributes of those taxa as they may infer functional changes in the plant–soil system.

Materials and methods

Field experiment design

The two field sites are located at Escanaba (ES), Michigan, USA (45°46′20″N, 87°11′43″W) and Rhinelander (RH), Wisconsin, USA (45°40′13″N, 89°12′45″W). The soil at the RH site is a loamy sand developed on well-sorted glacial outwash parent material, whereas the soil at ES is a fine sandy loam developed from limestone-derived glacial till. The RH site had been used for production of corn and other row crops until 2005, when it was converted to a cool-season grass hayfield. In contrast, the ES site was last cultivated in 1968, converted to cattle pasture in the 1970s, and had been used for hay production for at least the last 20 years. The 30-year mean annual precipitation at the ES site is 728 mm, while it is 675 mm at the RH site (Table S1). The experimental design is described in detail in Palmer *et al.* (2014). Briefly, identical experimental designs were initiated at both sites with 12 replicate plots randomly assigned to three treatments: (i) planted to willow, (ii) planted to hybrid poplar, or (iii) control plots, which were maintained in the prior land use (unplowed hayfield). Each plot was 40 × 40 m, with all samples collected from the inner 20 × 20 m to avoid edge effects and provide a buffer between plots.

In May 2010, all the existing vegetation was killed with glyphosate in the plots to be planted to poplar and willow, and then, these plots were cultivated with a moldboard plow and disked (Palmer *et al.*, 2014). Hybrid poplar cultivar NM6 (*Populus maximowiczii × Populus nigra*) was planted at a standard density of 1900 stems ha⁻¹. Willow cultivar Fish Creek (*Salix purpurea*) was planted at a standard density of 14 000 stems ha⁻¹. No fertilizer was applied to any plot during this study. Control plots were left untouched as uncultivated mixed species grass hayfields, which represents the baseline, preconversion condition at each site.

Soil sampling and DNA extraction

In 2010, soil samples were collected immediately before conversion on May 11 at the ES site and on May 13 at the RH site. After three growing seasons, samples were collected on August 29, 2012, at the ES site and on September 17, 2012, at the RH site. Five bulk soil cores were randomly collected from each plot to a 10-cm depth (surface soil ~1 cm was removed before sampling) using a soil probe. Soil from the cores was placed in plastic bags and kept on ice after collection and during transportation, and was stored at −20 °C in the laboratory until further processing. Five grams of soil from each core sample within the same plot was sieved (2-mm-mesh sieve), combined, and mixed thoroughly to make one sample to represent each plot. DNA was extracted from 0.5 g of the mixed soil sample using the Powersoil DNA Extraction kit (MOBIO Laboratories, Carlsbad, CA, USA) following the manufacturer's instructions.

Amplification and 454 pyrosequencing

Primer sets 454–577f adapter-mid-AYTGGGYDTAAAGNG and 454–926r adapter-mid-CCGTCAATTCMTTTRAGT were used to amplify partial 16S rRNA genes, and primers LR3 mid-CCGTGTTTCAAGACGGG and LR0R mid-ACCCGCTGAACT-TAAGC were used to amplify the 28S rRNA genes for all samples according to previously published protocols (Sul *et al.*, 2011; Penton *et al.*, 2013, respectively), where mid refers to a unique mid sequence used for sample sorting. The 28S rRNA amplicons were adapter ligated and bi-directionally sequenced on the 454 Life Sciences Titanium platform, using Lib-L kits (454 Life Sciences Corporation, Branford, CT, USA). All 16S rRNA samples were unidirectionally sequenced using Lib-A kits (454 Life Sciences Corporation). Sequencing was performed at Center for Integrated BioSystems, Utah State University, USA.

Measurement of soil characteristics

Soil characteristics, including pH, total nitrogen, ammonium, nitrate, total carbon, and C/N ratios, were determined on separate 10-cm core samples collected within 2 weeks of the initial and final sampling dates for soil microbial communities. Inorganic soil nitrate and ammonium concentrations were determined by extracting a 10-g subsample with 2 M potassium chloride for 1 h. Soil extracts were analyzed for ammonium and nitrate colorimetrically according to procedures described by Sinsabaugh *et al.* (2000) and Doane & Horwáth (2003), respectively. Total carbon, total nitrogen, and C/N ratios were determined on oven-dried soil via dry combustion on an elemental analyzer (Costech ECS 4010, Valencia, CA, USA). Soil pH was determined with a glass electrode in a 1 : 2 slurry of air-dried soil in 0.01 M CaCl$_2$.

Plant biomass

Leaf, stem, and fine-root biomass were determined at ES and RH sites for 2011 and 2012. Two 3.8 cm diameter × 10 cm deep soil cores were collected from each plot at 2–3 week intervals throughout the growing season. Fine roots (2 mm diameter) from these cores were hand sorted. Because there was significant within-plot variability that could potentially skew results with only two cores per plot per sample date, data were averaged across sampling times between June 1 and August 30 to calculate a growing-season average for each plot. Leaf and stem biomass production for poplar and willow plots were calculated by measuring tree

heights and diameters, together with allometric biomass equations developed for our sites as described by Palmer *et al.* (2014). Because poplar and willow are deciduous species, we make the assumption that leaf biomass in a given year is a suitable proxy for leaf litterfall. Finally, aboveground herbaceous biomass from control plots was measured by collecting all live plant materials from a 0.37-m^2 sampling frame in mid-July of each year.

Sequence processing

Raw sequences were sorted according to barcode and processed through the RDP pyrosequencing pipeline (http://pyro.cme.msu.edu) with low quality (Q score <20) and short reads (length <200) removed. Chimeric reads were filtered using UChime (Edgar, 2010) followed by alignment and clustering at 97% and 95% nucleotide identity for 16S and 28S rRNA reads, respectively (Krüger *et al.*, 2012; Shen *et al.*, 2014). All the reads were classified using RDP naive Bayesian classifier (Wang *et al.*, 2007). For all samples, the cluster data were randomly resampled to the sequence depth of the sample with lowest number of reads (2814 and 3637 reads for 16S and 28S rRNA genes, respectively) for downstream analysis. All sequences were deposited in the NCBI Sequence Read Archive (SRA) database (Accession numbers: SRX483129 and SRX483122).

Data analysis

Downstream analyses were performed in R (version 3.1.2; https://www.r-project.org) with package VEGAN (version 2.3-0; https://github.com/vegandevs/vegan) and PHEATMAP (1.0.2; https://cran.r-project.org/web/packages/pheatmap). Data were log or square root of relative abundance transformed to ensure normality and equal variance. Alpha-diversity (Shannon diversity and Chao richness) was calculated according to the re-sampled cluster data. Nonmetric multidimensional scaling (NMDS) was performed to illustrate the beta-diversity (Bray–Curtis distances) between treatments. Similarity percentage (SIMPER) analysis based on the estimation of average contribution of each individual microbial genus to the overall Bray–Curtis distances coupled with *t*-test was conducted to identify the indicator microbial groups between treatments (Warwick *et al.*, 1990). Only the genera that showed significant difference ($P < 0.05$) between treatments were considered as indicators. Heatmap plots were generated to illustrate microbial community shifts between treatments. The data were log-transformed and sorted by the maximum value of each genus among all treatments at each site, of which the top 30 genera were used to generate heatmap plots. Zero was transformed to −6 instead of log-transformation, which was smaller than the lowest value after transformation for all samples. Variance partition analysis was conducted to estimate the proportion of the microbial community variance that was explained by environmental factors. Significant community differences were tested using analysis of similarity (ANOSIM) (Clarke, 1993). Mantel tests were performed to identify the correlation between microbial groups and environmental factors (Smouse *et al.*, 1986). To determine the significance of differences, two-tailed, unpaired *t*-tests were performed on individual microbial groups between treatments. One-way ANOVA followed by the Tukey's HSD test was performed to compare the means in R (version 3.1.2) with AGRICOLAE package (version 1.2-1; https://cran.r-project.org/web/packages/agricolae).

Results

General information from sequencing data

In total, 187 546 bacterial 16S and 281 888 fungal 28S rRNA gene reads passed quality trimming and chimera check with an average 5861 ± 2098 and 8809 ± 5041 reads per sample for 16S and 28S rRNA genes, respectively. After re-sampling, the average number of OTUs per sample was 1033 ± 113 and 1070 ± 209 for 16S and 28S rRNA, respectively. All 16S rRNA reads were classified into 23 phyla, 57 classes, and 504 genera with the average number of unclassified reads at $13.7 \pm 4.0\%$ at the phylum level, $18.6 \pm 5.3\%$ at the class level, and $33.5 \pm 5.3\%$ at the genus level. All 28S rRNA reads were classified into 8 phyla, 29 classes, and 538 genera with

the average unclassified number of reads at $13.2 \pm 7.5\%$ at the phylum level, $18.8 \pm 9.5\%$ at the class level, and $39.4 \pm 11.2\%$ at the genus level. The top 10 most abundant bacterial classes across all samples in rank order were Alphaproteobacteria, Acidobacteria-Gp6, Gemmatimonadetes, Actinobacteria, Gammaproteobacteria, Acidobacteria-Gp4, Betaproteobacteria, Deltaproteobacteria, Acidobacteria-Gp3, and Planctomycetacia (Table S2). The top 10 most abundant fungal classes in rank order were Sordariomycetes, Agaricomycetes, Leotiomycetes, Dothideomycetes, Eurotiomycetes, Pezizomycetes, Chytridiomycetes, Tremellomycetes, Fungi incertae sedis, and Ustilaginomycetes (Table S3).

Patterns in microbial community compositions

The bacterial community at the ES site was compared between samples taken in May 2010 immediately prior to conversion and in August 2010 after the tree seedlings had begun to grow, but no differences were found in the communities (Fig. S1). As the trees were only seedlings and had no chance to impact the soil, further

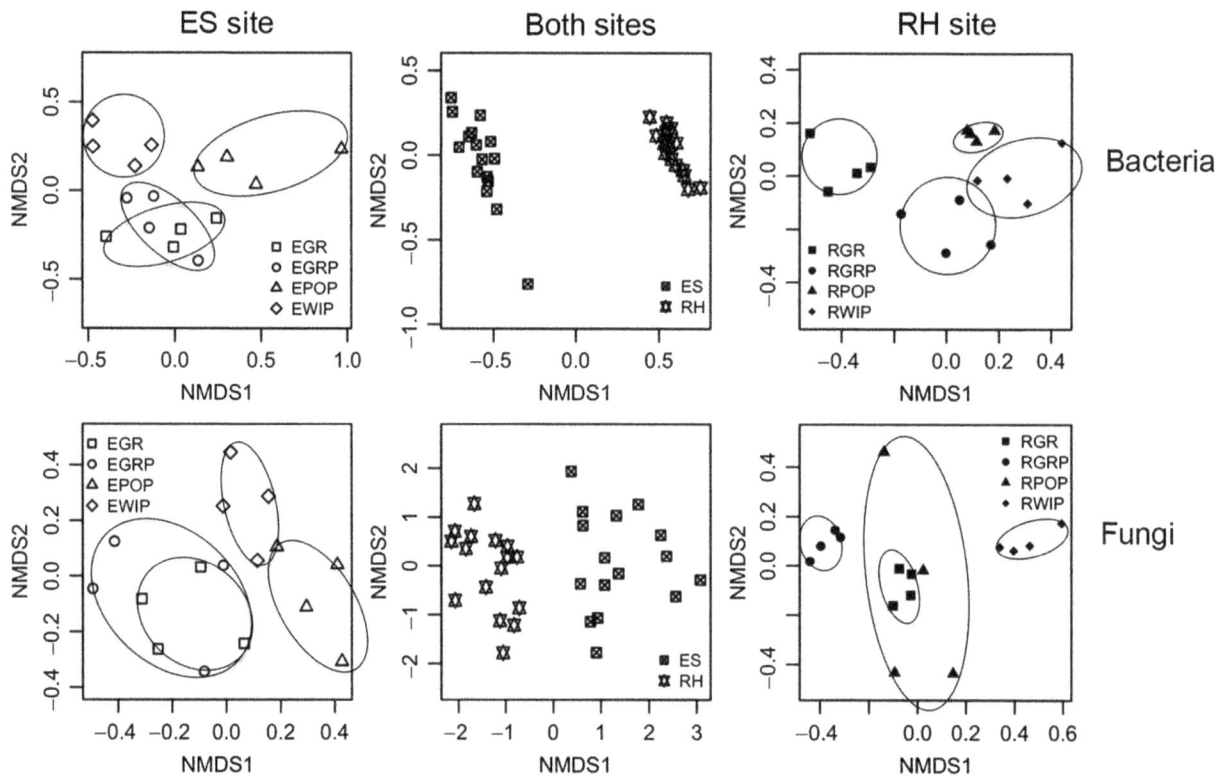

Fig. 1 Nonmetric multidimensional scaling (NMDS) ordination plots based on Bray–Curtis dissimilarity of Escanaba (ES) site samples, all samples, and Rhinelander (RH) site samples for bacteria and fungi. EGR and RGR are preconversion control plots at ES and RH sites, respectively. EGRP and RGRP are postconversion control plots at ES and RH sites, respectively. EPOP and RPOP are postconversion poplar plots at ES and RH sites, respectively. EWIP and RWIP are postconversion willow plots at ES and RH sites, respectively. ES represents all samples at Escanaba site, while RH represents all samples at Rhinelander site. NMDS stress values for bacterial communities are 0.079 (ES), 0.069 (both sites), and 0.105 (RH). NMDS stress values for fungal communities are 0.152 (ES), 0.148 (both sites), and 0.108 (RH).

sampling for community profiling was delayed until after the third season when the tree vegetation had more potential impact.

After three growing seasons postconversion, both the bacterial and fungal communities were altered at both sites (Fig. 1). All samples formed two groups based on location, and for each site, all samples could be further classified into three groups based on crop: grass (plots sampled at both pre- and postconversion times), willow, and poplar (Fig. 1). ANOSIM test revealed that converting grassland to willow and poplar significantly shifted both the bacterial and fungal communities into two distinct patterns based on vegetation (Table S4). Both bacterial and fungal community structures also differed in the RH control plots after three growing seasons, but not in the ES control plots (Table S4).

Conversion effect on microbial diversity

Changes in soil microbial diversities were influenced by both conversion and vegetation, but yielded different patterns at the two sites. At the ES site, converting grassland to poplar had no effect on fungal diversity, but it decreased bacterial diversity and richness significantly (Tukey's HSD test, $P < 0.05$) (Table 1). At the RH site, converting to poplar increased the bacterial diversity and richness (Tukey's HSD test, $P < 0.05$), while fungal diversity decreased after conversion (Tukey's HSD test, $P < 0.05$) (Table 1). In contrast, converting grassland to willow had no effect on bacterial diversity and richness with only a decrease in fungal diversity at the RH site (Tukey's HSD test, $P < 0.05$) (Table 1).

Conversion effect on microbial community composition

The effect of converting grassland to SRWCs on soil microbial composition was evaluated by comparing control plots with poplar and willow plots after the third growing season.

Poplar. Conversion altered both the bacterial and fungal communities from the genus to phylum levels. At the bacterial phylum level, the proportion of Acidobacteria increased at both sites after conversion (Tukey's HSD test, $P < 0.05$) with Proteobacteria, Planctomycetes, Gemmatimonadetes, and Actinobacteria as the most abundant bacterial phyla at both sites. After conversion, Sordariomycetes was the most abundant fungal class at both sites followed by Dothideomycetes, Agaricomycetes, Leotiomycetes, and Pezizomycetes. SIMPER analysis revealed that Acidobacteria groups, *Gemmatimonas*, and *Bradyrhizobium* were the most abundant bacterial contributing genera of conversion at both sites, with top 10 contributing genera comprising 58.9% and 52.6% of the total Bray–Curtis dissimilarity and 44.6% and 43.6% of the sequences at ES and RH site, respectively (Table S5). For fungi, only *Gibberella* was the shared indicator genus of conversion for the two sites. In the ES site poplar plots, *Hygrocybe* was the top contributing genus to the overall Bray–Curtis dissimilarity followed by *Schizothecium*, *Hebeloma*, and *Phaeodothis* with top 10 contributing genera comprising 56.4% of the total dissimilarity and 27.9% of the reads (Table S5). For RH, *Gaeumannomyces* was the top contributing genus to the dissimilarity followed by *Lycoperdon*, *Thelebolus*, and *Gibberella*, with top 10 contributing genera comprising 58.1% of the overall dissimilarity and 36.0% of the reads (Table S5).

Willow. Acidobacteria followed by Planctomycetes, Proteobacteria, Firmicutes, and Actinobacteria were the most abundant bacterial phyla in the ES site willow plots, while Proteobacteria followed by Acidobacteria, Actinobacteria, Bacteroidetes, and Gemmatimonadetes were the most abundant bacterial phyla at the RH site. Agaricomycetes was the most abundant fungal class at both sites followed by Sordariomycetes, Pezizomycetes, Leotiomycetes, and Dothideomycetes. For both sites, SIMPER analysis for this comparison was weighted heavily

Table 1 Measures of bacterial and fungal Chao richness and Shannon diversity for all treatments

Site	Crop	Conversion	Chao richness		Shannon diversity	
			Bacteria	Fungi	Bacteria	Fungi
Escanaba (ES)	Grass	Pre*	1748 ± 508^b	1389 ± 217^{ab}	6.51 ± 0.16^{ab}	6.13 ± 0.34^{abc}
ES	Grass	Post*	1707 ± 333^b	1568 ± 696^{ab}	6.48 ± 0.14^{abc}	6.04 ± 0.54^{abc}
ES	Poplar	Post	1201 ± 233^c	1464 ± 81^{ab}	6.25 ± 0.18^d	6.14 ± 0.25^{abc}
ES	Willow	Post	1516 ± 136^{bc}	1333 ± 354^{ab}	6.47 ± 0.09^{bc}	6.15 ± 0.47^{abc}
Rhinelander (RH)	Grass	Pre	1473 ± 158^{bc}	1541 ± 198^{ab}	6.44 ± 0.11^{bcd}	6.52 ± 0.15^{ab}
RH	Grass	Post	1444 ± 109^{bc}	1798 ± 108^a	6.31 ± 0.03^{cd}	6.68 ± 0.13^a
RH	Poplar	Post	2119 ± 69^a	1487 ± 384^{ab}	6.66 ± 0.07^a	6.02 ± 0.74^{bc}
RH	Willow	Post	1371 ± 114^c	1110 ± 233^b	6.26 ± 0.21^d	5.77 ± 0.35^c

*Pre and Post represent preconversion and postconversion, respectively.
Numbers in the same column with the same letter are not significantly different (Tukey's HSD test, $P > 0.05$, $n = 4$).

Table 2 Relative abundance of arbuscular mycorrhizal fungi genera detected in all treatments

Site	Crop	Conversion	Entrophospora (%)	Pacispora (%)	Glomus (%)	Paraglomus (%)
Escanaba (ES)	Grass	Pre*	ND	ND	0.01 ± 0.02b	0.01 ± 0.02a
ES	Grass	Post*	ND	ND	0.31 ± 0.28a	ND
ES	Poplar	Post	0 ± 0.01a	0.01 ± 0.01a	0.07 ± 0.12ab	0.02 ± 0.02a
ES	Willow	Post	ND	0.06 ± 0.12a	0.13 ± 0.17ab	0.02 ± 0.03a
Rhinelander (RH)	Grass	Pre	0.01 ± 0.01a	ND	0.02 ± 0.04ab	ND
RH	Grass	Post	ND	0.01 ± 0.01a	0.04 ± 0.07ab	0 ± 0.01a
RH	Poplar	Post	ND	ND	0.01 ± 0.01b	ND
RH	Willow	Post	0.01 ± 0.01a	ND	0.01 ± 0.01b	ND

*Pre and Post represent preconversion and postconversion, respectively.
Numbers in the same column with the same letter are not significantly different (Tukey's HSD test, $P > 0.05$, $n = 4$). ND means not detected. '0' is <0.005.

toward a few bacterial genera such as *Acidobacteria* groups, *Gemmatimonas*, and *Bradyrhizobium* with top 10 contributing genera comprising 53.0% and 54.7% of the total Bray–Curtis dissimilarity and 43.6% and 48.4% of the sequences at ES and RH sites, respectively (Table S6). For fungi, at the ES site, *Hygrocybe* contributed most (13.2%) to the Bray–Curtis dissimilarity followed by *Hebeloma, Sorocybe,* and *Clavulinopsis* with top 10 contributing genera comprising 60.9% of the overall dissimilarity and 26.7% of the reads. At the RH site *Hebeloma* contributed most (16.4%) to the Bray–Curtis dissimilarity followed by *Minimedusa, Thelebolus,* and *Peziza* with top 10 contributing genera comprising 63.1% of the overall dissimilarity and 36.9% of the reads (Table S6).

Arbuscular mycorrhizal fungi. In total, 1.3% of the total reads per sample were classified as arbuscular mycorrhizal fungi (AMF). Four AMF genera including *Entrophospora, Pacispora, Glomus,* and *Paraglomus* were detected, of which *Glomus* was consistently detected in all treatments and was the most dominant AMF genus (Table 2). Compared to the RH site, more AMF genera were detected in SRWC plots at the ES site. At both sites, the relative abundance of *Glomus* increased in control plots after three growing seasons (Table 2). Compared to control plots, the relative abundance of *Glomus* is lower in SRWC plots, though the difference is insignificant (Table 2).

Comparisons between bacteria and fungi. At both sites, conversion to SRWCs led to alterations in fungal com-

position with transitions occurring in both the rare (relative abundance <0.01%) and dominant populations (relative abundance >1%), for example, fungal groups such as *Hygrocybe, Hebeloma, Geopora, Cortinariaceae,* and *Inocybe.* In contrast, there was only a slight though significant variation in bacterial composition with no apparent shift between rare and dominant bacterial groups (Fig. 2). Furthermore, the majority of bacterial indicator genera such as *Bradyrhizobium, Gemmatimonas,* and *Nitrobacter* showed contrasting patterns at the two sites, while most of the fungal indicators such as *Hebeloma, Schizothecium, Gibberella, Geopora,* and *Peziza* exhibited the same shift at the two sites (Tables S5 and S6, Fig. 2).

Biomass

Measured or estimated leaf, stem, and fine-root biomass are presented in Table 3. For both sites, the control plots had the highest leaf biomass followed by willow and poplar in 2011 (Tukey's HSD test, $P < 0.05$) and 2012 (Tukey's HSD test, $P < 0.05$). Due to decreased leaf biomass in control plot (ES: Tukey's HSD test, $P > 0.05$; RH: Tukey's HSD test, $P < 0.05$) and a nonsignificant increase of leaf biomass in poplar and willow plots (Tukey's HSD test, $P > 0.05$) in 2012, the difference in leaf biomass between control and willow plots was insignificant. In addition, at the ES site, no significant difference of leaf biomass between poplar and willow was observed (Tukey's HSD test, $P > 0.05$), while willow had significantly higher leaf biomass than poplar in 2011 and 2012 at the RH site

Fig. 2 Heatmap displaying the relative abundances of top 30 bacterial and fungal genera for all treatments. The key from white to black represents the least abundant to most abundant, and the numbers represent log-transformed relative abundances of the microbial community. EGR and RGR are preconversion control plots at Escanaba (ES) and Rhinelander (RH) sites, respectively. EGRP and RGRP are postconversion control plots at ES and RH sites, respectively. EPOP and RPOP are postconversion poplar plots at ES and RH sites, respectively. EWIP and RWIP are postconversion willow plots at ES and RH sites, respectively.

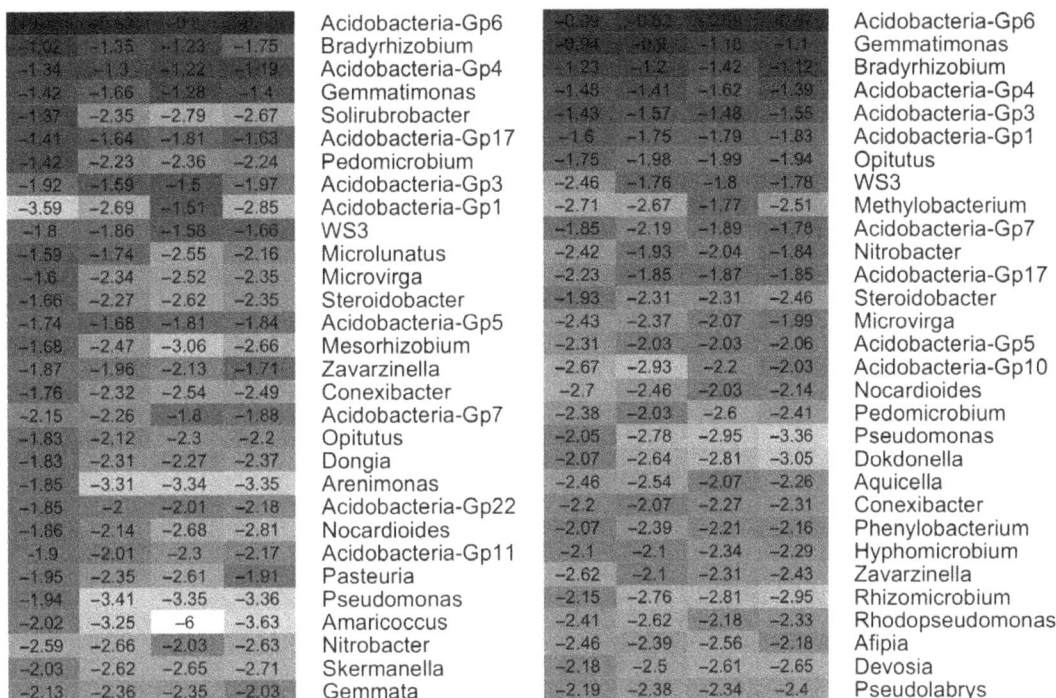

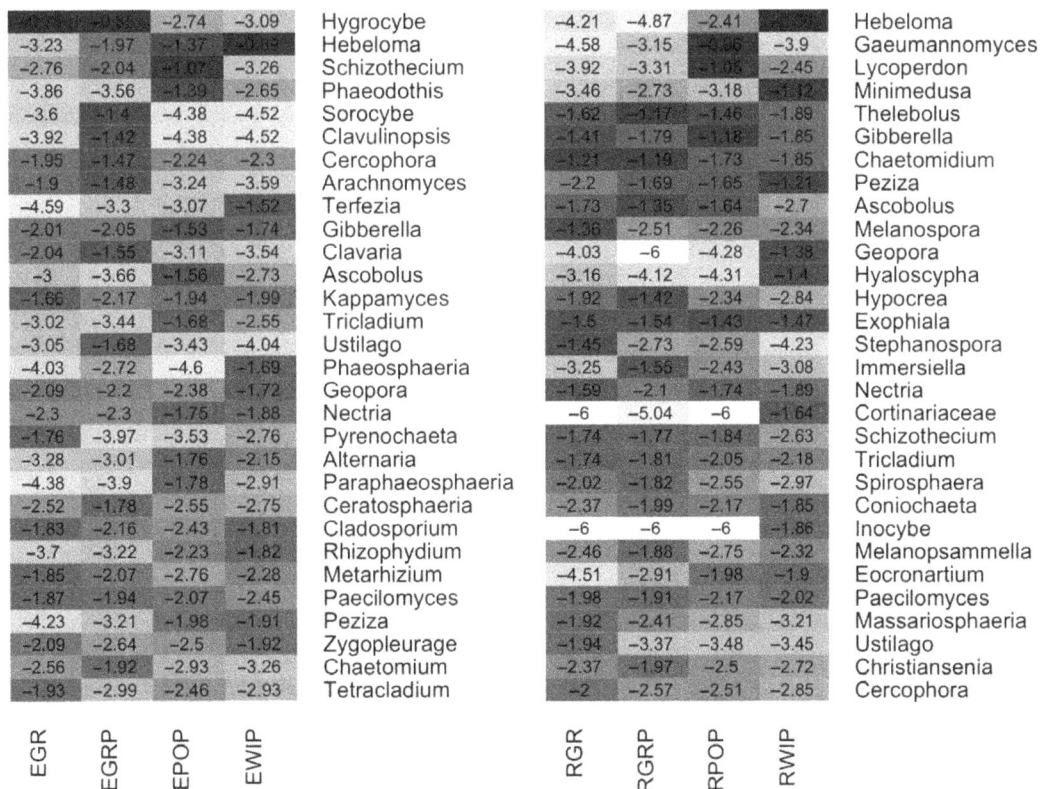

Table 3 Biomass of leaf, stem, and fine roots in 2011 and 2012 at Escanaba (ES) and Rhinelander (RH) sites

Site	Crop	Leaf		Stem		Root	
		2011	2012	2011	2012	2011	2012
ES	Grass	1.55 ± 0.38b	1.48 ± 0.35b	NA	NA	4.22 ± 0.75b	5.92 ± 0.93b
ES	Poplar	0.12 ± 0.06c	0.18 ± 0.1c	0.47 ± 0.32b	0.81 ± 0.51bc	0.79 ± 0.18c	1.28 ± 0.68c
ES	Willow	0.52 ± 0.33c	0.68 ± 0.34c	1.51 ± 0.98b	2.81 ± 1.7b	0.97 ± 0.50c	1.97 ± 0.77c
RH	Grass	3.49 ± 0.56a	2.45 ± 0.28a	NA	NA	10.09 ± 2.51a	9.00 ± 2.35a
RH	Poplar	0.19 ± 0.05c	0.54 ± 0.14c	0.83 ± 0.3b	3.29 ± 1b	2.42 ± 0.28bc	2.54 ± 0.70c
RH	Willow	2.15 ± 0.73b	2.25 ± 0.51a	5.68 ± 1.33a	10.67 ± 1.83a	1.53 ± 1.29bc	2.93 ± 1.00c

Numbers in the same column with the same letter are not significantly different (Tukey's HSD test, $P > 0.05$, $n = 4$). NA means data not available. Leaf and stem biomass for willow and poplar were calculated from tree diameter and height measurements as described by Palmer *et al.* (2014), whereas grass biomass in control plots was directly measured. The 2012 grass leaf data may under-represent the yield because it is from an adjacent area as the control plot was inadvertently mowed.

(Tukey's HSD test, $P < 0.05$). Willow had significantly higher stem biomass than poplar at both sites in both years (Tukey's HSD test, $P < 0.05$) except for the ES site in 2011. As found for the leaf biomass, the control plots had the highest fine-root biomass followed by willow and poplar at both sites in both years (Tukey's HSD test, $P < 0.05$), while no significant differences between willow and poplar were observed at both sites in both years (Tukey's HSD test, $P > 0.05$). In addition, higher fine-root biomass was observed in the drier year (2011 for the RH site and 2012 for the ES site) in control plots at both sites, while the willow and poplar fine-roots biomass was higher in 2012 at both sites; however, the differences were insignificant (Tukey's HSD test, $P > 0.05$).

Soil characteristics and microbial community variation

Soil pH ranged from 5.18 to 5.54, except the preconversion ES site (6.14), which was significantly higher (Tukey's HSD test, $P < 0.05$) (Table 4). Before conversion, the ES site had higher total nitrogen than the RH site (Tukey's HSD test, $P < 0.05$) though total nitrogen did not differ after conversion. Compared to the control plots, conversion to SRWCs only altered the total nitrogen in the willow plots at the ES site for which total nitrogen decreased 18% (Tukey's HSD test, $P < 0.05$). Nitrate did not change in control plots at the two sites over the three growing seasons. Conversion to SRWCs significantly increased nitrate from 3.2- to 11.6-fold (Tukey's HSD test, $P < 0.05$), except in the RH willow plots, for which the average is a nonsignificant 1.4-fold increase over the control plots. Before conversion, the ES site had higher total carbon than the RH site. After three growing seasons, total carbon decreased from 24.4 to 19.8 mg g^{-1} in the ES control plots (Tukey's HSD test, $P < 0.05$), while it increased in the control plots at the RH site from 18.6 to

22.1 mg g^{-1} (Tukey's HSD test, $P < 0.05$). Compared to the control plots, converting to SRWCs did not influence total carbon at the RH site. However, total carbon decreased significantly at the ES site after conversion to SRWCs (Tukey's HSD test, $P < 0.05$). At the RH site, C/N was higher before conversion and after three growing seasons, and C/N dropped significantly in the control plots at both sites (Tukey's HSD test, $P < 0.05$). Conversion to SRWCs did not change the soil C/N, except in willow plots at the ES site, in which the C/N decreased (Tukey's HSD test, $P < 0.05$).

Crop type, year (includes climate differences), and soil characteristics were evaluated for their contribution to microbial community variation. These factors explained ~59% of the community variation (Table 5). Crop type was identified as the main factor that influenced both the bacterial and fungal community shifts (Table 5). Only for the RH site did the bacterial community exhibit a significant correlation with time ($r = 0.67$, $P = 0.002$), agreeing with the previous ordinations. Total nitrogen had strong correlation with bacterial community structure (ES: $r = 0.31$, $P = 0.014$; RH: $r = 0.37$, $P = 0.007$), but not with fungal community structure. Nitrate ($r = 0.24$, $P = 0.048$) and ammonium ($r = 0.25$, $P = 0.026$) were correlated with bacterial community structures at the ES site, but not at the RH site. Total carbon ($r = 0.24$, $P = 0.024$) and C/N ($r = 0.31$, $P = 0.019$) exhibited a correlation with the bacterial community structure at the ES and RH sites, respectively. Soil pH did not show any correlation with microbial community structure, probably because differences were minor. The bacterial phylum Nitrospirae significantly increased in SRWC plots (Turkey's HSD test, $P < 0.05$). Its relative abundance was negatively correlated to soil total nitrogen ($r = -0.48$, $P = 0.005$), but positively correlated to soil nitrate ($r = 0.4$, $P = 0.023$).

Table 4 Soil characteristics of all treatments at Escanaba (ES) and Rhinelander (RH) sites

Site	Crop	Conversion	C/N (mass/mass)	Total C (%)	Total N (%)	pH	NH_4^+ (mg kg^{-1})	NO_3^- (mg kg^{-1})
ES	Grass	Pre*	12.94 ± 0.69^b	2.44 ± 0.08^a	0.19 ± 0.01^a	6.14 ± 0.92^a	1.19 ± 0.36^c	0.19 ± 0.03^c
ES	Grass	Post*	11.41 ± 1.30^c	1.98 ± 0.29^{bc}	0.17 ± 0.01^{ab}	5.18 ± 0.34^b	1.4 ± 0.28^{bc}	0.20 ± 0.20^c
ES	Poplar	Post	10.43 ± 1.00^{cd}	1.59 ± 0.26^{de}	0.15 ± 0.02^{bc}	5.43 ± 0.22^b	0.77 ± 0.31^c	2.32 ± 1.20^a
ES	Willow	Post	10.19 ± 0.29^d	1.40 ± 0.24^e	0.14 ± 0.02^{cd}	5.18 ± 0.15^b	1.07 ± 0.40^c	1.42 ± 1.08^{ab}
RH	Grass	Pre	14.77 ± 0.79^a	1.86 ± 0.17^{cd}	0.13 ± 0.02^d	5.37 ± 0.32^b	1.03 ± 0.19^c	0.27 ± 0.29^c
RH	Grass	Post	12.84 ± 0.47^b	2.21 ± 0.26^{ab}	0.17 ± 0.01^{ab}	5.47 ± 0.08^b	2.35 ± 0.88^{ab}	0.55 ± 0.13^{bc}
RH	Poplar	Post	12.56 ± 0.59^b	1.92 ± 0.15^{bc}	0.15 ± 0.01^{bc}	5.54 ± 0.13^b	1.52 ± 0.47^{bc}	1.77 ± 0.68^a
RH	Willow	Post	13.36 ± 0.56^b	2.11 ± 0.28^{bc}	0.16 ± 0.02^{bc}	5.38 ± 0.05^b	2.64 ± 1.62^a	0.76 ± 0.12^{bc}

*Pre and Post represent preconversion and postconversion, respectively.
Numbers in the same column with the same letter are not significantly different (Tukey's HSD test, $P > 0.05$, $n = 4$).

Table 5 Variance partition analysis and mantel test between microbial community and environmental factors

	Escanaba site						Rhinelander site					
	Bacteria			Fungi			Bacteria			Fungi		
Factor	Explained* (%)	r†	P†	Explained (%)	r	P	Explained (%)	r	P	Explained (%)	r	P
Crop	12.79	**0.43**	0.002	16.14	**0.29**	0.005	13.72	**0.39**	0.006	21.71	**0.55**	0.001
Year	5.40	0.07	0.282	6.60	0.04	0.353	6.38	**0.67**	0.002	6.86	−0.08	0.721
Total N	6.21	**0.31**	0.014	7.05	0.02	0.407	6.12	**0.37**	0.007	6.18	−0.17	0.922
NO_3^-	6.42	**0.24**	0.048	8.24	0.11	0.159	6.06	0.02	0.429	5.16	−0.07	0.647
NH_4^+	6.27	**0.25**	0.026	8.93	**0.33**	0.002	6.20	−0.02	0.502	4.19	−0.06	0.646
Total C	6.06	**0.24**	0.024	6.92	0.06	0.275	6.16	0.04	0.314	6.21	−0.01	0.51
C/N	6.29	0.08	0.230	6.70	0.13	0.103	6.17	**0.31**	0.019	6.02	0.01	0.477
pH	5.74	−0.05	0.592	3.91	0.01	0.402	6.14	0.05	0.323	3.21	−0.26	0.954

*Proportion of variance could be explained returned by variance partition analysis.
†R and P values returned by mantel test. R-value in bold means significantly correlated ($P < 0.05$, $n = 16$).

Discussion

Microbial community and environmental factors

Although the two experimental sites were relatively close to each other (~200 km apart, nearly identical latitude) with similar climate, soil type, and forested history during soil development and had identical experimental designs, the inherent site characteristics led to, in some cases, opposite shifts in patterns of soil microbial diversity and composition. These differences are likely due to the different compositions of the initial soil microbial communities that may have fundamentally influenced the different shift patterns observed at the two sites (Wakelin et al., 2008). There is also a difference in soil nutrients; ammonium and nitrate were significantly correlated to the microbial community structure at the ES site, but not at the RH site.

Although site differences were a factor, crop type was the main factor that influenced microbial community composition at each site. This agreed with previous studies on microbial communities in bulk and rhizosphere soils of three typical *Verticillium* host plants (Smalla et al., 2001), in forests in Britain (Grayston & Prescott, 2005), France (Lejon et al., 2005), and Malaysia (Ushio et al., 2008) and in a pasture to woody biomass conversion in Brazil (Rachid et al., 2015). This may be due to allelopathy, different chemical resources (e.g., root exudates) provided by the new roots and litter, and host selection for symbionts (Broeckling et al., 2008; Turner et al., 2013). In this study, roots more so than litter likely provided the new carbon for microbial growth as the soil core sampled the top 10 cm soil with the surface soil (~1 cm) removed and the litter impact was only at the surface and from the 2011 season as leaves had not fallen at the time of our 2012 sampling.

Soil microbial community structure can also be responsive to seasonality or larger climate differences (Cruz-Martínez et al., 2009; Andersson et al., 2010). In our case, both sites experienced wetter than normal growing seasons in the planting year of 2010 and drier than normal growing seasons in 2011. In 2012, RH rebounded to near normal precipitation, whereas ES was even drier (Table S1), which severely restricted stem and root growth of both willow and poplar despite more fertile soil at ES (Table 3). The effect of this second consecutive dry season at ES is consistent with significant variation being observed only at the RH site for both bacteria and fungi.

Historically, soil and vegetation at the ES site had remained undisturbed for 42 years before the start of our experiment. After four decades of adaptation to these conditions, the microbial community at the ES site was likely more resilient, thus possibly explaining the absence of temporal variation in control plots at this site. In contrast, the RH was under active cultivation until only 5 years prior to the start of the experiment. Thus, the significant temporal variation in control plots at this site suggests that the soil microbial community was still responding to the earlier conversion from row crops to grass. In support of this finding, a previous study reported that land-use history had a stronger impact on soil microbial community composition than aboveground vegetation and soil properties (Jangid et al., 2011). Thus, the combination of land-use history and climate prior to the initiation of the experiment may have impacted the temporal variability among these sites.

Conversion effect on soil microbial compositions

In the following discussion, we comment on potential function(s) of identified bacterial and fungal groups based on information from group members that have been studied. However, the extent to which these traits reflect the function of the whole group will vary with trait, group, and resolution of the marker gene, and as such should be considered only as potential ecological roles.

Conversion of grassland to SRWCs resulted in microbial community shifts at both sites. This was discernable by the genera associated with grasses and SRWC trees that were identified as contributing to the majority of the dissimilarity by SIMPER. The bacterial indicator genera were similar for all treatments at the two experimental sites. Acidobacteria groups, the most abundant bacterial groups in SRWCs plots, were the main indicators for postconversion. Acidobacteria are widespread in nature and have been reported as the most abundant bacterial group in poplar and willow rhizospheres (De Cárcer et al., 2007; Gottel et al., 2011). Other shared top indicator genera, such as *Gemmatimonas*, *Nocardioides*, and *Bradyrhizobium* are also widespread. *Gemmatimonas*, specifically *G. aurantiaca*, was reported as a polyphosphate-accumulating bacterium that could be used for biological phosphorus removal (Zhang, 2003). *Nocardioides* was reported to accumulate

in soil after poplar was planted (Hur et al., 2011) and has been described as an endophytic bacteria in poplar (Ulrich et al., 2008). *Nocardioides* has also been reported as enriched in soils with persistent organic pollutants (Golovleva et al., 1990; Iwabuchi & Harayama, 1997; Cho et al., 2000), which suggests that it may have the capacity to degrade complex organic compounds, such as those found in tree litter. *Bradyrhizobium* is a well-known N_2-fixing bacterial genus, reported as one of the dominant bacterial groups in the rhizo- and endosphere of poplar (Gottel et al., 2011). However, these indicator genera did not show consistent shift patterns at both sites after conversion. Deforestation was reported to alter soil microbial populations from *Fibrobacter* and *Syntrophomonas* assemblages in forest soil to *Burkholderia* and *Rhizobium–Agrobacterium* assemblages when the soil was converted to pasture (Nusslein & Tiedje, 1999). In our study, *Burkholderia* and *Rhizobium* both decreased in abundance after conversion and *Fibrobacter* and *Syntrophomonas* were not detected in any of the plots. These results indicate that *Burkholderia* and *Rhizobium* might be more dominant in grassland. Lastly, the bacterial phylum Nitrospirae was found significantly increased in poplar and willow plots after conversion, which is negatively correlated to soil total nitrogen ($r = -0.48$, $P < 0.05$) and positively correlated to nitrate ($r = 0.4$, $P < 0.05$). One confounding factor in these analyses involves the temporal scale and the resulting comparisons between SRWC plots to the control. As there was no sprayed and plowed control (vs. the native grassland control), there is the potential for changes in microbial community composition due to these pretreatments. However, the majority of community changes after 3 years appear to be most related to vegetation type and the direct comparisons between SRWC's remain valid, although a portion of the shared microbial community between SRWC's may be have been shaped by the initial spray/plow treatment.

For fungi, the grassland contributing genera reflect the characteristics of the site. At the ES site, *Hygrocybe* is a leading contributor, the dominant fungal group in grass plots. These gilled fungi are associates of mosses and are characteristic of old, unimproved grasslands (Rotheroe et al., 1996). Another contributor of the grassland at ES site was the genus *Clavulinopsis*. This genus belongs to the Clavariaceae, described as part of the CHEG (Clavariaceae, Hygrocybe, Entolomataceae, and Geoglossaceae) community that have been associated with old, undisturbed, unimproved grasslands (Russell, 2005; Moore et al., 2008). As a likely endophyte of the native grasses, *Cercophora* (specifically *coprophila*) has been identified as an endophyte of *Ammophila arenaria*, a coastal grass (Sánchez Márquez et al., 2008), and *Stipa grandis* (needlegrass) (Su et al., 2010). Species within the genus *Clavaria* have close physiological relationships to roots that possess *Ericoid* mycorrhizae (Englander & Hull, 1980; Petersen & Litten, 1989). Members of the genus *Ustilago* are smut fungi that are parasitic on grasses (Kirk *et al.*, 2008).

At the RH site, *Trichoderma* was a contributing genus. As the anamorph of *Hypocrea*, the species *hamatum*, *harzianum*, and *koningii* have been isolated from roots of wheat and rye grass (Dewan & Sivasithamparam, 1988) and thus may be associated with grasses present at this site. Interestingly, some *Trichoderma* species have been found to inhibit root rot by *Gaeumannomyces graminis var. tritici*, which was found in reduced abundance in the control plot at this site. *Chaetosphaeria* have been isolated as endophytes of sand couch grass (*Elymus farctus*) (Sánchez Márquez *et al.*, 2008). Little is known of *Thelebolus* except that it has been isolated as an endophyte of *Scandinavian* small reed (*C. phragmitoides*).

Among the top 20 fungal genera, *Hebeloma*, *Gibberella*, and *Peziza* showed a consistent shift after conversion to either poplar or willow at both sites. *Basidiomycota*, associated with high-lignin litter degradation (Blackwood *et al.*, 2007) and very common in forest soils, were higher in abundance in the SRWC plots. Lignin degradation-associated fungi belonging to the genera *Hebeloma* and *Peziza* have been identified on rotted alder, willow, and poplar (Beug *et al.*, 2014). These all increased abundance in SRWC plots at both sites, and *Hebeloma* was the overall most abundant Basidiomycota group in SRWC plots. *Hebeloma* (sp. *fastibile*) and *Peziza* (e.g., *Peziza ostrcoderma*) have also been observed to form ectomycorrhiza with Balsam poplar (*Populus balsamea*; Siemens & Zwiazek, 2008) and *Salix* (willow) (Baum *et al.*, 2009; Danielsen *et al.*, 2012). *Gibberella* (telemorph *Fusarium*), a potential pathogenic lineage in poplar, has been found in high abundance of diseased willow plants and during first year growth in previously arable land (Corredor *et al.*, 2012). In this study, *Gibberella* increased four times in poplar plots and two times in willow plots after conversion. However, not all species are pathogenic and this lineage is poorly resolved with the 28S rRNA gene.

Differences between bacteria and fungi

Conversion significantly altered fungal compositions leading to transitions between rare fungal groups and the dominant ones. However, conversion did not cause a dramatic change in the dominant bacterial genera and no apparent shift between rare and dominant bacterial groups. In addition, fungal communities had more consistent shift patterns at the two sites, indicating that fungi appear more responsive to con-version and the community is more specific to the aboveground vegetation than the bacterial community. This is in agreement with a previous study on afforestation of pastures with *Pinus radiate* (Macdonald *et al.*, 2009) where the soil fungal community was more affected by land conversion than the bacterial community. In addition, Tedersoo *et al.* (2014) found large differences in fungal communities between forested and treeless ecosystems and that fungal diversity was linked to plant diversity. Agricultural tillage is also known to negatively affect fungi more than bacteria (Stahl *et al.*, 1999; Kubicek & Druzhinina, 2007), but those studies are with repeated tillage, not the one time tillage three growing seasons earlier as was used here. Nonetheless, the early tillage event was a major disturbance and would have allowed some populations to gain an early advantage, which may have persisted, especially if those populations were effective competitors for tree leaf litter or root carbon. AMF is widely acknowledged as beneficial fungal group for plants (Bainard *et al.*, 2012a). In this study, four AMF genera were detected and only *Glomus* was consistently detected in both grass and SRWC plots. The failure of detection of other AMF groups is likely due to (i) the use of a universal fungal primer in this study, not an AMF specific primer set, and/or (ii) low numbers of non-*Glomus* AMF groups in agroecosystems (Bainard *et al.*, 2012b). As such, deeper sequencing may be required to detect these rare AMF groups. *Glomus* was in higher relative abundance in the dry year than the wetter year in control plots at both sites, in agreement with previous studies that AMF would benefit the plant in water uptake in drought (Doubková *et al.*, 2013; Jayne & Quigley, 2014). Overall, SRWC conversion decreased the relative abundance of *Glomus*, compared to the control. Conversely, some ectomycorrhizal fungi (ECM), namely Pezizales, Hebeloma, Cortinariaceae, Ceratobasidiaceae, Cantharellales, Atheliales, Tricholomataceae, and Thelephorales, increased in SRWC plots, though others decreased (Table S7). Overall, more ECM fungi increased in relative abundance in SRWC plots. Hence, there was the expected switch from AM fungi to ECM fungi with change in vegetation, which is important to tree productivity on these marginal lands (Smith & Read, 2010; Zuccaro *et al.*, 2014).

Soil microbial community and soil nitrogen

If nitrate is present in excess of plant demand, there is potential for leaching of nitrate and N_2O emissions (Galloway *et al.*, 2003). In our previous study at these sites (Palmer *et al.*, 2013), we observed a decrease in ammonium and large spikes in soil extractable NO_3^-, NO_3^-

leaching losses, and soil N_2O emissions in the first 2 years following conversion from grassland to poplar or willow plantations. This trend is consistent with a decrease in ammonium due to plant uptake and nitrification, the latter resulting in nitrate accumulation and N_2O production. The bacterial phylum Nitrospirae, a well-known nitrite-oxidizing bacterial group, significantly increased in abundance in the SRWC plots. Glyphosate application alone, associated with the SRWC conversion, has been shown to reduce Nitrospirae abundance (Barriuso et al., 2011), but in our case, the tillage would have enhanced N mineralization providing substrate for their growth and producing nitrate. Their increase was negatively correlated to soil total nitrogen, but positively correlated to soil nitrate, which is consistent with our previous findings on increased N_2O emissions at the SRWC sites (Palmer et al., 2013).

Acknowledgements

This project was funded by the US Department of Energy (DOE) Great Lakes Bioenergy Research Center (DOE BER Office of Science DE-FC02-07ER64494 and DOE OBP Office of Energy Efficiency and Renewable Energy DE-AC05-76RL01830) to Michigan State University and University of Wisconsin-Madison, by US Department of Energy award # DE-EE-0000280, and by the US Department of Agriculture, Agriculture and Food Research Initiative, Sustainable Bioenergy Program (2010-03866).

We thank Michael Cook, Christopher Johnson, Marin Palmer, and Paul Irving for assistance with the fieldwork and soil analyses, and Eliane Aparecida Gomes, Mark Charbonneau, Adam Dingens, and Julian Yu for help with the DNA-based work.

References

Andersson AF, Riemann L, Bertilsson S (2010) Pyrosequencing reveals contrasting seasonal dynamics of taxa within Baltic Sea bacterioplankton communities. The ISME Journal, 4, 171–181.

Aylott MJ, Casella E, Tubby I, Street NR, Smith P, Taylor G (2008) Yield and spatial supply of bioenergy poplar and willow short-rotation coppice in the UK. The New Phytologist, 178, 358–370.

Bainard LD, Koch AM, Gordon AM, Klironomos JN (2012a) Growth response of crops to soil microbial communities from conventional monocropping and tree-based intercropping systems. Plant and Soil, 363, 345–356.

Bainard LD, Koch AM, Gordon AM, Klironomos JN (2012b) Temporal and compositional differences of arbuscular mycorrhizal fungal communities in conventional monocropping and tree-based intercropping systems. Soil Biology and Biochemistry, 45, 172–180.

Barriuso J, Marín S, Mellado RP (2011) Potential accumulative effect of the herbicide glyphosate on glyphosate-tolerant maize rhizobacterial communities over a three-year cultivation period. PLoS One, 6, e27558.

Baum C, Toljander YK, Eckhardt K-U, Weih M (2009) The significance of host-fungus combinations in ectomycorrhizal symbioses for the chemical quality of willow foliage. Plant and Soil, 323, 213–224.

Beug M, Bessette AE, Bessette AR (2014) Ascomycete Fungi of North America: A Mushroom Reference Guide. University of Texas Press, Austin, TX.

Bever JD, Platt TG, Morton ER (2012) Microbial population and community dynamics on plant roots and their feedbacks on plant communities. Annual Review of Microbiology, 66, 265–283.

Blackwood CB, Waldrop MP, Zak DR, Sinsabaugh RL (2007) Molecular analysis of fungal communities and laccase genes in decomposing litter reveals differences among forest types but no impact of nitrogen deposition. Environmental Microbiology, 9, 1306–1316.

Broeckling CD, Broz AK, Bergelson J, Manter DK, Vivanco JM (2008) Root exudates regulate soil fungal community composition and diversity. Applied and Environmental Microbiology, 74, 738–744.

Cai X, Zhang X, Wang D (2011) Land availability for biofuel production. Environmental Science & Technology, 45, 334–339.

Cho Y-G, Rhee S-K, Lee S-T (2000) Influence of phenol on biodegradation of p-nitrophenol by freely suspended and immobilized Nocardioides sp. NSP41. Biodegradation, 11, 21–28.

Clarke KR (1993) Non-parametric multivariate analyses of changes in community structure. Austral Ecology, 18, 117–143.

Compant S, Clément C, Sessitsch A (2010) Plant growth-promoting bacteria in the rhizo- and endosphere of plants: their role, colonization, mechanisms involved and prospects for utilization. Soil Biology and Biochemistry, 42, 669–678.

Corredor AH, Van Rees K, Vujanovic V (2012) Changes in root-associated fungal assemblages within newly established clonal biomass plantations of Salix spp. Forest Ecology and Management, 282, 105–114.

Cruz-Martínez K, Suttle KB, Brodie EL, Power ME, Andersen GL, Banfield JF (2009) Despite strong seasonal responses, soil microbial consortia are more resilient to long-term changes in rainfall than overlying grassland. The ISME Journal, 3, 738–744.

Danielsen L, Thürmer A, Meinicke P et al. (2012) Fungal soil communities in a young transgenic poplar plantation form a rich reservoir for fungal root communities. Ecology and Evolution, 2, 1935–1948.

Davis SC, Dietze M, DeLucia E et al. (2012) Harvesting carbon from Eastern US forests: opportunities and impacts of an expanding bioenergy industry. Forests, 3, 370–397.

De Cárcer DA, Martín M, Karlson U, Rivilla R (2007) Changes in bacterial populations and in biphenyl dioxygenase gene diversity in a polychlorinated biphenyl-polluted soil after introduction of willow trees for rhizoremediation. Applied and Environmental Microbiology, 73, 6224–6232.

Demirbas A (2008) The importance of bioethanol and biodiesel from biomass. Energy Sources, Part B: Economics, Planning, and Policy, 3, 177–185.

Dewan MM, Sivasithamparam K (1988) Identity and frequency of occurrence of Trichoderma spp. in roots of wheat and rye-grass in Western Australia and their effect on root rot caused by Gaeumannomyces graminis var. tritici. Plant and Soil, 109, 93–101.

Doane TA, Horwáth WR (2003) Spectrophotometric determination of nitrate with a single reagent. Analytical Letters, 36, 2713–2722.

Doubková P, Vlasáková E, Sudová R (2013) Arbuscular mycorrhizal symbiosis alleviates drought stress imposed on Knautia arvensis plants in serpentine soil. Plant and Soil, 370, 149–161.

Edgar RC (2010) Search and clustering orders of magnitude faster than BLAST. Bioinformatics (Oxford, England), 26, 2460–2461.

EISA (2007) Energy Independence and Security Act. Public Law, pp. 110–140.

Englander L, Hull RJ (1980) Reciprocal transfer of nutrients between ericaceous plants and a Clavaria sp. New Phytologist, 84, 661–667.

Galloway JN, Aber JD, Erisman JW, Seitzinger SP, Howarth RW, Cowling EB, Cosby BJ (2003) The nitrogen cascade. BioScience, 53, 341.

Garbeva P, van Veen JA, van Elsas JD (2004) Microbial diversity in soil: selection microbial populations by plant and soil type and implications for disease suppressiveness. Annual Review of Phytopathology, 42, 243–270.

Ghatak HR (2011) Biorefineries from the perspective of sustainability: feedstocks, products, and processes. Renewable and Sustainable Energy Reviews, 15, 4042–4052.

Glick BR (1995) The enhancement of plant growth by free-living bacteria. Canadian Journal of Microbiology, 41, 109–117.

Golovleva LA, Pertsova RN, Evtushenko LI, Baskunov BP (1990) Degradation of 2,4,5-Trichlorophenoxyacetic acid by a Nocardioides simplex culture. Biodegradation, 1, 263–271.

Gopalakrishnan G, Negri MC, Wang M, Wu M, Snyder SW, LaFreniere L (2009) Biofuels, land, and water: a systems approach to sustainability. Environmental Science & Technology, 43, 6094–6100.

Gottel NR, Castro HF, Kerley M et al. (2011) Distinct microbial communities within the endosphere and rhizosphere of Populus deltoides roots across contrasting soil types. Applied and Environmental Microbiology, 77, 5934–5944.

Grayston SJ, Prescott CE (2005) Microbial communities in forest floors under four tree species in coastal British Columbia. Soil Biology and Biochemistry, 37, 1157–1167.

Hansen EA (1991) Poplar woody biomass yields: a look to the future. Biomass and Bioenergy, 1, 1–7.

Hanson CA, Fuhrman JA, Horner-Devine MC, Martiny JBH (2012) Beyond biogeographic patterns: processes shaping the microbial landscape. Nature Reviews. Microbiology, 10, 497–506.

Hill J, Nelson E, Tilman D, Polasky S, Tiffany D (2006) Environmental, economic, and energetic costs and benefits of biodiesel and ethanol biofuels. Proceedings of the National Academy of Sciences, USA, 103, 11206–11210.

Hur M, Kim Y, Song H-R, Kim JM, Choi YI, Yi H (2011) Effect of genetically modified poplars on soil microbial communities during the phytoremediation of waste mine tailings. *Applied and Environmental Microbiology*, **77**, 7611–7619.

Iwabuchi T, Harayama S (1997) Biochemical and genetic characterization of 2-carboxybenzaldehyde dehydrogenase, an enzyme involved in phenanthrene degradation by *Nocardioides* sp. strain KP7. *Journal of Bacteriology*, **179**, 6488–6494.

Jangid K, Williams MA, Franzluebbers AJ, Schmidt TM, Coleman DC, Whitman WB (2011) Land-use history has a stronger impact on soil microbial community composition than aboveground vegetation and soil properties. *Soil Biology and Biochemistry*, **43**, 2184–2193.

Jayne B, Quigley M (2014) Influence of arbuscular mycorrhiza on growth and reproductive response of plants under water deficit: a meta-analysis. *Mycorrhiza*, **24**, 109–119.

Kirk M, Cannon P, Minter D, Stalpers J (2008) *Dictionary of the Fungi* (10th edn). CABI, Wallingford.

Krüger D, Kapturska D, Fischer C, Daniel R, Wubet T (2012) Diversity measures in environmental sequences are highly dependent on alignment quality-data from ITS and new LSU primers targeting basidiomycetes. *PLoS One*, **7**, e32139.

Kubicek CP, Druzhinina IS (2007) *Environmental and Microbial Relationships*, Vol 4, pp. 47–68. Springer, Berlin Heidelberg, Berlin.

Labrecque M, Teodorescu TI (2005) Field performance and biomass production of 12 willow and poplar clones in short-rotation coppice in southern Quebec (Canada). *Biomass and Bioenergy*, **29**, 1–9.

Lauber CL, Hamady M, Knight R, Fierer N (2009) Pyrosequencing-based assessment of soil pH as a predictor of soil bacterial community structure at the continental scale. *Applied and Environmental Microbiology*, **75**, 5111–5120.

Lejon DPH, Chaussod R, Ranger J, Ranjard L (2005) Microbial community structure and density under different tree species in an acid forest soil (Morvan, France). *Microbial Ecology*, **50**, 614–625.

Loon LC (2007) Plant responses to plant growth-promoting rhizobacteria. *European Journal of Plant Pathology*, **119**, 243–254.

Lugtenberg B, Kamilova F (2009) Plant-growth-promoting rhizobacteria. *Annual Review of Microbiology*, **63**, 541–556.

Macdonald CA, Thomas N, Robinson L, Tate KR, Ross DJ, Dando J, Singh BK (2009) Physiological, biochemical and molecular responses of the soil microbial community after afforestation of pastures with *Pinus radiata*. *Soil Biology and Biochemistry*, **41**, 1642–1651.

Mladenoff DJ, Sahajpal R, Johnson CP, Rothstein DE (2015) Recent land cover change to agriculture in the Lake States, USA: implications for perennial cellulosic biomass production. *Ecosphere*, In Press.

Moore D, Nauta MM, Evans SE, Rotheroe M (2008) *Fungal Conservation: Issues and Solutions*. Cambridge University Press, New York.

Nacke H, Thürmer A, Wollherr A et al. (2011) Pyrosequencing-based assessment of bacterial community structure along different management types in German forest and grassland soils. *PLoS One*, **6**, e17000.

Nikièma P, Rothstein DE, Miller RO (2012) Initial greenhouse gas emissions and nitrogen leaching losses associated with converting pastureland to short-rotation woody bioenergy crops in northern Michigan, USA. *Biomass and Bioenergy*, **39**, 413–426.

Nusslein K, Tiedje JM (1999) Soil bacterial community shift correlated with change from forest to pasture vegetation in a tropical soil. *Applied and Environmental Microbiology*, **65**, 3622–3626.

Palmer MM, Forrester JA, Rothstein DE, Mladenoff DJ (2013) Conversion of open lands to short-rotation woody biomass crops: site variability affects nitrogen cycling and N$_2$O fluxes in the US Northern Lake States. *GCB Bioenergy*, **6**, 1–15.

Palmer MM, Forrester JA, Rothstein DE, Mladenoff DJ (2014) Establishment phase greenhouse gas emissions in short rotation woody biomass plantations in the Northern Lake States, USA. *Biomass and Bioenergy*, **62**, 26–36.

Penton CR, St Louis D, Cole JR et al. (2013) Fungal diversity in permafrost and tallgrass prairie soils under experimental warming conditions. *Applied and Environmental Microbiology*, **79**, 7063–7072.

Petersen RH, Litten W (1989) A new species of *Clavaria* fruiting with *Vaccinium*. *Mycologia*, **81**, 325–327.

Rachid CTCC, Balieiro FC, Fonseca ES, Peixoto RS, Chaer GM, Tiedje JM, Rosado AS (2015) Intercropped silviculture systems, a key to achieving soil fungal community management in eucalyptus plantations. *PLoS One*, **10**, e0118515.

Rotheroe M, Newton A, Evans S, Feehan J (1996) Waxcap-grassland survey. *Mycologist*, **10**, 23–25.

Rousk J, Bååth E, Brookes PC et al. (2010) Soil bacterial and fungal communities across a pH gradient in an arable soil. *The ISME Journal*, **4**, 1340–1351.

Russell P (2005) Grassland fungi and the management history of St. Dunstan's farm. *Field Mycology*, **6**, 85–91.

Sánchez Márquez S, Bills GF, Zabalgogeazcoa I (2008) Diversity and structure of the fungal endophytic assemblages from two sympatric coastal grasses. *Fungal Diversity*, **33**, 87–100.

Schweitzer JA, Bailey JK, Fischer DG, LeRoy CJ, Lonsdorf EV, Whitham TG, Hart SC (2008) Plant–soil–microorganisms interactions: heritable relationship between plant genotype and associated soil microorganisms. *Ecology*, **89**, 773–781.

Shen Z, Wang D, Ruan Y, Xue C, Zhang J, Li R, Shen Q (2014) Deep 16S rRNA pyrosequencing reveals a bacterial community associated with banana Fusarium Wilt disease suppression induced by bio-organic fertilizer application. *PLoS One*, **9**, e98420.

Siemens JA, Zwiazek JJ (2008) Root hydraulic properties and growth of balsam poplar (*Populus balsamifera*) mycorrhizal with *Hebeloma crustuliniforme* and *Wilcoxina mikolae* var. *mikolae*. *Mycorrhiza*, **18**, 393–401.

Sinsabaugh R, Reynolds H, Long T (2000) Rapid assay for amidohydrolase (urease) activity in environmental samples. *Soil Biology and Biochemistry*, **32**, 2095–2097.

Smalla K, Wieland G, Buchner A et al. (2001) Bulk and rhizosphere soil bacterial communities studied by denaturing gradient gel electrophoresis: plant-dependent enrichment and seasonal shifts revealed. *Applied and Environmental Microbiology*, **67**, 4742–4751.

Smith SE, Read DJ (2010) *Mycorrhizal Symbiosis*. Academic Press, London.

Smouse PE, Long JC, Sokal RR (1986) Multiple regression and correlation extensions of the mantel test of matrix correspondence. *Systematic Zoology*, **35**, 627–632.

Stahl PD, Parkin TB, Christensen M (1999) Fungal presence in paired cultivated and uncultivated soils in central Iowa, USA. *Biology and Fertility of Soils*, **29**, 92–97.

Su Y-Y, Guo L-D, Hyde KD (2010) Response of endophytic fungi of *Stipa grandis* to experimental plant function group removal in Inner Mongolia steppe, China. *Fungal Diversity*, **43**, 93–101.

Sul WJ, Cole JR, Jesus EdaC, Wang Q, Farris RJ, Fish JA, Tiedje JM (2011) Bacterial community comparisons by taxonomy-supervised analysis independent of sequence alignment and clustering. *Proceedings of the National Academy of Sciences, USA*, **108**, 14637–14642.

Tang Y, Xie J-S, Geng S (2010) Marginal land-based biomass energy production in China. *Journal of Integrative Plant Biology*, **52**, 112–121.

Tedersoo L, Bahram M, Polme S et al. (2014) Global diversity and geography of soil fungi. *Science*, **346**, 1256688.

Thoms C, Gattinger A, Jacob M, Thomas FM, Gleixner G (2010) Direct and indirect effects of tree diversity drive soil microbial diversity in temperate deciduous forest. *Soil Biology and Biochemistry*, **42**, 1558–1565.

Turner TR, Ramakrishnan K, Walshaw J et al. (2013) Comparative metatranscriptomics reveals kingdom level changes in the rhizosphere microbiome of plants. *The ISME Journal*, **7**, 2248–2258.

Ulrich K, Ulrich A, Ewald D (2008) Diversity of endophytic bacterial communities in poplar grown under field conditions. *FEMS Microbiology Ecology*, **63**, 169–180.

Ushio M, Wagai R, Balser TC, Kitayama K (2008) Variations in the soil microbial community composition of a tropical montane forest ecosystem: does tree species matter? *Soil Biology and Biochemistry*, **40**, 2699–2702.

Wakelin SA, Macdonald LM, Rogers SL, Gregg AL, Bolger TP, Baldock JA (2008) Habitat selective factors influencing the structural composition and functional capacity of microbial communities in agricultural soils. *Soil Biology and Biochemistry*, **40**, 803–813.

Wang Q, Garrity GM, Tiedje JM, Cole JR (2007) Naive Bayesian classifier for rapid assignment of rRNA sequences into the new bacterial taxonomy. *Applied and Environmental Microbiology*, **73**, 5261–5267.

Warwick RM, Platt HM, Clarke KR, Agard J, Gobin J (1990) Analysis of macrobenthic and meiobenthic community structure in relation to pollution and disturbance in Hamilton Harbour, Bermuda. *Journal of Experimental Marine Biology and Ecology*, **138**, 119–142.

Yang C-H, Crowley DE (2000) Rhizosphere microbial community structure in relation to root location and plant iron nutritional status. *Applied and Environmental Microbiology*, **66**, 345–351.

Zalesny RS, Donner DM, Coyle DR, Headlee WL (2012) An approach for siting poplar energy production systems to increase productivity and associated ecosystem services. *Forest Ecology and Management*, **284**, 45–58.

Zhang H (2003) *Gemmatimonas aurantiaca* gen. nov., sp. nov., a Gram-negative, aerobic, polyphosphate-accumulating micro-organism, the first cultured representative of the new bacterial phylum Gemmatimonadetes phyl. nov. *International Journal of Systematic and Evolutionary Microbiology*, **53**, 1155–1163.

Zhuang D, Jiang D, Liu L, Huang Y (2011) Assessment of bioenergy potential on marginal land in China. *Renewable and Sustainable Energy Reviews*, **15**, 1050–1056.

Zuccaro A, Lahrmann U, Langen G (2014) Broad compatibility in fungal root symbioses. *Current Opinion in Plant Biology*, **20**, 135–145.

10

Impacts of biogenic CO_2 emissions on human health and terrestrial ecosystems: the case of increased wood extraction for bioenergy production on a global scale

ROSALIE VAN ZELM[1], PATIENCE A. N. MUCHADA[1], MARIJN VAN DER VELDE[2], GEORG KINDERMANN[2], MICHAEL OBERSTEINER[2] and MARK A. J. HUIJBREGTS[1]

[1]Department of Environmental Science, Institute for Water and Wetland Research, Radboud University Nijmegen, P.O. Box 9010, Nijmegen, GL 6500, the Netherlands, [2]International Institute of Applied Systems Analysis, Ecosystem Services and Management Program, Laxenburg A-2361, Austria

Abstract

Biofuels are a potentially important source of energy for our society. Common practice in life cycle assessment (LCA) of bioenergy has been to assume that any carbon dioxide (CO_2) emission related to biomass combustion equals the amount absorbed in biomass, thus assuming no climate change impacts. Recent developments show the significance of contributions of biogenic CO_2 emissions during the time they stay in the atmosphere. The goal of this article is to develop a global, spatially explicit method to quantify the potential impact on human health and terrestrial ecosystems of biogenic carbon emissions coming from forest wood extraction for biofuel production. For this purpose, changes in aboveground carbon stock (ΔC_{forest}) due to an increase in wood extraction via changes in rotation time are simulated worldwide with a 0.5° × 0.5° grid resolution. Our results show that both impacts and benefits can be obtained. When the extraction increase is reached by creating a longer rotation time, new growth is allowed resulting in carbon benefits. In a case study, we assessed the life cycle impacts of heat production via wood to determine the significance of including biogenic CO_2 emissions due to changes in forest management. Impacts of biogenic CO_2 dominate the total climate change impacts from a wood stove. Depending on the wood source country, climate change impacts due to heat production from wood either have an important share in the overall impacts on human health and terrestrial ecosystems, or allow for a large additional CO_2 sink.

Keywords: carbon balance, characterization factor, ecosystems quality, forestry rotation time, global scale modelling, human health impact, life cycle impact assessment, spatially explicit

Introduction

Biofuels are a potentially important source of energy for our society in an era of increasing demand for energy, progressive depletion of fossil fuels, and increasing societal risks from climate change. Wood is the oldest and still one of the main forms of renewable energy (Demirbas, 2001). Global wood usage for both heat and electric energy production has been increasing (FAO, 2010), and advances have been made towards increased production of cellulose-based biofuels, such as bioethanol from wood biomass (Nieminen *et al.*, 2012). This increased focus on renewable energy sources has encouraged a strong debate on how to derive a scientifically valid greenhouse gas (GHG) life cycle balance related to the production and use of biofuels (Schlamadinger *et al.*,

1997; Lal, 2004a; Davidson & Janssens, 2006; Cherubini, 2010; Cherubini & Strømman, 2011; Cherubini *et al.*, 2011a; Levasseur *et al.*, 2012; Brandão *et al.*, 2013). Several GHG sources and sinks need to be considered for bioenergy production, including changes in the sequestered stock of carbon due to biomass extraction, contribution of land use and land use change related emissions, as well as emissions from use of fossil fuels during planting, maintenance and harvesting of the biomass (Cherubini *et al.*, 2011b; Levasseur *et al.*, 2012; Brandão *et al.*, 2013).

Common practice in life cycle assessment (LCA) of bioenergy has been to assume that any carbon dioxide (CO_2) emission related to biomass combustion equals the amount of CO_2 absorbed in biomass, thus assuming a carbon neutral system with no climate change impacts (Cherubini *et al.*, 2011a). Therefore, most LCA studies on biofuels do not include impacts of wood extractions nor emissions of biogenic CO_2 (van der Voet *et al.*, 2010;

Correspondence: Rosalie van Zelm
e-mail: R.vanZelm@science.ru.nl

Cherubini & Strømman, 2011). Lal (2004b) and Ostle & Ward (2012), however, showed that biogenic CO_2 emissions such as those resulting from changes in soil carbon due to biogenic crop cultivation, are a significant flow in the carbon cycle. Next to the fact that CO_2 release from biogenic sources may not be taken up at the original site of release, biogenic CO_2 spends time in the atmosphere before being captured by biomass regrowth, which can possibly lead to climate change related impacts.

Some recent studies proposed ways to improve bioenergy GHG accounting. Müller-Wenk & Brandão (2010) proposed a method to estimate the CO_2 emissions due to transformation of land from one use to another. They focused on the potential carbon stock of a specific land use type compared with the other and estimated the net release to air and the average duration of carbon stay in the atmosphere before taken up elsewhere (e.g. other land and oceans). The Global Warming Potential (GWP) has been used as a climate metric for greenhouse gas emission comparisons (IPCC, 2007). It is defined as the ratio of the integrated radiative forcing of a component in question relative to that of CO_2 over a given future time horizon, after a 1 kg pulse release to the atmosphere. Several other metrics have been studied as well in recent years (Tanaka et al., 2010; Peters et al., 2011; Reisinger et al., 2013). The most prominent is the Global Temperature change Potential (GTP), which is defined as the ratio between the global mean surface temperature change at a given future time horizon following an emission (pulse or sustained) of a compound relative to CO_2 (Shine et al., 2005). Cherubini et al. (2011a, 2012) added a biogenic component to the GWP. Their GWP_{bio} is based on the modelled atmospheric decay of biogenic CO_2 by local and global removal mechanisms. Their method is world generic and primarily depends on the rotation time of the crop of concern and the time horizon selected. In another study, Cherubini et al. (2012) further explored the GWP_{bio}s looking at various forest stands in Northern America and Norway. Kilpeläinen et al. (2011) developed an LCA tool to calculate the net carbon exchange of forest bioenergy production taking into account uptake of carbon into biomass, the decomposition of litter and humus, emissions from forest management operations and carbon released from the combustion of biomass and degradation of wood-based products. The tool has been applied to Finnish case studies (Kilpeläinen et al., 2012; Routa et al., 2012) for which many data are available. Subsequently, Kilpeläinen et al. (2012) determined the radiative forcing associated with net CO_2 emissions in Finnish boreal conditions.

To the best of our knowledge, there is currently no study that quantifies the potential climate change related impacts on humans or the environment, resulting from biogenic CO_2 emissions from forest wood extraction in a spatially explicit way on a global scale, to be applied in LCA case studies. The goal of this article is therefore to develop a global, spatially explicit method to quantify the impacts on human health and terrestrial ecosystems per amount of wood extracted for LCA. For this, effects on the carbon balance due to increases in forest wood extraction for bioenergy through changes in forest rotation time are determined. We link this change in forest carbon sequestration to a change in carbon release to the atmosphere, which in turn alters atmospheric temperature through changes in CO_2's concentration and associated radiative forcing. The resultant change in global mean temperature may lead to impacts on human health and terrestrial ecosystems (De Schryver et al., 2009). The significance of including biogenic emissions due to wood extraction in LCA is tested for the case of heat production from wood extracted from specific countries, compared with heat production from anthracite coal.

Materials and methods

Framework

In LCIA, characterization factors are determined that quantify the impact of emissions to impacts to selected areas of protection such as human health and ecosystem quality. We followed the cause-effect pathway outlined in Figure 1 to determine the effects of changes in wood resource extraction on carbon stock changes and subsequent impacts on human health and terrestrial ecosystems.

When there is an increase in wood extraction for biofuel use, the amount of aboveground carbon stock in the forest changes, followed by a change in the quantity of atmospheric carbon. Changes in the related CO_2 concentration in the air can result in a temperature change worldwide and cause global warming due to changes in radiative forcing. Changes in global mean temperature cause impacts on human health, which can be expressed as Disability Adjusted Life Years (DALY), and on terrestrial ecosystem quality in terms of potentially disappeared fraction of species (PDF). This pathway translates into a characterization model that quantifies the impact per unit of extraction in terms of a characterization factor (CF):

$$CF_e = \frac{-\Delta C_{forest}}{\Delta R} \cdot \frac{\Delta T}{-\Delta C_{forest}} \cdot \frac{\Delta I_e}{\Delta T} \qquad (1)$$

where ΔR is the change in wood resource extraction ($m^3 \ yr^{-1}$), $-\Delta C_{forest}$ is the reduction in the stock of carbon in the forest (tC), ΔT is the change in global mean temperature (°C) and ΔI_e is the change in impact at endpoint e, i.e. humans or ecosystems. CFs were determined for a change in PDF of terrestrial species, including birds, butterflies, mammals and plants, caused by a change in wood extraction in managed forests (PDF yr m^2 m^{-3} of wood extracted) and the change in impacts

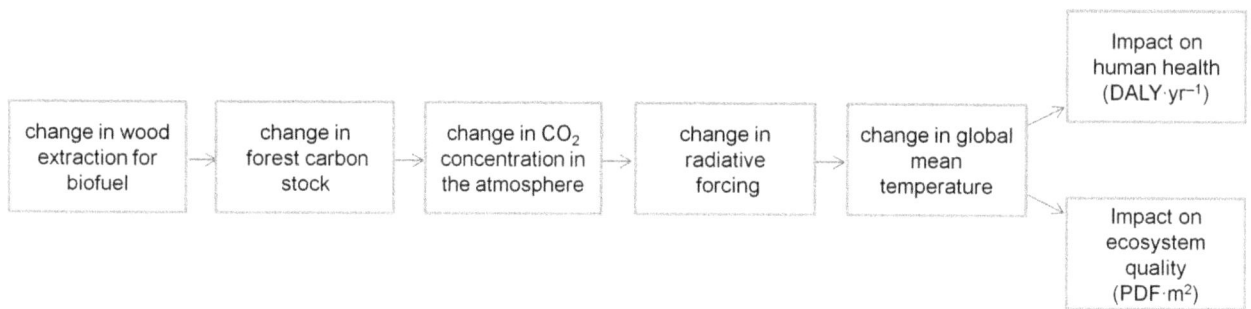

Fig. 1 Cause-effect pathway used in determining the effect of increased wood resource extraction on human health and terrestrial ecosystem quality (adapted from De Schryver *et al.* (2009)).

on human health in DALYs related to a number of climate sensitive diseases and health risks (malaria, malnutrition, drowning, diarrhoea and cardio-vascular diseases) (DALY m^{-3} of wood extracted).

CFs were determined for a time horizon of 100 years, following the IPCC (Intergovernmental Panel on Climate Change) default time horizon in the calculation of GWPs. Since most process inventory data (e.g. wood extracted per source location) are available on country level, country-specific CFs were determined as well. For this, grid-specific factors were aggregated based on the G4M simulated extraction rates.

Forest growth and management simulation

To determine the change in carbon stock due to a change in wood resource extraction ($-\Delta C_{\mathrm{forest}}/\Delta R$), the Global forestry model (G4M) was applied. G4M simulates the processes of growth and management of forests on a global scale (Gusti *et al.*, 2008; Kindermann *et al.*, 2013). It calculates aboveground forest biomass and amounts extracted on a 0.5° × 0.5° resolution in a yearly time step. Model inputs include global land cover, which outlines the forest area and carbon stock. The forest carbon stock was determined based on Net primary productivity (NPP) data and biomass quantities. NPP data were taken from Cramer *et al.* (1999) and used to describe the site productivity of the forest. The current biomass quantity was determined based on the forest standing biomass, which is reported by FAO statistics on a country level and downscaled as described by Kindermann *et al.* (2008). The forests modelled by G4M consist of generic trees, i.e. individual species are not distinguished. For the forest carbon stock we considered the aboveground biomass carbon only, as was done by Cherubini *et al.* (2011a).

In G4M, extraction rates depend on availability of extractable wood, which is largely determined by the forest rotation time. The rotation time is in turn dependent on the site productivity of the forest. Fast growing stands typically have a shorter rotation length, while slow growing sites have a longer one (Kindermann *et al.*, 2006). The wood extracted is an estimated potential based on the site productivity, stocking biomass and age structure. The model assumes wood harvesting in a forest by means of continuous-cover forestry with selective harvesting to keep or move to a continuous extraction and constant

rotation time. For this study, we simulated a change in wood extraction in every grid by employing a 5% change in rotation time of current managed forests. The response of the G4M model is approximately linear for such a relatively small change. This change resulted in either an increase or decrease in wood extraction per grid. From an LCA perspective we are interested in an increase in wood extraction from the forest, therefore we proceeded with the grids in which we observed an increase in wood extraction, either caused by a decrease in rotation time, or an increase. Calculations were performed and analysed using ArcGIS 9.2 and presented at 0.5° × 0.5° grid resolution.

Environmental impacts

The negative change in forest carbon stock is linked to global mean temperature change via three subsequent steps. First, the carbon released to the atmosphere ($-\Delta C_{\mathrm{forest}}$) was converted into ppb CO_2 via the conversion factor 5.17×10^{-7} ppb $CO_2 \cdot$t C^{-1} (De Schryver *et al.*, 2009). Second, a change in concentration results in a change in the radiative forcing (W m^{-2}) by a factor of 1.4×10^{-5} W m$^{-2} \cdot$ppb^{-1} (Forster *et al.*, 2007). Third, the change in radiative forcing causes a change in the global mean temperature of 0.48 °C W^{-1} m^2 for a 100 year time horizon as calculated by De Schryver *et al.* (2009) using the IMAGE model (Eickhout *et al.*, 2004).

For the change in impacts on human health and terrestrial ecosystems caused by a change in global mean temperature ($\Delta I_e/\Delta T$), we directly used the so-called damage factors from De Schryver *et al.* (2009). We followed the hierarchical perspective, of which the choices are based on the level of (scientific) consensus. This perspective is most frequently used by organizations like the International Organization for Standardization (ISO), and is generally applied in LCA (De Schryver, 2011). For human health impacts, value choices regarding a future scenario, certainty about climate-related health impacts, age weighting and discount rate are of importance in the DALY calculations. We assumed a baseline future scenario with no age weighting and discount rate at 3% in the DALY-calculations. An intermediate number of diseases was included, most dominantly malnutrition, resulting in a $\Delta I_e/\Delta T$ of 6.12×10^6 yr yr$^{-1} \cdot$ °C^{-1}. For ecosystem impacts, the species' resilience to climate change according to dispersal ability and level of protection based on the IUCN Red list classification (IUCN, 2001) are of importance. Inclusion of impacts to all species and a dispersal of species resulted in a $\Delta I_e/\Delta T$ of 0.06 PDF·°C^{-1} over 10.8×10^{13} m^2.

Case study

We performed a case study on heat production to (i) test the significance of including the biogenic carbon emissions due to wood extraction in LCA, and (ii) compare the impacts of using wood biomass for heat production to the impact of using fossil fuel, in this case coal anthracite. We calculated human health and ecosystem quality impacts for a functional unit (FU) of "1 MJ heat production". The systems compared were a 6 kW wood heater (assuming hardwood as input) with 75% efficiency and a 5–15 kW coal anthracite furnace with 70% efficiency. Emission data were taken from Eco-invent v2.2 (Hischier *et al.*, 2010). We assumed that about 30% of the extracted wood is left on the field and not harvested (Kindermann *et al.*, 2006). The energy content of the wood was estimated to be 9.6×10^3 MJ m^{-3} based on values from Werner *et al.* (2007). For the coal anthracite furnace, world average CFs were used to estimate biogenic CO$_2$ impacts. The impact of biogenic CO$_2$ emissions due to wood sourcing were determined for wood sourced from the most wood fuel producing country of each continent and of the European Union. These countries are the United States of America (USA), Russia, Germany, India, Ethiopia, Australia and Brazil and were identified from the FAO database's latest (i.e. year 2011) figures for total wood fuel production (http://faostat.fao.org/site/626/default.aspx#ancor).

Characterization factors for greenhouse gas emissions from fossil energy sources were taken from De Schryver *et al.* (2009). Next to CO$_2$, CH$_4$ and N$_2$O emissions contributed to global warming. Other impacts included in the case study were, for human health impacts (DALY kg^{-1}), fine particulate matter formation, photochemical oxidant formation, human toxicity, ozone depletion and ionizing radiation. Impacts to terrestrial ecosystems (PDF m^2 yr kg^{-1} or PDF m^2 yr m^{-2} yr^{-1}) included were acidification, land use (change) and ecotoxicity. These

endpoint CFs were taken from the ReCiPe methodology, hierarchist perspective (Goedkoop *et al.*, 2009).

Results

Characterization factors

Figure 2 and Figure 3 show grid-specific characterization factors for impacts on human health (DALY m^{-3} wood) and ecosystem quality (PDF m^2 yr m^{-3} wood) due to increased wood extraction. Country average CFs are in Table 1. Only countries are included that contain at least one grid cell of which forestry is the dominant land cover. Grids in grey are areas where wood extraction is at its maximum and therefore an increase is assumed not realistic. Grids without CF values, for example in northern Africa, do not contain forest cover.

Both positive CFs, i.e. impacts to humans and the environment, and negative CFs, i.e. benefits to human health and the environment, were obtained. Positive CFs were mainly observed in tropical climate regions, i.e. Central and South Africa, Malaysia, and Mid– and South America. Also, central and Eastern Europe showed negative CFs. Northern and South Europe, Russia, Australia and most of Northern America showed negative CFs. In Southern America, also many places showed negative CFs.

Impacts due to heat energy production

Figure 4 shows the contribution of each impact category to the total impact on human health (DALY) caused by the production of 1 MJ of heat. Wood combustion has slightly smaller impacts on human health as coal anthracite combustion, if the contribution of biogenic carbon changes in aboveground forest biomass is neglected. Accounting for biogenic carbon changes, the

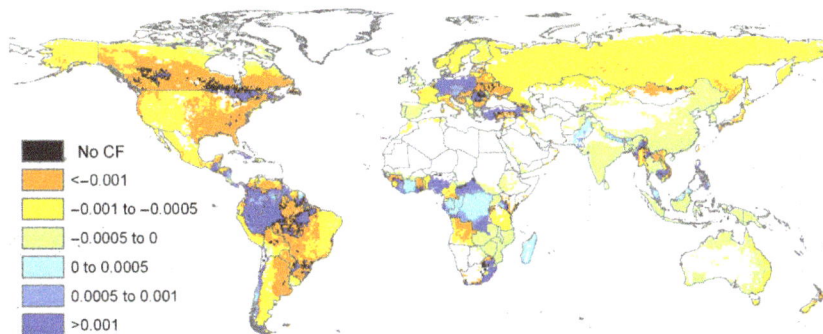

Fig. 2 Characterization factors for impacts on human health due to forest wood extraction for bioenergy production [disability adjusted life years (DALY) m^{-3} of wood] at grid level. Grids in dark grey are areas where wood extraction is already close to its optimum with the current rotation time therefore changes in rotation time only result in decreased harvest. Grids with no data, shown in white, represent areas with no forest cover.

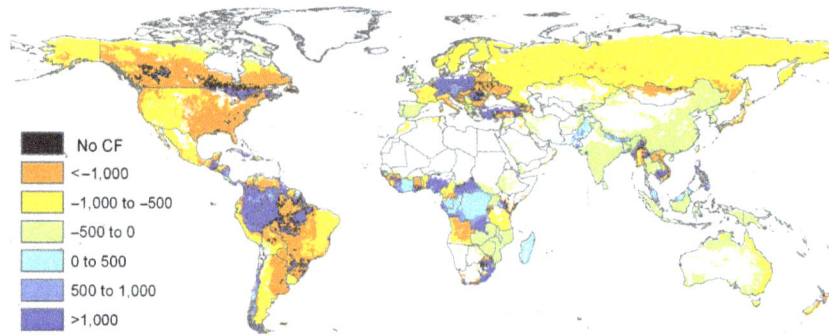

Fig. 3 Characterization factors for impacts on ecosystem quality due to forest wood extraction for bioenergy production (PDF m^2 yr m^{-3} of wood) at grid level. Grids in dark grey are areas where wood extraction is already close to its optimum with the current rotation time therefore changes in rotation time only result in decreased harvest. Grids with no data, shown in white, represent areas with no forest cover.

total human health impact by heat production from wood from Brazil and Germany was larger than the human health impacts resulting from use of coal anthracite. For the other countries, there are biogenic carbon storage benefits.

Figure 5 shows the contribution per impact category to the total impact on ecosystem quality caused by production of 1 MJ of heat (PDF m^2 yr). Unlike in the case of human health, wood combustion has impacts that are twice as large as coal anthracite combustion, if the contribution of biogenic carbon changes in aboveground forest biomass is neglected. Accounting for biogenic carbon changes, the total impact on terrestrial ecosystems by heat production from wood from Brazil and Germany even becomes four times larger than the impacts resulting from use of coal anthracite. Again, for the other countries, there are biogenic carbon storage benefits.

Discussion

We discuss here the main limitations in our global scale modelling of carbon changes in forests and subsequent climate change related impacts on human health and terrestrial ecosystems. We then provide an interpretation of our presented results followed by a discussion on their application in LCAs of bioenergy production.

Model uncertainty

Simplifications are inherent to global scale modelling, due to necessary generalizations and data limitations. Here, we outline the main model uncertainties. One limitation is that we modelled only aboveground biomass carbon without considering potential other pools of forest carbon, such as belowground biomass, soil organic carbon (SOC) and dead wood. The carbon content in belowground living biomass such as roots is much lower than the carbon content in aboveground biomass, and although SOC is a significant component of total forest carbon, a review by Jandl et al. (2007) shows that continuous-cover forestry, including selective harvesting, causes much less SOC losses compared with clear-cut harvesting. Changes in SOC do depend on harvesting techniques, but effects when not clear-cutting are generally small (Johnson & Curtis, 2001).

We assumed that the carbon that is no longer in the forest contributes to climate change. However, only the carbon that stays in the air will contribute to a temperature change. Next to the uptake in local forests, there is uptake in oceans and terrestrial biosphere worldwide. Cherubini et al. (2011a) included this uptake in their GWP_{bio} calculations, i.e. following a pulse emission of carbon. They applied an impulse response function, integrating uptake in oceans, terrestrial biosphere and local vegetation. To include the uptake in oceans and terrestrial biosphere in our flux approach, a more integrated model is needed, including both local and global carbon uptake.

We were not able to include albedo effects in the current study. The albedo of a forested landscape is generally lower when snow is lying. Decreased albedo exerts a positive radiative forcing on climate. This can lead to underestimations of the effects in regions that experience seasonal snow cover, like many boreal forests (Betts, 2000; Cherubini et al., 2012). It would therefore be interesting to include albedo effects in future damage assessment work.

Although our study shows a large spatial variability, this is still underestimated. First, because the forest modelled in G4M is a generic forest, identical for the whole globe, modelled with an assumed equal share area for all age classes. In reality, forests differ in terms of e.g. tree species or wood density. Some studies

Table 1 Country-specific characterization factors for human health and ecosystems

Country	Human health DALY m^3	Ecosystems PDF m^2 yr m^{-3}	Country	Human health DALY m^3	Ecosystems PDF m^2 yr m^{-3}
Afghanistan	−9.6E-05	−1.0E+02	Lao People's Democratic Republic (Laos)	−1.6E-03	−1.7E+03
Albania	−8.4E-04	−8.9E+02	Latvia	−9.2E-03	−9.7E+03
Algeria	−8.5E-04	−9.0E+02	Lebanon	−9.4E-05	−9.9E+01
Angola	−2.0E-03	−2.2E+03	Lesotho	9.1E-05	9.6E+01
Argentina	−1.6E-03	−1.6E+03	Liberia	8.4E-04	8.8E+02
Armenia	−1.1E-03	−1.1E+03	Lithuania	−9.8E-04	−1.0E+03
Australia	−5.1E-04	−5.3E+02	Luxembourg	2.3E-03	2.4E+03
Austria	7.7E-04	8.1E+02	Madagascar	1.8E-04	1.9E+02
Azerbaijan	−9.9E-03	−1.0E+04	Malawi	−3.7E-04	−3.9E+02
Bangladesh	−2.5E-04	−2.6E+02	Malaysia	3.9E-04	4.1E+02
Belarus	−1.2E-03	−1.2E+03	Mali	−9.9E-05	−1.0E+02
Belgium	−2.5E-03	−2.6E+03	Mexico	−1.1E-03	−1.2E+03
Belize	−1.4E-04	−1.5E+02	Moldova	−7.4E-04	−7.9E+02
Benin	−9.0E-04	−9.6E+02	Mongolia	−7.4E-03	−7.9E+03
Bhutan	3.8E-04	4.0E+02	Montenegro	−5.6E-04	−5.9E+02
Bolivia	−2.6E-03	−2.7E+03	Morocco	−6.6E-04	−7.0E+02
Bosnia and Herzegovina	−2.1E-03	−2.2E+03	Mozambique	−2.4E-04	−2.5E+02
Botswana	−3.1E-04	−3.3E+02	Myanmar (Burma)	3.2E-03	3.4E+03
Brazil	2.1E-03	2.2E+03	Namibia	−2.9E-04	−3.0E+02
Brunei	5.6E-04	5.9E+02	Nepal	3.3E-04	3.5E+02
Bulgaria	−1.7E-03	−1.7E+03	Netherlands	−1.0E-03	−1.1E+03
Burkina Faso	−4.8E-04	−5.1E+02	New Zealand	−2.6E-03	−2.7E+03
Burundi	−1.4E-03	−1.4E+03	Nicaragua	6.7E-04	7.1E+02
Cambodia	2.8E-03	3.0E+03	Niger	−4.7E-04	−4.9E+02
Cameroon	−1.6E-03	−1.7E+03	Nigeria	1.2E-03	1.3E+03
Canada	−4.9E-03	−5.2E+03	Norway	−7.7E-04	−8.2E+02
Central African Republic	2.2E-03	2.3E+03	Oman	−3.1E-02	−3.3E+04
Chad	−3.0E-03	−3.2E+03	Pakistan	2.5E-04	2.7E+02
Chile	5.9E-04	6.2E+02	Panama	8.7E-04	9.2E+02
China	−1.9E-04	−2.0E+02	Papua New Guinea	−3.1E-04	−3.2E+02
Colombia	1.2E-03	1.2E+03	Paraguay	−1.5E-03	−1.6E+03
Congo	1.8E-04	1.9E+02	Peru	1.5E-03	1.6E+03
Costa Rica	−3.0E-04	−3.2E+02	Philippines	1.2E-03	1.3E+03
Côte D'Ivoire	2.0E-04	2.1E+02	Poland	3.4E-03	3.6E+03
Croatia	−1.0E-02	−1.1E+04	Portugal	−1.2E-04	−1.3E+02
Cuba	2.6E-03	2.7E+03	Qatar	1.6E-03	1.7E+03
Czech Republic	6.8E-04	7.2E+02	Republic of Korea (South Korea)	−1.9E-04	−2.1E+02
Democratic People's Republic of Korea (North Korea)	−2.3E-04	−2.4E+02	Romania	1.7E-02	1.8E+04
Democratic Republic of the Congo	5.6E-04	5.9E+02	Russian federation	−7.5E-04	−8.0E+02
Denmark	−4.6E-04	−4.9E+02	Rwanda	−3.2E-03	−3.3E+03
Dominican Republic	−3.2E-04	−3.4E+02	Saudi Arabia	−2.5E-04	−2.7E+02
Ecuador	1.6E-03	1.7E+03	Senegal	−4.6E-04	−4.8E+02
Egypt	7.4E-04	7.8E+02	Serbia	−5.5E-04	−5.8E+02
El Salvador	−4.9E-03	−5.2E+03	Sierra Leone	−1.5E-03	−1.6E+03
Equatorial Guinea	−5.4E-04	−5.8E+02	Slovakia	1.4E-03	1.4E+03
Eritrea	−6.2E-04	−6.5E+02	Slovenia	1.2E-03	1.3E+03
Estonia	−6.4E-03	−6.8E+03	Somalia	1.9E-03	2.0E+03
Ethiopia	−7.7E-05	−8.2E+01	South Africa	5.1E-03	5.3E+03

(continued)

Table 1 (continued)

Country	Human health DALY m^3	Ecosystems PDF m^2 yr m^{-3}	Country	Human health DALY m^3	Ecosystems PDF m^2 yr m^{-3}
Falkland Islands (Islas Malvinas)	−8.4E-04	−8.9E+02	Spain	−1.2E-04	−1.3E+02
Finland	−4.9E-04	−5.2E+02	Sri Lanka	−8.3E-05	−8.8E+01
France	−9.0E-04	−9.6E+02	Sudan	−1.2E-04	−1.2E+02
French Guiana	7.3E-04	7.7E+02	Suriname	−1.1E-03	−1.1E+03
Gabon	5.3E-04	5.6E+02	Swaziland	−3.3E-04	−3.5E+02
Gambia, The	4.0E-02	4.2E+04	Sweden	−6.1E-04	−6.5E+02
Georgia	1.1E-02	1.2E+04	Switzerland	8.5E-04	9.0E+02
Germany	1.5E-03	1.6E+03	Syrian Arab Republic	−5.6E-04	−5.9E+02
Ghana	−8.1E-03	−8.6E+03	Taiwan	−1.4E-03	−1.5E+03
Greece	−1.1E-04	−1.2E+02	Tajikistan	−1.1E-04	−1.1E+02
Guatemala	4.8E-03	5.1E+03	Tanzania, United Republic of	−7.7E-04	−8.1E+02
Guinea	−1.8E-03	−1.9E+03	Thailand	−2.6E-04	−2.8E+02
Guinea-Bissau	−2.2E-04	−2.3E+02	The former Yugoslav Republic of Macedonia	−1.3E-04	−1.4E+02
Guyana	6.1E-03	6.4E+03	Togo	−1.1E-03	−1.1E+03
Haiti	−4.8E-04	−5.1E+02	Trinidad and Tobago	1.8E-03	2.0E+03
Honduras	−3.8E-03	−4.1E+03	Tunisia	−7.6E-05	−8.0E+01
Hungary	−1.6E-02	−1.7E+04	Turkey	2.0E-03	2.1E+03
India	−2.8E-04	−2.9E+02	Turkmenistan	−9.3E-05	−9.9E+01
Indonesia	−3.4E-04	−3.6E+02	Uganda	−1.7E-04	−1.8E+02
Iran	−4.6E-04	−4.9E+02	Ukraine	−1.5E-02	−1.6E+04
Iraq	−7.1E-04	−7.5E+02	United Kingdom	−3.0E-04	−3.2E+02
Ireland	−1.6E-04	−1.7E+02	United States of America	−3.4E-03	−3.6E+03
Israel	−2.4E-04	−2.6E+02	Uruguay	−5.3E-04	−5.7E+02
Italy	−1.2E-03	−1.3E+03	Uzbekistan	−5.8E-05	−6.2E+01
Japan	−1.9E-03	−2.0E+03	Venezuela	2.4E-02	2.5E+04
Jordan	−3.1E-04	−3.2E+02	Vietnam	−2.9E-03	−3.1E+03
Kazakhstan	−6.8E-04	−7.2E+02	West Bank	−6.2E-04	−6.5E+02
Kenya	2.0E-04	2.1E+02	Yemen	−1.7E-04	−1.8E+02
Kuwait	−1.5E-03	−1.6E+03	Zambia	−1.5E-04	−1.6E+02
Kyrgyzstan	−1.3E-04	−1.4E+02	Zimbabwe	−2.7E-04	−2.9E+02

showed the effect of tree species type on carbon stock dynamics. Kaipainen *et al.* (2004), for example, found that growth of spruce tree is more age-dependent than growth of pine tree. Also the yield tables from Marschall (1975) suggest that even for the same species in a relatively small area, the growth dependency with age varies. This shows that region-specific differences can even be larger than indicated in our simulations. Decisions on forestry management activities such as fertilization and harvesting technique also differ between or within regions (González-García *et al.*, 2009; Michelsen *et al.*, 2012). However, including more region-specific detail of the forest is currently not possible on a global scale due to data limitations. Furthermore, the type of harvesting varies around the globe as well, leading to different carbon cycles, such as when forests are clear-cut (Cherubini *et al.*, 2011b). Second, the grid-specific

biomass stock was downscaled from FAO country estimates by using the relationship between biomass, NPP and human impact (Kindermann *et al.*, 2008). This method highly depends on data input reported by countries and assumed relationships between NPP and aboveground biomass (Kindermann *et al.*, 2008). Third, no differences in forest management was assumed. In reality, some forests are clear-cut at their specific rotation time and the time-dependent carbon cycle can be nonlinear.

Interpretation

Our CFs quantify the contribution to global warming and resultant impact on human health and terrestrial ecosystems of biogenic CO_2 emissions experienced worldwide due to forest wood extraction for bioenergy.

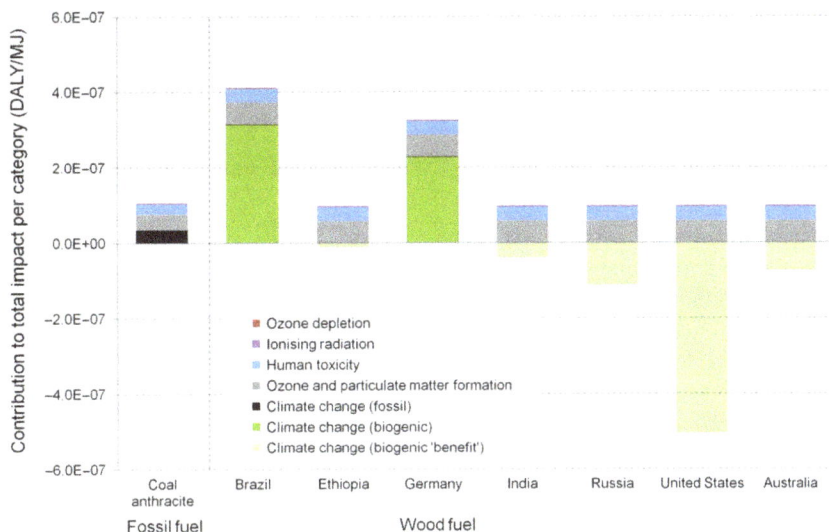

Fig. 4 Impact on human health due to production of 1MJ heat energy [disability adjusted life years (DALY)/MJ] using coal anthracite or wood sourced from the indicated countries.

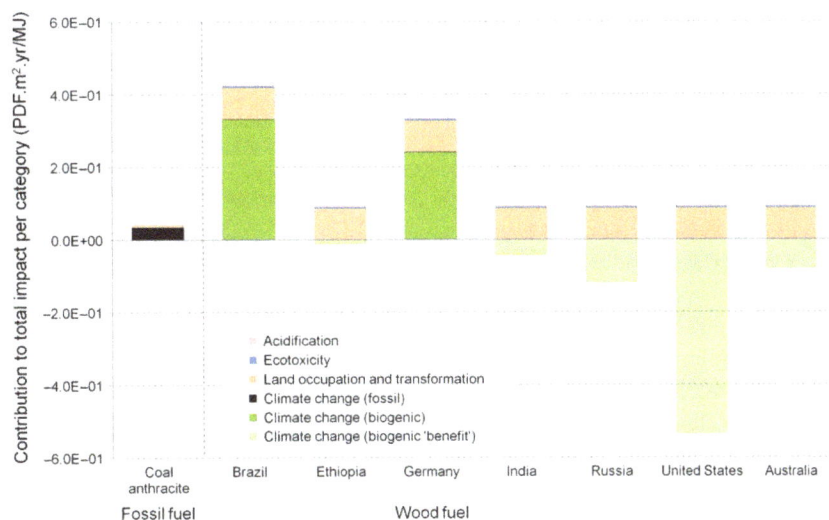

Fig. 5 Impact on ecosystem quality due to production of 1 MJ heat energy (PDF m^2 yr/MJ) using coal anthracite or wood sourced from the indicated countries.

Negative CFs for impacts on human health should not be interpreted as an addition of years to human life. Similarly, the negative CFs for impacts on terrestrial ecosystems do not mean introduction of new species in the other parts. They only indicate that the applied management activity, in this case increased wood extraction due to changes in rotation time, results in an increase in carbon sequestration in the forest. This is because the extraction increase is reached by creating a longer rotation time to allow for new growth to reach a new steady state. It thus allows for extra CO_2 uptake in

the forest, that can also be fossil CO_2 emissions. This is especially the case for longer rotation times, as they allow for older forests, which have a larger carbon stock than young ones (Davidson & Janssens, 2006; Hergoualc'h & Verchot, 2011; Zhang et al., 2012). An increase in the forest carbon stock implies a decrease in atmospheric carbon pool, thereby slowing down global mean temperature rise and the resultant impact on humans and ecosystems. In an LCA case study, such as the heat production in this study, we propose that climate benefits due to wood extraction can only compensate climate impacts from fossil-related activities, and not from other environmental stressors, such as toxicants and fine particulate matter.

Application

Impacts caused by forest wood extraction are dependent on a number of site-specific aspects, such as biomass stock, rotation times, and climatic conditions. Site-specific contributions to global warming are important in the determination of the environmental impacts of a product taking into account the origin of the wood. We recommend to apply the CFs on the smallest scale possible, to capture the detailed information they are based on. However, the chosen resolution for the CFs needs to be applicable in LCA, i.e. it needs to match up the spatial scale of the inventory data. Our starting point was the smallest scale currently possible, $0.5° \times 0.5°$. These grid level CFs can be aggregated to a larger spatial scale, such as a country, but also to ecoregions or biomes, when necessary. Such manual aggregations, however, result in loss of spatial detail. A large forested country as Brazil, for example, has country-specific CFs of 2.1×10^{-3} DALY m^{-3} and 2.2×10^{3} PDF·m^2 yr m^{-3}. These CFs vary, however, throughout the country from -0.5 to 1.2 DALY m^{-3} and -5.0×10^{5} to 1.3×10^6 PDF·m^2 yr m^{-3}. Another example is Russia, with country-specific CFs of -7.6×10^{-4} DALY m^{-3} and -8.0×10^{2} PDF·m^2 yr m^{-3}, but ranges from -5.8×10^{-3} to -1.7×10^{-4} DALY m^{-3} and -6.1×10^{3} to -1.8×10^{2} PDF m^2 yr m^{-3}. Care should therefore be taken when applying country-specific averages, and preferably ranges should be indicated as well. Mutel *et al.* (2012) proposed the application of geostatistical analysis tools such as autocorrelation analysis towards defining an optimum spatial scale for impact characterization. However, most life cycle inventory data are currently available at a minimum scale of country level.

Our CFs are based on the total amount that is extracted from a forest, i.e. the wood that is cut. However, about 30% of this extracted wood is on average left on the field and not harvested (Kindermann *et al.*, 2006). To combine our CFs with LCI data reporting harvested amounts, these amounts first need to be converted to extracted wood. When no data is available on the amount left on the field we recommend to use the generic 30% from Kindermann *et al.* (2006).

Results from the heat production case study depict the influence of spatial factors in determining the effect of global wood combustion. This is evidenced by the differences in the sign and magnitude of the biogenic emission impacts due to wood extraction from one country to another. The results show that in current LCAs of bioenergy, climate change impacts due to heat production from wood are miscalculated when biogenic emissions due to wood extraction are disregarded.

Acknowledgements

This research was partly funded by the European Commission under the 7th framework program on environment; ENV.2008.3.3.2.1: PROSUITE - Sustainability Assessment of Technologies, grant agreement number 227078, and ENV.2009.3.3.2.1: LC-IMPACT – Improved Life Cycle Impact Assessment methods (LCIA) for better sustainability assessment of technologies, grant agreement number 243827. We thank two anonymous reviewers for their helpful comments on the previous version of this document.

References

Betts RA (2000) Offset of the potential carbon sink from boreal forestation by decreases in surface albedo. *Nature*, **408**, 187–190.

Brandão M, Levasseur A, Kirschbaum MUF *et al.* (2013) Key issues and options in accounting for carbon sequestration and temporary storage in life cycle assessment and carbon footprinting. *International Journal of Life Cycle Assessment*, **18**, 230–240.

Cherubini F (2010) GHG balances of bioenergy systems – Overview of key steps in the production chain and methodological concerns. *Renewable Energy*, **35**, 1565–1573.

Cherubini F, Strømman AH (2011) Life cycle assessment of bioenergy systems: state of the art and future challenges. *Bioresource technology*, **102**, 437–451.

Cherubini F, Peters GP, Berntsen T, Strømman AH, Hertwich E (2011a) CO$_2$ emissions from biomass combustion for bioenergy: atmospheric decay and contribution to global warming. *Global Change Biology Bioenergy*, **3**, 413–426.

Cherubini F, Stromman AH, Hertwich E (2011b) Effects of boreal forest management practices on the climate impact of CO$_2$ emissions from bioenergy. *Ecological Modelling*, **223**, 59–66.

Cherubini F, Bright RM, Stromman AH (2012) Site-specific global warming potentials of biogenic CO$_2$ for bioenergy: contributions from carbon fluxes and albedo dynamics. *Environmental Research Letters*, **7**, 045902.

Cramer W, Kicklighter D, Bondeau A *et al.* (1999) Comparing global models of terrestrial net primary productivity (NPP): overview and key results. *Global Change Biology*, **5**, 1–15.

Davidson EA, Janssens IA (2006) Temperature sensitivity of soil carbon decomposition and feedbacks to climate change. *Nature*, **440**, 165–173.

De Schryver AM (2011) *Value Choices in Life Cycle Impact Assessment*. PhD thesis, Radboud University, Nijmegen, The Netherlands.

De Schryver AM, Brakkee KW, Goedkoop MJ, Huijbregts MAJ (2009) Characterization factors for global warming in life cycle assessment based on damages to humans and ecosystems. *Environmental science & technology*, **43**, 1689–1695.

Demirbas A (2001) Biomass resource facilities and biomass conversion processing for fuels and chemicals. *Energy Conversion and Management*, **42**, 1357–1378.

Eickhout B, De Elzen M, Kreileman G (2004) *The Atmosphere-Ocean System of IMAGE 2.2*. National Institute for Human Health and the Environment (RIVM), Bilthoven, The Netherlands.

FAO (2010) *Global Forest Resources Assessment 2010*. Food Agriculture Organization of the United Nations, Rome, Italy.

Forster P, Ramaswamy V, Artaxo P *et al.* (2007) Changes in atmospheric constituents and in radiative forcing. In: *Climate Change 2007: The Physical Science Basis. Contribution of Working Group I to the Fourth Assessment Report of the Intergovernmental Panel on Climate Change.* (eds Solomon S, Qin D, Manning M, Chen Z, Marquis M, Averyt K, Tignor M, Miller H), pp. 129–234. Cambridge University Press, Cambridge, United Kingdom.

Goedkoop M, Heijungs R, Huijbregts M, De Schryver A, Struijs J, Van Zelm R (2009) *ReCiPe 2008: A life cycle impact assessment method which comprises harmonised category indicators at the midpoint and the endpoint level.* VROM–Ruimte en Milieu, Ministerie van Volkshuisvesting, Ruimtelijke Ordening en Milieubeheer.

González-García S, Berg S, Feijoo G, Moreira M (2009) Environmental impacts of forest production and supply of pulpwood: Spanish and Swedish case studies. *International Journal of Life Cycle Assessment*, **14**, 340–353.

Gusti M, Havlik P, Obersteiner M (2008) *Technical description of the IIASA model cluster*. International Institute for Applied Systems Analysis (IIASA), Laxenburg.

Hergoualc'h K, Verchot LV (2011) Stocks and fluxes of carbon associated with land use change in Southeast Asian tropical peatlands: a review. *Global Biogeochemical Cycles*, **25**, GB2001. doi: 2010.1029/2009GB003718.

Hischier R, Weidema B, Althaus H-J *et al.* (2010) *Implementation of Life Cycle Impact Assessment Methods. Data v2.2.*, Swiss Centre for Life Cycle Inventories, St. Gallen, Switzerland.

IPCC (2007) Climate change 2007: The physical science basis. *Contribution of Working Group I to The Fourth Assessment Report of the Intergovernmental Panel on Climate Change.* (eds Solomon S, Qin D, Manning M, Chen Z, Marquis M, Averyt KB, M T, Miller HL). IPCC, Cambridge, New York.

IUCN (2001) *IUCN Red List Categories and Criteria: Version 3.1.* IUCN–The World Conservation Union Gland, Switzerland.

Jandl R, Lindner M, Vesterdal L *et al.* (2007) How strongly can forest management influence soil carbon sequestration? *Geoderma*, **137**, 253–268.

Johnson DW, Curtis PS (2001) Effects of forest management on soil C and N storage: meta analysis. *Forest Ecology and Management*, **140**, 227–238.

Kaipainen T, Liski J, Pussinen A, Karjalainen T (2004) Managing carbon sinks by changing rotation length in European forests. *Environmental Science & Policy*, **7**, 205–219.

Kilpeläinen A, Alam A, Strandman H, Kellomäki S (2011) Life cycle assessment tool for estimating net CO_2 exchange of forest production. *Global Change Biology Bioenergy*, **3**, 461–471.

Kilpeläinen A, Kellomäki S, Strandman H (2012) Net atmospheric impacts of forest bioenergy production and utilization in Finnish boreal conditions. *Global Change Biology Bioenergy*, **4**, 811–817.

Kindermann GE, Obersteiner M, Rametsteiner E, McCallum I (2006) Predicting the deforestation-trend under different carbon-prices. *Carbon Balance and Management*, **1**, 15.

Kindermann GE, McCallum I, Fritz S, Obersteiner M (2008) A global forest growing stock, biomass and carbon map based on FAO statistics. *Silva Fennica*, **42**, 387–396.

Kindermann G, Schorghuber S, Linkosalo T, Sanchez A, Rammer W, Seidl R, Lexer M (2013) Potential stocks and increments of woody biomass in the European Union under different management and climate scenarios. *Carbon Balance and Management*, **8**, 2.

Lal R (2004a) Soil carbon sequestration impacts on global climate change and food security. *Science*, **304**, 1623–1627.

Lal R (2004b) Soil carbon sequestration to mitigate climate change. *Geoderma*, **123**, 1–22.

Levasseur A, Lesage P, Margni M, Brandao M, Samson R (2012) Assessing temporary carbon sequestration and storage projects through land use, land-use change and forestry: comparison of dynamic life cycle assessment with ton-year approaches. *Climatic Change*, **115**, 759–776.

Marschall J (1975) *Hilfstafeln für die Forsteinrichtung*. Österreichischer Agrarverlag, Vienna, Austria.

Michelsen O, Cherubini F, Strømman AH (2012) Impact assessment of biodiversity and carbon pools from land use and land use changes in life cycle assessment, exemplified with forestry operations in Norway. *Journal of Industrial Ecology*, **16**, 231–242.

Müller-Wenk R, Brandão M (2010) Climatic impact of land use in LCA—carbon transfers between vegetation/soil and air. *International Journal of Life Cycle Assessment*, **15**, 172–182.

Mutel CL, Pfister S, Hellweg S (2012) GIS-based regionalized life cycle assessment: how big is small enough? Methodology and case study of electricity generation. *Environmental science & technology*, **46**, 1096–1103.

Nieminen K, Robischon M, Immanen J, Helariutta Y (2012) Towards optimizing wood development in bioenergy trees. *New Phytologist*, **194**, 46–53.

Ostle N, Ward S (2012) Climate change and soil biotic carbon cycling. In: *Soil Ecology and Ecosystem Services* (ed. Wall D), pp. 241–255. Oxford University Press, Oxford, United Kingdom.

Peters GP, Aamaas B, Berntsen T, Fuglestvedt JS (2011) The integrated global temperature change potential (iGTP) and relationships between emission metrics. *Environmental Research Letters*, **6**, 044021.

Reisinger A, Havlik P, Riahi K, van Vliet O, Obersteiner M, Herrero M (2013) Implications of alternative metrics for global mitigation costs and greenhouse gas emissions from agriculture. *Climatic Change*, **117**, 677–690.

Routa J, Kellomäki S, Strandman H (2012) Effects of forest management on total biomass production and CO_2 emissions from use of energy biomass of Norway spruce and Scots pine. *BioEnergy Research*, **5**, 733–747.

Schlamadinger B, Apps M, Bohlin F *et al.* (1997) Towards a standard methodology for greenhouse gas balances of bioenergy systems in comparison with fossil energy systems. *Biomass and Bioenergy*, **13**, 359–375.

Shine KP, Fuglestvedt JS, Hailemariam K, Stuber N (2005) Alternatives to the global warming potential for comparing climate impacts of emissions of greenhouse gases. *Climatic Change*, **68**, 281–302.

Tanaka K, Peters GP, Fuglestvedt JS (2010) Multicomponent climate policy: why do emission metrics matter? *Carbon Management*, **1**, 191–197.

van der Voet E, Lifset RJ, Luo L (2010) Life-cycle assessment of biofuels, convergence and divergence. *Biofuels*, **1**, 435–449.

Werner F, Althaus H, Künninger T, Richter K, Jungbluth N (2007) *Life Cycle Inventories of Wood as Fuels and Construction Material*. Final report ecoinvent 2000 No. 9. EMPA Dübendorf, Swiss Centre for Life Cycle Inventories, Dübendorf, CH.

Zhang X, Zhao Y, Ashton M, Lee X (2012) Measuring carbon in forests. In: *Managing Forest Carbon in a Changing Climate*. (eds Ashton M, Tyrrell M, Spalding D, Gentry B), pp. 139–164. Springer, Netherlands.

11

Importance of biophysical effects on climate warming mitigation potential of biofuel crops over the conterminous United States

PENG ZHU[1], QIANLAI ZHUANG[1], JOO EVA[2] and CARL BERNACCHI[2,3]

[1]Department of Earth, Atmospheric, and Planetary Sciences, Purdue University, West Lafayette, IN 47907, USA, [2]Department of Plant Biology, University of Illinois, Urbana, IL, USA, [3]Global Change and Photosynthesis Research Unit, USDA-ARS, Urbana, IL, USA

Abstract

Current quantification of climate warming mitigation potential (CWMP) of biomass-derived energy has focused primarily on its biogeochemical effects. This study used site-level observations of carbon, water, and energy fluxes of biofuel crops to parameterize and evaluate the community land model (CLM) and estimate CO_2 fluxes, surface energy balance, soil carbon dynamics of corn (*Zea mays*), switchgrass (*Panicum virgatum*), and miscanthus (*Miscanthus × giganteus*) ecosystems across the conterminous United States considering different agricultural management practices and land-use scenarios. We find that neglecting biophysical effects underestimates the CWMP of transitioning from croplands and marginal lands to energy crops. Biogeochemical effects alone result in changes in carbon storage of -1.9, 49.1, and 69.3 g C m^{-2} y^{-1} compared to 20.5, 78.5, and 96.2 g C m^{-2} y^{-1} when considering both biophysical and biogeochemical effects for corn, switchgrass, and miscanthus, respectively. The biophysical contribution to CWMP is dominated by changes in latent heat fluxes. Using the model to optimize growth conditions through fertilization and irrigation increases the CWMP further to 79.6, 98.3, and 118.8 g C m^{-2} y^{-1}, respectively, representing the upper threshold for CWMP. Results also show that the CWMP over marginal lands is lower than that over croplands. This study highlights that neglecting the biophysical effects of altered surface energy and water balance underestimates the CWMP of transitioning to bioenergy crops at regional scales.

Keywords: agricultural management, biofuel crops, biophysical effect, carbon sequestration, community land model, marginal land

Introduction

Biomass energy has been widely considered a major renewable and sustainable energy source for increase energy security while contributing to mitigating climate change (Field *et al.*, 2008; Beringer *et al.*, 2011). Bioenergy from crop-based biofuels is currently a promising biomass feedstock for replacing fossil fuels, and its demand is expected to continually increase to meet the mandate targets for biofuel production (US Congress, 2007). However, traditional crop-based biofuels have many unintended consequences for feedstock availability, food security, environmental sustainability, and societal welfare. For example, converting lands occupied by natural ecosystems to managed ecosystems for biofuel production could contaminate water quality with agricultural pollutants and converting food crop

ecosystems for biofuel production could potentially threaten food supplies (Clifton-Brown *et al.*, 2007; Field *et al.*, 2008; Gibbs *et al.*, 2008).

Recently, perennial grasses such as switchgrass and miscanthus have been favored as a better alternative to traditional row crops because they have higher productivity and water use efficiency (Hickman *et al.*, 2010; Vanloocke *et al.*, 2010, 2012; Zeri *et al.*, 2013). They also accumulate and sequestrate carbon into the soil, enhancing soil organic matter storage (Clifton-Brown *et al.*, 2007; Anderson-Teixeira *et al.*, 2009; Qin *et al.*, 2012; Valentine *et al.*, 2012). Meanwhile, these grasses could provide abundant biomass but require relatively less nutrient than conventional food crops (Lewandowski *et al.*, 2003; Heaton *et al.*, 2004; Clifton-Brown *et al.*, 2007; Stewart *et al.*, 2009; Zeri *et al.*, 2013). Therefore, they can grow on degraded agricultural land, that is marginal land, including idle or fallow cropland, abandoned or degraded cropland, and abandoned pastureland, where most food crops may not survive due to

Correspondence: Qianlai Zhuang
e-mail: qzhuang@purdue.edu

poor soil or climate conditions (Cai *et al.*, 2010; Gopalakrishnan *et al.*, 2012; Bandaru *et al.*, 2013), which could avoid competing with food crops for land.

It has been widely recognized that perennial biofuel grasses could mitigate climate change by sequestrating carbon, although the extent to which carbon can be sequestered will depend on the amount of carbon removed from the ecosystem and management practices employed (Zeri *et al.*, 2011; Anderson-Teixeira & Delucia, 2011; Liska *et al.*, 2014). Biophysical effects of land-use change are also critical to consider due to the potential for altered surface energy budgets which may feedback on local climate (Loarie *et al.*, 2011; He *et al.*, 2014; Peng *et al.*, 2014; Zhang *et al.*, 2014). Land management and land conversion each can impact surface temperature at comparable magnitudes (Luyssaert *et al.*, 2014). The direct climatic effect can be significant to climate warming mitigation and has been investigated in the field of deforestation and afforestation (Lee *et al.*, 2011; Loarie *et al.*, 2011; Peng *et al.*, 2014), but little research has been conducted within the framework of biofuel lifecycle analysis under different scenarios (Anderson-Teixeira *et al.*, 2012).

Presently, many crop models have been developed to estimate regional or global scale biomass production and greenhouse gas (GHG) emissions for biofuel crops (Nair *et al.*, 2012; Qin *et al.*, 2014; Surendran *et al.*, 2012; Thomas *et al.*, 2013). However, there are still large uncertainties in the simulated carbon and water balance such as biomass production, GHG emissions, and water demand (Liu *et al.*, 2015). These uncertainties are due to the different model parameterizations including feedstock chosen, cultivation practices, harvesting dates, fertilizer application, and land-use conversion pattern (Hudiburg *et al.*, 2015). A fully coupled earth system model can provide a comprehensive evaluation of both biogeochemical and biophysical effects due to land cover change on climate. However, significant challenges to these models exist and particularly related to computational resources need to run these coupled models. In contrast, most ecosystem models are sufficient to quantify carbon balance of biofuel ecosystems but often cannot accurately capture the high-frequency variation of surface energy due to their simplified surface energy balance schemes. Thus, land surface models which have a higher time frequency and detailed carbon and surface energy parameterization scheme are a more favorable compromise.

Using data collected at the University of Illinois Energy Farm, we parameterize and validate an advanced version land surface model CLM4.5 to evaluate carbon flux, biomass production, and surface energy balance of switchgrass and miscanthus. We then conducted an explicit spatial estimation of biogeochemical and biogeophysical effects for corn, switchgrass, and miscanthus across the conterminous United States. Model simulations were conducted to quantify responses of surface energy and carbon balance to different land-use scenarios and management practices compared to current land-use patterns. The surface energy and carbon balance changes were then integrated into calculations of CWMP. We hypothesize, at the regional scale, that (1) compared to maize and annual C_3/C_4 grasses, switchgrass and miscanthus will have higher productivity and sequester more carbon into soils, (2) CWMP of planting biofuels will be enhanced when accounting for evaporative cooling effects, and (3) agricultural management practice such as fertilization and irrigation will result in higher total carbon uptake, higher below ground biomass, and substantial evaporative cooling due to the sufficient water supply, consequently yielding a higher CWMP.

Materials and methods

Site description

The observational data were obtained at University of Illinois Energy Farm located in central Illinois (40.064°N, 88.197°W, ~220 m above sea level). This experiment consists of four ecosystems of 4 Ha (200 m × 200 m) each instrumented with eddy covariance and micrometeorological instrumentation at the center of each plot (a full site description is provided in Zeri *et al.*, 2011). In 2008, four ecosystems: corn–soybean rotation, miscanthus, switchgrass, and a mix of native prairie species were planted, with a replantation of miscanthus in 2010 after poor establishment, to examine bioenergy production and the associated environmental services. Above-ground biomass of each ecosystem was determined from harvested dry biomass at the end of the growing season. Leaf area index (LAI) of all species was measured optically (LAI-2200; LI-COR Biosciences, Lincoln, NE, USA) at weekly intervals during the period of active canopy development. The eddy covariance systems were established with a three-dimensional sonic anemometer (model 81000 V; R.M. Young Company, Traverse City, MI, USA) and an infrared gas analyzer (model LI-7500; LI-COR Biosciences, Lincoln, NE, USA) and were adjusted in height following plant growth. This system collected high-frequency data (10 Hz) of wind speed, and fluxes of CO_2 and H_2O analyzed using the Alteddy software package. The high-frequency data were corrected for coordinate alignment, humidity effects of the temperature measurements by the sonic anemometer, and density fluctuations of the infrared gas analyzer. A double-rotation scheme was used to align the coordinate system to the main wind direction and make the average vertical velocity zero (Kaimal & Finnigan, 1994). Data were also corrected for high-frequency data losses due to sensor separation (Moore, 1986). Data collected at low turbulence conditions were removed from the dataset and filtered by the u*-threshold (Aubinet *et al.*, 2001; Foken *et al.*, 2005). The footprint model of (Hsieh *et al.*, 2000) was applied to identify periods when the fluxes were outside the edges of the plots and records removed if less than 70% of cumulative flux came from within the plot area. Quality control of the data filtered out unreasonable fluxes (Zeri *et al.*, 2011). Missing data were gap-filled and fluxes partitioned from net ecosystem exchange into ecosystem respiration (R_{eco}) and gross primary production (GPP) (Reichstein *et al.*, 2005; Zeri *et al.*, 2011). Other essential meteorological variables to drive the gap-filling and partitioning model, including solar radiation (shortwave and longwave, both incoming and outgoing components; CNR1, Kipp & Zonen, the Netherlands), precipitation, air temperature, pressure, and relative humidity (HMP-45C; Campbell Scientific, Logan, UT, USA), were also collected at the center of each plot. A full site description with details about data analysis and quality control has been published previously (Zeri *et al.*, 2011, 2013; Joo *et al.*, 2016). The data collected in 2011 were used for model parameterization and evaluation.

Model description and improvement

Model simulations were performed using CLM4.5 to simulate the effects of climate, land-use change and agricultural management on carbon and surface energy budgets in bioenergy ecosystems. CLM was initially developed by concurrent effort at NCAR, merging community-developed land model focusing on biogeophysics to expand NCAR Land Surface Model (Bonan, 1996). CLM was incorporated with a number of biophysical processes for different plant functional types (PFT) including stomatal physiology, photosynthesis, energy and momentum fluxes with vegetation canopy and soil, heat transfer in soil and snow, and hydrology of canopy, soil, and snow. Carbon allocation and developmental stages are based on temperature thresholds and the accumulation of growing degree-days which is dynamic throughout the growing season. Soil organic carbon (SOC) is estimated from the turnover of soil organic matter pools, which change with decomposition rate. Version CLM4.5 was released as the land surface component of Community Earth System Model (CESM) with many improvements, including a revised canopy radiation scheme and canopy scaling of leaf processes, colimitations on photosynthesis and updated photosynthetic parameters (Bonan *et al.*, 2011). In CLM4.5, there is already a crop submodel, inherited from Agro-IBIS (Foley *et al.*, 1996; Kucharik *et al.*, 2000) to represent the role of agriculture in land surface processes. Processes of land management such as crop type, planting, harvesting, fertilization, and irrigation were added. In this study, the two major agricultural management practices, fertilization, and irrigation are accounted for, because these two management practices are considered to be crucial in determining carbon sequestration potentials of biofuel crops (Elshout *et al.*, 2015). The irrigation parameterization scheme is based loosely on the implementation of Ozdogan *et al.* (2010). This parameterization did not account for timing and background climate conditions, and it responds dynamically to climate. Irrigation can significantly influence the surface water and energy balances partition in the model and thus has an evident biophysical effect (Ozdogan *et al.*, 2010). Thus, water can be added to soil through irrigation so that a target soil moisture is

reached. Interactive fertilization is also enabled in this version, and nitrogen is added directly into the soil mineral nitrogen pool to meet crop demands. Total nitrogen fertilizer amounts are 150 kg N ha^{-1} for maize, 80 kg N ha^{-1} for temperate cereals, and 25 kg N ha^{-1} for soybean, representative of central US annual fertilizer application amounts. For biofuel crops, 100 kg N ha^{-1} is applied based on previous field experiments (Fike *et al.*, 2006; Heaton *et al.*, 2008; Propheter *et al.*, 2010; Nikièma *et al.*, 2011).

To reach our research goal, a new parameterization scheme for CLM is necessary for those perennial grasses including switchgrass and miscanthus, which have different physiological traits. Unlike annual crops, perennial grasses allocate a large amount of resources to belowground organs such as rhizomes (Anderson-Teixeira *et al.*, 2009; Atkinson, 2009). The new scheme was calibrated by adjusting relevant model parameters based on observations of switchgrass and miscanthus in 2011 to compare simulation results against observations. Several key parameters and their corresponding values (Table 1) in switchgrass and miscanthus parameterization were incorporated into the model. These parameters can be generally grouped into parameters controlling photosynthesis capacity including Vcmax25, Q, and slatop; phenology parameters including lfemerg, hybgdd, mxmat, baset, min_NH_planting_-date, min_planting_temp; and allocation parameters including Astem, Aroot, fleafi, Cnleaf. We combined the carbon allocated to rhizome with those to roots to minimize the change of the original model structure. Meanwhile, as perennial biomass crops usually needs 2–5 year to reach full maturity, we first run the model for 5 years without harvesting to allow carbon allocation to rhizomes to stabilize. The harvest frequency is one time per year. The current model does not consider recultivation and tillage that may be necessary for long-term bioenergy production. At site level, the model was run at a half-hour intervals to correspond with the eddy covariance data. The collected meteorological forcing data during 2011 is used to drive the model. At least 500 years of model spin-up is established to allow soil carbon pools to reach equilibrium.

Regional experiments under various land-use and management scenarios

Regional simulations were run at half-hourly time step from 2000 to 2010 at 0.5° × 0.5° spatial resolution. This recent 10-year time period was selected to capture the effects of interannual variations in climate. The regional spin-up procedure was the same as the single site and used current vegetation map for each grid cell. In the control run (cntl), each grid cell is initialized with a distribution of plants from current vegetation maps generated from the International Geosphere Biosphere Programme's 1-km DISCover (IGBP) land cover dataset (Loveland & Belward, 1997). For the remaining 12 simulations, the marginal land distribution utilizes the map estimated from Cai *et al.* (2010). In their study, global marginal lands were classified according to the marginal agricultural productivity based on land suitability indicators such as topography, climate conditions, and soil fertility. The first scenario in Cai *et al.* (2010) was used in this study. This scenario included marginal, abandoned, mixed crops, and vegetation land yet does not sacrifice

Table 1 New parameter values for switchgrass and miscanthus calibrated from site observational data

Parameter name	Description	Switchgrass	Miscanthus
Vcmax25	Maximum rubisco activity at 25 °C at top of canopy (μmol m^{-2} s^{-1})	75	92
Q	Intrinsic quantum efficiency (dimensionless)	0.04	0.04
Slatop	Specific leaf area (m^2 g C^{-1}) at top of canopy	31	70
Laimx	Maximum leaf area index (LAI) allowed (m^2 m^{-2})	6.5	8.5
hybgdd	Maximum growing degree-days (base 0 °C) required for physiological maturity	3700	3820
mxmat	Maximum number of days allowed past planting for physiological maturity to be reached	260	260
Fleafi	Fraction of assimilated carbon allocated to leaves	0.6	0.7
Astem	Fraction of assimilated carbon allocated to stems	0.2	0.2
Aroot	Fraction of assimilated carbon allocated to roots	0.15	0.12
Cnleaf	C:N ratio of leaf biomass	100	80
Baset	Base temperature for GDD calculation	0	0
min_planting_temp	Average 5 day daily minimum temperature needed for planting (K)	274.1	275
min_NH_planting_date	Minimum planting date for the Northern Hemisphere	301	301
lfemerg	Leaf emergence parameter	0.02	0.03

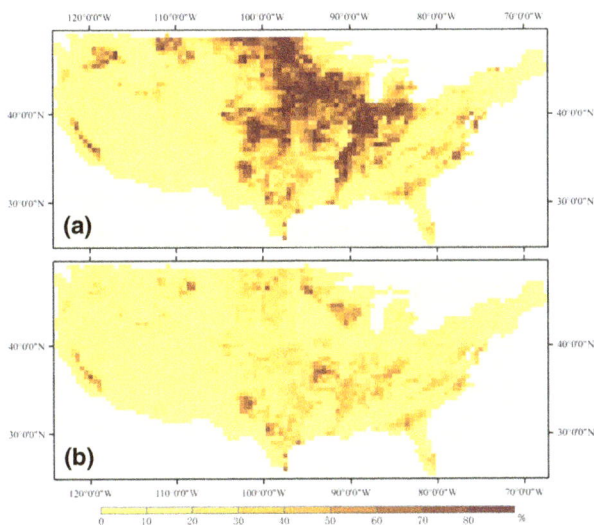

Fig. 1 Two scenarios showing of land conversion percentages for the study domain: (a) both marginal lands and croplands are converted, (b) only marginal lands are converted. Numbers in the color bar represent the proportion (%) of each grid cell that is converted to biofuel crops.

large amounts of current crop and natural ecosystems (forest and grassland) to bioenergy production. This scenario was considered as baseline land-use conditions and was used here to represent the spatial distribution of marginal lands in the United States. The data in Cai et al. (2010) were aggregated to 0.5° × 0.5° spatial resolution, and then, two land conversion scenarios were generated according to the proportion of marginal lands and croplands in each grid: One scenario (Fig. 1a) in which both marginal lands and croplands are converted and one scenario (Fig. 1b) in which only marginal lands are converted. The darker pixels in the figure represent higher fractions of convertible land. Compared to the first scenario (Fig. 1a), most of croplands in the second scenario remain

unchanged and only the scattered marginal lands are converted (Fig. 1b). Soil texture and soil color class for each 0.5° grid cell are based on the Harmonized World Soil Database (HWSD, Wieder et al., 2014) and are used by CLM4.5 to determine soil hydraulic and thermal properties. The climate data needed to drive simulations at the half-hourly time steps were obtained from CLM4.5 standard atmospheric forcing data sets CRUNCEP (Viovy, 2011), which is a combination of two existing datasets: the CRU TS3.2 0.5°×0.5° monthly data covering the period 1901 to 2002 (Mitchell & Jones, 2005) and the NCEP reanalysis 2.5°×2.5° 6-hourly data covering the period 1948 to 2010. For all model runs, harvest is set to once per year and without recultivation over the length of the experiment. The full perennial crop lifetime is determined by GDD and the related parameters have been calibrated according to observed LAI. Since the analytical time frame is 50 years, analysis focused on the period after the perennial crops reached maturity, which can minimize the albedo effect of the first year of cultivation in the energy balance analysis.

Twelve experiments were conducted to assess climate warming mitigation potential under different combinations of land conversion scenarios and agricultural management practices (Table 2). Applying extensive agricultural management, particularly irrigation, is not practical, even risky, considering its environmental impacts. However, the management scenarios present an upper boundary of reachable climate warming mitigation relative to the baseline. In addition, the proportion of crop residues removal could have a noticeable impact on soil carbon pool (Liska et al., 2014). In the control run, in addition to total crop grain harvest, 20% of residue was removed to represent SOC loss by soil disturbance from cultivation, which is neglected in CLM (Levis et al., 2014). For the remaining 12 experiments, 70% of above-ground biomass was removed to simulate harvest for lignocellulosic biofuel crops. This removal rate is considered to maintain sustainable utilization while maximizing yields. Across all of the 13 simulations, natural and crop ecosystems in each grid cell were modeled separately and then aggregated based on their fractions within each grid

Table 2 A list of model experiments allowing for variation in biofuel crop types, land conversion scenarios, and altered management practices associated with irrigation and fertilization

Experiments	Biofuel type	Land conversion scenarios	Management practices
corn1	Corn	Marginal land and cropland	No
corn2	Corn	Marginal land	No
corn3	Corn	Marginal land and cropland	Yes
corn4	Corn	Marginal land	Yes
sw1	Switchgrass	Marginal land and cropland	No
sw2	Switchgrass	Marginal land	No
sw3	Switchgrass	Marginal land and cropland	Yes
sw4	Switchgrass	Marginal land	Yes
mx1	Miscanthus	Marginal land and cropland	No
mx2	Miscanthus	Marginal land	No
mx3	Miscanthus	Marginal land and cropland	Yes
mx4	Miscanthus	Marginal land	Yes

cell. Comparisons of CWMP were based on the differences of 10-year average for the 12 experiments from the control.

Climate mitigation potential metrics

CWMP of growing biofuel crops was often quantified using net GHG fluxes and SOC change, both are important in the life cycle analysis of biofuel carbon balance. However, the contribution of biophysical effects to CWMP was overlooked in previous research (Qin *et al.*, 2012, 2015; Albanito *et al.*, 2015). Here, we combine carbon fluxes, soil carbon pool changes, evaporative cooling effects, and net radiation (Rn is the balance between incoming and outgoing long-wave and short-wave radiation, mainly determined by albedo) changes to construct a synthetic CWMP metric using carbon as the currency. Both biophysical effects and biogeochemical effects can be converted to radiative forcing effects, that is biogeochemical effects influence the capacity of absorbing long-wave radiation while biophysical effects concerns short-wave radiation and latent heat flux:

$$T \frac{\Delta E}{S} = \frac{\Delta C_e / M_c}{A} R_e \qquad (1)$$

where ΔE is the surface energy change (W m^{-2}; $\Delta E = \Delta LE - \Delta Rn$). ΔC_e is the equivalent carbon change. $A = 1.78 \times 10^{20}$ mol is the moles of air in the atmosphere. $R_e = 1.4 \times 10^4$ nW m^{-2} ppb^{-1} is the effective radiative forcing efficiency of CO_2. $S = 5.1 \times 10^{14}$ m^2 is the global surface area, here acting as scale factor to convert the local ΔE to global radiative forcing effects. Mc is the molar mass of carbon. As radiative forcing of CO_2 has cumulative effect, here T is multiplied as the time frame to balance the two sides. We choose T to be 50 years as used previously (Anderson-Teixeira *et al.*, 2012) to account for the residence time of CO_2 in atmosphere. An additional time frame is

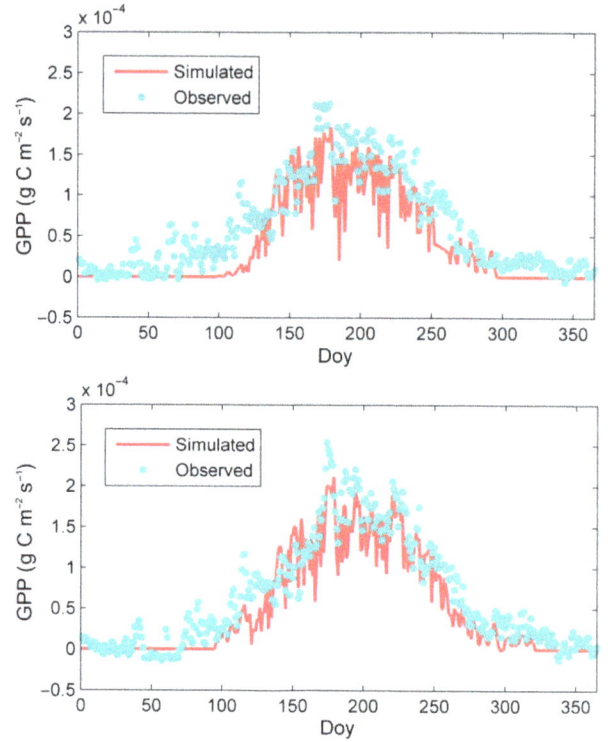

Fig. 2 Simulated vs. observed daily gross primary productivity (GPP) over the 2011 growing season for switchgrass (upper panel, model = $0.92 \times$ obs + 0.000017, $R^2 = 0.71$, RMSE = 4.47×10^{-6} g C m^{-2} s^{-1}) and miscanthus (lower panel, model = $0.94 \times$ obs + 0.000013, $R^2 = 0.75$, RMSE = 3.78×10^{-6} g C m^{-2} s^{-1}).

needed to allow for SOC changes be comparable with the change of annual net carbon fluxes and also set to 50 years. Thus, CWMP can be defined as:

$$CWMP = \Delta C_e + \Delta NEP + \frac{\Delta SOC}{50} \qquad (2)$$

According to Eqn (1), the surface energy change of 1 W m^{-2} is roughly equal to 6 g C m^{-2} over a 50-year time span. More technical details of these conversions could be found in Anderson-Teixeira *et al.* (2012). The CWMP of each grid cells occurring in biofuel crops expansion is finally aggregated based on land conversion rate.

Results

Model evaluation at site

The simulated GPP compared well with measurements for both switchgrass and miscanthus with a slight underestimation during the maximum carbon uptake period in 2011 (Fig. 2). Simulated GPP captures the annual variation in productivity over the whole growing season, including the initial increase after leaf emergence, the timing of peak values, and decline after leaf senescence (Fig. 2). For switchgrass, simulated timing occurs later than observations for leaf onset but leaf senescence matches better with the observations. The model performs well for miscanthus in capturing the timing of leaf onset and senescence. The model explained 71% and 75% of observed GPP for switchgrass and miscanthus, respectively. Miscanthus showed a longer growing season due to its later leaf senescence date, leading to a higher annual GPP of 2.88 kg C m^{-2} relative to 2.34 kg C m^{-2} for switchgrass.

The timing and magnitude of simulated latent heat (LE) matched well with the observed values at a half-hour time step. Simulated NEE matched the eddy covariance measurements by capturing the transition from winter dormancy to spring uptake to summer maximum uptake (Figs 3 and 4). Compared to eddy covariance measurements, the simulated LE and NEE were slightly overestimated. The annual LE differences between simulation and observation were 4.7 W m^{-2} and 4.1 W m^{-2}, while the NEE differences were 32.2 g C m^{-2} and 24.3 g C m^{-2} for switchgrass and miscanthus, respectively. All of these differences were within the 10% of the annual observation.

Model projections of biofuel crop carbon and energy balance

Carbon balance for growing biofuel crops. Simulating corn grown for harvesting both grain and stover for biofuel production showed that soils acted as a carbon source when no management practices were applied, primarily owing to the higher rate of residue removal for biofuel production (Fig. 5). Higher productivity and a longer growing season led to increased soil litter inputs for switchgrass and miscanthus and correspond to accumulation in soil carbon despite above-ground biomass removal (Figs 6 and 7). Corn agroecosystem showed higher soil C accumulation in the north, while switchgrass and miscanthus tended to gain more SOC in the south, consistent with previous results (Miguez *et al.*, 2012). There was a substantial increase in SOC when the arid areas (e.g., western United States) were fertilized and irrigated. Areas planted with corn had a moderate increase in net carbon fluxes relative to natural vegetation and miscanthus had the largest carbon sequestration potential, followed with switchgrass and corn (Fig. 8). All of the three biofuel crops showed increased carbon sequestration with increased management to 49.3, 66.0, and 84.9 g C m^{-2} y^{-1} compared with -1.9, 49.1, and 69.3 g C m^{-2} y^{-1} without management for corn, switchgrass, and miscanthus, respectively. Carbon sequestration capacity was generally larger for crops than for marginal lands due to nutrient limitation in marginal lands. These results suggest that for a given mitigation target, more marginal lands are required for conversion to bioenergy compared with croplands. However, converting marginal lands will not interfere

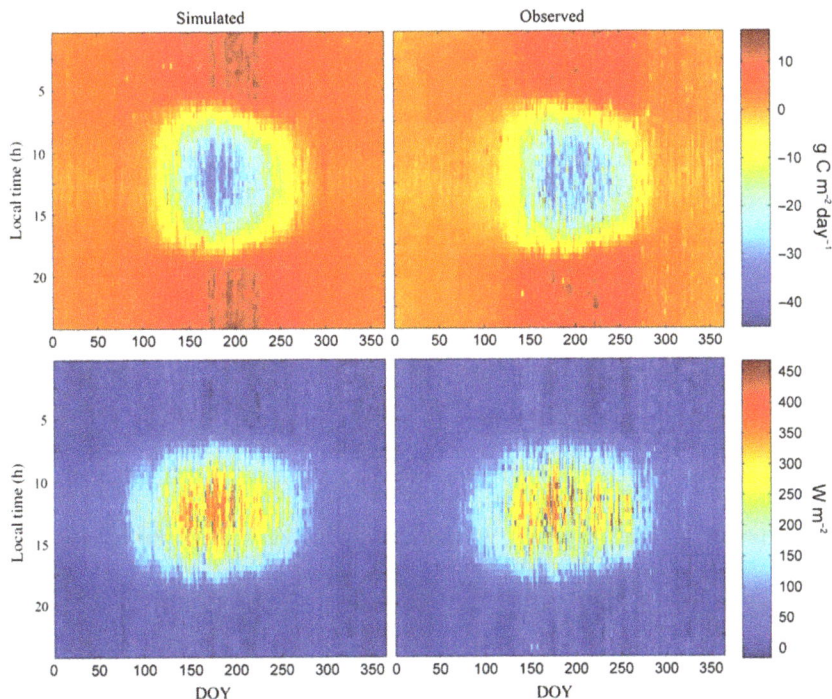

Fig. 3 Observed (left column) and simulated (right column) net ecosystem exchange (NEE g C m^{-2} day^{-1}, top row) and latent heat flux (LE W m^{-2}, bottom row) at half-hour intervals for mature switchgrass in 2011.

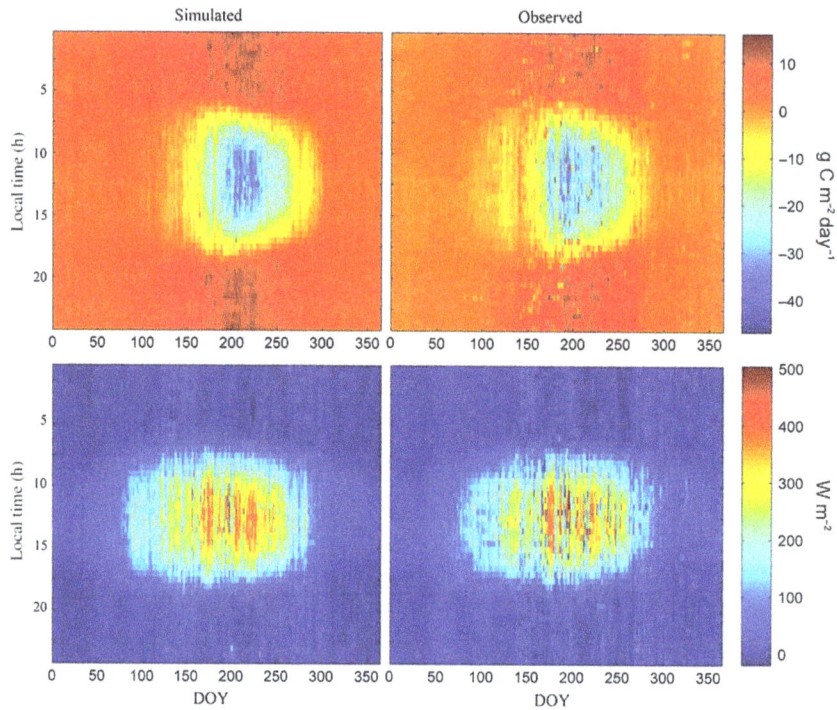

Fig. 4 Observed (left column) and simulated (right column) NEE (g C m^{-2} day^{-1} top row) and LE (W m^{-2} bottom row) at half-hour interval for mature miscanthus in 2011.

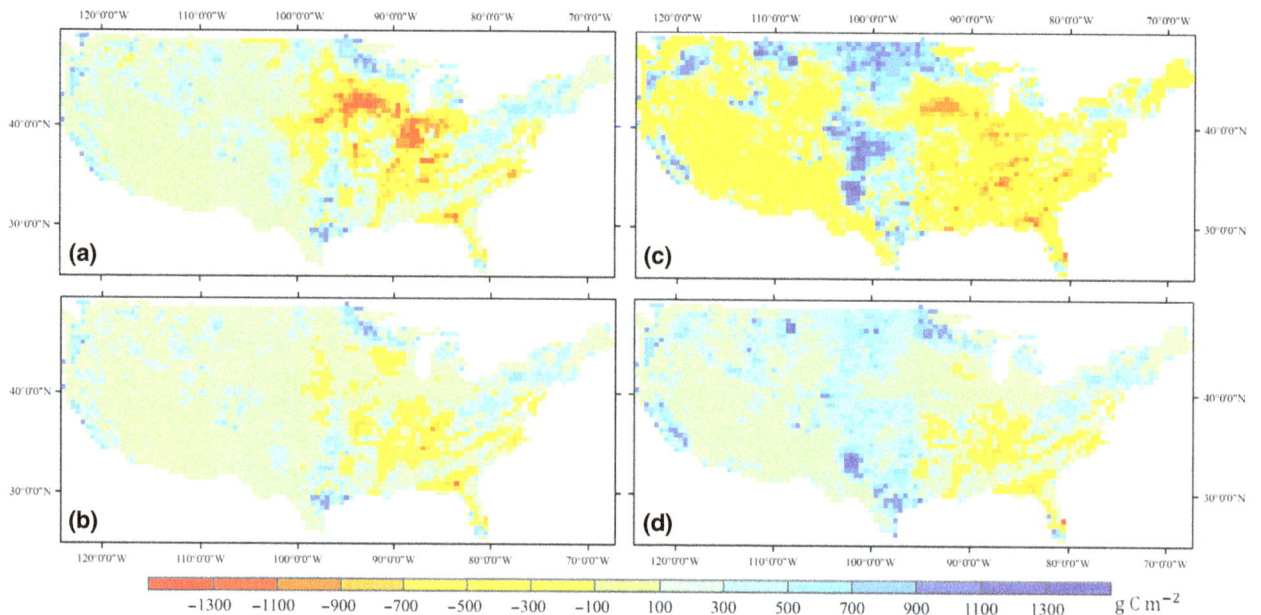

Fig. 5 Simulated differences in SOC (g C m^{-2}) based on 10-year (2000–2010) climate forcing data when the soil carbon pool reaches equilibrium for corn1-cntl (a), corn2-cntl (b), corn3-cntl (c), corn4-cntl (d).

with food production. Our simulations were generally consistent with previous findings (Qin *et al.*, 2012, 2015; Elshout *et al.*, 2015), suggesting that switchgrass and miscanthus could increase carbon sequestration.

Changes of energy balance. The spatial pattern in simulated Rn and LE was generally consistent with previous modeling results, indicating that annual cumulative ET for switchgrass and miscanthus was larger than corn

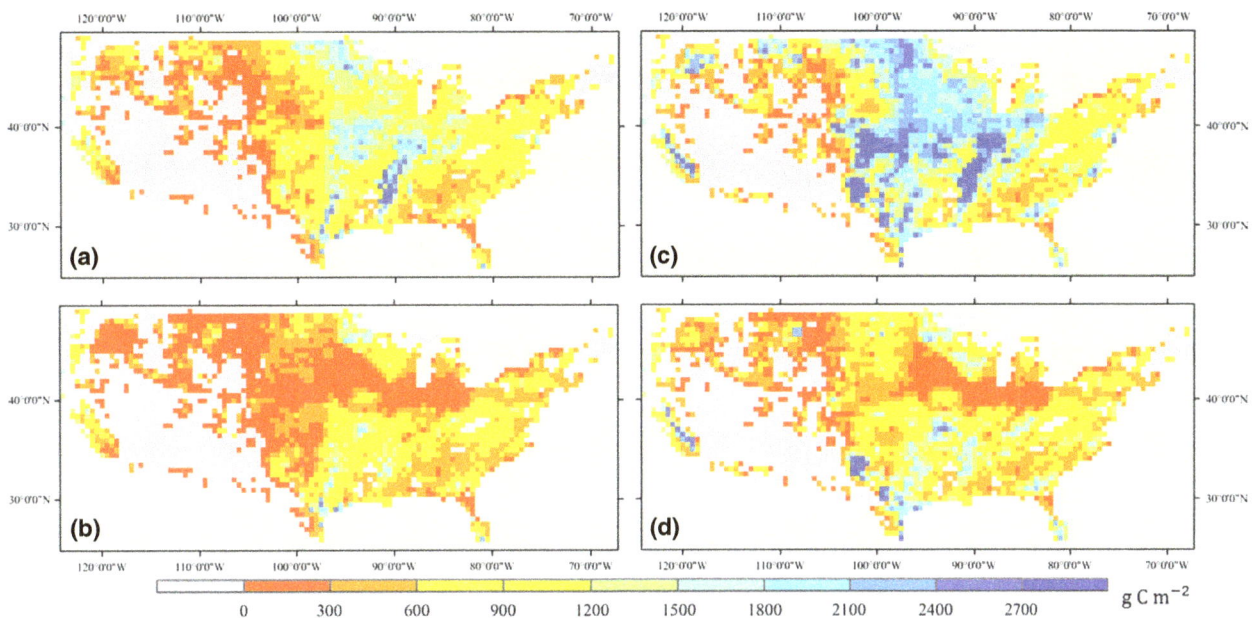

Fig. 6 The simulated difference of SOC (g C m^{-2}) based on 10-year (2000–2010) climate forcing data when the soil carbon pool reaches equilibrium for sw1-cntl (a), sw2-cntl (b), sw3-cntl (c), sw4-cntl (d).

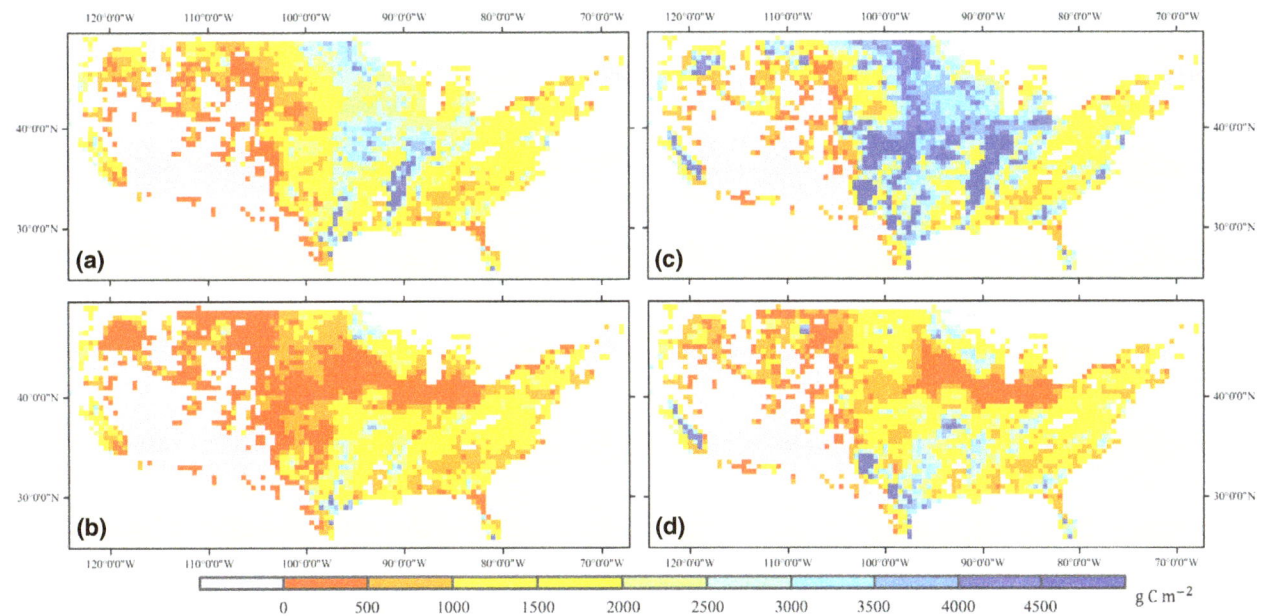

Fig. 7 The simulated difference of SOC (g C m^{-2}) based on 10-year (2000–2010) climate forcing data when the soil carbon pool reaches equilibrium for mx1-cntl (a), mx2-cntl (b), mx3-cntl (c), mx4-cntl (d).

due to their longer growing season and higher rate of evapotranspiration (Hickman *et al.*, 2010; Vanloocke *et al.*, 2010; Zeri *et al.*, 2013; Joo *et al.* submitted to Plant Physiology). The distribution of ΔLE showed a similar spatial pattern to carbon flux, implying there was a tight nexus between carbon and energy exchanges (Fig. 9). The three biofuel crops had larger LE than existing vegetation due to their higher plant transpiration.

Switchgrass and miscanthus showed a higher net radiation, indicating a lower albedo due to higher LAI. ΔLE of corn, switchgrass, and miscanthus growing on marginal lands and croplands without management are 3.8 W m^{-2}, 5.2 W m^{-2}, and 5.2 W m^{-2}, respectively. Maximum ΔLE of switchgrass and miscanthus were 8.8 W m^{-2} and 9.3 W m^{-2} in the southeast of the United States, which corresponded with a higher ΔRn

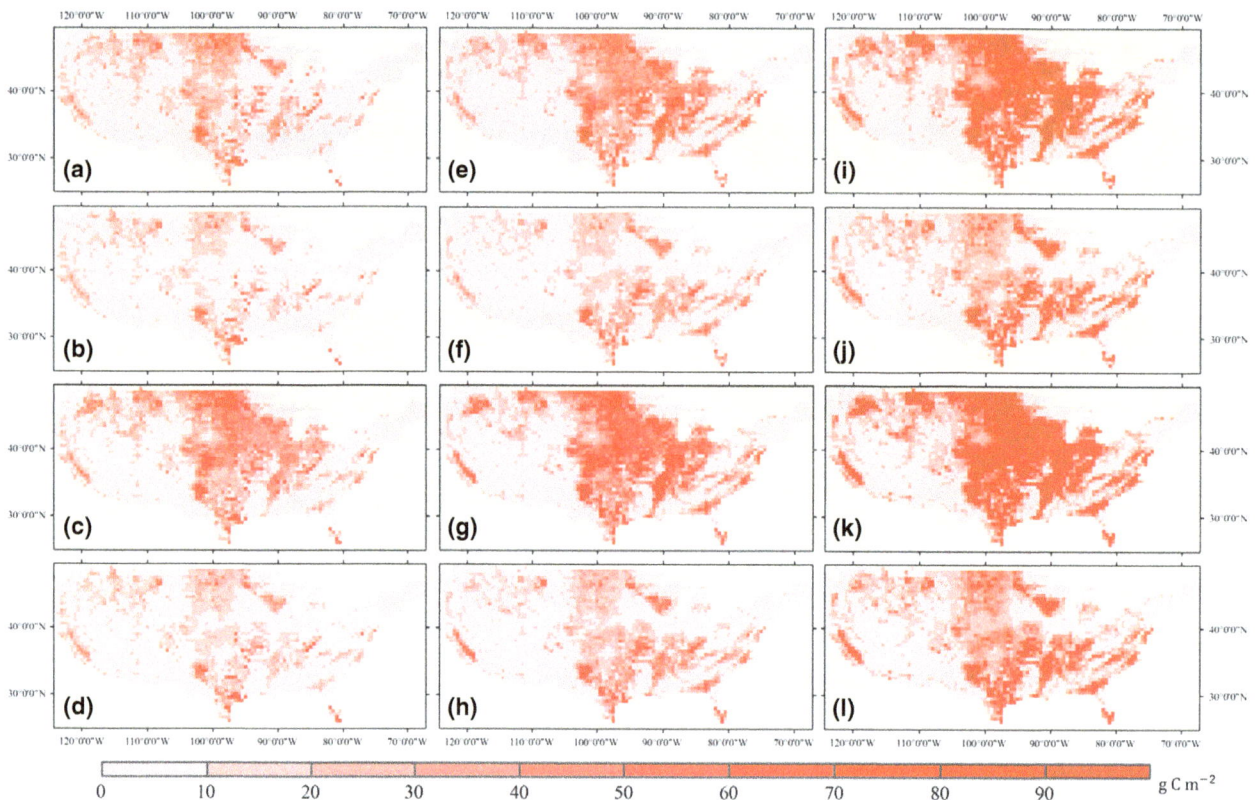

Fig. 8 The simulated difference of annual net carbon flux (g C m^{-2}) based on 10-year (2000–2010) climate forcing data among each experiments (a–l) corresponds to the difference between corn1, corn2, corn3, corn4, sw1, sw2, sw3, sw4, mx1, mx2, mx3, mx4, and cntl, respectively.

Table 3 Simulated CWMP change based on 10-year (2000–2010) means under various modeled scenarios

Experiment	Carbon flux (g C m^{-2})	SOC (g C m^{-2})	LE (W m^{-2})	Rn (W m^{-2})	CWMP (g C m^{-2})
corn1	24.4	−26.3	4.3	0.30	20.5
corn2	21.5	−8.3	3.8	0.26	33.0
corn3	37.5	11.8	5.8	0.38	79.6
corn4	32.1	8.2	4.9	0.32	65.9
sw1	30.6	18.5	5.6	0.35	78.5
sw2	28.7	16.3	5.2	0.32	71.3
sw3	39.2	26.8	6.2	0.42	98.3
sw4	34.1	22.2	5.8	0.37	86.7
mx1	38.5	30.8	5.7	0.67	96.2
mx2	32.4	27.2	5.1	0.55	85.1
mx3	47.2	37.7	6.9	0.87	118.8
mx4	43.5	32.4	6.3	0.68	107.3

(Figs 9 and 10). The spatial variation of Rn was similar for switchgrass and miscanthus as both are perennial grasses and had relatively similar physiological and phenological traits compared with corn. However, the

mean value of switchgrass was lower than miscanthus. In most regions covered by biofuel crops, ΔLE typically outweighed ΔRn such that the biophysical effects of land conversion are dominated by localized evaporative cooling. When agricultural management was applied, the increase of LE was much greater than Rn, leading to a higher cooling effect. This could be attributed to (1) irrigation maintaining high soil moisture and/or (2) fertilization leading to higher LAI and thus increased transpiration. The spatial pattern of LE change showed larger enhancement in the southern United States for the three biofuel crops, which was possibly attributed to the higher evaporative demand.

CWMP under various alternative scenarios. Our simulated annual CWMP under various alternatives (Table 3) indicated that CWMP could be significantly improved when biophysical effects were added. The corn ecosystem changed from carbon source to sink in the corn1 experiment, which affirmed the previous research that biophysical effects of bioenergy crops can be even larger than biogeochemical effects at regional scales (Georgescu *et al.*, 2011; Anderson-Teixeira *et al.*, 2012). This improvement can be mainly explained by an increase in LE for biofuel crops leading to cooling effects that contribute to

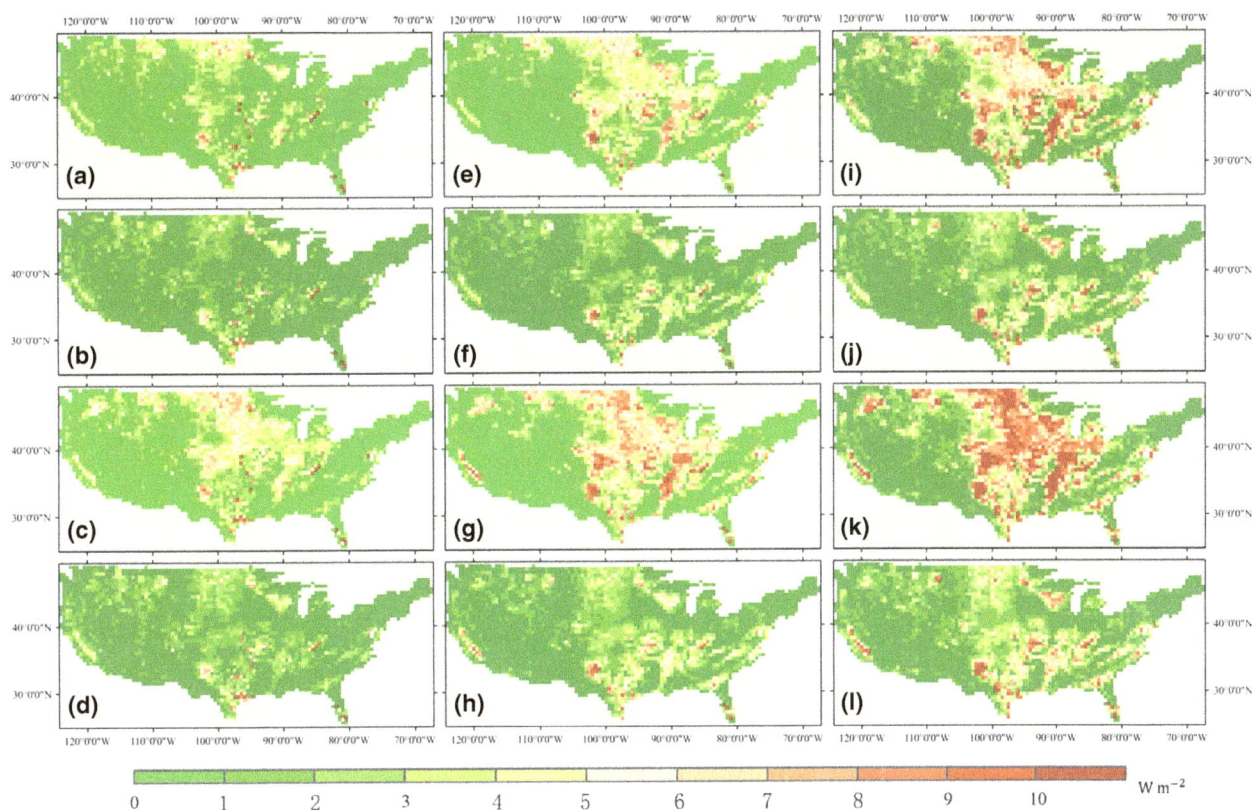

Fig. 9 The simulated difference of annual mean LE (W m^{-2}) based on 10-year (2000–2010) climate forcing data among each experiments, (a–l) corresponds to the difference between corn1, corn2, corn3, corn4, sw1, sw2, sw3, sw4, mx1, mx2, mx3, mx4, and cntl, respectively.

climate warming mitigation. The synergistic effect of fertilization and irrigation significantly improves the CWMP of biofuel crops, especially for corn, similar to previous studies (Lee *et al.*, 2012). This management triples CWMP of corn from 20.5 g C m^{-2} to 79.6 g C m^{-2} in the scenario where both croplands and marginal lands were converted to bioenergy crops. This result is consistent with previous research and confirmed high input could reduce carbon payback time of crop-based biofuel (Elshout *et al.*, 2015). If biofuel crops were planted only on marginal lands with no management, their CWMP ranged from 33.0 g C m^{-2} to 85.1 g C m^{-2} while this range shifts to 20.5 g C m^{-2} to 96.2 g C m^{-2} when cropland is also converted, implying CWMP over marginal lands is lower than that over croplands. The highest CWMP of 118.8 g C m^{-2} is achieved by mx3, which is ~50% higher than if biogeochemical effects are considered alone (84.9 g C m^{-2}). The simulated CWMP of switchgrass lies between corn and miscanthus.

Discussion

In this study, we used the revised land surface model, CLM4.5, to evaluate the climate regulation service of the

grain and cellulosic crops across conterminous United States over a multiyear time frame. The results show that harvesting corn grain and residue for biofuel production under a scenario without any agricultural management will progressively deplete the soil carbon pool. Previous research concluded that cultivation of switchgrass and miscanthus increased SOC on average 10–100 g C m^{-2} per year in the top 30 cm (Anderson-Teixeira *et al.*, 2009). Our modeled SOC change of 16.3–37.7 g C m^{-2} per year fell within this range. Results confirmed that cellulosic crops, which normally had higher nutrient use efficiency and higher water use efficiency, sequester more carbon and produce more biomass for bioenergy feedstocks (Davis *et al.*, 2011; Vanloocke *et al.*, 2012; Jones *et al.*, 2015). This suggests that these biofuel crops are more promising in areas that extend beyond current croplands. The results demonstrate high spatial variation in carbon sequestration ability controlled by the climatic and soil conditions as well as the type of land being replaced. Previous research demonstrated that the conversion of tropical and temperate forests, savannahs, and peatland for biofuel production could cause net carbon emissions because of the large amount of stored carbon released

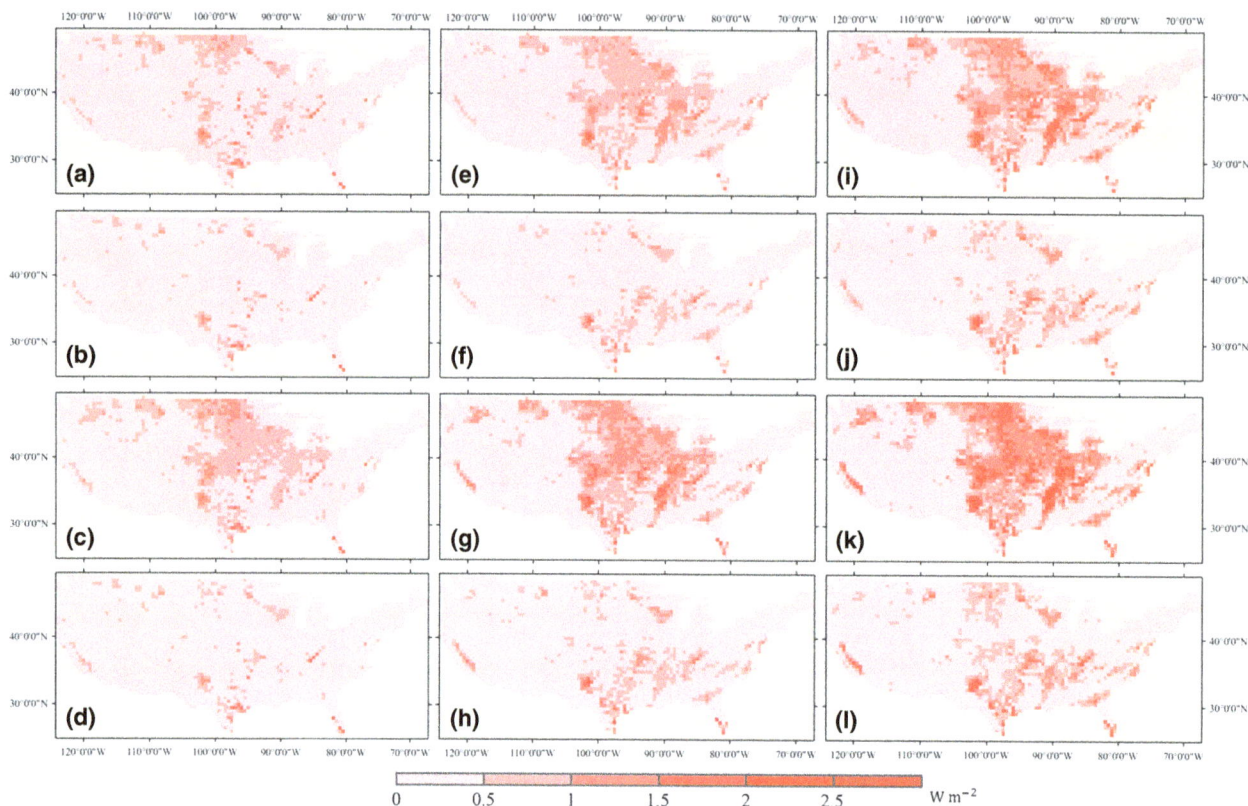

Fig. 10 The simulated difference of annual mean Rn (W m^{-2}) based on 10-year (2000–2010) climate forcing data among each experiments, (a–l) corresponds to the difference between corn1, corn2, corn3, corn4, sw1, sw2, sw3, sw4, mx1, mx2, mx3, mx4, and cntl, respectively.

(Farglone et al., 2008; Elshout et al., 2015). In this study, only marginal land and cropland were taken into account for land conversion. These scenarios therefore present a practical approach based on previous experimental conclusion that cultivation of biofuels on marginal land can enhance productivity while minimizing environmental degradation (Bhardwaj et al., 2011). This and previous research highlights that marginal land is less fertile and can sustain lower carbon sequestration capacity (Gelfand et al., 2013); therefore, more marginal land is needed to achieve mitigation targets. Our model simulation only considered 100% land conversion fraction, which can be viewed as an upper limit. Lower fraction of land conversion will dampen the CWMP of biofuels crops observed here, similar to previous results (Vanloocke et al., 2010).

The proposed CWMP here covered carbon fluxes, carbon storage, and surface energy change, which presents a complete perspective of evaluating climate mitigation of biofuel crops (Anderson et al., 2010; Knoke et al., 2012). CWMP of both cellulosic crops and corn significantly increased when accounting for both biogeochemical and biophysical, rather than just biogeochemical, effects. One biophysical component, Rn which is

dominated by albedo, increased when current land cover was displaced by biofuel crops. Evaporative cooling of miscanthus and switchgrass, therefore, is augmented by higher albedo, consistent with previous results (Anderson-Teixeira et al., 2012). However, observed experiment data indicated that miscanthus and switchgrass have a higher albedo than corn during the growing season, thereby reducing Rn in bioenergy crops (Miller et al., 2015). This discrepancy likely originates from the current model being unable to predict albedo of these perennial grasses through improper parameterization of leaf transmittance and reflectance, leaf angle, and canopy structure (Lawrence et al., 2011). Thus, further efforts are needed to improve surface energy processes of these biofuel crops. The albedo effects, however, are secondary to the effects of crop type on LE, which is a major component of the water cycle. Both crop productivity and LE are strongly influenced by irrigation, particularly in arid environments (Roncucci et al., 2014). Experiments suggest miscanthus has larger transpiration due to the higher stomatal conductance to support its high carbon assimilation rate (Dohleman et al., 2009), which is consistent with our results. Higher LE induced by the expansion of biofuel crops are likely to impact the hydrological cycle although this influence spatially variable based on a range of factors (Vanloocke *et al.*, 2010; Abraha *et al.*, 2015).

While this study indicates that both biogeochemical and biophysical feedbacks need to be considered in evaluating biofuel crops, several limitations to the analysis still exist. First, this model neglects other important GHGs from agroecosystems particularly N_2O. Higher N_2O emissions are expected with increased fertilization. However, previous results show that perennial systems leak less N_2O through leaching and denitrification (Smith *et al.*, 2013; Hudiburg *et al.*, 2015). Thus, our estimates of the climate change mitigation potential are likely conservative. Our results also neglect environmental impact of increasing nitrate leaching induced by fertilization application (e.g., Chamberlain *et al.*, 2011). Second, soil carbon storage is heavily dependent on crop residue remove rate (Smith *et al.*, 2012; Liska *et al.*, 2014), and we set crop residue remove rate as a constant across the United States. More realistic and/or flexible removal rates should be introduced in the future research. Third, the irrigation in CLM4.5 is automatically triggered based on soil water status. Although irrigation is shown to improve CWMP of biofuel crops and might save more lands, its possible threat to local water resource is not accounted. Recent research highlights the need to institute policies to balance the water and land requirements during bioenergy production (Bonsch *et al.*, 2014). Finally, we used land surface energy change to represent total cooling effects of growing biofuel crops on the climate. It is desirable to use dynamic climate models to examine how these land-use change and management scenarios affect the climate in terms of air temperature and precipitation. For instance, altered evapotranspiration due to growing biofuel crops will impact atmospheric vapor pressure deficit and its consequences on plant physiology, meteorology, and climate (Lobell *et al.*, 2009; Puma & Cook, 2010). For example, an increase in cloud formation will affect short-wave radiation and air temperature, the processes that current land surface models neglect.

Previous research has demonstrated that integrating the ideal farming management practices (e.g., improved harvesting techniques, harvest timing, organic matter amendments, reduced-till, and/or rotating cereals with grain legumes) can reduce GHG emissions and improve soil carbon sequestration capacity and soil quality. This can lead to environmental protection and biodiversity conservation (Gan *et al.*, 2014; Cheng *et al.*, 2014; Hudiburg *et al.* 2015; Davis *et al.*, 2013). Our research confirmed the importance of agricultural management in enhancing CWMP especially when accounting for biophysical effects. In addition to climate mitigation, improving current farming practices can lead to improved

ecosystem services while maximizing bioenergy production. Faced with increasing land-use pressures driven by growing population, our spatially explicit results accounting both biophysical and biogeochemical effect enable policymakers to make wiser decisions on the landscape planning of biofuel crops expansion to accomplish climate mitigation target (Campbell *et al.*, 2010).

Acknowledgements

This research is partially supported by funding to Q.Z. through a NSF project (DEB- #0919331), the NASA Land Use and Land Cover Change program (NASA-NNX09AI26G), Department of Energy (DE-FG02-08ER64599), and the NSF Division of Information & Intelligent Systems (NSF-1028291).

References

Abraha M, Chen J, Chu H *et al.* (2015) Evapotranspiration of annual and perennial biofuel crops in a variable climate. *GCB Bioenergy*, **7**, 1344–1356.

Albanito F, Beringer T, Corstanje R, Poulter B, Stephenson A, Zawadzka J, Smith P (2015) Carbon implications of converting cropland to bioenergy crops or forest for climate mitigation: a global assessment. *GCB Bioenergy*, **8**, 81–95.

Anderson RG, Canadell JG, Randerson JT *et al.* (2010) Biophysical considerations in forestry for climate protection. *Frontiers in Ecology and the Environment*, **9**, 174–182.

Anderson-Teixeira KJ, DeLucia EVAN (2011) The greenhouse gas value of ecosystems. *Global Change Biology*, **17**, 425–438.

Anderson-Teixeira KJ, Davis SC, Masters MD, Delucia EH (2009) Changes in soil organic carbon under biofuel crops. *GCB Bioenergy*, **1**, 75–96.

Anderson-Teixeira KJ, Snyder PK, Twine TE, Cuadra SV, Costa MH, DeLucia EH (2012) Climate-regulation services of natural and agricultural ecoregions of the Americas. *Nature Climate Change*, **2**, 177–181.

Atkinson C (2009) Establishing perennial grass energy crops in the UK: a review of current propagation options for Miscanthus. *Biomass and Bioenergy*, **33**, 752–759.

Aubinet M, Chermanne B, Vandenhaute M, Longdoz B, Yernaux M, Laitat E (2001) Long term carbon dioxide exchange above a mixed forest in the Belgian Ardennes. *Agricultural and Forest Meteorology*, **108**, 293–315.

Bandaru V, Izaurralde RC, Manowitz D, Link R, Zhang X, Post WM (2013) Soil carbon change and net energy associated with biofuel production on marginal lands: a regional modeling perspective. *Journal of Environmental Quality*, **42**, 1802–1814.

Beringer TIM, Lucht W, Schaphoff S (2011) Bioenergy production potential of global biomass plantations under environmental and agricultural constraints. *GCB Bioenergy*, **3**, 299–312.

Bhardwaj AK, Zenone T, Jasrotia P, Robertson GP, Chen J, Hamilton SK (2011) Water and energy footprints of bioenergy crop production on marginal lands. *GCB Bioenergy*, **3**, 208–222.

Bonan GB (1996) Land surface model (LSM version 1.0) for ecological, hydrological, and atmospheric studies: technical description and users guide. Technical note (No. PB–97-131494/XAB; NCAR/TN–417-STR). National Center for Atmospheric Research, Boulder, CO, USA. Climate and Global Dynamics Div.

Bonan GB, Lawrence PJ, Oleson KW *et al.* (2011) Improving canopy processes in the Community Land Model version 4 (CLM4) using global flux fields empirically inferred from FLUXNET data. *Journal of Geophysical Research*, **116**, 1–22.

Bonsch M, Humpenöder F, Popp A *et al.* (2014) Trade-offs between land and water requirements for large-scale bioenergy production. *GCB Bioenergy*, **8**, 11–24.

Cai X, Zhang X, Wang D (2010) Land availability for biofuel production. *Environmental Science & Technology*, **45**, 334–339.

Campbell JE, Block E (2010) Land-use and alternative bioenergy pathways for waste biomass. *Environmental Science and Technology*, **44**, 8665–8669.

Chamberlain JF, Miller SA, Frederick JR (2011) Using DAYCENT to quantify on-farm GHG emissions and N dynamics of land use conversion to N-managed switchgrass in the Southern US. *Agriculture, Ecosystems & Environment*, **141**, 332–341.

Cheng K, Ogle SM, Parton WJ, Pan G (2014) Simulating greenhouse gas mitigation potentials for Chinese Croplands using the DAYCENT ecosystem model. *Global Change Biology*, **20**, 948–962.

Clifton-Brown JC, Breuer J, Jones MB (2007) Carbon mitigation by the energy crop, Miscanthu. *Global Change Biology*, **13**, 2296–2307.

Davis SC, Parton WJ, Grosso SJD, Keough C, Marx E, Adler PR, DeLucia EH (2011) Impact of second-generation biofuel agriculture on greenhouse-gas emissions in the corn-growing regions of the US. *Frontiers in Ecology and the Environment*, **10**, 69–74.

Davis SC, Boddey RM, Alves BJ et al. (2013) Management swing potential for bioenergy crops. *GCB Bioenergy*, **5**, 623–638.

Dohleman FG, Heaton EA, Leakey ADB, Long SP (2009) Does greater leaf-level photosynthesis explain the larger solar energy conversion efficiency of Miscanthus relative to switchgrass? *Plant, Cell & Environment*, **32**, 1525–1537.

Gan Y, Liang C, Chai Q, Lemke RL, Campbell CA, Zentner RP (2014) Improving farming practices reduces the carbon footprint of spring wheat production. *Nature Communications*, **5**, 5012.

Elshout PMF, van Zelm R, Balkovic J et al. (2015) Greenhouse-gas payback times for crop-based biofuels. *Nature Climate Change*, **5**, 604–610.

Fargione J, Hill J, Tilman D, Polasky S, Hawthorne P (2008) Land clearing and the biofuel carbon debt. *Science*, **319**, 1235–1238.

Field CB, Campbell JE, Lobell DB (2008) Biomass energy: the scale of the potential resource. *Trends in Ecology & Evolution*, **23**, 65–72.

Fike JH, Parrish DJ, Wolf DD, Balasko JA, Green JT, Rasnake M, Reynolds JH (2006) Long-term yield potential of switchgrass-for-biofuel systems. *Biomass and Bioenergy*, **30**, 198–206.

Foken T, Göockede M, Mauder M, Mahrt L, Amiro B, Munger W (2005) Post-field data quality control. In: *Handbook of Micrometeorology* (eds Lee X, Massman W, Law B), pp. 181–208. Springer, Netherlands.

Foley JA, Prentice IC, Ramankutty N, Levis S, Pollard D, Sitch S, Haxeltine A (1996) An integrated biosphere model of land surface processes, terrestrial carbon balance, and vegetation dynamics. *Global Biogeochemical Cycles*, **10**, 603–628.

Gelfand I, Sahajpal R, Zhang X, Izaurralde RC, Gross KL, Robertson GP (2013) Sustainable bioenergy production from marginal lands in the US Midwest. *Nature*, **493**, 514–517.

Georgescu M, Lobell DB, Field CB (2011) Direct climate effects of perennial bioenergy crops in the United States. *Proceedings of the National Academy of Sciences*, **108**, 4307–4312.

Gibbs HK, Johnston M, Foley JA, Holloway T, Monfreda C, Ramankutty N, Zaks D (2008) Carbon payback times for crop-based biofuel expansion in the tropics: the effects of changing yield and technology. *Environmental Research Letters*, **3**, 034001.

Gopalakrishnan G, Negri C, Salas W (2012) Modeling biogeochemical impacts of bioenergy buffers with perennial grasses for a row-crop field in Illinois. *GCB Bioenergy*, **4**, 739–750.

He F, Vavrus SJ, Kutzbach JE, Ruddiman WF, Kaplan JO, Krumhardt KM (2014) Simulating global and local surface temperature changes due to Holocene anthropogenic land cover change. *Geophysical Research Letters*, **41**, 623–631.

Heaton E, Voigt T, Long SP (2004) A quantitative review comparing the yields of two candidate C 4 perennial biomass crops in relation to nitrogen, temperature and water. *Biomass and Bioenergy*, **27**, 21–30.

Heaton EA, Dohleman FG, Long SP (2008) Meeting US biofuel goals with less land: the potential of Miscanthus. *Global Change Biology*, **14**, 2000–2014.

Hickman GC, Vanloocke A, Dohleman FG, Bernacchi CJ (2010) A comparison of canopy evapotranspiration for maize and two perennial grasses identified as potential bioenergy crops. *GCB Bioenergy*, **2**, 157–168.

Hsieh C-I, Katul G, Chi T-W (2000) An approximate analytical model for footprint estimation of scalar fluxes in thermally stratified atmospheric flows. *Advances in Water Resources*, **23**, 765–772.

Hudiburg TW, Davis SC, Parton W, Delucia EH (2015) Bioenergy crop greenhouse gas mitigation potential under a range of management practices. *GCB Bioenergy*, **7**, 366–374.

Jones MB, Finnan J, Hodkinson TR (2015) Morphological and physiological traits for higher biomass production in perennial rhizomatous grasses grown on marginal land. *GCB Bioenergy*, **7**, 375–385.

Joo E, Hussain MZ, Zeri M et al. (2016) The influence of drought and heat stress on long term carbon fluxes of bioenergy crops grown in the Midwestern US. *Plant, Cell & Environment*, doi: 10.1111/pce.12751.

Kaimal JC, Finnigan JJ (1994). Atmospheric boundary layer flows: their structure and measurement.

Knoke T, Román-Cuesta RM, Weber M, Haber W (2012) How can climate policy benefit from comprehensive land-use approaches? *Frontiers in Ecology and the Environment*, **10**, 438–445.

Kucharik CJ, Foley JA, Delire C et al. (2000) Testing the performance of a dynamic global ecosystem model: water balance, carbon balance, and vegetation structure. *Global Biogeochemical Cycles*, **14**, 795–825.

Lawrence DM, Oleson KW, Flanner MG et al. (2011) Parameterization improvements and functional and structural advances in version 4 of the Community Land Model. *Journal of Advances in Modeling Earth Systems*, **3**, doi: 10.1029/2011MS00045.

Lee X, Goulden ML, Hollinger DY et al. (2011) Observed increase in local cooling effect of deforestation at higher latitudes. *Nature*, **479**, 384–387.

Lee J, Pedroso G, Linquist BA, Putnam D, Kessel C, Six J (2012) Simulating switchgrass biomass production across ecoregions using the DAYCENT model. *GCB Bioenergy*, **4**, 521–533.

Levis S, Hartman MD, Bonan GB (2014) The Community Land Model underestimates land-use CO_2 emissions by neglecting soil disturbance from cultivation. *Geoscientific Model Development*, **7**, 613–620.

Lewandowski I, Scurlock JM, Lindvall E, Christou M (2003) The development and current status of perennial rhizomatous grasses as energy crops in the US and Europe. *Biomass and Bioenergy*, **25**, 335–361.

Liska AJ, Yang H, Milner M et al. (2014) Biofuels from crop residue can reduce soil carbon and increase CO_2 emissions. *Nature Climate Change*, **4**, 398–401.

Liu Y, Zhuang Q, Miralles D et al. (2015) Evapotranspiration in Northern Eurasia: impact of forcing uncertainties on terrestrial ecosystem model estimates. *Journal of Geophysical Research: Atmospheres*, **120**, 2647–2660.

Loarie SR, Lobell DB, Asner GP, Mu Q, Field CB (2011) Direct impacts on local climate of sugar-cane expansion in Brazil. *Nature Climate Change*, **1**, 105–109.

Lobell D, Bala G, Mirin A, Phillips T, Maxwell R, Rotman D (2009) Regional differences in the influence of irrigation on climate. *Journal of Climate*, **22**, 2248–2255.

Loveland TR, Belward AS (1997) The IGBP-DIS global 1 km land cover data set, DISCover: first results. *International Journal of Remote Sensing*, **18**, 3289–3295.

Luyssaert S, Jammet M, Stoy PC et al. (2014) Land management and land-cover change have impacts of similar magnitude on surface temperature. *Nature Climate Change*, **4**, 389–393.

Miguez FE, Maughan M, Bollero GA, Long SP (2012) Modeling spatial and dynamic variation in growth, yield, and yield stability of the bioenergy crops Miscanthus × giganteus and Panicum virgatum across the conterminous United States. *GCB Bioenergy*, **4**, 509–520.

Miller JN, VanLoocke A, Gomez-Casanovas N, Bernacchi CJ (2015) Candidate perennial bioenergy grasses have a higher albedo than annual row crops. *GCB Bioenergy*, doi: 10.1111/gcbb.12291.

Mitchell TD, Jones PD (2005) An improved method of constructing a database of monthly climate observations and associated high-resolution grids. *International Journal of Climatology*, **25**, 693–712.

Moore CJ (1986) Frequency response corrections for eddy correlation systems. *Boundary-Layer Meteorology*, **37**, 17–35.

Nair S, Sujithkumar SK, Zhang X et al. (2012) Bioenergy crop models: descriptions, data requirements, and future challenges. *GCB Bioenergy*, **4**, 620–633.

Nikièma P, Rothstein DE, Min DH, Kapp CJ (2011) Nitrogen fertilization of switchgrass increases biomass yield and improves net greenhouse gas balance in northern Michigan, USA. *Biomass and Bioenergy*, **35**, 4356–4367.

Ozdogan M, Rodell M, Beaudoing HK, Toll DL (2010) Simulating the effects of irrigation over the United States in a land surface model based on satellite-derived agricultural data. *Journal of Hydrometeorology*, **11**, 171–184.

Peng S-S, Piao S, Zeng Z et al. (2014) Afforestation in China cools local land surface temperature. *Proceedings of the National Academy of Sciences*, **111**, 2915–2919.

Propheter JL, Staggenborg SA, Wu X, Wang D (2010) Performance of annual and perennial biofuel crops: yield during the first two years. *Agronomy Journal*, **102**, 806–814.

Puma MJ, Cook BI (2010) Effects of irrigation on global climate during the 20th century. *Journal of Geophysical Research*, **115**, D16120, doi: 10.1029/2010JD014122.

Qin Z, Zhuang Q, Chen M (2012) Impacts of land use change due to biofuel crops on carbon balance, bioenergy production, and agricultural yield, in the conterminous United States. *GCB Bioenergy*, **4**, 277–288.

Qin Z, Zhuang Q, Zhu X (2014) Carbon and nitrogen dynamics in bioenergy ecosystems: 1. Model development, validation and sensitivity analysis. *GCB Bioenergy*, **6**, 740–755.

Qin Z, Zhuang Q, Zhu X (2015) Carbon and nitrogen dynamics in bioenergy ecosystems: 2. Potential greenhouse gas emissions and global warming intensity in the conterminous United States. *GCB Bioenergy*, **7**, 25–39.

Reichstein M, Falge E, Baldocchi D et al. (2005) On the separation of net ecosystem exchange into assimilation and ecosystem respiration: review and improved algorithm. *Global Change Biology*, **11**, 1424–1439.

Roncucci N, Di Nasso NO, Bonari E, Ragaglini G (2014) Influence of soil texture and crop management on the productivity of Miscanthus (*Miscanthus × giganteus* Greef et Deu.) in the Mediterranean. *GCB Bioenergy*, **7**, 998–1008.

Smith WN, Grant BB, Campbell CA, McConkey BG, Desjardins RL, Kröbel R, Malhi SS (2012) Crop residue removal effects on soil carbon: measured and inter-model comparisons. *Agriculture, Ecosystems & Environment*, **161**, 27–38.

Smith CM, David MB, Mitchell CA, Masters MD, Anderson-Teixeira KJ, Bernacchi CJ, DeLucia EH (2013) Reduced nitrogen losses after conversion of row crop agriculture to perennial biofuel crops. *Journal of Environmental Quality*, **42**, 219–228.

Stewart J, Toma YO, Fernandez FG, Nishiwaki AYA, Yamada T, Bollero G (2009) The ecology and agronomy of *Miscanthus sinensis*, a species important to bioenergy crop development, in its native range in Japan: a review. *GCB Bioenergy*, **1**, 126–153.

Surendran Nair S, Kang S, Zhang X *et al.* (2012) Bioenergy crop models: descriptions, data requirements, and future challenges. *GCB Bioenergy*, **4**, 620–633.

Thomas AR, Bond AJ, Hiscock KM (2013) A multi-criteria based review of models that predict environmental impacts of land use-change for perennial energy crops on water, carbon and nitrogen cycling. *GCB Bioenergy*, **5**, 227–242.

US Congress (2007) The energy independence and security act of 2007 (H.R. 6). Available at: https://www.gpo.gov/fdsys/pkg/PLAW-110publ140/pdf/PLAW-110publ140.pdf

Valentine J, Clifton-Brown J, Hastings A, Robson P, Allison G, Smith P (2012) Food vs. fuel: the use of land for lignocellulosic 'next generation' energy crops that minimize competition with primary food production. *GCB Bioenergy*, **4**, 1–19.

Vanloocke A, Bernacchi CJ, Twine TE (2010) The impacts of *Miscanthus × giganteus* production on the Midwest US hydrologic cycle. *GCB Bioenergy*, **2**, 180–191.

Vanloocke A, Twine TE, Zeri M, Bernacchi CJ (2012) A regional comparison of water use efficiency for Miscanthus, switchgrass and maize. *Agricultural and Forest Meteorology*, **164**, 82–95.

Viovy N (2011) CRU-NCEPv4. CRUNCEP dataset. See http://dods.extra.cea.fr/data/p529viov/cruncep/readme.htm.

Wieder WR, Boehnert J, Bonan GB (2014) Evaluating soil biogeochemistry parameterizations in Earth system models with observations. *Global Biogeochemical Cycles*, **28**, 211–222.

Zeri M, Anderson-Teixeira K, Hickman G, Masters M, DeLucia E, Bernacchi CJ (2011) Carbon exchange by establishing biofuel crops in Central Illinois. *Agriculture, Ecosystems & Environment*, **144**, 319–329.

Zeri M, Hussain MZ, Anderson-Teixeira KJ, DeLucia E, Bernacchi CJ (2013) Water use efficiency of perennial and annual bioenergy crops in central Illinois. *Journal of Geophysical Research: Biogeosciences*, **118**, 581–589.

Zhang M, Lee X, Guirui Y *et al.* (2014) Response of surface air temperature to small-scale land clearing across latitudes. *Environmental Research Letters*, **9**, 034002.

12

Contribution of N from green harvest residues for sugarcane nutrition in Brazil

DANILO A. FERREIRA[1], HENRIQUE C. J. FRANCO[1], RAFAEL OTTO[2], ANDRÉ C. VITTI[3], CAIO FORTES[4], CARLOS E. FARONI[4], ALAN L. GARSIDE[5] and PAULO C. O. TRIVELIN[4]

[1]Brazilian Bioethanol Science and Technology Laboratory – CTBE/CNPEM, Rua Giuseppe Máximo Scolfaro, 10.000, CP. 13083-970 Campinas, SP, Brazil., [2]Soil Science department – ESALQ/USP, University of Sao Paulo, Av. Pádua Dias, 11, CP. 13418-900 Piracicaba, SP, Brazil, [3]Sao Paulo's Agency for Agribusiness Technology, Center South Pole, Rodovia SP 127, Km 30, CP. 13400-970 Piracicaba, SP, Brazil, [4]Stable Isotopes Laboratory - CENA/USP, University of Sao Paulo, Av. Centenário, 303, CP. 13400-970 Piracicaba, SP, Brazil, [5]Visiting Researcher, Brazilian Bioethanol Science and Technology Laboratory – CTBE/CNPEM, Rua Giuseppe Máximo Scolfaro, 10.000, CP. 13083-970 Campinas, SP, Brazil

Abstract

Brazil is recognized as a prominent renewable energy producer due to the production of ethanol from sugarcane. However, in order for this source of energy to be considered truly sustainable, conservation management practices, such as harvesting the cane green (without burning) and retaining the trash in the field, need to be adopted. This management practice affects mostly the nitrogen (N) cycle through the effect of trash on immobilization–mineralization of N by soil microorganisms. The aim of the experiments reported here was to evaluate N recovery from trash (trash-N) by sugarcane during three ratoon crop seasons: 2007, 2008 and 2009. Two field experiments were carried out, one in Jaboticabal and the other in Pradopolis, in the state of Sao Paulo, Brazil. The experiments were set up in a randomized block design with four replications. Within each plot, microplots were installed where the original trash was replaced by trash labelled with ^{15}N, and maintained up to the fourth crop cycle. Trash-N recovery was higher in the Jaboticabal site, the most productive one, than in the Pradópolis site. The average trash-N recovery across the two sites after three crop cycles was 7.6 kg ha^{-1} (or 16.2% of the initial N content in trash), with the remaining trash-N being incorporated into soil organic matter reserves. While these results indicate that the value of trash for sugarcane nutrition is limited in the short term, maintaining trash on the field will serve as a long-term source of N and C for the soil.

Keywords: ^{15}N, mineralization, nitrogen, *Saccharum spp*, sustainability, trash

Introduction

Based on its positive energy balance, the sugarcane crop has been highlighted globally as an important feedstock for biofuel production (Renouf *et al.*, 2008; Cavalett *et al.*, 2011). With almost 9 million ha producing over 6.5 million tons of cane annually, Brazil is the largest producer of sugarcane in the world (CONAB, 2013).

In the main Brazilian region for production of sugarcane (Center South), there has been a steady increase in mechanical harvesting without prior burning resulting in large amounts of cane trash (straw, residue) being returned to the field. This practice of harvesting without prior burning is referred to as green cane trash blanketing (GCTB). During the crop season of 2013/2014, green harvesting was estimated to account for 85% of the cane harvested in Center South region.

The amount of trash generated by GCTB harvesting of sugarcane can range from 10 to 20 Mg ha^{-1} of dry matter (Trivelin *et al.*, 1995, 1996; Vitti *et al.*, 2011; Fortes *et al.*, 2012) and has the potential to increase soil organic matter content and release nutrients into the soil (Wood, 1991; Razafimbelo *et al.*, 2006; Robertson & Thorburn, 2007b; de Luca *et al.*, 2008). According to Carvalho *et al.* (2013), the main inputs of soil organic carbon (SOC) in sugarcane fields are derived from aboveground part of the plant (dry leaves and tops) rather than roots. These authors calculate that the total allocation to SOC (mean of four harvests from three areas) from above- and belowground compartments was about 1.1 Mg C ha^{-1} yr^{-1}, of which 33% was from root system and 67% from trash.

However, sugarcane trash represents an important energy source, which can be used for electricity generation, steam production for the boilers, and in the future, as raw material for the second generation of ethanol production by means of enzymatic hydrolysis. For these

Correspondence: Danilo A. Ferreira
e-mail: danilo.ferreira@bioetanol.org.br

reasons, understanding the benefits of returning sugarcane trash to the field and knowing its effect on soil sustainability is important for the assessment of how much trash can be removed for energy production without causing further soil degradation.

Many benefits have been identified as a result of sugarcane trash being maintained in the field. For example, the enhancement of soil physical quality (Sparovek & Schnug, 2001), reduction in greenhouse gas (GHG) emissions (Figueiredo & La Scala, 2011), increase in carbon stocks (Cerri et al., 2011) and of biological activity (Souza et al., 2012), improvement of soil fertility (Vitti et al., 2011), better nutrient cycling (Franco et al., 2007) and higher yield and lifespan of sugarcane ratoons (Gava et al., 2001) are all important issues that result from trash maintenance.

Among the nutrients released during trash decomposition, nitrogen (N) has been reported as one of the most important elements to be studied, due to its complex dynamics in plant–soil system (Trivelin & Franco, 2011). The availability of N from trash mineralization is dependent on factors such as temperature, moisture, soil aeration and chemical composition of trash, especially the carbon: nitrogen ratio (C : N). Thus, there are various interacting factors that control the rate of transformation of organic N to inorganic N (Janssen, 1996).

Sugarcane trash has typical carbon and nitrogen contents ranging from 390 to 450 g kg^{-1} (carbon) and 4.6 to 6.5 g kg^{-1} (nitrogen) (Ng Kee Kwong et al., 1987; Oliveira et al., 1999), thus having a C : N ratio of around 100 : 1. This high C : N ratio promotes substantial immobilization of N in the soil and results in only small amounts of nitrogen being mineralized for the following crop season (Gava et al., 2001; Robertson & Thorburn, 2007a; Fortes et al., 2012). Usually, N immobilization takes place in plant residues (trash) when N content is than 18 g kg^{-1} and the C : N ratio is higher than 20 (Smith & Douglas, 1971; White, 1984; Bengtsson et al., 2003). Thus, in the short term, the nitrogen contribution from sugarcane trash is expected to be minor as immobilization will be substantial. Under such conditions, competition for the small amount of N available will occur between the roots and soil microorganisms (Jingguo & Bakken, 1997).

The contribution of N from trash is insignificant for nutritional purposes with only 4% of N being recovered from trash during the first ratoon under Brazilian field conditions (Vitti et al., 2005). Further, Vitti et al. (2005) suggested that the majority of uptake of this trash-N occurred during the last months before sugarcane harvest. Thus, a significant contribution of N from sugarcane trash may occur during subsequent ratoons because more than 90% of trash-N remains in the soil. Robertson & Thorburn (2007b) verified that almost 80%

of trash-N still remained in the soil after 6 years of implementation of green harvest sugarcane management.

In the research reported in this study, the contribution to sugarcane nutrition of N from green harvest residues during three consecutive crop seasons in different environmental conditions is evaluated. Such studies are important because few data are available under the conditions prevailing in Sao Paulo state, Brazil. The use of data from other areas that experience different growing conditions to establish models and parameters about trash decomposition, N mineralization from trash and N-trash uptake by sugarcane may be inappropriate.

Materials and methods

Sites description

This study was carried out at two sites located near Jaboticabal city in the state of Sao Paulo, Brazil, on sugarcane fields that were harvested green without prior burning. Site 1 was located at the Santa Adelia Bioenergy farm (21°19′, 48°19′W, 600 m asl), on a Typic Kandiudox (TK) with a medium texture (Soil Survey Staff, 2011). Site 2 was located at the Sao Martinho Bioenergy farm (21°17′S, 48°12′W, 580 m asl), on a Rhodic Eutrudox soil (RE) with clayey texture (Soil Survey Staff, 2011). The climate in both sites is classified as Aw (tropical or savannah), according to Köppen classification (Rolim et al., 2007). Chemical soil attributes were determined in samples collected at 0–0.25 and 0.25–0.5 m, in order to determine lime, gypsum and fertilizer requirements to avoid nutritional limitations.

In Site 1, the tillage practices adopted were as follows: herbicide applications over the previous sugarcane ratoon, deep ploughing (depth of 0.4 m) to incorporate previous crop residues and 2 Mg ha^{-1} of limestone into the soil, soil disking (depth of 0.2 m) for final soil preparation before furrow opening (depth of 0.35 m) and planting. In Site 2, the practice adopted was reduced tillage by means of herbicide application over the old ratoon, soil subsoiling (depth of 0.35 m) followed by furrow opening (depth of 0.35 m) and planting.

The planting was performed in February 2005 in Site 2 and in April 2005 in Site 1. Sugarcane setts of SP813250 variety with 15 buds per metre of row were planted. A total of 80 kg ha^{-1} of N as Urea, 120 kg ha^{-1} of P_2O_5 as single superphosphate and 120 kg ha^{-1} of K_2O as potassium chloride were applied at the bottom of each furrow. During the experimental period, weather data were measured by automatic weather stations located near to each experimental site. Climatological water balance was calculated according to the Penman–Monteith approach made by Allen et al. (1998) (Fig. 1).

Experimental design

After plant cane harvest (performed in July and August 2006, respectively, for the sites 1 and 2), the trials were set up in a randomized block design with four replications. The experimental

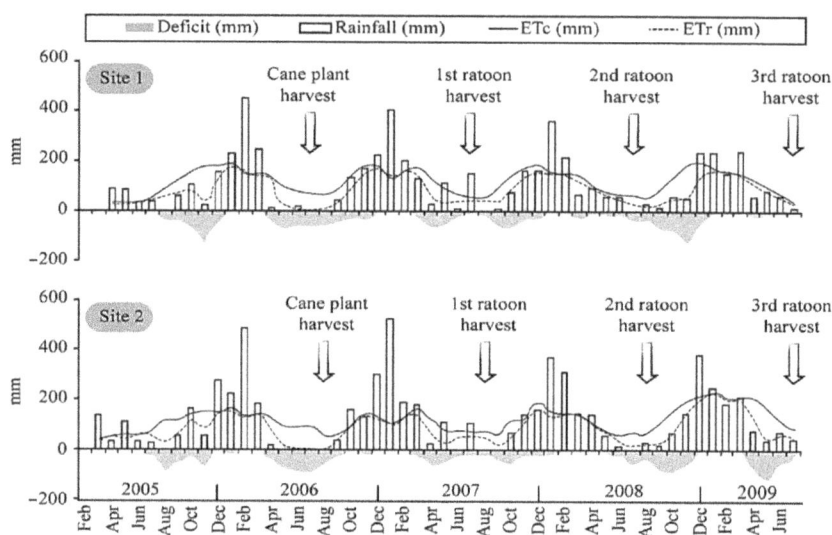

Fig. 1 Climatological water balance at sites 1 and 2 during the crop cycles. ETc, crop evapotranspiration; ETr, real evapotranspiration.

plots had 12 rows of sugarcane (1.5 m apart and 15 m long). In the centre of each plot, a microplot (3 m^2) was established (2 m long and 1.5 m wide). The trash present in each microplot was replaced by the same quantity of trash labelled with ^{15}N. The trash-^{15}N (10 Mg ha^{-1} of dry matter) was deposited over the soil surface in October 2006. The trash-^{15}N material, composed of dry leaves and tops of sugarcane, was obtained from another field experiment, where aboveground part of sugarcane was labelled by spray application of ^{15}N-urea solution according to the method of Faroni *et al.* (2007). Quantitative analysis performed on the trash material before its distribution at each site showed a biomass enrichment of 0.83 and 1.00% ^{15}N of atoms, and 51 and 41 kg ha^{-1} N for the sites 1 and 2, respectively.

Measurements

The harvest of the first ratoon took place in July (Site 1) and August 2007 (Site 2); the subsequent crop harvests were performed in July (2008 and 2009) for both sites.

During crop harvest, all the plants located in the microplots and in both adjacent rows were collected manually. The procedure adopted was the same described by Trivelin *et al.* (1994). The plants were separated into dry leaves, tops and stalks. The fresh biomass (kg) of each component was obtained directly in the field. The samples were then chopped in a forage chopper and homogenized, and a subsample of each component was weighed before being dried at 65 °C for 72 h after which they were reweighed. The subsample was then further ground in a knives mill. Measurements of total N (%) and ^{15}N isotopic abundance (atom % ^{15}N) were taken in a mass spectrometer coupled with a N analyser, model ANCA-GSL, from Sercon Co., Crewe, UK (Barrie & Prosser, 1996).

After microplot evaluation, the entire plot was harvested mechanically to obtain industrial stalk yield. Following harvest, the microplots were maintained, and the 'new' trash

(unlabelled) was deposited over the trash-^{15}N. This procedure was also adopted in the subsequent crop cycle. Nitrogen fertilization was not performed on the ratoon crops, to isolate the effect of the trash-^{15}N recovery by sugarcane plants.

Nitrogen uptake and recovery

The N in plant from trash (NPFT) and recovery of ^{15}N-trash (RNT) by sugarcane aboveground biomass were estimated by the isotopic dilution approach using the following equations:

$$\text{NPFT} = [(A - C)/(B - C)].\text{NT} \qquad (1)$$

$$\text{RNT}\,(\%) = (\text{NPFT}/\text{NAF}).100 \qquad (2)$$

where NPFT = N in the plant from ^{15}N-trash (kg ha^{-1}); RNT – recovery of ^{15}N-trash by sugarcane (%); A – abundance of ^{15}N (% of atoms) in the plant; B – abundance of ^{15}N (% of atoms) in the source (^{15}N-trash); C – natural abundance of ^{15}N (0.366% of atoms); NT – N content in the plant (kg ha^{-1}); NAF – amount of N-trash applied (kg ha^{-1}).

Statistical analyses

The results were submitted to analysis of variance (ANOVA), using F-test at $P < 0.05$ level, the averages being compared by Tukey test at 5% of probability.

Results

Biomass and nitrogen accumulation

Comparing the two experimental locations, Site 1 showed higher aboveground biomass and N accumulation in 2008, 2009 and in the sum of three seasons

(2007/2009) (Fig. 2). However, there was no difference between sites for these parameters for the first year (2007). Similar results were measured for stalks biomass and N accumulation, which is not surprising as the majority of aboveground biomass would be comprised of stalks.

For the aggregated data from both sites, the total aboveground biomass production was lower in 2008 than those measured in 2007 and 2009, in which biomass yields were similar. On the other hand, N accumulation was higher in 2009, followed by 2007 and 2008 (Table 1). Stalks biomass showed an increase from the second to third year (2008–2009), which could be related to favourable seasonal conditions during the 2009 [third season when adequate rainfall was distributed throughout the growing period and especially for the months preceding sugarcane maturation (Fig. 1)].

Trash-N uptake and recovery by sugarcane

The trash-N uptake (NPFT) by aboveground biomass was higher in Site 1 compared with Site 2 for 2008, 2009

and in the sum of three seasons (2007/2009) (Fig. 3). For RNT, the same trend was verified only in 2008 and with the combined data across 2007–2009. In Site 1, the NPFT in aboveground part of sugarcane was 3.7 (2007), 1.3 (2008) and 4.4 (2009) kg ha^{-1}, corresponding to 7.3, 2.5 and 8.7% RNT. In Site 2, NPFT for the same period was 2.4, 0.7 and 2.6 kg ha^{-1} (5.9, 1.6 and 6.3% RNT).

For the aggregate data, differences were observed in aboveground relating to NPFT and RNT in 2007 and 2009 (Table 2) and the same trend was apparent for the stalk component. At the end of the experimental period, the total NPFT was 7.6 kg ha^{-1}, corresponding to 16.2% RNT, which represents only 2.1% of total N accumulated by the crop.

Discussion

The RNT varied from 5 to 8% at the end of the first ratoon cycle (2007) at both sites (Fig. 3), similar to those presented by Ng Kee Kwong *et al.* (1987) and Gava *et al.* (2003). The NPFT on tops was higher at Site 1 than Site 2, although the biomass and N accumulation were the

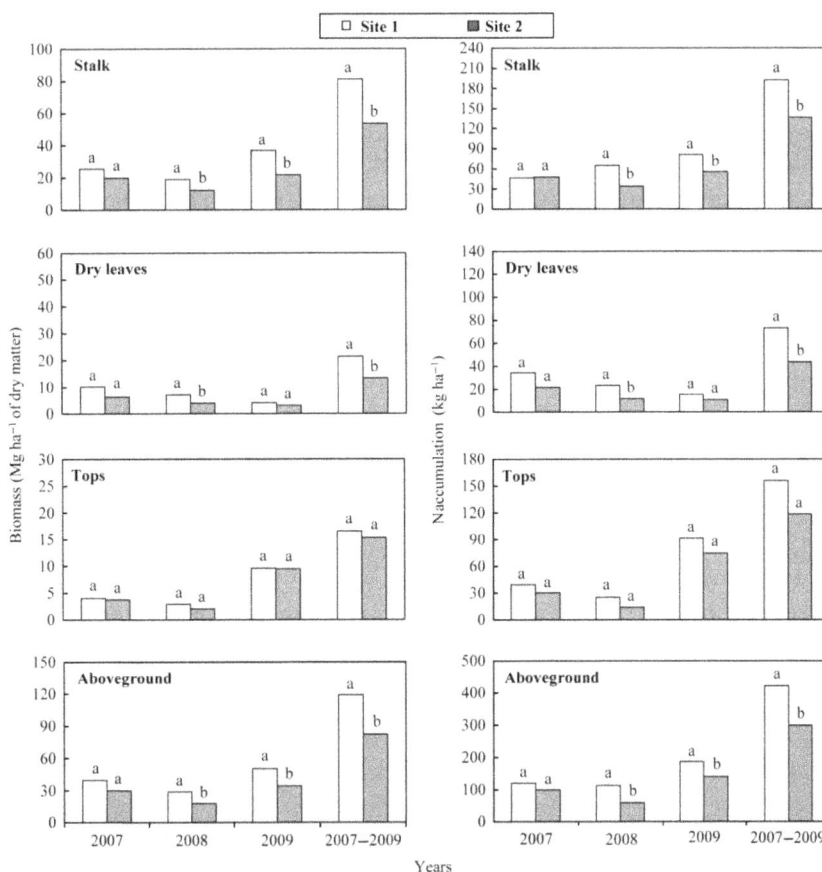

Fig. 2 Biomass and N accumulation in sugarcane components (stalk, dry leaves, tops and aboveground part) throughout three crop seasons in two experimental sites in Brazil (1 and 2). Same letters indicate no difference between sites in each year according to Tukey's test (P < 0.05).

Table 1 Biomass and N accumulation in parts of sugarcane plant during three crop season. (All data are the average across sites*)

Year	Biomass				N accumulation			
	Stalk	Dry leaves	Tops	Aboveground	Stalk	Dry leaves	Tops	Aboveground
	Mg ha^{-1} of dry matter				kg ha^{-1}			
2007	22.6 ab	8.3 a	3.9 b	34.8 a	34.8 a	28.1 a	34.8 b	110.0 b
2008	15.5 b	5.5 ab	2.5 c	23.6 b	23.6 b	17.3 b	19.6 c	86.3 b
2009	29.2 a	3.6 b	9.5 a	42.3 a	42.3 a	12.9 b	82.8 a	163.9 a
2007–2009	67.4	17.4	15.9	100.7	164.6	58.3	137.2	360.5
LSD	9.4	3.3	0.5	11.2	17.8	9.5	7.7	24.2
CV (%)	14.5	19.7	3.5	11.4	11.4	16.9	5.8	6.9

LSD, least significant difference; CV, coefficient of variation.

*Values followed by same small letters in column indicate no difference between years accordingly the Tukey' test ($P < 0.05$).

same. No differences were detected between sites for all the other parameters evaluated. Among the plant components, stalks showed greater recovery when compared to tops and dry leaves. The higher N recovery in stalks was probably associated with the period of measurement, carried out during the sugarcane maturation phase, when stalks make up the bulk of the biomass (Table 1). At this stage, leaves constitute only 10% of plant biomass (Robertson et al., 1996; Franco et al., 2013). Furthermore, later in the season, part of the N from the senescent leaves is expected to be remobilized to the active parts of the plant (stalks).

In the second ratoon season (2008), even though differences occurred between sites for NPFT and RNT on aboveground biomass, it is possible to observe a sharp decrease in the amount of trash-N recovery by plants, especially in Site 2 (Fig. 2) which may be explained by the lower dry biomass and N accumulation obtained at this site in 2008 when compared with the previous season (Fig. 2) which was most likely associated with less favourable seasonal conditions.

The rainfall during 2nd ratoon (2008) at Site 2 was almost 350 mm lower than for the first ratoon (1729 mm in 2007 against 1371 mm in 2008), which affected the biomass production and ^{15}N-trash recovery. Further, the reduction in ^{15}N-trash recovery may also be associated with low temperature average during 2008 compared to 2007 (25.7 in 2007 against 23.3 °C in 2008). Temperature has important implications for determining microorganism activities, and therefore trash mineralization (Stanford et al., 1973; Katterer et al., 1998). Stanford et al. (1973) reported that, in the temperature range from 5 to 35 °C, the rate of N mineralization doubles for each 10 °C increase in temperature. In other work, Lara Cabezas et al. (2004) showed that high temperature and moisture levels in the soil favour crop residues mineralization, while Torres et al. (2005) reported a decrease in the decomposition rate and N release from cover crops residues under low temperature and moisture.

The unfavourable water balance in the 2nd ratoon (2008) has also affected the sugarcane yield in Site 1 (Fig. 2). Nevertheless, biomass and nitrogen accumulation reduction were not as remarkable as those observed in Site 2. However, during the 2009 season (3rd ratoon), better seasonal conditions for plant growth resulted in higher biomass production, N accumulation and hence the increases in NPFT and RNT at both sites (Fig. 3). Furthermore, the rate of 7.5% of RNT in the third year may indicate a decrease in the C : N ratio of trash, considering that at the start of the experiment, this ratio was around 100 : 1, which would certainly cause immobilization of N by microorganisms (Jadhav, 1996; Bengtsson et al., 2003). However, Fortes et al. (2012) found a reduction in the C : N ratio from 108 : 1 to 24 : 1 during the years of evaluation (three ratoons), at the Site 2 experiment, resulting in 31% of trash-N being released to the soil.

Interestingly in 2009, significant differences between the sites were measured only with the tops, which showed higher NPFT and RNT at Site 1. This resulted in the same trend with aboveground trash-N recovery as in the previous seasons (Fig. 3). The influence of tops on aboveground NPFT is due to the huge increase in tops biomass in 2009 when compared with 2008. Tops biomass in 2009 represented 22.5% of total aboveground biomass (Table 1).

At the end of the experimental period, the differences found represents 3.7 kg ha^{-1} more NPFT by plants at Site 1 than Site 2, which represents 24.9% more N from trash (Fig. 3). These results are due to the higher biomass production in Site 1 than Site 2 (119 vs. 82 tonness ha^{-1}) (Fig. 2), resulting in 29% more N accumulation in Site 1. The lower biomass production at Site 2 may be attributed to the reduced tillage adopted prior the sugarcane planting and to water deficit during the ratoon crop cycle. Regardless reduced tillage provides an effective means to conserve overall soil fertility, even though it may initially increase the need for N fertilizer

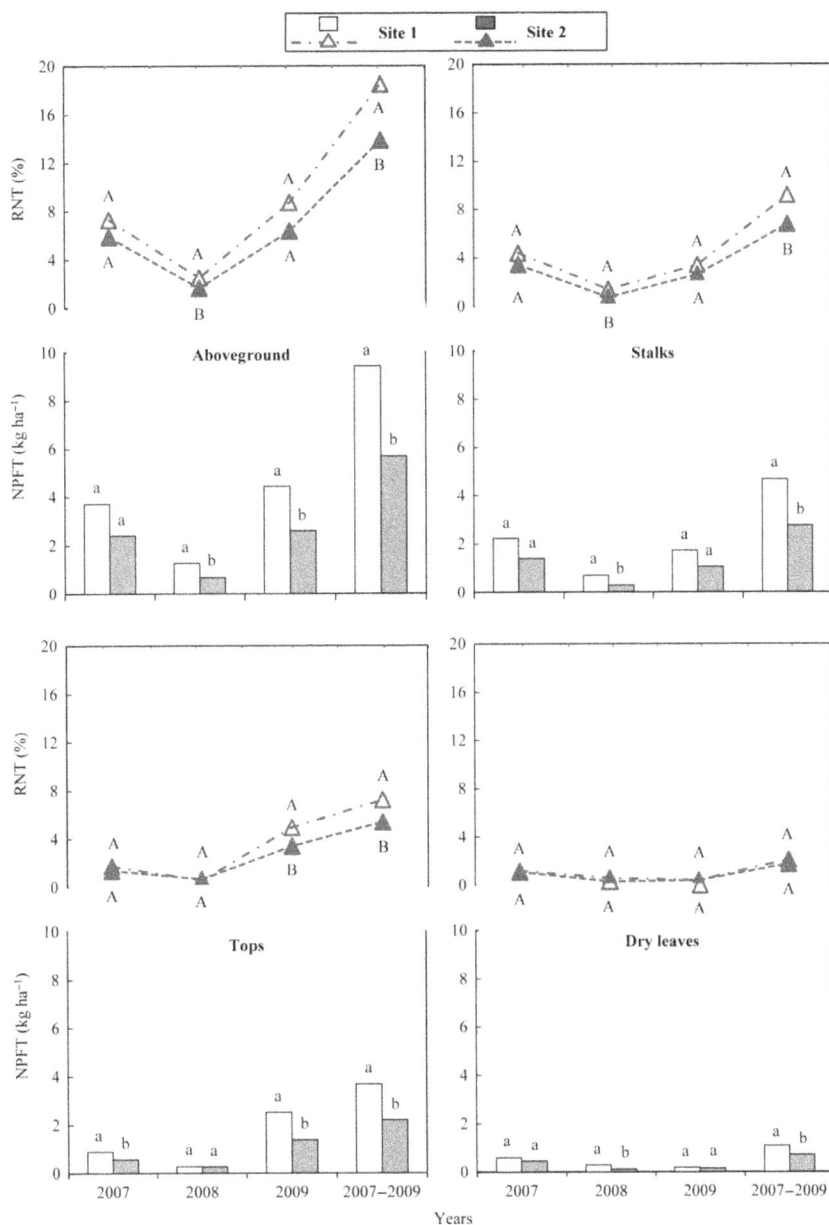

Fig. 3 Recovery of N from trash (RNT, %) and nitrogen in plant from trash (NPFT, kg ha^{-1}) in sugarcane components (stalk, dry leaves, tops and aboveground part) throughout three crop seasons in two experimental sites in Brazil (1 and 2). Same letters indicate no difference between sites in each year according to Tukey's test ($P < 0.05$).

(Maltas *et al.*, 2013), particularly if the trash has a high C : N ratio, as is the case with sugarcane trash. Further, in these studies N fertilization was not carried out with the ratoon crops.

Averaged across the two sites after three crop cycles, only 16.2% of N-trash was recovered by sugarcane (Table 2), which represents a small contribution to crop nutrition (2.1% of total N needs) in the short term. The remainder was most likely located in the various nitrogen pools known to exist in the soil/plant continuum associated with the dynamics of nitrogen movement

between different pools in the soil. Myers *et al.* (1994) reported that, during the decomposition of the crop residues, there is partitioning of N into the compartments of mineral N (soil and fertilizer), humic N and immobilized N due to the soil microbial biomass, with a continuous turnover of this N among the compartments.

If the initial trash composition for Site 2 (C : N ratio of 108) described by Fortes *et al.* (2012) is taken into account and calculations proposed by Robertson & Thorburn (2007b) are applied, 98 and 120 kg ha^{-1} N

Table 2 Nitrogen in plant from trash (NPFT) and recovery of N from trash (RNT) in each part of sugarcane over the experimental period. (All data are the average across sites*)

Year	NPFT				RNT			
	Stalk	Dry leaves	Tops	Aboveground	Stalk	Dry leaves	Tops	Aboveground
	kg ha^{-1}				%			
2007	1.8 a	0.5 a	0.7 b	3.1 a	3.9 a	1.1 a	1.5 b	6.6 a
2008	0.5 b	0.2 b	0.3 c	1.0 b	1.1 b	0.4 b	0.6 c	2.1 b
2009	1.4 a	0.2 b	1.9 a	3.5 a	3.0 a	0.4 b	4.2 a	7.5 a
2007–2009	3.7	0.9	2.9	7.6	8.0	1.9	6.3	16.2
LSD	0.8	0.2	0.2	1.0	1.9	0.4	0.5	2.4
CV (%)	22.0	21.4	8.4	14.3	24.2	23.4	8.5	15.1

LSD, least significant difference; CV, coefficient of variation.
*Values followed by same small letters in column indicate no difference between years accordingly the Tukey' test ($P < 0.05$).

would be immobilized to achieve trash decomposition in sites 1 and 2, respectively. This means 57–69 kg ha^{-1} N needs to be supplied by soil organic N or fertilizers inputs. Thus, in order to ensure high sugarcane yield and avoid soil fertility depletion, the rate of fertilizer N needs to be increased at least for the early years of sugarcane green harvest as an increase N immobilization will most certainly occur due the high C : N ratio of the trash.

However, although sugarcane trash only provides a small amount of N for the crop during the first years of green harvest establishment, the trash deposition on soil surface by successive harvests should contribute towards a greater accumulation of organic N into the soil. According to Robertson & Thorburn (2007b), around 79% of the trash-N may be retained in the soil in the long term under sugarcane green harvesting systems. Simulation studies performed in Australia indicate that after 20 years of sugarcane trash retention, it is possible to save around 40 kg ha^{-1} N per year in fertilizer N due to long-term N mineralization of sugarcane (Vallis et al., 1996). Simulations performed for Brazilian conditions resulted in similar findings (Trivelin et al., 2013). However, it is important to note that all of the data currently available (Vallis et al., 1996 included) are based on trash incorporation between sugarcane cycles and not on continuous deposition onto the soil surface. Adopting the latter approach may well result in substantially more N accumulation over time.

In our experiments, sugarcane absorbed 7.6 kg ha^{-1} of N (means of two sites) from trash after 3 years of maintenance in the field (representing 16.2% of the initial N content of trash) (Table 2). The results indicate that the most part of trash-N remains in the soil and serves as a long-term source of N to the crop, as well as an important source of soil organic carbon. This finding is important considering the interest of farmers in removing trash from fields for energy purposes. Based

on our results, as the trash only provides 2.1% of total N needs, its removal will not affect sugarcane nitrogen nutrition in the short term, but is likely to show a potential negative effect in the long term for sugarcane nutrition and soil C sequestration.

Acknowledgements

To 'FAPESP' (Thematic Project 2002/10534-8) and 'Conselho Nacional de Desenvolvimento Científico e Tecnológico' for financial support; 'Centro de Tecnologia Canavieira (CTC)', São Martinho and Santa Adélia sugarcane mills for the sites and field work support.

References

Allen RG, Pereira LS, Raes D, Smith M (1998) *Crop Evapotranspiration: Guidelines for Computing Crop Water Requirements.* (FAO Irrigation and Drainage Paper, 56). Rome: FAO. pp. 1–15.

Barrie A, Prosser SJ (1996) Automated analysis of light-element stable isotopes by isotope ratio mass spectrometry. In: *Mass Spectrometry of Soils* (eds Boutton TW, Yamashi S), pp. 1–46. Marcel Dekker, New York, NY.

Bengtsson G, Bengtson P, Mansson KF (2003) Gross nitrogen mineralization, immobilization, and nitrification rates as a function of soil C/N ratio and microbial activity. *Soil Biology & Biochemistry*, **35**, 143–154.

Carvalho JLN, Otto R, Franco HCJ, Trivelin PCO (2013) Input of sugarcane post-harvest residues into the soil. *Scientia Agricola*, **70**, 336–344.

Cavalett O, Cunha MP, Junqueira TL et al. (2011) Environmental and economic assessment of bioethanol, sugar and bioelectricity production from sugarcane. *Chemical Engineering Transactions*, **25**, 1007–1012.

Cerri CC, Galdos MV, Maia SMF, Bernoux M, Feigl BJ, Powlson D, Cerri CEP (2011) Effect of sugarcane harvesting systems on soil carbon stocks in Brazil: an examination of existing data. *European Journal of Soil Science*, **62**, 23–28.

COMPANHIA NACIONAL DE ABASTECIMENTO – CONAB. Available at: http://www.conab.gov.br/OlalaCMS/uploads/arquivos/13_04_09_10_29_31_boletim_cana_portugues_abril_2013_1o_lev.pdf (accessed: 17 may 2013).

Faroni CE, Trivelin PCO, da Silva PH, Bologna IR, Vitti AC, Franco HCJ (2007) Marcação de fitomassa de cana-de-açúcar com aplicação de solução de uréia marcada com 15N. *Pesquisa Agropecuária Brasileira*, **42**, 851–857.

Figueiredo EB, La Scala N (2011) Greenhouse gas balance due to the conversion of sugarcane areas from burned to green harvest in Brazil. *Agriculture, Ecosystems & Environment*, **141**, 77–85.

Fortes C, Trivelin PCO, Vitti AC (2012) Long-term decomposition of sugarcane harvest residues in Sao Paulo state, Brazil. *Biomass and Bioenergy*, **42**, 189–198.

Franco HCJ, Vitti AC, Faroni CE, Cantarella H, Trivelin PCO (2007) Estoque de nutrientes em resíduos culturais incorporados ao solo na reforma de áreas com cana-de-açúcar. STAB. *Sociedade dos Técnicos Açucareiros e Alcooleiros do Brasil*, **25**, 32–36.

Franco HCJ, Pimenta MTB, Carvalho JLN, Rossell CEV, Braunbeck OA, Kolln OT, Rossi Neto J (2013) Assessment of sugarcane trash for agronomic and energy purposes in Brazil. *Scientia Agrícola*, **70**, 305–312.

Gava GJC, Trivelin PCO, Oliveira MW, Penatti CP (2001) Crescimento e acúmulo de nitrogênio em cana-de-açúcar cultivada em solo coberto com palhada. *Pesquisa Agropecuária Brasileira*, **36**, 1347–1354.

Gava GJC, Trivelin PCO, Vitti AC, Oliveira MW (2003) Recuperação do nitrogênio (15N) da uréia e da palhada por soqueira de cana-de-açúcar (*Saccharum* spp.). *Revista Brasileira de Ciência do Solo*, **27**, 621–630.

Jadhav SB (1996) *Effect of incorporation of sugarcane trash on cane productivity and soil fertility*. In: International society of sugar cane technologists congress 22., Cartagena, 1995. Proceeding. Cali: Tecnicanã.

Janssen BH (1996) Nitrogen mineralization in relation to C: N ratio and decomposability of organic materials. *Plant and Soil*, **181**, 39–45.

Jingguo W, Bakken LR (1997) Competition for nitrogen during mineralization of plant residues in soil: microbial response to C and N availability. *Soil Biology & Biochemistry*, **29**, 163–170.

Katterer T, Reichstein M, Andrén O, Lomander A (1998) Temperature dependence of organic matter decomposition: a critical review using data analyzed with different models. *Biology and Fertility of Soils*, **27**, 258–262.

Lara Cabezas WAR, Alves BJR, Caballero SSU, Santana DC (2004) Influência da cultura antecessora a da adubação nitrogenada na produtividade de milho em sistema de plantio direto e solo preparado. *Ciência Rural*, **34**, 1005–1013.

de Luca EF, Feller C, Cerri CC, Barthès B, Chaplot V, Campos DC, Manechini C (2008) Avaliação de atributos físicos e estoques de carbono e nitrogênio em solos com queima e sem queima do canavial. *Revista Brasileira de Ciência do Solo*, **32**, 789–800.

Maltas A, Charles R, Jeangros B, Sinaj S (2013) Effect of organic fertilizers and reduced-tillage on soil properties, crop nitrogen response and crop yield: results of a 12-year experiment in Changins, Switzerland. *Soil and Tillage Research*, **126**, 11–18.

Myers RJK, Palm CA, Cuevas E, Gunatilleke IUN, Brossard M (1994) The synchronization of nutrient mineralization and plant nutrient demand. In: *The Biological Management of Tropical Soil Fertility* (eds Woomer PL, Swift MJ), pp. 81–116. Wiley-Sayce Publications, New York, NY.

Ng Kee Kwong KF, Deville J, Cavalot PC, Riviere V (1987) Value of cane trash in nitrogen nutrition of sugarcane. *Plant and Soil*, **102**, 79–83.

Oliveira MW, Trivelin PCO, Penatti CP, Piccollo MC (1999) Decomposição e liberação de nutrientes da palhada de cana-de-açúcar em campo. *Pesquisa Agropecuária Brasileira*, **34**, 2359–2362.

Razafimbelo T, Barthès B, Larré-Larrouy MC, Luca EF, Laurent JY, Cerri CC, Feller C (2006) Effect of sugarcane residue management (mulching versus burning) on organic matter in a clayey Oxisol from southern Brazil. *Agriculture, Ecosystems & Environment*, **115**, 285–289.

Renouf MA, Wegener MK, Nielsen LK (2008) An environmental life cycle assessment comparing Australian sugarcane with US corn and UK sugar beet as producers of sugars for fermentation. *Biomass and Bioenergy*, **32**, 1144–1155.

Robertson FA, Thorburn PJ (2007a) Decomposition of harvest residue in different climatic zones. *Australian Journal of Soil Research*, **45**, 1–11.

Robertson FA, Thorburn PJ (2007b) Management of sugarcane harvest residues: consequences for soil carbon and nitrogen. *Australian Journal of Soil Research*, **45**, 13–23.

Robertson MJ, Wood AW, Muchow RC (1996) Growth of sugarcane under high input conditions in tropical Australia.1. Radiation use, biomass accumulation and partitioning. *Field Crops Research*, **48**, 11–25.

Rolim GS, Camargo MBP, Lania DG, Moraes JFL (2007) Climatic classification of Köppen and Thornthwaite systems and their applicability in the determination of agroclimatic zoning for the State of São Paulo, Brazil. *Bragantia*, **66**, 711–720.

Smith JH, Douglas CL (1971) Wheat straw decomposition in the field. *Soil Science Society of America, Proceedings*, **35**, 269–272.

Soil Survey Staff (2011) *Keys to Soil Taxonomy. 2010*, 11th edn. USDA-Natural Resources Conservation Service, Washington, DC. pp. 241–256.

Souza RA, Telles TS, Machado W, Hungria M, Tavares Filho J, Guimarães MF (2012) Effects of sugarcane harvesting with burning on the chemical and microbiological properties of the soil. *Agriculture, Ecosystems & Environment*, **155**, 1–6.

Sparovek G, Schnug E (2001) Soil tillage and precision agriculture: a theoretical case study for soil erosion control in Brazilian sugar cane production. *Soil & Tillage Research*, **61**, 47–54.

Stanford G, Frere MH, Shwaninger DH (1973) Temperature coefficient of soil nitrogen mineralization. *Soil Science*, **115**, 321–323.

Torres JLR, Pereira MG, Andrioli I, Polidoro JC, Fabian AJ (2005) Decomposição e liberação de nitrogênio de resíduos culturais de plantas de cobertura em um solo de cerrado. *Revista Brasileira de Ciência do Solo*, **29**, 609–618.

Trivelin PCO, Franco HCJ (2011) Adubação nitrogenada e a sustentabilidade de agrossistemas. *Tópicos em Ciência do Solo*, **7**, 193–219.

Trivelin PCO, Lara Cabezas WAR, Victoria RL, Reichardt K (1994) Evaluation of 15N plot design for estimating plant recovery of fertilizer nitrogen applied to sugar cane. *Scientia Agricola*, **51**, 226–234.

Trivelin PCO, Victoria RL, Rodrigues JCS (1995) Aproveitamento por soqueira de cana-de-açúcar de final de safra do nitrogênio da aquamônia-15N e uréia-15N aplicado ao solo em complemento à vinhaça. *Pesquisa Agropecuária Brasileira*, **30**, 1375–1385.

Trivelin PCO, Rodrigues JCS, Victoria RL (1996) Utilização por soqueira de cana-de-açúcar de início de safra do nitrogênio da aquamônia-15N e uréia-15N aplicado ao solo em complemento à vinhaça. *Pesquisa Agropecuária Brasileira*, **31**, 89–99.

Trivelin PCO, Franco HCJ, Otto R *et al.* (2013) Impact of sugarcane trash on fertilizer requirements for São Paulo, Brazil. *Scientia Agrícola*, **70**, 345–352.

Vallis I, Parton WJ, Keating BA, Wood AW (1996) Simulation of the effects of trash and N fertilizer management on soil organic matter levels and yields of sugarcane. *Soil & Tillage Research*, **38**, 115–132.

Vitti AC, Trivelin PCO, Gava GJC, Penatti CP, Piedade SMS (2005) Recuperação do N fertilizante-15N e da palha-15N no sistema solo-planta-palha em cana-de-açúcar. International Nuclear Atlantic Conference.

Vitti AC, Franco HCJ, Trivelin PCO, Ferreira DA, Otto R, Fortes C, Faroni CE (2011) Nitrogênio proveniente da adubação nitrogenada e resíduos culturais na nutrição da cana-planta. *Pesquisa Agropecuária Brasileira*, **46**, 287–293.

White PJ (1984) Effects of crop residues incorporation on soil properties and growth of subsequent crops. *Australian Journal of Experimental Agriculture*, **24**, 219–235.

Wood AW (1991) Management of crop residues following green harvesting of sugarcane in north Queensland. *Soil & Tillage Research*, **20**, 69–85.

13

Ecosystem model parameterization and adaptation for sustainable cellulosic biofuel landscape design

JOHN L. FIELD[1,2], ERNIE MARX[1], MARK EASTER[1], PAUL R. ADLER[3] and KEITH PAUSTIAN[1,4]

[1]Natural Resource Ecology Laboratory, Colorado State University, 1499 Campus Delivery, Fort Collins, CO 80523, USA, [2]Department of Mechanical Engineering, Colorado State University, 1374 Campus Delivery, Fort Collins, CO 80523, USA, [3]Pasture Systems and Watershed Management Research Unit, United States Department of Agriculture-Agricultural Research Service, University Park, PA 16802, USA, [4]Department of Soil and Crop Science, Colorado State University, 1170 Campus Delivery, Fort Collins, CO 80523, USA

Abstract

Renewable fuel standards in the US and elsewhere mandate the production of large quantities of cellulosic biofuels with low greenhouse gas (GHG) footprints, a requirement which will likely entail extensive cultivation of dedicated bioenergy feedstock crops on marginal agricultural lands. Performance data for such systems is sparse, and non-linear interactions between the feedstock species, agronomic management intensity, and underlying soil and land characteristics complicate the development of sustainable landscape design strategies for low-impact commercial-scale feedstock production. Process-based ecosystem models are valuable for extrapolating field trial results and making predictions of productivity and associated environmental impacts that integrate the effects of spatially variable environmental factors across diverse production landscapes. However, there are few examples of ecosystem model parameterization against field trials on both prime and marginal lands or of conducting landscape-scale analyses at sufficient resolution to capture interactions between soil type, land use, and management intensity. In this work we used a data-diverse, multi-criteria approach to parameterize and validate the DayCent biogeochemistry model for upland and lowland switchgrass using data on yields, soil carbon changes, and soil nitrous oxide emissions from US field trials spanning a range of climates, soil types, and management conditions. We then conducted a high-resolution case study analysis of a real-world cellulosic biofuel landscape in Kansas in order to estimate feedstock production potential and associated direct biogenic GHG emissions footprint. Our results suggest that switchgrass yields and emissions balance can vary greatly across a landscape large enough to supply a biorefinery in response to variations in soil type and land-use history, but that within a given land base both of these performance factors can be widely modulated by changing management intensity. This in turn implies a large sustainable cellulosic biofuel landscape design space within which a system can be optimized to meet economic or environmental objectives.

Keywords: biogeochemistry, carbon sequestration, cellulosic biofuel, ecosystem model parameterization, ecosystem modeling, feedstock, greenhouse gas emissions, landscape design, marginal land, nitrous oxide

Introduction

The Energy Independence and Security Act of 2007 expanded the US renewable fuel standard to require the use of large quantities of 'advanced' and 'cellulosic' biofuels with lifecycle greenhouse gas (GHG) emissions reductions of 50% and 60%, respectively, relative to a conventional gasoline baseline (110th Congress of the United States, 2007). This mandate is predicated on the wide availability of biomass feedstocks with low direct environmental impacts and causing minimal disruption to agricultural commodity markets, which could lead to indirect leakage effects (Searchinger et al., 2008). A variety of low-impact feedstock provisioning strategies have been envisioned including the collection of agricultural residues, forestry residues, and municipal wastes, as well as the cultivation of dedicated woody and herbaceous crops on marginal or non-agricultural lands (Campbell et al., 2008; Robertson et al., 2008; Tilman et al., 2009). In particular, perennial grasses such as switchgrass, Miscanthus, bioenergy-optimized sugarcane, and mixed prairie species have been identified as promising cellulosic feedstock crops due to their high

Correspondence: John L. Field
e-mail: John.L.Field@colostate.edu

yield potential, relatively low input requirements, high nitrogen use efficiency, and ability to sequester carbon through soil organic matter formation (Vogel *et al.*, 2002; Tilman *et al.*, 2006; Heaton *et al.*, 2008; Walter *et al.*, 2014). A comprehensive review of US feedstock availability by the Department of Energy suggests that widespread cultivation of high-yielding dedicated perennial grasses will be necessary to achieve the target of displacing 30% of U.S. petroleum consumption with biofuels (U.S. Department of Energy, 2011), and it has been estimated that biofuel supply chains based on such feedstocks will have highly favorable lifecycle GHG impacts as compared to first-generation biofuel technologies (Schmer *et al.*, 2008; Davis *et al.*, 2012; Wang *et al.*, 2012).

Despite their promise, agronomic experience with perennial grass feedstock crops is still relatively limited, and questions around the best management practices to balance the often-competing goals of maintaining high yields while minimizing environmental impacts have not yet been resolved. The debate between the relative merits of low-intensity cultivation over large areas (often referred to as 'land-sharing') vs. more intensive production on a more limited land base ('land-sparing') is still being waged (Anderson-Teixeira *et al.*, 2012). Process-based ecosystem modeling can play an important role in extrapolating limited existing perennial grass field trial results to make more general estimates of productivity, environmental impacts, and optimal management and landscape design strategies.

Modeling cellulosic feedstock yields & environmental impacts

The use of crop models for assessing management-environment interactions and predicting bioenergy feedstock crop productivity was thoroughly reviewed by Nair *et al.* (2012). Crop models such as APSIM, BioCro, and ALMANAC have been applied at regional or national scales to assess the productivity of first- and second-generation feedstock crops as affected by broad-scale climate-soil associations (Bryan *et al.*, 2010; Miguez *et al.*, 2012; Behrman *et al.*, 2014). Bioenergy system design is not just a question of yield, however, and understanding the biogeochemical cycling of carbon, nitrogen, and water through these systems is essential for quantitatively evaluating their sustainability (Robertson *et al.*, 2011). Perennial grass feedstock crops are often associated with a high potential for soil carbon sequestration. One recent meta-analysis suggests that switchgrass increases soil organic carbon (SOC) levels at a median rate of ca. 0.7 t C ha^{-1} yr^{-1} when cultivated on carbon-depleted agricultural soils, though performance is more neutral on pastureland or areas that

were not previously cropped (Qin *et al.*, 2015a). Application of nitrogen fertilizers is typically required to replace losses during harvest and maintain yield levels, but emissions of nitrous oxide (N_2O, a byproduct of soil microbial activity and a potent GHG) increase non-linearly with increasing N rate (Hoben *et al.*, 2011; Shcherbak *et al.*, 2014) and can threaten the overall lifecycle GHG footprint of any bioenergy system based on feedstocks with inefficient nitrogen cycling (Crutzen *et al.*, 2008). The biogeochemical cycling of C, N, and H_2O and associated fluxes of biogenic GHGs are tightly linked in all agroecosystems by fundamental mechanisms including plant tissue stoichiometry, photosynthetic pathway (C_3 vs. C_4 vs. CAM), stomatal conductance, and microbial mineralization/immobilization, and are sensitive to local climate, soil type, and land-use history.

Detailed reviews of process-based biogeochemical model use to capture these interactions in the context of bioenergy system sustainability assessment are provided by Thomas *et al.* (2013) and Robertson *et al.* (2015). The CENTURY model was among the first to be applied to bioenergy sustainability assessment, and it and its derivative DayCent model have been widely used to evaluate corn grain production, corn stover removal, and the dedicated cultivation of switchgrass and Miscanthus from the level of individual sites to national scales (Sheehan *et al.*, 2003; Kim & Dale, 2005; Chamberlain *et al.*, 2011; Davis *et al.*, 2012; Lee *et al.*, 2012; Duval *et al.*, 2013). The Environmental Policy Integrated Climate model has also been applied extensively to bioenergy feedstocks in the context of economic analyses (Jain *et al.*, 2010; Egbendewe-Mondzozo *et al.*, 2011) and environmental sustainability assessments (Gelfand *et al.*, 2013) at scales from regional (Zhang *et al.*, 2010) to global (Kang *et al.*, 2014).

Experience with using models to assess the productivity and biogeochemical implications of fine-scale variations in environmental factors such as soil type or topography across landscapes or individual farms is more limited. Studies of switchgrass cultivation in the southeastern US using crop production models have come to different conclusions as to whether yield is sensitive (Woli, 2012) or insensitive (Persson *et al.*, 2011) to underlying soil type. At the finest spatial scale, a group at Idaho National Laboratory has used corn yield data from a precision agriculture system in Iowa to drive DayCent and models of soil erosion and identify areas of low nitrogen use efficiency to target for conversion to switchgrass (Abodeely *et al.*, 2013). Many studies conduct large numbers of fine-scale simulations to make regional-scale estimates of feedstock productivity and environmental performance, though they typically do not report on soil-climate-management interactions explicitly but rather emphasize more aggregate descriptions of landscape performance (e.g. Davis *et al.*, 2012; Gelfand *et al.*, 2013).

Challenges in bioenergy ecosystem modeling

The use of ecosystem models to assess different bioenergy landscape design strategies is complicated by challenges around the representation of marginal lands, adequate bioenergy crop parameterization, and selection of the most appropriate spatial resolution and agronomic management practices for simulation. While biogeochemical process models are increasingly used to simulate conversion of marginal agricultural lands to bioenergy feedstock cultivation (Bandaru *et al.*, 2013; Qin *et al.*, 2015b), these scenarios are particularly challenging from a modeling perspective. The definition of 'marginal' land itself is not straightforward or consistent across studies, and the term is applied to both land with low productivity potential and to land with ample productivity but vulnerable to long-term degradation (erosion, loss of soil organic matter, etc.) under conventional cropping systems. In many bioenergy assessment studies the marginal designation has been based on unfavorable biophysical properties as judged using land suitability ratings (Gelfand *et al.*, 2013) or remote sensing techniques (Cai *et al.*, 2011). Another more direct basis for the designation considers past transitions in an out of agricultural production as inferred from land-use datasets (Campbell *et al.*, 2008), remote sensing (Wright & Wimberly, 2013), or sector-level economic modeling (Swinton *et al.*, 2011). When considering lands designated as marginal based on their low productivity potential, bioenergy feedstock crops are often not immune to the factors that make such lands challenging for conventional crops. Recent perennial grass field trials purposely conducted on marginal sites indicate reduced productivity relative to performance on the prime lands typically encountered at many agricultural field stations (Mooney *et al.*, 2009; Shield *et al.*, 2012; Boyer *et al.*, 2013). From an ecosystem modeling perspective, accurate assessment is only possible to the extent that underlying biophysical limitations on productivity (e.g. unfavorable climates, soil texture extremes, shallow soils, low soil organic matter levels, site drainage problems, slope, vulnerability to erosion, etc.) are represented directly or indirectly in model data inputs and the processes simulated.

It is well understood that process-based ecosystem models require proper parameterization specific to the agroecosystems being simulated in order to achieve good performance, and that such models have limited predictive power when extrapolated far beyond their parameterization scope (Thomas *et al.*, 2013). However, many bioenergy assessment studies are based on models parameterized under prime conditions and then extrapolated to highly marginal sites, or lacking an explicit independent validation of performance (e.g. Gelfand *et al.*, 2013; Kang *et al.*, 2014). In the case of the DayCent model, parameterization typically involves adjusting study site and crop parameters by hand in order to match observed real-world performance for a small number of field trial cases for which extensive data are available (Del Grosso *et al.*, 2011). While this approach can often yield a high degree of fidelity across a range of performance criteria for the sites in question (Hudiburg *et al.*, 2015), the very large number of empirically-determined crop and site parameters in the model makes the process vulnerable to over-parameterization. In such cases, model fit to the training dataset is improved via mechanisms that lack broader underlying ecological significance, reducing the generality of the resulting model for other geographic areas, environmental conditions, or management regimes (Necpálová *et al.*, 2015). It is also possible to introduce bias with the selection of the parameterization cases themselves, if the researcher gravitates to focus on studies that confirm *a priori* assumptions of how a system 'should' perform.

There are additional challenges around the spatial resolution of landscape modeling and assumptions about crop agronomic management. Management factors including tillage intensity (Adler *et al.*, 2007), fertilizer application rate (Davis *et al.*, 2013), and rotation length (Pyörälä *et al.*, 2014) can potentially change the lifecycle GHG performance of a bioenergy system from positive to negative, an effect termed the 'management swing potential' (Davis *et al.*, 2013). Management recommendations for bioenergy crops are not always well defined. For example, nitrogen fertilizer recommendations for Miscanthus have been widely debated (Arundale *et al.*, 2014a), with important implications for the overall GHG footprint of production (Roth *et al.*, 2015). To the extent that there are interactions between best management practices and site-level ecosystem properties (soil texture, land-use history, etc.), assumptions about management used within an assessment study should ideally be implemented at the level of management decision-making, that is, the field-scale. Some landscape modeling studies have started to endogenize management intensity questions by simulating productivity and environmental impacts at different rates of N application (Gelfand *et al.*, 2013) or different levels of tillage (Zhang *et al.*, 2010) for a given simulation run, for example, accounting for realistic variations in best management practices across the study landscape.

Study goals

Our study used the DayCent biogeochemical process model to assess perennial grass productivity and associated biogenic GHG emissions as a function of land quality and management intensity. Implications for bioenergy landscape design were explored through a case study of switchgrass production around a newly constructed commercial-scale cellulosic biorefinery in an area with substantial heterogeneity in soils and land use. This investigation expands on previous work in two main ways:

1. We conducted an extensive model parameterization and validation effort based on a data-diverse, multi-criteria approach, using a large parameterization dataset collected from the literature spanning wide gradients of climate, soil texture, and management intensity.

2. We evaluated the impacts of management intensification at the full spatial resolution of the assessment, estimating optimal levels of nitrogen fertilizer application for each point on the landscape in order to either maximize yield or minimize biogenic GHG emissions.

Our objective was to develop a rigorous, well-validated spatially explicit biogeochemical modeling capability that can serve as the basis for future integrated assessment and landscape optimization efforts.

Materials and methods

Case study introduction

We performed a landscape assessment case study simulating the cultivation of switchgrass (*Panicum virgatum*) to supply biomass to the Abengoa cellulosic biorefinery located outside the town of Hugoton in southwestern Kansas (Peplow, 2014), which began operations in fall 2014. While the plant will initially produce 25 million gallons of ethanol per year using corn stover as the primary lignocellulosic feedstock, switchgrass has been mentioned as an advanced cellulosic feedstock of interest and Biomass Crop Assistance Program Project Area 7 is sponsored by the company and targets the establishment of 20 000 acres of switchgrass production in the area (U.S. Department of Agriculture FSA, 2011). The case study region has long been at the center of agricultural sustainability and energy issues, having been deeply affected by the Dust Bowl in the 1930s (Kansas Historical Society, 2014) and being the site of the earliest hydraulic fracturing trials in the U.S. (Borowski, 2012).

Today, the surrounding Stevens County is a highly diverse and productive agricultural area. In 2012, 21.4% of the county was dedicated to highly-productive irrigated corn cultivation (average yield of 12.1 Mg ha^{-1}, or 192 bushels acre^{-1}) and 11% to dryland wheat (1.1 Mg ha^{-1}, or 18 bu acre^{-1}), with smaller fractions devoted to other crops and pasture/rangeland,

supporting an inventory of 45 500 head of cattle including calves (U.S. Department of Agriculture NASS, 2014). In this study we investigated the biogeochemical implications of converting non-irrigated cropland and rangeland in Stevens County and its six neighboring counties in southwestern Kansas and the Oklahoma panhandle to switchgrass cultivation (see Fig. 1), examining tradeoffs between productivity and associated biogenic GHG emissions as a function of underlying soil type and management intensity, specifically nitrogen fertilizer application rate. Issues of land ownership, conservation easement status, and other land-use policy limitations are excluded here, but explored in a subsequent publication dedicated to bioenergy landscape optimization.

The DayCent model

Productivity and net fluxes of biogenic CO_2 and N_2O from soils under switchgrass cultivation were modeled with the DayCent biogeochemistry model (Parton *et al.*, 1998; Del Grosso *et al.*, 2011). DayCent is a semi-empirical process-based model that simulates cycling of C, N, and water in natural and agroecosystems based on site-specific biophysical factors, land-use history, and management practices (e.g. tillage, fertilizer application, irrigation, etc.). The spatial and temporal scope of the model lies in between that of dedicated crop growth models (Miguez *et al.*, 2012) and generalized earth climate system models (Anderson *et al.*, 2013; Hallgren *et al.*, 2013). DayCent has been used extensively to predict yields and environmental impacts of switchgrass cultivation (Adler *et al.*, 2007; Chamberlain *et al.*, 2011; Davis *et al.*, 2012; Lee *et al.*, 2012) and is also used to predict agricultural soil GHG emissions for the annual Inventory of U.S. Greenhouse Gas Emissions and Sinks (U.S. Environmental Protection Agency, 2014a).

DayCent computes soil temperature and moisture for different layers of the soil profile (resolved separately based on soil texture, bulk density, and pH) using daily climate data inputs. Crop growth (net primary productivity, or NPP) is simulated using species-specific parameters describing photosynthetic efficiency, tissue C/N ratio limits, above- and below-ground C partitioning, and phenology, many of which are determined empirically using model parameterization datasets as described previously. Daily biomass growth potential (NPP_{pot}) is derived from top-of-atmosphere radiation (srad), corrected for atmospheric transmission losses and multiplied by a series of 0–1 factors that represent deviations from ideal temperature (f_{temp}) or soil moisture (f_{H_2O}) or limitations due to canopy immaturity or self-shading (f_{canopy}). This potential growth is then adjusted down if the available soil mineral nitrogen supply is limiting, as determined based on a maximum incremental biomass C : N ratio at the given level of plant maturity:

$$NPP = \max(NPP_{pot}, NPP_{Nlim}) \text{ where:}$$

$$NPP_{pot} = srad \cdot transmission \cdot f_{temp} \cdot f_{H_2O} \cdot f_{canopy}$$

$$NPP_{Nlim} = N_{avail} \cdot (C/N)_{max}$$

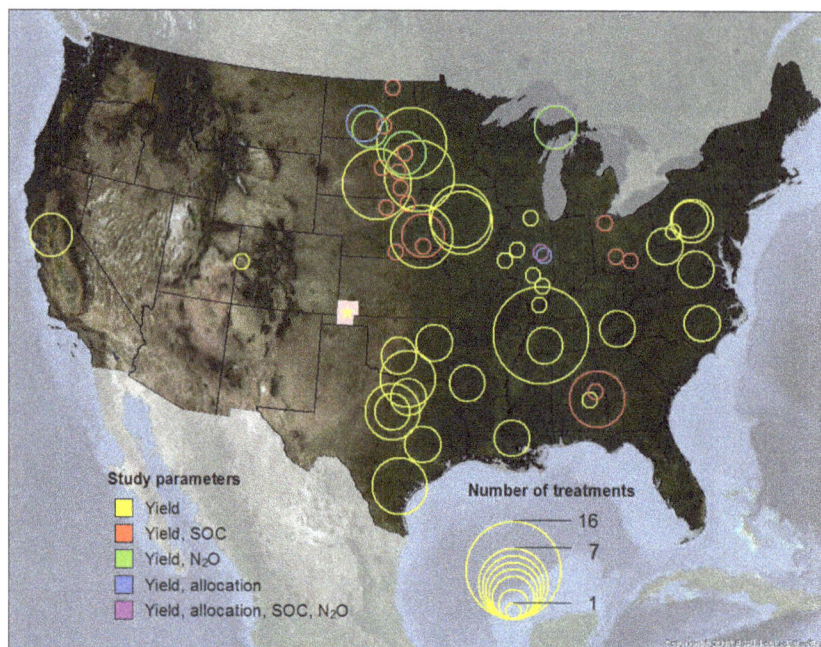

Fig. 1 Map of all switchgrass field trial sites included in the full model parameterization and validation dataset prior to ecotype/latitude filtering. Ring size indicates the number of experimental treatments (e.g. different ecotypes or nitrogen fertilizer application rates) conducted at that site, and color represents the types of data available. The 7-county case study area is highlighted in pink, with a star designating the biorefinery location.

In addition to plant productivity DayCent also estimates soil carbon and nitrogen cycling including net changes in SOC levels and N_2O emissions, the major constituents of the agricultural soil GHG balance. Carbon dynamics are simulated for soil surface pools and the top soil layer (0–20 cm by default), with organic matter represented by two litter pools (metabolic and structural) and three SOC pools- an 'active' pool representing microbial biomass and associated microbial products with a rapid turnover rates, and two others representing chemically/physically stabilized carbon with decadal- ('slow' pool) and century-scale ('passive' pool) turnover times. Nitrogen mineralization and immobilization rates for each pool are controlled by the maximum and minimum permissible C : N ratios for each pool, soil temperature, soil moisture, and microbial efficiency as a function of soil texture (Parton et al., 1987). The soil nitrogen balance considers synthetic N fertilizer addition, manure and other organic N amendments, atmospheric deposition, volatilization, leaching, plant uptake, and N mineralization and immobilization associated with soil organic matter transformations. The model simulates nitrification of ammonium (NH_4^+) to nitrate (NO_3^-) including NO_X and N_2O by-products, as well as denitrification of nitrate to gaseous products (N_2O, NO_X, N_2). Nitrification is simulated by multiplying available soil ammonium by a maximum potential nitrification rate adjusted based on soil temperature, water-filled pore space (WFPS), and pH limitations (Del Grosso et al., 2000; Parton et al., 2001). The overall rate of denitrification and the N_2O/N_2 ratio of its products are modeled based on the availability of nitrate and organic matter substrates (as inferred from heterotrophic respiration rate), and local soil micropore redox state and gas diffusion rates as inferred from WFPS and heterotrophic respiration rate.

Ecosystem model parameterization & validation

We undertook an extensive parameterization and validation of the DayCent switchgrass growth model to improve overall simulation accuracy and verify the model response across gradients of land quality and management intensity. A large dataset of switchgrass field trials for the continental United States was assembled from the peer-reviewed literature. Studies were included in the dataset provided they (a) specified the underlying soil in sufficient detail that a corresponding Natural Resource Conservation Service Soil Survey Geographic database (SSURGO; Ernstrom & Lytle, 1993) map unit could be identified using the Web Soil Survey (U.S. Department of Agriculture NRCS); (b) specified key management variables (switchgrass ecotype planted, N fertilizer application rate, harvest date) in sufficient detail to define key DayCent simulation parameters; and (c) reported disaggregated annual yield and GHG results (studies reporting results averaged across multiple sites or N treatments were excluded). All modeled-vs.-measured performance comparisons were made on the basis of time-averaged results across the duration of the field trial, so studies that report either annual yield data or treatment-averaged data were included. If multiple cultivars of the same ecotype were included in a study, results were averaged across cultivars into a single representative value for the ecotype for simplicity. Additional details on the parameterization and validation dataset are provided in the Appendix S1.

A total of 24 appropriate studies were identified, covering 573 annual biomass yield points across 147 unique combinations of site (soil, climate) and management treatments (N rate, harvest date, etc.). Initial exploratory analysis confirmed the need to exclude the first two seasons of yield data before switchgrass yields stabilize (Lesur *et al.*, 2013; Arundale *et al.*, 2014b), and to filter out treatments for ecotypes grown outside their typical latitude range (up to 54°N for lowland varieties and down to 34°N for upland, as per Casler, 2012). The remaining dataset was then randomly split at the level of individual studies 70 : 30 into a parameterization dataset and an independent validation dataset (Table 1). Several of these field trials included more detailed information on biomass partitioning or nitrogen content (Frank *et al.*, 2004; Dohleman *et al.*, 2012; Anderson-Teixeira *et al.*, 2013), long-term changes in SOC based on either repeat measurements or paired sites (Ma *et al.*, 2001; Liebig *et al.*, 2008; Follett *et al.*, 2012; Anderson-Teixeira *et al.*, 2013; Bonin & Lal, 2014), and/or time-resolved N_2O emissions (Nikièma *et al.*, 2011; Hong *et al.*, 2012; Schmer *et al.*, 2012; Smith *et al.*, 2013). There was insufficient data available to perform an independent validation of these model performance criteria, so all of these studies were included in the parameterization dataset. While the raw SOC dataset was very noisy, eliminating individual data points that were not reported as

statistically significant yielded a final reduced modeling dataset that behaved more predictably and was consistent with recent switchgrass SOC meta-analysis results (i.e. Qin *et al.*, 2015a).

The combined parameterization and validation dataset covers a wide range of latitudes and longitudes (Fig. 1) and temperature/precipitation regimes (Fig. S1), as well as a wide range of soil textures and Natural Resource Conservation Service land capability class (LCC) ratings (Helms, 1992; Fig. S2). Note that LCC ratings are reflective of a variety of land-use limitations, some of which are explicitly simulated in the Day-Cent model (e.g. dry climates, extreme textures, shallow soils) and some of which are not (e.g. erosion susceptibility, drainage class). The switchgrass crop parameterization was further informed with data from greenhouse or growth chamber experiments looking at productivity response across gradients of temperature (Balasko & Smith, 1971; Hsu *et al.*, 1985; Reddy *et al.*, 2008; Kandel *et al.*, 2013; Wagle & Kakani, 2014) or moisture (Xu *et al.*, 2006).

The parameterization process started with a default switchgrass crop parameter set based on previous work (Adler *et al.*, 2007; Davis *et al.*, 2012) and focused on refining parameters relating to productivity, temperature and moisture stress response, nitrogen management, shoot vs. root partitioning, and tissue death and turnover rates, with separate parameterizations

Table 1 DayCent switchgrass crop parameterization and validation and data sources

Parameter types	Data sources	# treatments
Parameterization		
Aboveground biomass yield	Ma *et al.* (2001), Fuentes & Taliaferro (2002), Vogel *et al.* (2002), Frank *et al.* (2004), Pearson (2004), Mulkey *et al.* (2006), Schmer *et al.* (2008, 2012), Nikièma *et al.* (2011), Dohleman *et al.* (2012), Follett *et al.* (2012), Hong *et al.* (2012), Kering *et al.* (2012), Anderson-Teixeira *et al.* (2013), Boyer *et al.* (2013), Bonin & Lal (2014) and Pedroso *et al.* (2014)	98 (67 after filtering)
Soil organic carbon (SOC) changes	Ma *et al.* (2001), Liebig *et al.* (2008), Follett *et al.* (2012), Anderson-Teixeira *et al.* (2013) and Bonin & Lal (2014)	18 (8 after filtering)
Soil nitrous oxide (N_2O) emissions	Nikièma *et al.* (2011), Hong *et al.* (2012), Schmer *et al.* (2012) and Smith *et al.* (2013)	11
Seasonal aboveground and belowground biomass accumulation, and/or C/N ratio	Frank *et al.* (2004), Dohleman *et al.* (2012) and Anderson-Teixeira *et al.* (2013)	3
Phenology	Sanderson (1992), Hopkins *et al.* (1995), Sanderson *et al.* (1997), Berdahl *et al.* (2005), Casler *et al.* (2007) and Wang *et al.* (2013)	NA
Productivity response to temperature	Balasko & Smith (1971), Hsu *et al.* (1985), Reddy *et al.* (2008), Kandel *et al.* (2013) and Wagle & Kakani (2014)	NA
Productivity response to soil moisture	Xu *et al.* (2006)	NA
Independent validation		
Yield	Staley *et al.* (1991), Muir *et al.* (2001), Cassida *et al.* (2005), Adler *et al.* (2006), Fike *et al.* (2006), Arundale *et al.* (2014b) and Wilson *et al.* (2014)	49 (44 after filtering)

for both upland and lowland ecotypes as appropriate. Initial exploratory analysis suggested that capturing differences in phenology between upland and lowland ecotypes was essential for accurate yield simulation, consistent with the current understanding of maturation based on photoperiod being a strong determinant of yield differences between different cultivars grown at a given latitude (Casler *et al.*, 2004). We set green-up dates uniformly for both ecotypes as function of latitude based on a variety of literature sources (Sanderson, 1992; Sanderson *et al.*, 1997; Wang *et al.*, 2013) as illustrated in Fig. S5. Peak biomass dates were predicted as a function of both latitude and ecotype as inferred from a variety of sources (Sanderson, 1992; Hopkins *et al.*, 1995; Sanderson *et al.*, 1997; Vogel *et al.*, 2002; Frank *et al.*, 2004; Berdahl *et al.*, 2005; Casler *et al.*, 2007; Anderson-Teixeira *et al.*, 2013; see Figs S6 and S7) and used to trigger plant senescence events within DayCent.

After crop phenology was set other parameter adjustments were implemented manually, using the model automation routine described in the next section to rapidly evaluate parameter changes against the 67-point parameterization dataset. The parameterization process focused on maximizing modeled-vs.-measured fidelity (based on visual inspection of the plotted data and calculation of modeled-vs-measured regression parameters, Pearson correlation coefficient, and root mean squared error; e.g. Smith *et al.*, 1996) for upland and lowland ecotype yields, changes in SOC, and growing season N_2O emissions, but also took into account time-resolved shoot : root and C : N ratio data where available for a multi-criteria evaluation of parameter set performance (e.g. Fig. S13). Once the parameterization process was complete, independent validation of upland and lowland yield performance was conducted based on the data held in reserve (i.e. using the holdout validation method).

Spatial data inputs, model initialization & automation

A variety of spatially explicit data inputs are necessary to initialize and run the DayCent model for a large-scale parameterization or landscape analysis, including data on climate, soil type, and land-use history. Data sources used in this analysis are summarized in Table 2. Soil texture, rock fraction, and pH

for different soil profile layers of the dominant soil component for each map unit were taken directly from the SSURGO database (Ernstrom & Lytle, 1993), and bulk density, field capacity, wilting point, and saturated hydraulic conductivity were computed using the Saxton equations (Saxton *et al.*, 1986). Climate data on a 32 km grid was derived from the North American Regional Reanalysis database (NARR; Mesinger *et al.*, 2006).

Land-use history and current land management practices were compiled from a variety of sources. Current land use was determined from the National Land Cover Database 2006 (NLCD; Wickham *et al.*, 2013), re-sampled from the native 30 m resolution to 240 m for ease of use, and re-classified into the simplified categories of annual agricultural lands ('cultivated crops', 'pasture/hay'), rangeland ('dwarf shrub', 'shrub/scrub', 'grassland/herbaceous', 'sedge/herbaceous'), and excluded (all other categories including forested and developed lands). Irrigated areas were identified using the MIrAD-US database (Pervez & Brown, 2010), and federally-owned lands were identified using the USGS Federal Lands of the United States data layer (U.S. Geological Survey, 2015) and excluded from further analysis (3.4% of the landscape, most part of Cimarron National Grassland). These five GIS layers were then intersected and small slivers were eliminated by merging all polygons smaller than 1 ha into the neighbor with which they shared the longest border. This yielded 39 320 polygons of a variety of sizes across the seven-county Hugoton case study area (Fig. S16), representing 3779 unique combinations of model inputs requiring individual simulation, which we refer to as DayCent modeling 'strata'.

For each strata, the DayCent model was pre-initialized using the same pre-settlement and historical land use assumptions as used in the EPA Inventory of U.S. Greenhouse Gas Emissions and Sinks (U.S. Environmental Protection Agency, 2014a) and described in detail by Ogle *et al.* (2010). Model initialization included an equilibrium run of several thousand years duration reflecting the natural state of the land prior to conversion to agriculture in order for all soil C and N pools to achieve steady-state values. Historical management between initial plow-out and the modern period was simulated with crop rotations and management practices compiled at regional scale from a variety of historical and modern sources (Ogle *et al.*,

Table 2 Summary of spatial data inputs

Spatial database	Data type	Year	Native resolution	URL
SSURGO	Soils	2012	1 : 12 000–1 : 63 360	http://www.nrcs.usda.gov/wps/portal/nrcs/detail/soils/survey/geo/?cid=nrcs142p2_053627
NARR	Daily weather	1979–2010	32 km	http://www.emc.ncep.noaa.gov/mmb/rreanl
NLCD	Land use	2006	30 m	http://www.mrlc.gov/nlcd2006.php
MIrAD-US	Irrigation extent	2007	250 m	http://earlywarning.usgs.gov/USirrigation/
Federal Lands of the United States	Federal land ownership	2005	640 acres/1 mi^2/1 : 2 000 000	http://nationalmap.gov/small_scale/mld/fedlanp.html

MIrAD-US, MODIS Irrigated Agriculture Dataset for the United States; NARR, North American Regional Reanalysis; NLCD, National Land Cover Database; SSURGO, Soil Survey Geographic database.

2010). The future switchgrass simulations were then executed across part of a 29-node, 288-processor cluster computing system at the Colorado State University Natural Resource Ecology Laboratory. Parallel execution was implemented in Python (http://www.python.org/) using forking operations to take advantage of multiple cores within a given node.

Landscape design analysis scenarios, results processing & sensitivity analysis

For the landscape analysis case study we simulated conversion of all non-irrigated, non-federally owned polygons within the seven-county case study area to rain-fed lowland switchgrass cultivation. We conducted 30-year forward simulations to assess long-term productivity and trends in soil C and N cycling, recycling the full range of the NARR historic weather record to represent future weather conditions. In order to assess the response of crop productivity and GHG performance to management intensity, seven different rates of nitrogen fertilizer application were simulated for each strata (0–150 kgN ha^{-1} in 25 kgN ha^{-1} increments). We assumed that switchgrass would be replanted every 10 years after field preparation consisting of chisel plow and field cultivator operations, and that the crop would be neither fertilized nor harvested the year of establishment in order to limit competition from weeds and ensure robust crop establishment, as per local

extension recommendations. These assumptions are highly conservative as switchgrass is often established in this region without tillage, the need for periodic replanting is widely debated, and first-year harvest can be possible if the crop achieves sufficient first-year productivity.

Switchgrass harvest yields, changes in SOC levels, and annual N_2O emissions were then averaged over the 30-year simulation period for each strata. Average annual N_2O emission values were converted into CO_2 equivalents using a 100-year global warming potential (GWP$_{100}$) value of 298 (Forster et al., 2007) then added to the CO_2 flux values associated with average annual net SOC changes for an estimate of total direct biogenic emissions. To determine the biogenic GHG intensity of production (Mg CO_2eq Mg^{-1} biomass harvested) total emissions per hectare were divided by the associated simulated switchgrass yield. Continuous functions of biomass yield and biogenic GHG intensity vs. nitrogen fertilizer application rate were developed for each strata by applying a cubic regression to the 30-year averaged simulation results for the different N rate simulations, and the yield-maximizing and GHG balance-minimizing N rates were then interpolated. The simulated yield and GHG intensity associated with these optimized N rates was then re-associated with the appropriate landscape polygons, and aggregated across the full landscape in order to develop curves illustrating total potential landscape productivity and biogenic GHG emissions balance when strata are managed for these different

Table 3 DayCent switchgrass crop parameter changes

Type of parameter change	Parameter name	Original value	Lowland value	Upland value
Productivity potential and temperature response:	PRDX(1)	2.75	4*	3.5*
Increase in productivity to reflect updated solar	PPDF(1)	30	30	30
radiation model and compensate for increased	PPDF(2)	45	44	44
belowground partitioning; differentiation in	PPDF(3)	1	0.75	0.75
productivity and temperature response between	PPDF(4)	2.5	2	2
upland and lowland ecotypes				
Growth response to moisture stress: Reduced	CWSCOEFF(1,1)	0.38	0.35	0.35
sensitivity to soil moisture stress	CWSCOEFF(1,2)	9	14	14
Belowground partitioning: Reduction in baseline	CFRTCN(1)	0.5	0.7	0.7
BG partitioning rate. BG partitioning in response	CFRTCN(2)	0.3	0.25	0.25
to moisture stress reduced, but response to	CFRTCW(1)	0.6	0.4	0.4
nutrient stress increased	CFRTCW(2)	0.3	0.25	0.25
Tissue N and lignin content: Root maximum	PRBMX(1,1)	55	50	50
allowable C : N ratio lowered slightly. Root lignin	FLIGNI(1,2)	0.26	0.06	0.06
content reduced. Small amount of N fixation	FLIGNI(1,3)	0.26	0.13	0.13
added to make growth under no-fertilizer	SNFXMX(1)	0	0.005	0.005
conditions more realistic				
Tissue death rates: Death rate and fall rate for	FSDETH(3)	0.05	0.075	0.075
shoots increased. Root maturation rate increased,	FALLRT	0.01	0.1	0.1
and turnover rate of both juvenile and mature	CMXTURN	0.12	0.3	0.3
roots increased	RDRJ	0.4	0.72	0.72
	RDRM	0.2	0.54	0.54
N conservation: Increased translocation of	CRPRTF(1)	0.15	0.43	0.43
nitrogen from shoots to roots during senescence				

*Change in the PRDX(1) values also reflects a recent change in the DayCent model to simulate atmospheric transmission losses of photosynthetically active radiation.

objectives. The sensitivity of these landscape results to key crop parameters, landscape characterization, and switchgrass cultivation scenario assumptions was assessed as detailed in Table 4 in order to determine the overall robustness of our conclusions. All results analysis routines were automated in Python through a combination of SQLite database operations (http://sqlite.org/) via the *sqlite3* module (http://docs.python.org/2/library/sqlite3.html), data manipulation in the native Python list data type, and figure generation using the *matplotlib.pyplot* module (http://matplotlib.org/api/pyplot_api.html).

Results

Ecosystem model parameterization & validation

A total of 79 different switchgrass parameter set iterations were ultimately developed and tested. Final upland and lowland ecotype parameter values that differ from the default DayCent switchgrass crop parameterization are detailed in Table 3. The most significant changes were:

- Increased plant potential NPP rate – The most recent version of DayCent explicitly models solar radiation atmospheric transmission losses, and the revised potential NPP parameter PRDX(1) must be adjusted higher relative to previously-published versions to reflect the new growth calculation on a canopy photosynthetically active radiation (PAR) basis rather than a top-of-atmosphere PAR basis. Further fine adjustments were made to PRDX(1) to optimize the observed yield difference between the different ecotypes and to offset slightly increased belowground C partitioning.

- Adjusted temperature and moisture stress response curves – Crop temperature (Fig. S8) and moisture stress (Fig. S9) response curves were set based on the greenhouse and growth chamber studies listed in Table 1. In the case of temperature response, fine adjustments to the edges of the curve where direct empirical data were lacking were implemented to improve overall modeled-vs.-measured yield performance across the full parameterization dataset. The same temperature and moisture response curves were used for both ecotypes. Comparison of measured and modeled yield ranges binned by site average growing degree day accumulation (Fig. S10) or annual precipitation (Fig. S11) verifies that the model accurately captures increasing switchgrass productivity at warmer, wetter sites.

- Increased belowground partitioning and root turnover – The default parameterization slightly underestimated belowground biomass, significantly underestimated observed SOC increases, and over-

predicted N_2O emissions. A small increase in belowground partitioning coupled with a large increase in root turnover rates resulted in more carbon being cycled into the soil and more mineral N being taken up by the plant, improving model performance on all three criteria.

Model parameterization and validation results for yields and soil GHG fluxes are shown in Fig. 2. Sufficient data was available to perform holdout method independent validation of yield predictions for both the upland and lowland ecotypes (Fig. 2a). The out-of-sample validation root mean square errors (RMSE) are 3.7 and 4.1 Mg ha^{-1} for the upland and lowland ecotypes, respectively, with minimal bias (mean difference of -0.2 and $+2.3$ Mg ha^{-1}, respectively). When these data are binned by nitrogen fertilizer application rate (Fig. 2c) we see that the model is able to capture the general trend toward increased switchgrass productivity with increasing management intensity. In contrast, yield response to land quality was more ambiguous. Neither measured nor modeled yields exhibited a clear relationship with soil texture across the full parameterization and validation dataset (Fig. 2d), and the weak trend towards lower yields at sites with higher LCC rating (more marginal) was not replicated in our model (Fig. S12).

Field trial data on soil GHG balance were sparser, and the results presented in Fig. 2b reflect within-sample model performance against the parameterization dataset rather than an independent validation. Observed annual changes in SOC under switchgrass were much larger than measured growing season N_2O emissions when compared in CO_2-equivalent terms. The within-sample RMSE value for the combined SOC and N_2O dataset is 0.69 MgCO$_2$ eq ha^{-1} yr^{-1} across the wide range of sites, climates, and nitrogen application rates represented in the underlying studies ($n = 19$, Fig. 1). Bias calculations show that the model tends to err in the direction of overestimating N_2O and underestimating SOC accumulation, so the resulting predictions of switchgrass GHG balance are somewhat conservative. Additional detail on SOC and N_2O performance is available in Figs S14 and S15.

Landscape design case study

Simulated lowland switchgrass yields as a function of nitrogen fertilizer application rate across the 3779 unique DayCent strata in the Hugoton case study area are shown in Fig. 3 for both crop land and rangeland conversion, color-coded to the underlying soil texture of each strata. The maximum attainable yield under arbitrarily well-fertilized conditions shows significant

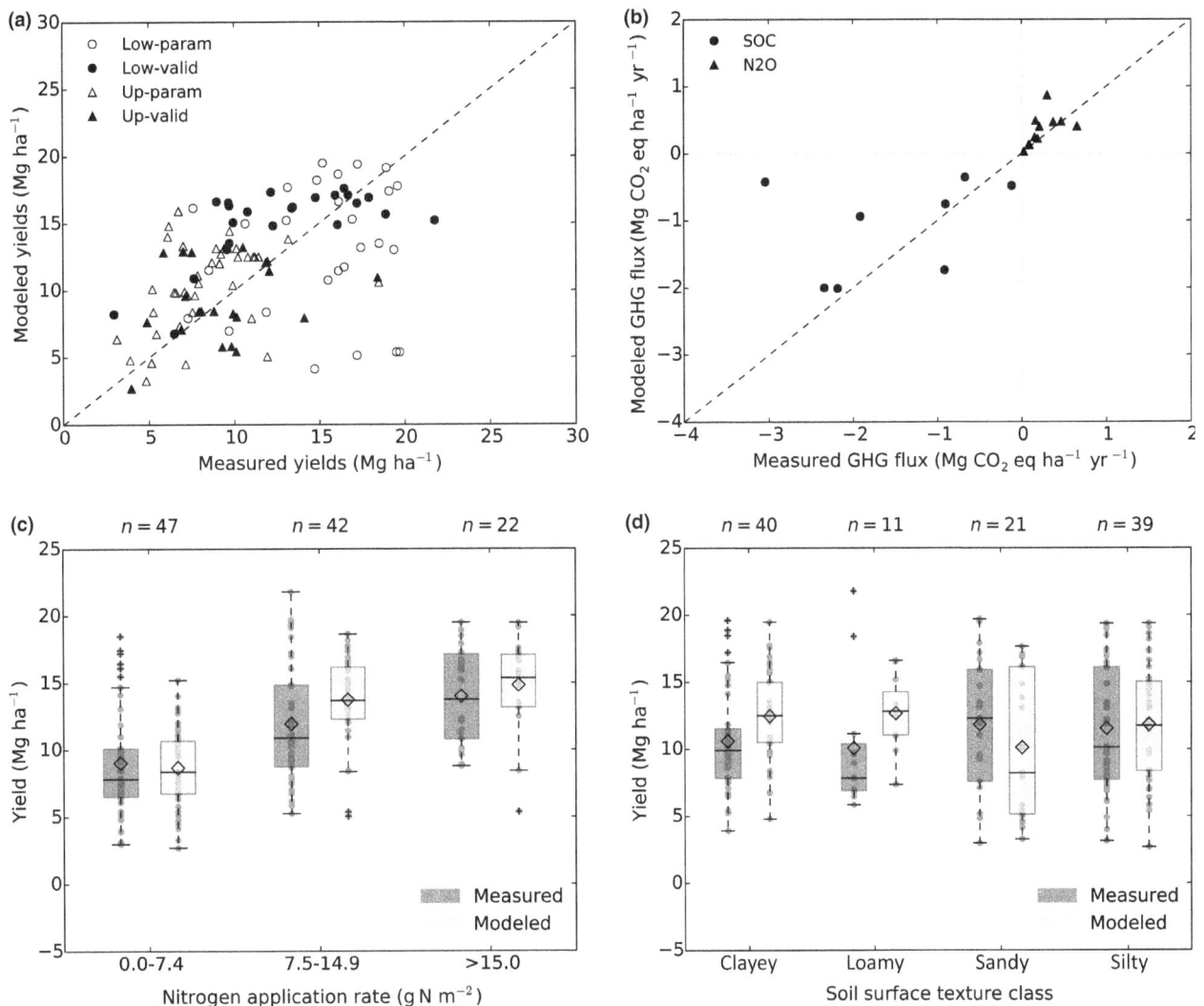

Fig. 2 Parameterization and validation results: (a) modeled vs. measured switchgrass biomass yield fits for lowland ecotype within sample ('Low-param', $r = 0.10$, root mean square error (RMSE) = 6.2 Mg ha^{-1}, mean difference (MD) = −2.1 Mg ha^{-1}) and out-of-sample ['Low-valid', $r = 0.66$, RMSE = 4.1, MD = +2.3] field trial results and upland ecotype within-sample ('Up-param', $r = 0.38$, RMSE = 4.0, MD = +2.1) and out-of-sample ('Up-valid', $r = 0.26$, RMSE = 3.7, MD = −0.2) results; (b) modeled vs. measured annualized changes in soil organic carbon ($r = 0.08$, RMSE = 1.05 MgCO2eq ha^{-1} yr^{-1}, MD = +0.42 MgCO2eq ha^{-1} yr^{-1}) and growing season N$_2$O emissions ($r = 0.54$, RMSE = 0.22, MD = +0.10) on a CO$_2$-equivalent basis (combined $r = 0.79$, RMSE = 0.69, MD = +0.24); (c) measured and modeled switchgrass yield ranges for different levels of nitrogen application across all parameterization and validation data points; and (d) measured and modeled switchgrass yield ranges for different soil surface texture classes across all parameterization and validation data points.

variation with soil texture, ranging from more than 10 Mg ha^{-1} in certain clay and sandy soils down to about 6 Mg ha^{-1} in the more moderate-textured silty soils. In this semi-arid climate of intermittent precipitation events soil moisture levels are often near wilting point, and simulated average yields reflect a tension between the greater total water holding capacity of finer-textured soils (an advantage during relatively wet years) vs. more effective infiltration and less surface soil evaporation in coarser soils (the so-called 'inverse texture effect', beneficial during dry years;

Noy-Meir, 1973; Epstein et al., 1997; Lane et al., 1998). Switchgrass yields on converted rangeland are generally less sensitive to fertilizer application rates than those on converted cropland, in some cases showing no response to increasing N rate. This is due to higher background mineral nitrogen levels from transient soil organic matter turnover following conversion in these areas. However, for most strata, full switchgrass yield potential is realized at N rates from 60 to 100 kgN ha^{-1}, as indicated with circular markers in the upper panels of Fig. 3.

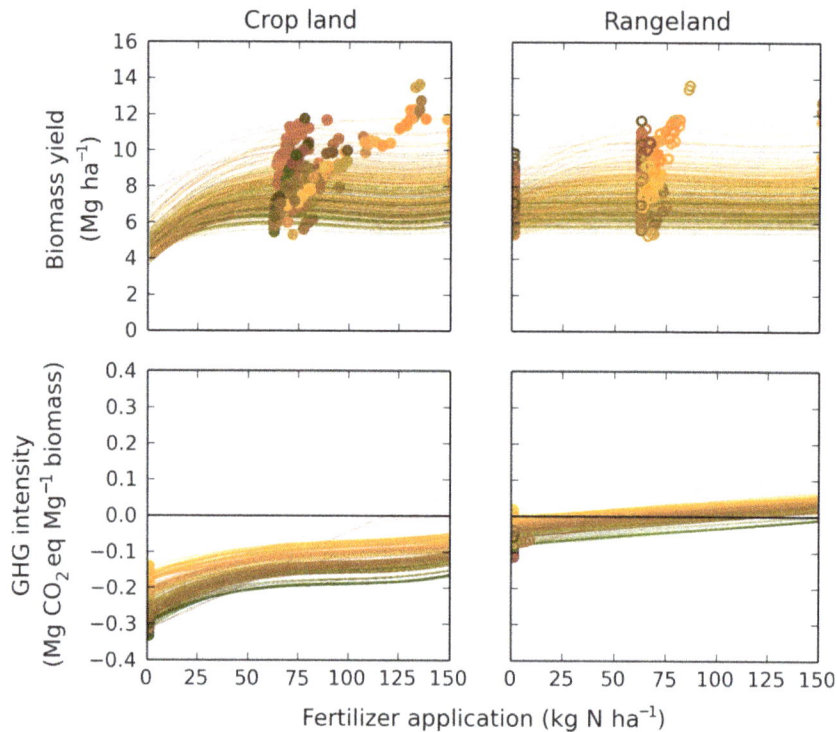

Fig. 3 Simulated yield and GHG intensity response to increasing nitrogen fertilizer rate for switchgrass production on former crop land and former rangeland across the 3779 distinct DayCent strata in the case study landscape. Interpolated yield maxima and GHG intensity minima are marked with solid markers for cropland conversion and open markers for rangeland conversion. The color of the lines and markers indicates the soil surface texture of the strata, with yellow = sand, red = clay, green = silt, and brown = loam.

When examining landscape assessment results under different levels of management intensity it is important to note a discrepancy between the metrics of *total* area GHG emissions or mitigation (MgCO$_2$eq per hectare per year) and the GHG *intensity* of biomass production (MgCO$_2$eq per Mg biomass grown). Biomass production GHG intensity results are shown as a function of nitrogen application rate across all simulation strata in the lower panels of Fig. 3. For most strata at most nitrogen fertilizer rates, 30 year-average soil carbon sequestration outweighs nitrous oxide emissions on a CO$_2$-equivalent basis. Biomass has the lowest (most negative) direct biogenic GHG footprint when cultivated with no nitrogen fertilizer on previously cropped fine-textured soils due to their high potential for SOC accumulation. This sequestration value is increasingly offset at higher N rates due to marginal soil N$_2$O emissions outpacing corresponding yield and SOC gains. In contrast, rangeland has higher initial SOC levels and thus less capacity for carbon sequestration after conversion to bioenergy feedstock cropping, which results in an overall net GHG intensity much closer to zero.

Total area biogenic GHG emissions are explored in relation to productivity in Fig. 4. Since switchgrass can

be managed either to maximize yields or to optimize soil GHG balance (i.e. to maximize soil carbon sequestration while minimizing N$_2$O emissions), we selected a random 10% of the landscape polygons, determined the nitrogen application rate that maximizes yield and the rate that maximizes GHG *mitigation* for each polygon, and aggregated the results for these different management strategies across our landscape sub-sample. The difference between the two curves is representative of the degree to which system productivity and GHG performance can be modulated by adjusting management on the same limited land base. Converting 10% of the landscape to switchgrass managed under either strategy results in >0.75 Mt biomass feedstock production annually, enough material to supply approximately two facilities the size of the Abengoa biorefinery, with the associated net GHG impact of sequestering >0.06 Mt (60 000 metric tons) of CO$_2$. Figure 5 illustrates that the optimal switchgrass management intensity for maximum GHG mitigation is related to the distribution of soil types and current land use across the case study area, factors that are also correlated with one another as conventional cropping tends to be concentrated on the moderate textured soils of this landscape.

Sensitivity analysis

Our modeled landscape productivity and GHG balance are highly sensitive to certain key crop model parameters, particularly optimal growth temperature, potential NPP, and response to moisture stress (Table 4), highlighting the importance of careful parameterization and validation. While the first two parameters were reasonably well constrained as a result of the switchgrass parameterization process, empirical data on switchgrass moisture stress in the literature is rare, and uncertainty around those parameters in the model should be considered large. Interestingly, the aggregate landscape simulation results show little sensitivity to the length of the switchgrass growing season in early spring and late fall, even though accurate characterization of the phenology proved essential for determining crop parameter sets with adequate out-of-sample validation performance. Accurate characterization of the landscape in terms of climate and land-use history is important to the integrity of simulation results as well, with GHG mitigation particularly sensitive to past land-use history. Landscape results were generally less sensitive to crop management scenario assumptions around tillage intensity and fertilizer timing.

Discussion

Challenges of model development and data-diverse parameterization

This study was grounded in an extensive model parameterization and validation effort using a data-diverse multi-criteria approach enabled by automation of all simulation runs and results analyses. The diversity of studies included in our switchgrass field trial dataset

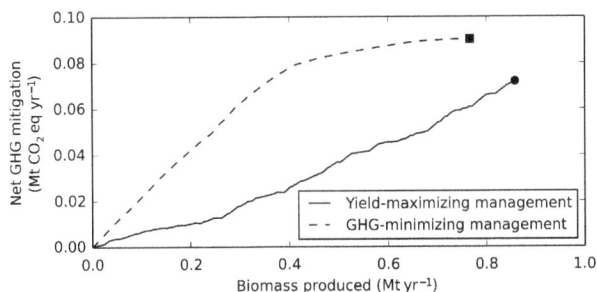

was intended to provide a highly general test of model performance independent of a single study, environment, or management practice, and the simultaneous consideration of different types of data (Table 1) was designed to ensure that accuracy of one model performance criteria was never improved at the expense of others. This approach proved useful for sorting out interrelated model responses, for example, the strong effects of belowground partitioning and root turnover rate parameters on harvested biomass yield, SOC changes, and N_2O rates that must be balanced in a systematic manner during model parameterization.

Our resulting model explains approximately half of the observed variability in yield and GHG observations in our dataset, and realistically captures climate effects and responses to management intensity. While soil type has been observed to have an effect on bioenergy grass productivity in semi-arid climates outside the U.S. (Di Virgilio et al., 2007; o Di Nasso et al., 2015; Roncucci et al., 2015), we did not observe a strong texture signal in our domestic switchgrass productivity dataset, consistent with a similar previous large-scale model parameterization effort (Wullschleger et al., 2010). Future modeling work on soil-climate interactions in the U.S. would greatly benefit from additional field trails like Wilson et al. (2014) that include paired trails across multiple soil types on a landscape, in order to tease apart soil-precipitation interactions without the other larger sources of variability present in composite datasets.

While we believe the approach presented here represents an improvement over more limited model parameterization efforts, it is still possible to over-parameterize to a dataset of this size and achieve good fits via unrealistic mechanisms that do not translate well out-of-sample. In our experience, reliable yield performance was only achieved once crop phenology was adequately captured. This parameterization dependency is somewhat challenging, as there are a limited number of studies in the literature reporting detailed phenology in the form required for this type of generalized modeling effort (i.e. dates for green-up and peak biomass or senescence as a function of crop cultivar and site latitude). Additionally, since maturation dates can vary significantly even for cultivars within a single ecotype (Frank et al., 2004), focusing on only two ecotype groupings introduces additional errors. Accurate representation of phenology can be superseded to a certain extent by narrowing crop temperature response curves to truncate early and late-season productivity, but our initial parameterization efforts in this direction performed very poorly on validation.

After several dozen parameter set iterations we were approaching the limit of what could be accomplished with setting parameter values manually, as model

Fig. 4 Cumulative total soil GHG mitigation vs. cumulative switchgrass biomass production for a random 10% of the case study landscape under different management goals: managing each land parcel to maximize switchgrass yields (dashed line, highest-productivity sites aggregated first), or managing each land parcel to maximize ecosystem GHG mitigation (solid line, strongest mitigation sites aggregated first).

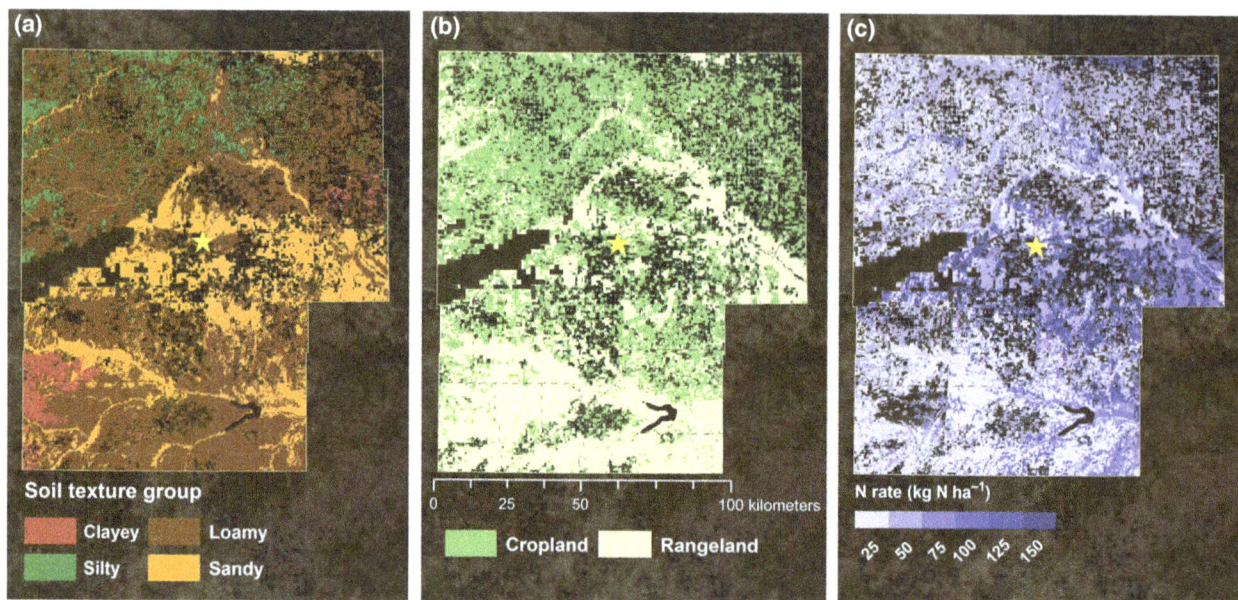

Fig. 5 Map illustrating (a) soil texture distribution, (b) current land use, and (c) the switchgrass nitrogen fertilizer application rates associated with maximum area GHG mitigation (as opposed to biomass GHG intensity explored in Fig. 3), for all non-federal non-irrigated cropland and rangeland across the Hugoton case study area. Aerial imagery is included as a background layer for scale, and the boundary of the 7-county case study area is shown in white with the biorefinery location marked with a yellow star.

responses became more subtle and antagonistic across the different performance criteria. This, coupled with the relatively large number of crop parameters in the DayCent model, suggests that future efforts could benefit greatly from systematic parameter optimization techniques such as inverse modeling and/or κ-fold cross validation. Such techniques provide a more systematic and transparent approach, facilitate maximum extraction of information from a field trial dataset, and help to identify and avoid over-parameterization issues. While inverse modeling techniques have been demonstrated to improve model performance against small datasets (Necpálová et al., 2015), they are not commonly applied to crop growth and biogeochemistry models used in bioenergy sustainability analysis at the scale of a full parameterization and validation effort such as the one presented in this study.

More broadly, model performance is dependent on underlying model structure in addition to parameterization, and DayCent model development efforts to improve its performance in perennial grass systems are ongoing. For example, accurate yield estimates depend on capturing crop phenology and canopy dynamics, and efforts are underway to represent early-season leaf canopy closure and late-season senescence dynamics more realistically. Similarly, while past model development efforts have largely focused on the upper 0–20 cm soil layer most relevant for annual crops, representation of soil carbon changes under perennial grasses may be improved through more explicit simulation of deep

rooting, dissolved organic carbon movement, and other aspects of soil organic matter dynamics deeper in the soil profile (Campbell, 2015).

Landscape design case study interpretation and climate accounting issues

The Hugoton case study site was selected because it featured the most heterogeneity in soils and current land use among the first three commercial-scale cellulosic biorefinery sites in the U.S. (the other two being located in the Iowa corn belt; Peplow, 2014). However, the dry climate at this site proved challenging from a modeling perspective, with landscape simulation results showing high sensitivity to interactions between crop moisture stress parameterization and soil texture, and with N_2O response to increasing N rate lower than what might be expected in a wetter climate. Overall, our landscape simulation results are similar to others in the literature that find the GHG balance of perennial grasses dominated by soil carbon sequestration immediately post-establishment, with only modest N_2O emissions (e.g. Gelfand et al., 2013). In field trials where baseline N_2O emission rates are reported and an Intergovernmental Panel on Climate Change (IPCC) Tier 1 estimate can be made, observed switchgrass N_2O emissions rates are often near or below the lower bound of the IPCC N_2O emissions factor range (0.3– 3% fertilizer N emitted as N_2O-N; Fig. S15). This suggests that switchgrass is an efficient nitrogen cycler, and that the critique of biofuel

Table 4 Sensitivity analysis

Parameter or assumption	Change	Change in total landscape productivity	Change in total landscape GHG mitigation
Baseline average landscape performance			
Yield = 7.24 Mg ha^{-1} yr^{-1}			
GHG mitigation = 0.76 Mg CO$_2$eq ha^{-1} yr^{-1}			
DayCent crop parameterization			
Optimal temperature	Increase PPDF(1) by 3 degrees	−21%	−42%
Productivity potential	Decrease PRDX(1) value by 10%	−9.4%	−18%
Moisture sensitivity	Switch to the default curve associated with annual crops	−1.1%	+8.6%
Root death rates	Decrease RDRJ and RDRM by 10%	+0.11%	+0.60%
Phenology	Decrease growing season length by 15 days	+0.49%	−0.41%
Site characterization			
Land-use history	Assume a uniform cropped history rather than a mix of cropped and grazed	+5.0%	+29%
Climate	Assume the NARR weather for Hugoton for the entire 7-county study area	−4.2	−9.6%
Agronomic management			
Fertilizer date	Assume fertilizer application 1 month earlier	+0.11%	+1.5%
Tillage intensity	Switch to no-till crop establishment	+0.13%	+0.97%

GHG, greenhouse gas; NARR, North American Regional Reanalysis.

GHG mitigation benefits being outweighed by direct and indirect N$_2$O emissions (Crutzen et al., 2008) is likely overstated for second-generation perennial grass feedstocks. Our predicted SOC sequestration rates after cropland conversion are very similar to median meta-analysis results in Qin et al. (2015a), while the finding of a small net positive sequestration with the conversion of grassland is consistent with, but on the optimistic edge of, their observed range.

Full quantification of the confidence interval around these landscape assessment results is highly challenging due to multiple levels of uncertainty in problem specification, landscape input data, model structure, and model parameterization (Walker et al., 2003). While previous work has shown that uncertainty in DayCent-estimated changes in SOC is dominated by model structural uncertainty (Ogle et al., 2010), repeating this empirical uncertainty assessment approach here is impractical as only one of the 145 treatments in our parameterization and validation dataset includes data for all three of the factors that determine biomass GHG intensity (yield, SOC change, and N$_2$O; Fig. 1). Here we rely on sensitivity analysis to identify priorities for future model improvement efforts. When discussing model sensitivity and uncertainty issues it is also important to be cognizant of issues in the underlying field

trial datasets. Field measurements of N$_2$O are very challenging due to emissions variability on extremely fine spatial (Li et al., 2013) and temporal (Jørgensen et al., 2012; van der Weerden et al., 2013) scales, and studies based on sampling at weekly or every-other-week frequency (as was the case for all N$_2$O studies in our dataset) are vulnerable to systematic biases of up to 20% and 60%, respectively (Parkin, 2008; Smemo et al., 2011). Additionally, small-scale field trials do not necessarily reflect imperfections in agronomic management (e.g. uneven fertilizer application) and real-world harvest losses, possibly introducing a systematic over-estimation of switchgrass productivity (Searle & Malins, 2014).

Finally, there are several issues around climate impact accounting relevant to this landscape study. While soil carbon sequestration will eventually attenuate even though N$_2$O emissions will persist for as long as N fertilizer is being applied (Sheehan et al., 2003; Adler et al., 2007), our assessment for this semi-arid system suggests it will take 60–80 years for annual sequestration and N$_2$O rates to reach parity (Fig. S17). A more dynamic climate impact accounting approach that takes transient forcing benefits into account (e.g. Holtsmark, 2015) would tend to further weight near-term SOC benefits against future N$_2$O emissions, though no standardized accounting approach is yet widely accepted (U.S. Environmental Protection Agency, 2014b). The current study is also somewhat limited in focusing on the climate impacts of biomass feedstock production from a purely biogeochemical perspective, ignoring potential biophysical impacts such as changes in surface albedo or evapotranspiration and water dynamics that are significant in some bioenergy production scenarios (Muñoz *et al.*, 2010; Georgescu *et al.*, 2011; Cherubini *et al.*, 2012; Caiazzo *et al.*, 2014), and ignoring any potential broader impacts on other ecosystem services (Chamberlain & Miller, 2012). Additionally, future changes to atmospheric CO_2 concentrations and climate are not considered here, though they may have large repercussions for landscape design (Bryan *et al.*, 2010). The ability of current ecosystem models to accurately extrapolate to such future conditions is an active area of investigation (De Kauwe *et al.*, 2013). Additional sensitivity analysis on these points would improve our understanding of bioenergy landscape performance potential, and highlight priorities for future research efforts.

Acknowledgements

This work was supported by a USDA/NIFA project 'Decision support tool for integrated biofuel greenhouse gas emission footprints' (grant # 2011-67009-30083), an NSF IGERT fellowship through the Multidisciplinary Approaches to Sustainable Bioenergy program at Colorado State University, and an NSF REU fellowship and graduate Chevron fellowship through the Colorado Center for Biorefining and Biofuels (C2B2). We would like to thank Jacqueline Marquez for her assistance in identifying and coding switchgrass field trial papers, Matthew Stermer for helping quality-controlling this dataset, Kendrick Killian and Ty Boyak for their advice on developing model automation and analysis code, Jill Schuler for her GIS contribution, Jeff Kent for help with DayCent growth submodel performance visualization, and Yao Zhang for his insights on crop model performance in dry climates.

References

110th Congress of the United States (2007) Energy Independence and Security Act of 2007.

Abodeely JM, Muth DJ, Koch JB, Bryden KM (2013) A model integration framework for assessing integrated landscape management strategies. In: *Environmental Software Systems. Fostering Information Sharing*, Vol 413 (eds Hřebíček J, Schimak G, Kubásek M, Rizzoli AE), pp. 121–128. Springer Berlin Heidelberg, Berlin, Heidelberg.

Adler PR, Sanderson MA, Boateng AA, Weimer PJ, Jung H-JG (2006) Biomass yield and biofuel quality of switchgrass harvested in fall or spring. *Agronomy Journal*, **98**, 1518–1525.

Adler PR, Del Grosso SJ, Parton WJ (2007) Life-cycle assessment of net greenhouse-gas flux for bioenergy cropping systems. *Ecological Applications*, **17**, 675–691.

Anderson CJ, Anex RP, Arritt RW, Gelder BK, Khanal S, Herzmann DE, Gassman PW (2013) Regional climate impacts of a biofuels policy projection. *Geophysical Research Letters*, **40**, 1217–1222.

Anderson-Teixeira KJ, Duval BD, Long SP, DeLucia EH (2012) Biofuels on the landscape: is "land sharing" preferable to "land sparing"? *Ecological Applications*, **22**, 2035–2048.

Anderson-Teixeira KJ, Masters MD, Black CK, Zeri M, Hussain MZ, Bernacchi CJ, DeLucia EH (2013) Altered belowground carbon cycling following land-use change to perennial bioenergy crops. *Ecosystems*, **16**, 508–520.

Arundale RA, Dohleman FG, Voigt TB, Long SP (2014a) Nitrogen fertilization does significantly increase yields of stands of *Miscanthus × giganteus* and *Panicum virgatum* in multiyear trials in Illinois. *BioEnergy Research*, **7**, 408–416.

Arundale RA, Dohleman FG, Heaton EA, Mcgrath JM, Voigt TB, Long SP (2014b) Yields of *Miscanthus × giganteus* and *Panicum virgatum* decline with stand age in the Midwestern USA. *GCB Bioenergy*, **6**, 1–13.

Balasko JA, Smith D (1971) Influence of temperature and nitrogen fertilization on the growth and composition of switchgrass (*Panicum virgatum* L.) and timothy (*Phleum pratense* L.) at anthesis. *Agronomy Journal*, **63**, 853–857.

Bandaru V, Izaurralde RC, Manowitz D, Link R, Zhang X, Post WM (2013) Soil carbon change and net energy associated with biofuel production on marginal lands: a regional modeling perspective. *Journal of Environmental Quality*, **42**, 1802–1814.

Behrman KD, Keitt TH, Kiniry JR (2014) Modeling differential growth in switchgrass cultivars across the Central and Southern Great Plains. *BioEnergy Research*, **7**, 1165–1173.

Berdahl JD, Frank AB, Krupinsky JM, Carr PM, Hanson JD, Johnson HA (2005) Biomass yield, phenology, and survival of diverse switchgrass cultivars and experimental strains in western North Dakota. *Agronomy Journal*, **97**, 549–555.

Bonin CL, Lal R (2014) Aboveground productivity and soil carbon storage of biofuel crops in Ohio. *GCB Bioenergy*, **6**, 67–75.

Borowski S (2012) Idea for "fracking" came from Civil War battlefield. *AAAS MemberCentral*. Available at: http://membercentral.aaas.org/blogs/scientia/idea-fracking-came-civil-war-battlefield (accessed 17 November 2015).

Boyer CN, Roberts RK, English BC, Tyler DD, Larson JA, Mooney DF (2013) Effects of soil type and landscape on yield and profit maximizing nitrogen rates for switchgrass production. *Biomass and Bioenergy*, **48**, 33–42.

Bryan BA, King D, Wang E (2010) Biofuels agriculture: landscape-scale trade-offs between fuel, economics, carbon, energy, food, and fiber. *GCB Bioenergy*, **2**, 330–345.

Cai X, Zhang X, Wang D (2011) Land availability for biofuel production. *Environmental Science & Technology*, **45**, 334–339.

Caiazzo F, Malina R, Staples MD, Wolfe PJ, Yim SHL, Barrett SRH (2014) Quantifying the climate impacts of albedo changes due to biofuel production: a comparison with biogeochemical effects. *Environmental Research Letters*, **9**, 024015.

Campbell EE (2015) Modeling soil organic matter: theory, development, and applications in bioenergy cropping systems. PhD Dissertation. Colorado State University, Fort Collins, CO, USA.

Campbell JE, Lobell DB, Genova RC, Field CB (2008) The global potential of bioenergy on abandoned agriculture lands. *Environmental Science & Technology*, **42**, 5791–5794.

Casler MD (2012) Switchgrass breeding, genetics, and genomics. In: *Green Energy and Technology* (ed Monti A), pp. 29–53. Springer London, London.

Casler MD, Vogel KP, Taliaferro CM, Wynia RL (2004) Latitudinal adaptation of switchgrass populations. *Crop Science*, **44**, 293–303.

Casler MD, Vogel KP, Taliaferro CM *et al.* (2007) Latitudinal and longitudinal adaptation of switchgrass populations. *Crop Science*, **47**, 2249–2260.

Cassida KA, Muir JP, Hussey MA, Read JC, Venuto BC, Ocumpaugh WR (2005) Biomass yield and stand characteristics of switchgrass in South Central U.S. Environments. *Crop Science*, **45**, 673–681.

Chamberlain JF, Miller SA (2012) Policy incentives for switchgrass production using valuation of non-market ecosystem services. *Energy Policy*, **48**, 526–536.

Chamberlain JF, Miller SA, Frederick JR (2011) Using DAYCENT to quantify on-farm GHG emissions and N dynamics of land use conversion to N-managed switchgrass in the Southern U.S. *Agriculture, Ecosystems & Environment*, **141**, 332–341.

Cherubini F, Bright RM, Strømman AH (2012) Site-specific global warming potentials of biogenic CO_2 for bioenergy: contributions from carbon fluxes and albedo dynamics. *Environmental Research Letters*, **7**, 045902.

Crutzen PJ, Mosier AR, Smith KA, Winiwarter W (2008) N_2O release from agro-biofuel production negates global warming reduction by replacing fossil fuels. *Atmospheric Chemistry and Physics*, **8**, 389–395.

Davis SC, Parton WJ, Del Grosso SJ, Keough C, Marx E, Adler PR, DeLucia EH (2012) Impact of second-generation biofuel agriculture on greenhouse-gas emissions in the corn-growing regions of the US. *Frontiers in Ecology and the Environment*, **10**, 69–74.

Davis SC, Boddey RM, Alves BJR *et al.* (2013) Management swing potential for bioenergy crops. *GCB Bioenergy*, **5**, 623–638.

De Kauwe MG, Medlyn BE, Zaehle S *et al.* (2013) Forest water use and water use efficiency at elevated CO2: a model-data intercomparison at two contrasting temperate forest FACE sites. *Global Change Biology*, **19**, 1759–1779.

Del Grosso SJ, Parton WJ, Mosier AR, Ojima DS, Kulmala AE, Phongpan S (2000) General model for N_2O and N_2 gas emissions from soils due to dentrification. *Global Biogeochemical Cycles*, **14**, 1045–1060.

Del Grosso SJ, Parton WJ, Keough CA, Reyes-Fox M, Ahuja LR, Ma L (2011) Special features of the DayCent modeling package and additional procedures for parameterization, calibration, validation, and applications. In: *Advances in Agricultural Systems Modeling* (eds Ahuja LR, Ma L), pp. 155–176. American Society of Agronomy, Crop Science Society of America, Soil Science Society of America, Madison, WI.

o Di Nasso NN, Lasorella MV, Roncucci N, Bonari E (2015) Soil texture and crop management affect switchgrass (*Panicum virgatum* L.) productivity in the Mediterranean. *Industrial Crops and Products*, **65**, 21–26.

Di Virgilio N, Monti A, Venturi G (2007) Spatial variability of switchgrass (*Panicum virgatum* L.) yield as related to soil parameters in a small field. *Field Crops Research*, **101**, 232–239.

Dohleman FG, Heaton EA, Arundale RA, Long SP (2012) Seasonal dynamics of above- and below-ground biomass and nitrogen partitioning in *Miscanthus × giganteus* and *Panicum virgatum* across three growing seasons. *GCB Bioenergy*, **4**, 534–544.

Duval BD, Anderson-Teixeira KJ, Davis SC, Keogh C, Long SP, Parton WJ, DeLucia EH (2013) Predicting greenhouse gas emissions and soil carbon from changing pasture to an energy crop. *PLoS ONE*, **8**, e72019.

Egbendewe-Mondzozo A, Swinton SM, Izaurralde CR, Manowitz DH, Zhang X (2011) Biomass supply from alternative cellulosic crops and crop residues: a spatially explicit bioeconomic modeling approach. *Biomass and Bioenergy*, **35**, 4636–4647.

Epstein HE, Lauenroth WK, Burke IC (1997) Effects of temperature and soil texture on ANPP in the U.S. Great Plains. *Ecology*, **78**, 2628–2631.

Ernstrom DJ, Lytle D (1993) Enhanced soils information systems from advances in computer technology. *Geoderma*, **60**, 327–341.

Fike JH, Parrish DJ, Wolf DD, Balasko JA, Green JT Jr, Rasnake M, Reynolds JH (2006) Switchgrass production for the upper southeastern USA: influence of cultivar and cutting frequency on biomass yields. *Biomass and Bioenergy*, **30**, 207–213.

Follett RF, Vogel KP, Varvel GE, Mitchell RB, Kimble J (2012) Soil carbon sequestration by switchgrass and no-till maize grown for bioenergy. *BioEnergy Research*, **5**, 866–875.

Forster P, Ramaswamy V, Artaxo P et al. (2007) Changes in atmospheric constituents and in radiative forcing. In: *Climate Change 2007: The Physical Science Basis. Contribution of Working Group I to the Fourth Assessment Report of the Intergovernmental Panel on Climate Change* (eds Solomon S, Qin D, Manning M et al.). Cambridge University Press, Cambridge, United Kingdom and New York, NY, USA.

Frank AB, Berdahl JD, Hanson JD, Liebig MA, Johnson HA (2004) Biomass and carbon partitioning in switchgrass. *Crop Science*, **44**, 1391–1396.

Fuentes RG, Taliaferro CM (2002) Biomass yield stability of switchgrass cultivars. In: *Trends in New Crop and New Uses* (eds Janick J, Whipkey A), pp. 276–282. ASHS Press, Alexandria, Virginia.

Gelfand I, Sahajpal R, Zhang X, Izaurralde RC, Gross KL, Robertson GP (2013) Sustainable bioenergy production from marginal lands in the US Midwest. *Nature*, **493**, 514–517.

Georgescu M, Lobell DB, Field CB (2011) Direct climate effects of perennial bioenergy crops in the United States. *Proceedings of the National Academy of Sciences*, **108**, 4307–4312.

Hallgren W, Schlosser CA, Monier E, Kicklighter D, Sokolov A, Melillo J (2013) Climate impacts of a large-scale biofuels expansion. *Geophysical Research Letters*, **40**, 1624–1630.

Heaton EA, Dohleman FG, Long SP (2008) Meeting US biofuel goals with less land: the potential of Miscanthus. *Global Change Biology*, **14**, 2000–2014.

Helms D (1992) *Readings in the History of the Soil Conservation Service*. Soil Conservation Service, Washington, DC, pp. 60–73.

Hoben JP, Gehl RJ, Millar N, Grace PR, Robertson GP (2011) Nonlinear nitrous oxide (N₂O) response to nitrogen fertilizer in on-farm corn crops of the US Midwest. *Global Change Biology*, **17**, 1140–1152.

Holtsmark B (2015) Quantifying the global warming potential of CO_2 emissions from wood fuels. *GCB Bioenergy*, **7**, 195–206.

Hong CO, Owens VN, Schumacher T, Clay D, Osborne S, Lehman M, Schumacher J (2012) Nitrogen losses from switchgrass as affected by nitrogen fertilizer rate. In: *Proceedings from Sun Grant National Conference: Science for Biomass Feedstock Production and Utilization*, pp. 183–190. New Orleans, LA. Available at: http://sungrant.tennessee.edu/NatConference/ConferenceProceedings/ (accessed 17 November 2015).

Hopkins AA, Vogel KP, Moore KJ, Johnson KD, Carlson IT (1995) Genotypic variability and genotype × environment Interactions among switchgrass accessions from the midwestern USA. *Crop Science*, **35**, 565–571.

Hsu FH, Nelson CJ, Matches AG (1985) Temperature effects on seedling development of perennial warm-season forage grasses. *Crop Science*, **25**, 249–255.

Hudiburg TW, Davis SC, Parton W, Delucia EH (2015) Bioenergy crop greenhouse gas mitigation potential under a range of management practices. *GCB Bioenergy*, **7**, 366–374.

Jain AK, Khanna M, Erickson M, Huang H (2010) An integrated biogeochemical and economic analysis of bioenergy crops in the Midwestern United States. *GCB Bioenergy*, **2**, 217–234.

Jørgensen CJ, Struwe S, Elberling B (2012) Temporal trends in N₂O flux dynamics in a Danish wetland – effects of plant-mediated gas transport of N₂O and O₂ following changes in water level and soil mineral-N availability. *Global Change Biology*, **18**, 210–222.

Kandel TP, Wu Y, Kakani VG (2013) Growth and yield responses of switchgrass ecotypes to temperature. *American Journal of Plant Sciences*, **4**, 1173–1180.

Kang S, Nair SS, Kline KL et al. (2014) Global simulation of bioenergy crop productivity: analytical framework and case study for switchgrass. *GCB Bioenergy*, **6**, 14–25.

Kansas Historical Society (2014) Dust clouds rolling over the prairies, Hugoton, Kansas. *Kansas Memory*.

Kering MK, Butler TJ, Biermacher JT, Guretzky JA (2012) Biomass yield and nutrient removal rates of perennial grasses under nitrogen fertilization. *BioEnergy Research*, **5**, 61–70.

Kim S, Dale BE (2005) Environmental aspects of ethanol derived from no-tilled corn grain: nonrenewable energy consumption and greenhouse gas emissions. *Biomass and Bioenergy*, **28**, 475–489.

Lane DR, Coffin DP, Lauenroth WK (1998) Effects of soil texture and precipitation on above-ground net primary productivity and vegetation structure across the Central Grassland region of the United States. *Journal of Vegetation Science*, **9**, 239–250.

Lee J, Pedroso G, Linquist BA, Putnam D, van Kessel C, Six J (2012) Simulating switchgrass biomass production across ecoregions using the DAYCENT model. *GCB Bioenergy*, **4**, 521–533.

Lesur C, Jeuffroy M-H, Makowski D et al. (2013) Modeling long-term yield trends of *Miscanthus × giganteus* using experimental data from across Europe. *Field Crops Research*, **149**, 252–260.

Li Y, Fu X, Liu X et al. (2013) Spatial variability and distribution of N₂O emissions from a tea field during the dry season in subtropical central China. *Geoderma*, **193–194**, 1–12.

Liebig MA, Schmer MR, Vogel KP, Mitchell RB (2008) Soil carbon storage by switchgrass grown for bioenergy. *BioEnergy Research*, **1**, 215–222.

Ma Z, Wood C, Bransby D (2001) Impact of row spacing, nitrogen rate, and time on carbon partitioning of switchgrass. *Biomass and Bioenergy*, **20**, 413–419.

Mesinger F, DiMego G, Kalnay E et al. (2006) North American regional reanalysis. *Bulletin of the American Meteorological Society*, **87**, 343–360.

Miguez FE, Maughan M, Bollero GA, Long SP (2012) Modeling spatial and dynamic variation in growth, yield, and yield stability of the bioenergy crops *Miscanthus × giganteus* and *Panicum virgatum* across the conterminous United States. *GCB Bioenergy*, **4**, 509–520.

Mooney DF, Roberts RK, English BC, Tyler DD, Larson JA (2009) Yield and breakeven price of "Alamo" switchgrass for biofuels in Tennessee. *Agronomy Journal*, **101**, 1234–1242.

Muir JP, Sanderson MA, Ocumpaugh WR, Jones RM, Reed RL (2001) Biomass production of "Alamo" switchgrass in response to nitrogen, phosphorus, and row spacing. *Agronomy Journal*, **93**, 896–901.

Mulkey VR, Owens VN, Lee DK (2006) Management of switchgrass-dominated Conservation Reserve Program lands for biomass production in South Dakota. *Crop Science*, **46**, 712–720.

Muñoz I, Campra P, Fernández-Alba AR (2010) Including CO₂-emission equivalence of changes in land surface albedo in life cycle assessment. Methodology and case study on greenhouse agriculture. *The International Journal of Life Cycle Assessment*, **15**, 672–681.

Nair SS, Kang S, Zhang X et al. (2012) Bioenergy crop models: descriptions, data requirements, and future challenges. *GCB Bioenergy*, **4**, 620–633.

Necpálová M, Anex RP, Fienen MN et al. (2015) Understanding the DayCent model: calibration, sensitivity, and identifiability through inverse modeling. *Environmental Modelling & Software*, **66**, 110–130.

Nikièma P, Rothstein DE, Min D-H, Kapp CJ (2011) Nitrogen fertilization of switchgrass increases biomass yield and improves net greenhouse gas balance in northern Michigan, U.S.A. *Biomass and Bioenergy*, **35**, 4356–4367.

Noy-Meir I (1973) Desert ecosystems: environment and producers. *Annual Review of Ecology and Systematics*, **4**, 25–51.

Ogle SM, Breidt FJ, Easter M, Williams S, Killian K, Paustian K (2010) Scale and uncertainty in modeled soil organic carbon stock changes for US croplands using a process-based model. *Global Change Biology*, **16**, 810–822.

Parkin TB (2008) Effect of sampling frequency on estimates of cumulative nitrous oxide emissions. *Journal of Environmental Quality*, **37**, 1390–1395.

Parton WJ, Schimel DS, Cole CV, Ojima DS (1987) Analysis of factors controlling soil organic matter levels in Great Plains grasslands. *Soil Science Society of America Journal*, **51**, 1173–1179.

Parton WJ, Hartman M, Ojima D, Schimel D (1998) DAYCENT and its land surface submodel: description and testing. *Global and Planetary Change*, **19**, 35–48.

Parton WJ, Holland EA, Grosso SJD et al. (2001) Generalized model for NOx and N$_2$O emissions from soils. *Journal of Geophysical Research*, **106**, 17403–17419.

Pearson CH (2004) Pasture grass species evaluation at fruita 1995–2001. In: *Colorado Forage Research 2003: Alfalfa, Irrigated Pastures, and Mountain Meadows* (eds Brummer JE, Pearson CH), pp. 65–70. Agricultural Experiment Station and Cooperative Extension Technical Bulletin TB04-01, Colorado State University, Fort Collins, CO.

Pedroso GM, van Kessel C, Six J, Putnam DH, Linquist BA (2014) Productivity, 15N dynamics and water use efficiency in low- and high-input switchgrass systems. *GCB Bioenergy*, **6**, 704–716.

Peplow M (2014) Cellulosic ethanol fights for life. *Nature*, **507**, 152–153.

Persson T, Ortiz BV, Bransby DI, Wu W, Hoogenboom G (2011) Determining the impact of climate and soil variability on switchgrass (*Panicum virgatum* L.) production in the south-eastern USA; a simulation study. *Biofuels, Bioproducts and Biorefining*, **5**, 505–518.

Pervez MS, Brown JF (2010) Mapping irrigated lands at 250-m scale by merging MODIS data and national agricultural statistics. *Remote Sensing*, **2**, 2388–2412.

Pyörälä P, Peltola H, Strandman H, Antti K, Antti A, Jylhä K, Kellomäki S (2014) Effects of management on economic profitability of forest biomass production and carbon neutrality of bioenergy use in Norway spruce stands under the changing climate. *BioEnergy Research*, **7**, 279–294.

Qin Z, Dunn JB, Kwon H, Mueller S, Wander MM (2015a) Soil carbon sequestration and land use change associated with biofuel production: empirical evidence. *GCB Bioenergy*, doi:10.1111/gcbb.12237.

Qin Z, Zhuang Q, Cai X (2015b) Bioenergy crop productivity and potential climate change mitigation from marginal lands in the United States: an ecosystem modeling perspective. *GCB Bioenergy*, **7**, 1211–1221.

Reddy KR, Matcha SK, Singh SK, Brand D, Seepaul R (2008) Quantifying the effects of temperature and nitrogen on switchgrass growth and development. In: *Presentation to the 38th Biological Systems Simulation Modeling Meeting Conference, Temple TX, USA*. Available at: http://www.spar.msstate.edu/Files/BSSG_2008_Reddy_KR.pdf (accessed 17 November 2015).

Robertson GP, Dale VH, Doering OC et al. (2008) Sustainable biofuels redux. *Science*, **322**, 49–50.

Robertson GP, Hamilton SK, Del Grosso SJ, Parton WJ (2011) The biogeochemistry of bioenergy landscapes: carbon, nitrogen, and water considerations. *Ecological Applications*, **21**, 1055–1067.

Robertson AD, Davies CA, Smith P, Dondini M, McNamara NP (2015) Modelling the carbon cycle of *Miscanthus* plantations: existing models and the potential for their improvement. *GCB Bioenergy*, **7**, 405–421.

Roncucci N, o Di Nasso NN, Bonari E, Ragaglini G (2015) Influence of soil texture and crop management on the productivity of miscanthus (*Miscanthus* × *giganteus* Greef et Deu.) in the Mediterranean. *GCB Bioenergy*, **7**, 998–1008.

Roth B, Finnan JM, Jones MB, Burke JI, Williams ML (2015) Are the benefits of yield responses to nitrogen fertilizer application in the bioenergy crop *Miscanthus* × *giganteus* offset by increased soil emissions of nitrous oxide? *GCB Bioenergy*, **7**, 145–152.

Sanderson MA (1992) Morphological development of switchgrass and kleingrass. *Agronomy Journal*, **84**, 415–419.

Sanderson MA, West CP, Moore KJ, Stroup J, Moravec J (1997) Comparison of morphological development indexes for switchgrass and bermudagrass. *Crop Science*, **37**, 871–878.

Saxton KE, Rawls WJ, Romberger JS, Papendick RI (1986) Estimating generalized soil-water characteristics from texture. *Soil Science Society of America Journal*, **50**, 1031–1036.

Schmer MR, Vogel KP, Mitchell RB, Perrin RK (2008) Net energy of cellulosic ethanol from switchgrass. *Proceedings of the National Academy of Sciences*, **105**, 464–469.

Schmer MR, Liebig MA, Hendrickson JR, Tanaka DL, Phillips RL (2012) Growing season greenhouse gas flux from switchgrass in the northern great plains. *Biomass and Bioenergy*, **45**, 315–319.

Searchinger T, Heimlich R, Houghton RA et al. (2008) Use of US croplands for biofuels increases greenhouse gases through emissions from land-use change. *Science*, **319**, 1238–1240.

Searle SY, Malins CJ (2014) Will energy crop yields meet expectations? *Biomass and Bioenergy*, **65**, 3–12.

Shcherbak I, Millar N, Robertson GP (2014) Global metaanalysis of the nonlinear response of soil nitrous oxide (N$_2$O) emissions to fertilizer nitrogen. *Proceedings of the National Academy of Sciences*, **111**, 9199–9204.

Sheehan J, Aden A, Paustian K, Killian K, Brenner J, Walsh M, Nelson R (2003) Energy and environmental aspects of using corn stover for fuel ethanol. *Journal of Industrial Ecology*, **7**, 117–146.

Shield IF, Barraclough TJP, Riche AB, Yates NE (2012) The yield response of the energy crops switchgrass and reed canary grass to fertiliser applications when grown on a low productivity sandy soil. *Biomass and Bioenergy*, **42**, 86–96.

Smemo KA, Ostrom NE, Opdyke MR, Ostrom PH, Bohm S, Robertson GP (2011) Improving process-based estimates of N$_2$O emissions from soil using temporally extensive chamber techniques and stable isotopes. *Nutrient Cycling in Agroecosystems*, **91**, 145–154.

Smith J, Smith P, Addiscott T (1996) Quantitative methods to evaluate and compare Soil Organic Matter (SOM) models. In: *Evaluation of Soil Organic Matter Models* (eds Powlson DS, Smith P, Smith JU), pp. 181–199. Springer Berlin Heidelberg, Berlin, Heidelberg.

Smith CM, David MB, Mitchell CA, Masters MD, Anderson-Teixeira KJ, Bernacchi CJ, DeLucia EH (2013) Reduced nitrogen losses after conversion of row crop agriculture to perennial biofuel crops. *Journal of Environmental Quality*, **42**, 219–228.

Staley TE, Stout WL, Jung GA (1991) Nitrogen use by tall fescue and switchgrass on acidic soils of varying water holding capacity. *Agronomy Journal*, **83**, 732–738.

Swinton SM, Babcock BA, James LK, Bandaru V (2011) Higher US crop prices trigger little area expansion so marginal land for biofuel crops is limited. *Energy Policy*, **39**, 5254–5258.

Thomas ARC, Bond AJ, Hiscock KM (2013) A multi-criteria based review of models that predict environmental impacts of land use-change for perennial energy crops on water, carbon and nitrogen cycling. *GCB Bioenergy*, **5**, 227–242.

Tilman D, Hill J, Lehman C (2006) Carbon-negative biofuels from low-input high-diversity grassland biomass. *Science*, **314**, 1598–1600.

Tilman D, Socolow R, Foley JA et al. (2009) Beneficial biofuels-the food, energy, and environment trilemma. *Science*, **325**, 270–271.

U.S. Department of Agriculture FSA (2011) Program Fact Sheets. Available at: http://www.fsa.usda.gov/FSA/newsReleases?area=newsroom&subject=landing&topic=pfs&newstype=prfactsheet&type=detail&item=pf_20110727_energ_en_bcap7.html (accessed 17 November 2015).

U.S. Department of Agriculture NASS (2014) Quick Stats Ad-hoc Query Tool. Available at: http://quickstats.nass.usda.gov/ (accessed 17 November 2015).

U.S. Department of Agriculture NRCS. Web Soil Survey. Available at: http://websoilsurvey.sc.egov.usda.gov/App/HomePage.htm (accessed 17 November 2015).

U.S. Department of Energy (2011) U.S. Billion-Ton Update: Biomass Supply for a Bioenergy and Bioproducts Industry (Leads Perlack RD, Stokes BJ), 227 pp. ORNL/TM-2011/224. Oak Ridge National Laboratory, Oak Ridge, TN.

U.S. Environmental Protection Agency (2014a) *DRAFT Inventory of U.S. Greenhouse Gas Emissions and Sinks: 1990–2012*. US Environmental Protection Agency, Washington, DC.

U.S. Environmental Protection Agency (2014b) *Framework for Assessing Biogenic CO2 Emissions from Stationary Sources*. Available at: http://www.epa.gov/climatechange/ghgemissions/biogenic-emissions.html (accessed 17 November 2015).

U.S. Geological Survey (2015) Federal Lands of the United States. Available at: http://nationalmap.gov/small_scale/mld/fedlanp.html (accessed 17 November 2015).

Vogel KP, Brejda JJ, Walters DT, Buxton DR (2002) Switchgrass biomass production in the Midwest USA. *Agronomy Journal*, **94**, 413–420.

Wagle P, Kakani VG (2014) Environmental control of daytime net ecosystem exchange of carbon dioxide in switchgrass. *Agriculture, Ecosystems & Environment*, **186**, 170–177.

Walker WE, Harremoës P, Rotmans J, van der Sluijs JP, van Asselt MBA, Janssen P, Krayer von Krauss MP (2003) Defining uncertainty: a conceptual basis for uncertainty management in model-based decision support. *Integrated Assessment*, **4**, 5–17.

Walter A, Galdos MV, Scarpare FV et al. (2014) Brazilian sugarcane ethanol: developments so far and challenges for the future. *Wiley Interdisciplinary Reviews: Energy and Environment*, **3**, 70–92.

Wang M, Han J, Dunn JB, Cai H, Elgowainy A (2012) Well-to-wheels energy use and greenhouse gas emissions of ethanol from corn, sugarcane and cellulosic biomass for US use. *Environmental Research Letters*, **7**, 045905.

Wang C, Hunt ER Jr, Zhang L, Guo H (2013) Phenology-assisted classification of C3 and C4 grasses in the U.S. Great Plains and their climate dependency with MODIS time series. *Remote Sensing of Environment*, **138**, 90–101.

van der Weerden T, Clough T, Styles T (2013) Using near-continuous measurements of N_2O emission from urine-affected soil to guide manual gas sampling regimes. *New Zealand Journal of Agricultural Research*, **56**, 60–76.

Wickham JD, Stehman SV, Gass L, Dewitz J, Fry JA, Wade TG (2013) Accuracy assessment of NLCD 2006 land cover and impervious surface. *Remote Sensing of Environment*, **130**, 294–304.

Wilson DM, Heaton EA, Schulte LA *et al.* (2014) Establishment and short-term productivity of annual and perennial bioenergy crops across a landscape gradient. *BioEnergy Research*, **7**, 885–898.

Woli P (2012) Soil and variety effects on the energy and carbon balances of switchgrass-derived ethanol. *Journal of Sustainable Bioenergy Systems*, **02**, 65–74.

Wright CK, Wimberly MC (2013) Recent land use change in the Western Corn Belt threatens grasslands and wetlands. *Proceedings of the National Academy of Sciences*, **110**, 4134–4139.

Wullschleger SD, Davis EB, Borsuk ME, Gunderson CA, Lynd LR (2010) Biomass production in switchgrass across the United States: database description and determinants of yield. *Agronomy Journal*, **102**, 1158–1168.

Xu B, Li F, Shan L, Ma Y, Ichizen N, Huang J (2006) Gas exchange, biomass partition, and water relationships of three grass seedlings under water stress. *Weed Biology and Management*, **6**, 79–88.

Zhang X, Izaurralde RC, Manowitz D *et al.* (2010) An integrative modeling framework to evaluate the productivity and sustainability of biofuel crop production systems. *GCB Bioenergy*, **2**, 258–277.

14

Bioethanol potential from miscanthus with low ILUC risk in the province of Lublin, Poland

SARAH J. GERSSEN-GONDELACH[1], BIRKA WICKE[1], MAGDALENA BORZĘCKA-WALKER[2], RAFAŁ PUDEŁKO[2] and ANDRE P. C. FAAIJ[3]

[1]*Copernicus Institute of Sustainable Development, Utrecht University, Heidelberglaan 2, 3584 CS Utrecht, The Netherlands,* [2]*Department of Agrometeorology and Applied Informatics, Institute of Soil Science and Plant Cultivation State Research Institute, 8 Czartoryskich Str., 24-100 Puławy, Poland,* [3]*Energy and Sustainability Research Institute, University of Groningen, Blauwborgje 6, 9747 AC Groningen, The Netherlands*

Abstract

Increasing production of biofuels has led to concerns about indirect land-use change (ILUC). So far, significant efforts have been made to assess potential ILUC effects. But limited attention has been paid to strategies for reducing the extent of ILUC and controlling the type of LUC. This case study assesses five key ILUC mitigation measures to quantify the low-ILUC-risk production potential of miscanthus-based bioethanol in Lublin province (Poland) in 2020. In 2020, a total area of 196 to 818 thousand hectare of agricultural land could be made available for biomass production by realizing above-baseline yield developments (95–413 thousand ha), increased food chain efficiencies (9–30 thousand ha) and biofuel feedstock production on underutilized lands (92–375 thousand ha). However, a maximum 203–269 thousand hectare is considered legally available (not protected) and biophysically suitable for miscanthus production. The resulting low-ILUC-risk bioethanol production potential ranges from 12 to 35 PJ per year. The potential from this region alone is higher than the national Polish target for second-generation bioethanol consumption of 9 PJ in 2020. Although the sustainable implementation potential may be lower, the province of Lublin could play a key role in achieving this target. This study shows that the mitigation or prevention of ILUC from bioenergy is only possible when an integrated perspective is adopted on the agricultural and bioenergy sectors. Governance and policies on planning and implementing ILUC mitigation are considered vital for realizing a significant bioenergy potential with low ILUC risk. One important aspect in this regard is monitoring the risk of ILUC and the implementation of ILUC mitigation measures. Key parameters for monitoring are land use, land cover and crop yields.

Keywords: case study, ILUC mitigation and prevention, land-use change, miscanthus × giganteus, policies and governance, technical biofuel potential

Introduction

From 2002 to 2012, the production of biofuels has expanded significantly in the EU (Observ'ER, 2015). This growth is largely policy driven, based on the idea that biofuels can play an important role in reducing GHG emissions and mitigating climate change (European Parliament and Council of the European Union, 2003). However, in recent years, this assumption has been widely debated. One of the main topics of concern is land-use change (LUC), and especially indirect land-use change (ILUC). Here, ILUC is defined as a change in land use that takes place if biofuel feedstock production displaces agricultural production of food, feed and fibers and this displacement results in 1) food, feed and

fibers being produced elsewhere to continue to meet the demand, or 2) more land being taken into agricultural production because of increased food prices (Searchinger *et al.*, 2008; Plevin *et al.*, 2010; Wicke *et al.*, 2012). When ILUC entails the conversion of high carbon stock lands, for example, forests or grasslands, this can lead to increased GHG emissions which reduces or even cancels out the GHG benefits of biofuels compared with fossil fuels (Searchinger *et al.*, 2008). Since the first publication on the negative effects of ILUC by Searchinger *et al.* (2008), multiple studies have attempted to model and quantify the extent of (I)LUC and the level of related GHG emissions caused by biofuel production (e.g., Al-Riffai *et al.*, 2010; EPA, 2010; Hertel *et al.*, 2010; Tyner *et al.*, 2010; Laborde, 2011). However, the modeling of LUC and (I)LUC-related GHG emissions is characterized by major limitations and challenges (Warner *et al.*, 2014). Results vary significantly between studies,

Correspondence: Sarah J. Gerssen-Gondelach
e-mail: s.j.gerssen-gondelach@uu.nl

and outcomes are expected to remain uncertain (Plevin *et al.*, 2010, 2015; Wicke *et al.*, 2012). Therefore, investigating how ILUC can be mitigated or prevented may be more important than assessing the scale of ILUC under current assumptions (Wicke *et al.*, 2012).

ILUC of biofuels can only be prevented when the direct LUC (DLUC) of the displaced activity is addressed as well. Therefore, it is necessary to take an integrated perspective on all land use, whether for food, feed, fiber and fuels. Previous research has identified the following key measures to reduce the extent of ILUC and control the type of land-use change: above-baseline yield development, improved integration of food and biofuel chains, increased chain efficiencies, biofuel feedstock production on underutilized lands and land zoning (van de Staaij *et al.*, 2012; Wicke *et al.*, 2012; Witcover *et al.*, 2013; Brinkman *et al.*, 2014). Very few studies, however, have investigated the potential of producing biofuels with low ILUC risk (van de Staaij *et al.*, 2012). For the assessment of low-ILUC-risk biofuel potentials, regional analyses are of great importance because of several reasons. First, a regional analysis considers the specific characteristics of a region, for example, biophysical conditions, agricultural practices and the socio-economic context. Such factors are needed to define a feasible and suitable biofuel target for the region and develop appropriate policy strategies for realizing this target and mitigating ILUC. Second, a regional analysis is important to assess the availability and quality of data and to translate this into parameters for monitoring the implementation of ILUC mitigation measures and ILUC risks. Monitoring is required for correct certification of low-ILUC-risk biofuels.

The aim of this case study was (1) to assess how much additional biofuel can be produced in 2020 by implementing ILUC mitigation measures (i.e., the low-ILUC-risk biofuel production potential), and (2) to identify parameters required for monitoring the risk of ILUC and the implementation of ILUC mitigation measures. The case study focuses on bioethanol production from miscanthus, in the Polish province of Lublin (Lubelskie voivodship). Lublin is located in the southeast of Poland. Diverse studies have shown that this province has a significant technical and economic potential for biomass production (Fischer *et al.*, 2010; de Wit & Faaij, 2010; Szymańska & Chodkowska-Miszczuk, 2011; Faber *et al.*, 2012; Pudełko *et al.*, 2012). In addition, the development level of agricultural systems and the agricultural yields in Eastern Poland are lower compared with Western regions (Eurostat, 2013; CSO, 2014a). This suggests that agricultural productivity can improve significantly and thereby make land available for bioenergy feedstock production without ILUC. The choice to conduct the case study at province level is based on the

good availability of data and regional differences in agricultural characteristics in Poland. Miscanthus is chosen because it has the potential to contribute to the development of the rural economy by the diversification of farms, which often enhances their economic resilience and profitability (Agricultural Sustainability Institute). In addition, crop diversity helps to maintain or improve the agroecosystem (Dauber *et al.*, 2010).

Methods and materials

The case study presented here is based on a report by Gerssen-Gondelach *et al.* (2014). The general method to quantify ILUC mitigation measures was developed by Brinkman *et al.* (2015). This section describes the main aspects of the method and provides case-specific details. For more details, the reader is referred to Gerssen-Gondelach *et al.* (2014) and Brinkman *et al.* (2015).

Assessment of low-ILUC-risk biofuel potential

The assessment of the low-ILUC-risk biofuel production potential is based on a combination of a top-down and bottom-up approach and distinguishes three main components, see Fig. 1. Below, these components are shortly described.

Step 1: Top-down assessment of agricultural production in the baseline and target scenario in 2020. From an economic model used to analyze ILUC factors (top-down approach), a biomass production baseline (without additional biofuels) and target (with a biofuel mandate) for the case study region in 2020 are established. The current study uses the outputs from the computable general equilibrium model MIRAGE-BioF (Modeling International Relationships in Applied General Equilibrium for Biofuel, hereafter referred to as MIRAGE) as generated for a study for DG Trade of the European Commission (Laborde, 2011). The baseline indicates the production of biomass for food, feed and fiber applications in the absence of the biofuels mandate (i.e., assuming current biofuel production to remain approximately constant). The target refers to the total biomass production when a biofuels mandate is implemented; it includes food, feed and fiber demand as well as the extra feedstocks for biofuels needed to meet the biofuels mandate. The difference between the target and baseline is the extra agricultural production induced by the mandate (whether directly caused by increased demand for meeting the mandate or induced by increased crop prices). In MIRAGE, this amount is projected to cause LUC (both direct and indirect).

The MIRAGE study (Laborde, 2011) includes two biofuel mandate scenarios which differ in their assumptions regarding future trade policy (business as usual or BAU vs. free trade). In the present study, the BAU scenario is applied, which means that all existing import tariffs on biofuels remain unchanged in 2020. The mandate includes first-generation biofuels: biodiesel from oil palm, rapeseed, soybean and sunflower and bioethanol from maize, wheat, sugar cane and sugar beet (Laborde, 2011).

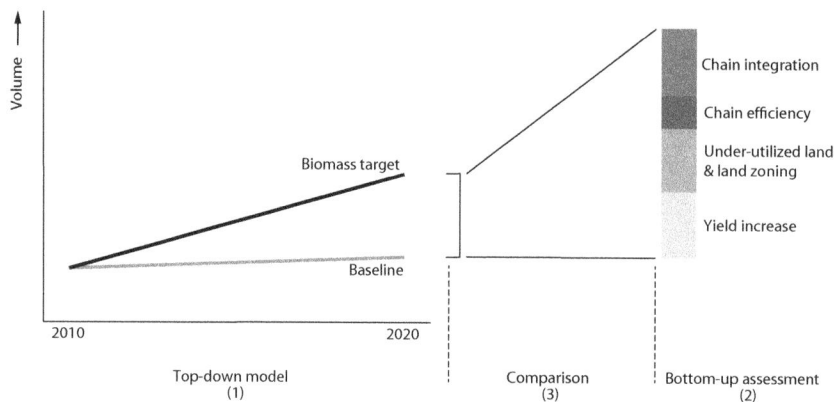

Fig. 1 General approach to analyze and quantify biomass production potential with low ILUC risks (Brinkman *et al.*, 2015).The three steps (1–3) of the analysis are described in the main text.

This study considers both crop and cattle (beef and milk) production. The MIRAGE model outputs for crop production volumes in the baseline and target scenario are only available on the EU27 level. Therefore, the outputs are disaggregated to the case study region, based on the current share of crop production in Lublin compared to the EU27 (see Brinkman *et al.*, 2015). The production of beef and milk in 2020 cannot be derived from the MIRAGE model. Therefore, the production in 2020 is estimated by assuming that the production trend will be in line with the recent trend in the European Union (1991–2012). It is assumed that the production is unaffected by a biofuel mandate and thus equal in the baseline and target scenario. In both the baseline and target scenario for 2020, the production volumes for crops and beef are projected to be lower than in 2010, see Data S2. The production volume of milk is projected to remain constant compared to 2010.

Step 2: Bottom-up assessment of low-ILUC-risk miscanthus-based ethanol production potential. It was shown that in MIRAGE, no mandate for miscanthus-based bioethanol is included. Therefore, the production of miscanthus-based ethanol further increases the biomass production volume above the level of the target scenario. Only when the total biomass production potential with low ILUC risk is higher than the production induced by the target scenario, production of low-ILUC-risk miscanthus-based ethanol is possible. To determine the potential to produce ethanol from miscanthus with low ILUC risk, it is assessed how much agricultural land can be made available for miscanthus cultivation by implementing ILUC mitigation measures. First, a baseline yield scenario is defined to determine the initial total agricultural land area required in 2020 for the projected biomass production volume in the target scenario. Then, it is assessed to what extent the different ILUC mitigation measures can contribute to reducing this land requirement and making the surplus land (i.e., land no longer needed for the targeted biomass production) available for bioenergy (section Assessment of ILUC mitigation measures). To this aim, this study takes an integrated view on all land uses and looks for synergies between agriculture, forestry and bioenergy. The following ILUC mitigation measures are assessed (bottom-up approach): above-baseline yield

development, improved chain integration, increased food chain efficiency, biofuel feedstock production on underutilized lands and land zoning. The latter measure, land zoning, is distinct from the first four measures. It does not reduce land requirements for agricultural production, but establishes constraints on future production areas to avoid the conversion of protected and biophysically unsuitable areas to miscanthus cultivation.

Finally, the total surplus land resulting from integrating all ILUC mitigation measures and the potential bioethanol production on this land are calculated. This potential is the technical potential, which takes into account the demand for land for food and feed production, legal requirements regarding environmental conservation and minimal biophysical requirements for miscanthus cultivation. In the present paper, this potential is called the low-ILUC-risk potential. Although the implementation of ILUC mitigation measures reduces the risk of ILUC, it not necessarily decreases the risk to zero. The ILUC risk will only be zero if it is guaranteed that biofuel feedstocks are only produced on land that is made available by one of the ILUC mitigation measures. This requires legislation and enforcement of regulations.

In the general approach, the baseline yield scenario for crops is derived from MIRAGE (Brinkman *et al.*, 2015). However, in this case study, the MIRAGE projections are not in line with recent yield trends in Lublin. As it is found that crop yields often follow a linear trend over time (Ray *et al.*, 2012; Gerssen-Gondelach *et al.*, 2015), the baseline scenario in this case study is based on linear extrapolation of historical yield trends (1999–2012) in Lublin. For cattle, the selected parameters for yield are the beef and milk productivity (beef or milk production per animal per year) and the cattle density on meadows and pastures. The productivity and density values in 2020 are defined similar to crop yields. Currently, the total agricultural area needed for the production of the selected crops and for cattle in Lublin covers 1224 thousand ha (87% of the utilized agricultural land area). Because of the increasing yields and the projected reduction in the total agricultural production volume in the target scenario compared to the level in 2010, the total land use reduces to 944 thousand ha in 2020 (see Data S2).

Step 3: Comparison of low-ILUC-risk bioethanol potential to biofuel production target. The low-ILUC-risk biofuel production potential is compared to the biofuel production target. In the general method, this target is derived from MIRAGE's biofuel mandate scenario. As MIRAGE includes no target for miscanthus-based bioethanol, the production potential can only be compared to MIRAGE's production target for first-generation bioethanol. The bioethanol production target from MIRAGE for the EU27 is disaggregated to the Polish national level based on the share of bioethanol production in Poland compared to the EU27 (Table 1). The target is not disaggregated to the level of Lublin province because no information is available about current biofuel production levels at the provincial level. In addition to the bioethanol production target from MIRAGE, the low-ILUC-risk biofuel potential is compared to the targets for biofuel consumption in 2020 as set in the Polish National Renewable Energy Action Plan (NREAP) for meeting the requirements of Directive 2009/28/EC (Table 1).

Assessment of ILUC mitigation measures

The contribution of the ILUC mitigation measures to the miscanthus-based ethanol production potential is investigated for three scenarios; *low, medium* and *high*. Each scenario includes all ILUC mitigation measures, and for each measure, assumptions are made about how this measure contributes to the generation of surplus land compared to the target scenario from step 2. For example, the scenarios assume more rapid developments in agricultural productivity and food chain efficiency compared to the baseline projections in the target scenario. The rates of development increase from the low to the high scenario to indicate the variability and uncertainty in the data and to test the effect on the low-ILUC-risk potential. The next subsections explain the assumptions per measure for the different scenarios. Where the methods to assess the ILUC mitigation measures deviate from the general approach (Brinkman *et al.*, 2015), this is also explained in these sections. Finally, it is described how the total low-ILUC-risk biofuel potential in each scenario is calculated by integrating the results of all individual measures.

Above-baseline yield development

Increases in crop yield, beef and milk productivity and cattle density above the baseline projection result in a reduction in agricultural land required for crop and livestock production (assuming the production volume remains constant). On the resulting surplus land area, biomass can be produced with low ILUC risk (see Brinkman *et al.* (2015) for a detailed description of the calculation). The baseline yield scenario was defined based on the finding that crop yields often follow a linear trend over longer terms (Ray *et al.*, 2012; Gerssen-Gondelach *et al.*, 2015). Over shorter time periods, however, higher yield increases are possible, especially when the yield gap is still large (Gerssen-Gondelach *et al.*, 2015). The average yields and management levels in Lublin are lower compared with regions in Western Poland and Germany (see Table S2 in Data S1). Therefore, measures such as scaling up of farms, mechanization and improved use of chemicals, as already applied in these other regions, can enable higher annual yield growth rates compared to the baseline. Also, when crop yields increase and less land is needed for production, the use of lower quality land for production is likely to decrease which has a positive effect on the average yield levels. The crop yields in Western Poland and Germany give an indication of what yields can be attained in Lublin, but also maximum attainable crop yields based on, for example, climate and land suitability are taken into account. Similarly, for cattle production, cattle density and productivity levels in Poland and Germany (CSO, 2014a; FAO, 2014) are assumed to be appropriate indicators of what improvements are attainable in Lublin. Data on historical and current crop-specific yields, cattle density and beef and milk productivity are collected or derived from the Central Statistical Office in Poland (CSO) (CSO, 2014a) and the FAO (FAO, 2014). The agro-ecological potential crop yield is derived from the Global Agro-Ecological Zones database (FAO and IIASA, 2014). As an example, Table 2 compares the current average wheat yield to the yield level in the baseline and the low, medium and high scenarios.

Improved chain integration

The production of second-generation bioethanol generates various by-products such as lignin, proteins and carbon dioxide released during fermentation. These by-products can be used to produce a variety of value-added co-products (see, e.g., Patton, n.d.). Depending on the potential uses of these co-products and following the principles of consequential LCA (see Ekvall & Weidema, 2004; Finnveden *et al.*, 2009; Reinhard & Zah,

Table 1 Current and targeted production of first- and second-generation bioethanol in Poland

	Current production*	2020 projected baseline production without mandate†	2020 production target with mandate†	2020 consumption target Poland‡
First-generation bioethanol				
Million liter	177.5	174.9	567	760
PJ	4.2	4.1	13.3	17.8
Second-generation bioethanol				
Million liter	–	n/a	n/a	376
PJ	–	n/a	n/a	8.8

*Current: average 2009–2011 (Observ'ER, 2011, 2012, 2013).

†Derived from MIRAGE output for EU27.

‡Targets as set in 2010 in the National Renewable Energy Action Plan (Ministry of Economy, 2010).

Table 2 Current and projected average wheat yields and losses in transport and storage. Projections are given for the baseline and the low, medium and high scenarios

	Average wheat yield Lublin (t ha^{-1} yr^{-1})	Wheat losses (%)
Current*	3.7	5.0
Baseline scenario 2020	4.1	5.0
Low scenario 2020	4.5	3.8
Medium scenario 2020	5.7	2.5
High scenario 2020	7.5	0.8

*Current wheat yield: average 2008–2012; current wheat losses: average 2008–2011.

2011; Brinkman et al., 2015), these co-products could be argued to reduce land demand and thereby help to mitigate ILUC. For example, when co-products can substitute livestock feed from crops, a certain amount of land can be freed up from crop cultivation. In this case study, no co-products are included that have the potential to generate surplus land in 2020 (see the biofuel chain design in Data S3). Therefore, this measure is not further investigated.

Increased food chain efficiency

This ILUC mitigation measure addresses the reduction in food losses in transport, storage, (un)loading, etc., such that a higher share of the produced goods reaches the consumer. Therefore, the less land is needed to deliver the same amount of goods [see Brinkman et al. (2015) for a detailed description of the calculation]. In Poland, 27% of all food losses and food waste take place in the stages between farms and consumers (Rutten et al., 2013). These losses are equal to 15% of the total national food production; for comparison, the average percentage in the EU27 is 7% (Rutten et al., 2013). To analyze the land-saving potential of food chain efficiency improvements, regional figures are not available. Therefore, national figures on food losses occurring during storage and transportation from FAO food balances are used (FAO, 2014). These figures give estimated volumes of food losses for each separate agricultural product. In Table 2, the example of projected food losses is given for wheat in the baseline and the low, medium and high scenarios.

Biofuel feedstock production on underutilized lands

Underutilized land includes set-aside land, abandoned land, marginal lands or degraded land, which often has lower productivity than conventional agricultural land. The share of this land type that does not provide other services (e.g., agriculture, biodiversity, high carbon stocks or other ecosystem services) can be used for the production of biomass with low risk of ILUC. The total area of underutilized land in 2020 depends on the current area of underutilized land and an increase in this area due to reduced agricultural land use from 2010 to 2020 as

projected in the biofuel target scenario based on MIRAGE (see step 2).

Regarding the amount of underutilized land currently available in the case study area, the use of spatially explicit data about the location and extent of these types of land, its current uses and functions, and its suitability for the biofuel feedstock investigated in the case study is ideal. For Lublin province, however, spatially explicit data about the location of underutilized lands is not available. Therefore, the current area of underutilized agricultural land is estimated based on statistical data about set-aside, fallow and marginal land from the Central Statistical Office of Poland (CSO, 2013, 2014a,b), Eurostat (Eurostat, 2012) and FAO (FAO, 2003). In addition, it is estimated what part of the agricultural land not under agricultural activity can be considered as abandoned land potentially available for miscanthus. This is based on statistics and own estimates (for more details, see Data S4). Additional abandoned land available in 2020 is estimated based on the area of agricultural land no longer required because of the projected reduction in agricultural production and increase in yields (see step 2).

Often, the bioenergy crop yield on underutilized lands is expected to be lower than average. However, not in all cases yields on underutilized land are actually lower than on agricultural land as it depends on the soil and climate conditions. As the location and biophysical characteristics of underutilized lands are unknown, the suitability and the attainable miscanthus yield on these lands cannot be assessed. Therefore, the impact of the yield level on the miscanthus production potential is assessed in a sensitivity analysis (see methods section Integrated analysis of overall low-ILUC-risk biofuel potential).

Land zoning

While the previously described measures attempt to mitigate ILUC, land zoning aims at reducing the impacts of LUC, here especially the associated biodiversity losses and GHG emissions. This study includes land zoning to prevent the conversion of protected areas, including (primary and secondary) forest and high conservation value areas, for the production of biomass. In addition, in this case study, this measure also considers the land suitability for miscanthus production.

The land not excluded by land zoning for protection purposes is referred to as *legally available land*. *Suitable land* refers to land that is biophysically suitable for miscanthus production, considering minimal climate and soil requirements. The calculation of the legally available and suitable agricultural land area for miscanthus cultivation is based on the method applied by Pudełko et al. (2012) to assess the technical potential of perennial energy crops in Poland. Spatial analyses for Lublin province are performed in the geographic information system (GIS), using the following data sets: agricultural soil suitability (IUNG, 1974), Corine land cover (Nunes de Lima, 2005), digital elevation model, hydrogeological map (Institute of Geology, 1957), annual rainfall based on the Agroclimate Model of Poland (Górski & Zaliwski, 2002) and protected areas (European Commission, 1992; Ministry of the Environment, 2003).

First, to determine the legally available land, the following criterion is applied:

- Soils located on protected areas of land are removed. Protected areas include all forests, national parks, landscape parks, nature reserves, strict protection areas, Natura 2000 sites and their buffer zone.

Second, to assess what share of the total agricultural area is biophysically suitable for miscanthus cultivation, the following criteria are applied:

- Miscanthus roots can extract water to a depth of approximately 2 m (Caslin et al., 2011). Therefore, the ground water level is set at a depth up to 2 m for all soils. The areas with a lower ground water table are excluded;
- The minimal average annual precipitation is 550 mm yr^{-1} for all soils [see, e.g., Kuś and Faber, 2009 in Sliz-Szkliniarz (2013)]. Areas where the precipitation did not exceed this minimum are removed;
- Boggy and wet areas are excluded because the accessibility of machinery to waterlogged sites is limited and can cause soil damage. Also, the release of carbon dioxide due to land conversion will negatively affect the GHG emission balance of the biofuel;
- Areas over 350 m above sea level are excluded because production and transportation conditions are hampered in these regions.

Land not complying with these suitability criteria is only very marginally suitable for miscanthus production. On these lands, miscanthus yields would be significantly lower than on suitable lands (see Data S3).

Finally, the criteria for legally available land and suitable land are combined, resulting in the total agricultural area legally available and suitable. Although considered suitable, the soil quality and degree of suitability of the areas included varies. Therefore, in the results, it is shown how land is distributed among suitability classes. This distribution is determined using the Polish classification system that distinguishes twelve soil suitability classes or complexes (Terelak & Witek, 1995). Nine of these classes apply to arable land and can be categorized into very good and good quality soils, lower quality soils and very weak soils. Three classes apply to grasslands (meadows and pastures) of various qualities.

Pudełko et al., (2012) excluded good and very good quality soils from their analysis, based on the guideline that bioenergy crops should not be cultivated on these lands. However, ILUC mitigation measures may free some areas that have good or very good quality soils while this would not result in displacement of crop production. Therefore, the present study includes all soil classes.

The land zoning criteria applied in this study do not include specific conditions on maximum carbon stocks to allow land-use conversion. However, the analysis excludes all areas that are prohibited by the Renewable Energy Directive (EU) to be used for biomass production because of high carbon stocks (i.e., wetlands, forested areas and peat land).

In the criteria described above, all protected areas currently under agricultural use (e.g., parts of the Natura 2000 network) are excluded from bioenergy production to ensure the conservation of biodiversity (European Parliament and Council of the European Union, 2009). However, some of the protected areas may actually be designated as legally available for miscanthus cultivation because miscanthus can have a positive impact on the biodiversity of agricultural land. The biodiversity in miscanthus fields is found to be higher compared with annual crops (Smeets et al., 2009; Dauber et al., 2010). This is potentially also true for grasslands, but the number of studies is limited yet and more research is needed (Dauber et al., 2010; Donnelly et al., 2011). In the medium and high scenarios, it is assumed that a part of the suitable agricultural areas with high conservation value can be made legally available for miscanthus cultivation. Areas with high carbon stocks are excluded in all scenarios.

Integrated analysis of overall low-ILUC-risk biofuel potential

Table 3 provides a summary of the scenario assumptions per ILUC mitigation measure. Having evaluated the individual measures, the total potential biomass production without ILUC is analyzed. This is an integrated assessment that accounts for the interactions and feedback between different measures. An example of this is a reduction in food losses that decreases the food production volume required for supplying the same amount of food, which influences the effect of above-baseline yield developments. The order in which the measures are considered in the integrated analysis influences the outcome of the assessment. In this study, the integration calculations are performed as follows:

i The agricultural land area required for food, feed, fuel and fiber production in 2020 as derived from the MIRAGE target scenario is taken as the initial land base.

ii The measure *increased food chain efficiency* is implemented: the biomass production volume required after a reduction in food losses is calculated. The surplus area generated by this measure is calculated using the baseline yield development scenario.

iii The measure *above-baseline yield increases* is applied: based on the required food production as determined in step ii, the additional surplus area generated through above-baseline yield developments is calculated.

iv The measure *use of underutilized land* is taken into account: the area of underutilized land is added to the total surplus land area from steps ii and iii.

v The measure *land zoning* is implemented: The total surplus land area from steps ii to iv is compared to the total land area suitable and legally available for miscanthus production. In the case that the surplus land area is larger than the area suitable and legally available, the use of surplus land for biomass production is limited by land zoning restrictions. The total surplus land area resulting from applying all five measures is presented for a low, medium and high integrated scenarios in which the low, medium and high scenarios of each measure are combined, respectively. In addition, a distinction in the results is made between the surplus area of cropland and of meadow and pastureland.

Table 3 Summary of scenario assumptions per ILUC mitigation measure and per scenario

	Baseline	Low	Medium	High
Above-baseline yield developments – crops*	Extrapolation of the historical linear yield trends in Lublin for the period 1999–2012 to 2020. The average annual yield increase is 1.8%, but varies between crops	Annual yield increase of 2.3% for a period of 10 years for all crops, based on the REFUEL projection for Central and Eastern European countries (de Wit et al., 2011)	Crop-specific yields are set to the current maximum yield level attained in Poland at the province level (average 2008–2012) (CSO, 2014a). The average annual yield increase is 3.2%, but varies between crops	Crop-specific yields are set to the current yield level attained in Germany (average 2008–2012) (FAO, 2014). The average annual yield increase is 7.6%, but varies between crops
Above-baseline yield developments – cattle	Extrapolation of the historical linear trends in cattle density and milk productivity in Lublin for the period 1999–2012 to 2020 (CSO, 2014a); In 2020, beef productivity attains current average productivity level of Poland in 2012 (CSO, 2014b)	Beef and milk productivity are set equal to baseline scenario, and cattle density is set equal to the current cattle density level in Germany (average 2008–2011) (FAO, 2014)	Both beef and milk productivity and cattle density are set equal to the current level attained in Germany (average 2008–2012 for beef and milk productivity; average 2008–2011 for cattle density) (FAO, 2014)	Both beef and milk productivity and cattle density are set equal to the current maximum level attained in Poland at the province level (average 2008–2012) (CSO, 2014a)
Improved food chain efficiency†	Product-specific food losses in 2020 are similar to average losses in Poland for the period 2008–2011 (FAO, 2014)	Product-specific food losses reduce with 25%	Product-specific food losses reduce with 50%	Product-specific food losses reduce to the 15th percentile‡ of the loss percentages of all EU countries (FAO, 2014)
Biomass production on underutilized land	n.a.	Low estimation of underutilized land area, based on statistics	Medium estimation of underutilized land area, based on statistics	High estimation of underutilized land area, based on statistics
Land zoning	n.a.	All protected areas are excluded	Miscanthus cultivation is possible on a limited area of protected land where miscanthus cultivation could support improvements in biodiversity§	Miscant Miscanthus cultivation is possible on a limited area of protected land (larger than in the medium scenario) where miscanthus cultivation could support improvements in biodiversity¶

*In the calculations, it is assumed that yields will not decrease compared with the current level. If a yield in one of the scenarios is lower than the current yield, the current yield will be considered instead. If a yield in one of the scenarios is higher than the agro-ecological potential yield, this agro-ecological potential yield will be considered instead. The agro-ecological potential yield is derived from the Global Agro-Ecological Zones database (FAO and IIASA, 2014). The data reflect Polish average maximum attainable yields. Data was not found for all crops; in case no data was available, the maximum potential yield was not taken into account.

†It is assumed that losses do not increase. In case losses would increase in a certain scenario, the loss is set equal to the current level (average Poland 2008–2011), that is, the reduction in losses is zero. In FAOSTAT, no food loss figures are given for rapeseed and beef. Therefore, estimations are made based on losses in Austria (rapeseed) and Hungary (beef); these are considered to be most comparable to levels in Poland.

‡15% of the EU countries attain this or a lower loss percentage, 85% of the EU countries attain a higher loss percentage.

§In the medium scenario, the following is assumed: Of the area that is suitable for miscanthus production but not legally available according to the applied protection criteria, 50% will be made legally available because miscanthus cultivation has a positive impact on biodiversity. Protected areas that are not considered to be suitable are excluded for miscanthus production.

¶In the high scenario, the following is assumed: Of the area that is suitable for miscanthus production but not legally available according to the applied protection criteria, 100% will be made legally available because miscanthus cultivation has a positive impact on biodiversity. Protected areas that are not considered to be suitable are excluded for miscanthus production.

vi For each integrated scenario, the potential miscanthus and bioethanol production on the total surplus land area is calculated. These potentials depend on the miscanthus yield and the biofuel chain efficiency. Therefore, to assess the impact of the value chain design, the total chain productivity is defined for a medium scenario and two sensitivity scenarios (low and high), see Table 4.

Results: ILUC mitigation potentials

Above-baseline yield development

Table 5 presents the land savings for the low, medium and high above-baseline yield developments in crop and cattle production compared to the target scenario. The saving potentials of crops are higher compared to cattle. This is in line with the fact that the cropland area is larger than the area of meadows and pastures (Table S1, Data S1). In all three scenarios, wheat yield improvements account for the largest area saved, followed by barley, triticale and rapeseed. For potatoes, sugar beets and apples, the yields in the low and medium above-baseline scenarios are actually lower compared with the baseline projection, because extrapolation of the recent yield trend results in a high yield increase. The additional area required for these crops compared to the target scenario is lower than the area saved by other crops, but reduces the total area saved. With regard to cattle, increasing the cattle density on meadows and pastures has a larger effect on the area saved than increasing the beef and milk productivity. The impacts of improvements in beef and milk productivity are comparable to each other.

Increased food chain efficiency

The agricultural area saved in the increased food chain efficiency scenarios is presented in Table 5. The potentials are significantly lower compared with the potential from above-baseline yield development. Improved chain efficiencies of crops result in considerably higher land saving compared with cattle. Similar to the above-baseline yield improvement scenarios, wheat has the highest land-saving potential, followed by oats and rapeseed.

Table 4 Components and productivity of the value chain for miscanthus-based biofuel production (for detailed explanation see Data S3)

Chain component	Assumptions	Parameter	Baseline	Sensitivity range (low–high)	References
Miscanthus cultivation and harvest	Spring yield, farming conditions are suboptimal and plantations have not reached plateau yields yet	Yield (t dm ha^{-1})	13	10–17	Stampfl et al. (2007); Borkowska & Molas (2013); Matyka & Kus (2011); van Dam et al. (2007)
Storage	On-farm storage of bales in the open air covered with plastic sheeting or storage in a silo or under a bale tarp	Biomass loss (% dry matter)	3%	1–5% dry matter	Monti et al. (2009); Shinners et al. (2010); Smeets et al. (2009)
Transport	Truck transport	Biomass loss (% dry matter)	0%*	–	–
Conversion	Biochemical conversion	Biomass-to-ethanol conversion efficiency (% HHV)	35%	35–40%†	Bansal et al. (2013); Hamelinck & Faaij (2006); Tao & Aden (2009); Aden et al. (2002); POET-DSM Advanced Biofuels, (2014); Abengoa Bioenergy, (2014)
Overall ethanol yield			84 GJ ha^{-1}	64–129 GJ ha^{-1}	Own calculation‡

*Biomass losses during transport are assumed to be negligible.
†The low conversion efficiency for the sensitivity analysis is equal to the baseline efficiency, because newly build plants already attain the baseline efficiency (Aden et al., 2002; Hamelinck & Faaij, 2006; Tao & Aden, 2009; Bansal et al., 2013; Abengoa Bioenergy, 2014; POET-DSM Advanced Biofuels, 2014).
‡Calculated from combining the miscanthus yield, storage and transportation losses and conversion efficiency.

Table 5 Land saved by crops and cattle in the above-baseline yield scenarios and increased food chain efficiency scenarios

| | Area saved (1000 ha) | | | | | |
| | Above-baseline yield scenarios | | | Increased food chain efficiency scenarios | | |
Product	Low	Medium	High	Low	Medium	High
Wheat	25	80	132	3.7	7.4	12.5
Rapeseed	12	16	26	0.7	1.5	2.0
Potatoes, sugar beets and apples	−6	−13	9	0.6	1.2	1.9
Other crops	35	67	126	3.4	6.7	11.4
Cattle*	30	74	132	0.8	1.7	2.8
Total	96	224	426	9.2	18.4	30.5

*The land saved by cattle is meadow and pastureland.

Table 6 Estimated underutilized land area available in 2020 in the low, medium and high scenarios

| | Area Lublin (1000 ha) | | | |
Scenario	Low	Medium	High	References
Set-aside and fallow land	45	60	75	Eurostat, (2012); FAO, (2003); CSO, (2014a)
Abandoned land not held for agricultural activity	5	15	20	CSO, (2014a) and own estimates
Abandoned land baseline scenario Of which	42	98	280	MIRAGE projection (Laborde, 2011) and own estimates
Cropland	34	80	229	
Grassland	8	18	51	
Marginal land	0	0	0	CSO, (2014b, 2013)
Total underutilized land	92	173	375	

Biofuel feedstock production on underutilized lands

The total area of set-aside and fallow land is estimated to be 45–75 thousand hectare (Table 6, see Data S4 for a more detailed explanation). In addition, the area of agricultural land that is held by owners who do not conduct agricultural activities and that could potentially be considered as abandoned land suitable for miscanthus production is estimated to be 5–20 thousand hectare (Data S4). Finally, according to the biofuel target scenario for 2020 based on MIRAGE and own estimates for cattle production, a total area of 280 thousand hectare will be abandoned compared to 2010 (see step 2 in Methods and materials). However, the projected rate of reduction in the agricultural land area is high compared to what is expected based on recent developments in Lublin and Poland. In addition, several factors could result in more land use than projected. Both issues are considered in more detail in the discussion. For this measure, the area of abandoned land in the low and medium scenarios is estimated to be significantly smaller than 280 thousand

hectare. In the high scenario, the total abandoned land area of 280 thousand hectare is included.

In statistics, only 202 ha of land was defined as degraded land (CSO, 2013, 2014b). Therefore, the share of marginal land in the total area of agricultural land is considered to be negligible. The resulting total estimated area of underutilized in each scenario is presented in Table 6.

Land zoning

In the methods, criteria were given to assess both the legal availability and the biophysical suitability of agricultural land for miscanthus production. When only applying the protection criterion, the total agricultural area that is legally available is 1267 thousand ha, see Table 7. When only considering the suitability criteria, the total agricultural area that is suitable for miscanthus production is 269 thousand ha. Other agricultural areas are considered very marginally suitable for miscanthus because of limited (soil) water availability, which is low in summer due to

Table 7 Agricultural area legally available and suitable for miscanthus production

| Criteria applied | Resulting land area | Area (1000 ha) | % of total agricultural area | Area by soil quality (1000 ha) Arable land | | | Grassland total |
				Very good & good	Lower	Very weak	
None	Total agricultural land area*	1745	100	885	499	75	316
Protection	Total area legally available	1267	73	631	391	49	195
Suitability	Total area suitable	269	15	55	134	0	80
Protection and suitability	Total area suitable and legally available for miscanthus	203	12	40	114	0	50

*Equal to average of agricultural land area in 2010 and 2012.

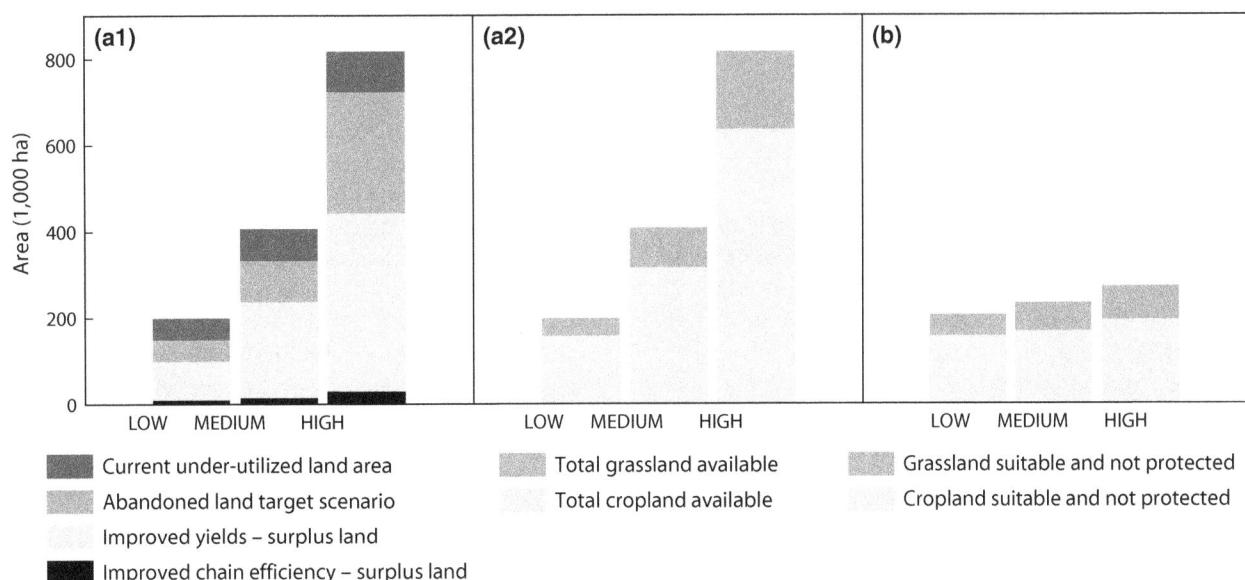

Fig. 2 Surplus area in the three integrated scenarios, breakdown by measures (a1) and agricultural cropland and grassland (a2); cropland and grassland area suitable for miscanthus cultivation and corrected for land zoning (b).

droughts (Mioduszewski, 2014). Although miscanthus has a good water-use efficiency compared with many other crops, it is found to be sensitive to water stress (Lewandowski *et al.*, 2000; Richter *et al.*, 2008). When combing the protection and suitability criteria, the total area suitable and legally available is 203 thousand hectare, which is equal to 12% of the total agricultural area. This value is used for the low scenario. Assuming that some protected areas could be made available for miscanthus production (as described in the methods section Land zoning), the suitable and legally available land area increases to 236 thousand ha in the medium scenario and 269 thousand ha in the high scenario.

Integrated analysis

Figure 2 presents the combined potential surplus land area generated by the measures *increased food chain*

efficiency, above-baseline yield development and *biofuel feedstock production on underutilized lands* and compares this land area to the area suitable and legally available for miscanthus based on the measure *land zoning*.

The ILUC mitigation measures *above-baseline yield development* and *biofuel feedstock production on underutilized lands* have the largest potential to make land available for biomass production. For the measure *biofuel feedstock production on underutilized lands*, a large share of the potential is related to the projected reduction in demand for agricultural land in the biofuel target scenario of MIRAGE compared to the current situation. The largest share of the area saved is considered to be cropland. The suitability and legal availability criteria for agricultural land limit the use of the arable land area saved in all scenarios and of the grassland area saved in the medium and high scenarios. The resulting ethanol production potential in each scenario is presented in

Fig. 3. This figure also presents the sensitivity of the biofuel potential to the miscanthus-ethanol chain productivity, as defined in the methods. The total bioethanol production potential ranges from 12.2 PJ per year in the case of a low ethanol yield in the low integrated scenario to 34.6 PJ per year in the case of a high ethanol yield in the high integrated scenario (i.e., 522–1479 million liter per year). The second-generation bioethanol consumption target for Poland as set in the NREAP is 8.8 PJ or 376 million liter in 2020. Thus, in all scenarios, the miscanthus-based ethanol production potential of only the province of Lublin is higher than this national target. In addition, in the NREAP, the total target for the national consumption of all first- and second-generation biofuels in Poland is 60.5 PJ (2582 million liter) in 2020 (Ministry of Economy, 2010). Thus, 20% to 57% of this target could be met by bioethanol production from miscanthus in the province of Lublin.

Monitoring ILUC and ILUC mitigation measures

The analysis shows that technically it is possible to produce large additional amounts of biofuel in Lublin with a low risk of causing ILUC. For certification, it needs to be verified that biofuel feedstock is indeed produced with low ILUC risk, that is, the risks of land conversion elsewhere or undesired land-use change in the case study region as a result of biofuel feedstock production are within certain thresholds. Also, to control and

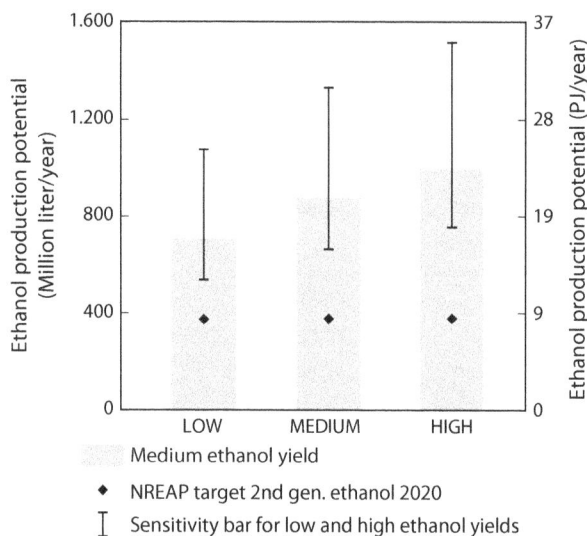

Fig. 3 Bioethanol production potentials in Lublin province in 2020 for medium ethanol yield and sensitivity bars for low ethanol yield and high ethanol yield, see Table 3. The National Renewable Energy Action Plan target is the Polish national consumption target for second-generation bioethanol of 8.8 PJ in 2020, as stated in the polish National Renewable Energy Action Plan.

manage the expansion of biofuel feedstock production, the implementation of the ILUC mitigation measures should be monitored.

For this case study, several parameters are identified that are important for monitoring ILUC risk (Table 8). First, the observation of land use (e.g., for agricultural or bioenergy production, or for forestry) and land-use change over time is vital. For this, it is required to frequently compose land use and land cover maps. This can be carried out using remote sensing (satellite monitoring), supplemented with field data for validation. Important is the detail of the land use and land cover maps. This means that maps should differentiate between, for example, forest and agricultural land, land under agricultural activity and abandoned or set-aside land, and between agricultural and bioenergy crops. In addition, appropriate spatial and temporal resolutions should be chosen. As farms in Lublin are often small, the spatial resolution should be high to enable the identification of differences in land use and land cover. It is especially important to observe areas that are excluded from bioenergy production through land zoning regulations. When land-use change would occur in these areas or their buffer zone, this is a sign for potential ILUC risk. The observation of land use and land-use change could be supported by monitoring land management, as this is a good indicator of how and for what purpose(s) land is used. Aspects of land management include, among others, the type (e.g., tillage or no tillage), intensity (e.g., full tillage or reduced tillage) and timing of management. However, collecting this type of data could be very time-consuming. Second, changes in food and trade balances could be an indicator for increasing ILUC risk. For example, when agricultural production volumes increase at a higher rate than expected, the land area required for food, feed and bioenergy production is potentially larger than the area available without ILUC. In addition, changes in imports or exports of agricultural products might indicate growing production demand in the region or relocation of local production to other regions, which can both cause ILUC. Small changes compared to the projected production and trade volumes, however, should be considered to be within the uncertainty range of the projection. It should therefore be assessed what an appropriate threshold would be.

The key indicators to monitor the implementation of the ILUC mitigation measures are as follows. First, for above-baseline yield development, the most important parameter is the crop-specific annual yield. As yields fluctuate over time, it is recommended to monitor the 5-year moving average yield. In addition, the targeted yield should be defined as a range within which the average yield should be in a certain year. Statistics on

Table 8 Parameters for monitoring ILUC risk and implementation of ILUC mitigation measures and data availability and quality for these parameters

Monitoring	Parameter for monitoring	Availability of data	Quality of available data
ILUC risk	Land use and land cover	Good (Nunes de Lima, 2005)	It needs to be assessed whether the spatial and temporal resolution of the data is appropriate for monitoring
	Land management	Some aggregated data on provincial level, for example, about amount of machinery and level of fertilizer use (CSO, 2014a)	Potentially data for individual farmers*
	Food balance	Available at provincial and national level (CSO, 2014a; FAO, 2014)†	To be assessed
	Trade balance	Available at national level (FAO, 2014)‡	To be assessed
Above-baseline yield development	Annual yield	Good for most important products, statistics are updated annually (CSO, 2014a). For other products, data is lacking or only provided on aggregated level and/or for selected years (CSO, 2014a,b)	Only provincial averages, no data for individual farmers§.
	Farm size and management	Only aggregated data on provincial level (CSO, 2014a)	Potentially data for individual farmers*
Increased food chain efficiency	Food losses in supply chain	Available per agricultural product, but only at national level and not specified per process step in the supply chain (FAO, 2014)	Poor, only estimates are provided and specification per process step is lacking
Land zoning and use of underutilized land	Land use and land cover	Good for protected areas (European Commission, 1992; Ministry of the Environment, 2003), no spatially explicit data available for underutilized lands	It needs to be assessed if the temporal resolution of the data is appropriate for monitoring protected areas.

*A national agricultural census is taken every 6–8 years and includes data about agricultural machinery (CSO, 2014a). Currently, the data is only provided at aggregated levels (CSO, 2014a). It should be further assessed whether data for individual farmers can be made available and used for monitoring.
†On provincial level, data is only available for production volumes and for the most important agricultural products (CSO, 2014a).
‡Trade figures on province level were not found during this study. It is recommended to further investigate whether such data is already collected and available and how missing data can be collected in the future.
§A national agricultural census is taken every 6–8 years, but does not include yields (CSO, 2014a).

provincial average crop yields are generally made available annually (CSO, 2014a). However, information about the performance of individual farmers is lacking. This information is useful to identify where efforts and investments for yield improvements are needed most. In addition, monitoring developments in farm size and management (e.g., the level and efficiency of fertilizers use) is valuable to assess whether subsidies and other stimulating policies are effective, and whether advances in farm management are substantial enough to realize the expected or targeted yield improvements. Second, monitoring of food chain efficiency requires data on food losses in the whole supply chain, specified per agricultural product and per process in the chain. In this study, data availability and quality are poor (see Table 8). More and better data need to be collected periodically to set targets for chain efficiency and monitor whether developments are in line with these targets. Third, land zoning and the use of underutilized lands can be mainly monitored by periodically assessing land use and land cover (see also above). When remote sensing is used, the ability to differentiate between miscanthus and other crops is very important, because miscanthus may be cultivated in areas where the production of other crops is undesired or prohibited. As the information on the location and size of underutilized lands is much more limited compared to protected areas, remote sensing and improved field data collection are important to set a baseline for monitoring this measure.

Discussion

Potential surplus land area

This case study assessed the production potential of miscanthus-based bioethanol with low ILUC risk in the Polish province of Lublin in 2020. Five measures have been analyzed that reduce the extent of ILUC and control the type of land-use change. The total potential of these measures has been investigated for a low, medium and high scenarios that refers to developments above the baseline projections. In 2020, a total area of 196 to 818 thousand hectare of agricultural land could become available for biomass production. This is equal to 11% to 47% of the total agricultural area in Lublin. The largest potential to generate surplus land comes from *above-baseline yield developments* (95–413 thousand hectare). Increasing especially wheat yields adds significantly to the total potential of this measure. Also, the projected area of underutilized land, 92–375 thousand hectare, is considerable. The large effect of these two ILUC mitigation measures illustrates the importance of improving land management. This finding is supported by assessments of land availability in Eastern Romania, Hungary and North-East Kalimantan (Indonesia) (Wicke *et al.*, 2015) and in Brazil (Woods *et al.*, 2015).

The potentials differ substantially between the scenarios, and also the feasibility and likelihood of the scenarios vary significantly. For example, the yields applied in the high scenario are considered to be feasible based on existing farming practices in Germany. But it is questionable whether the adoption of these practices can take place in the limited timeframe to 2020. Second, based on the disaggregation of results from the MIRAGE model, the production of crops in Lublin is projected to decline in both the baseline and biofuel target scenarios. This reduction strongly affects land use. The decline in crop production is primarily caused by a reduction in the cultivation of potatoes and cereals (except wheat and maize) as projected by MIRAGE for the EU27. These crops account for a significant part of the agricultural production in Lublin. Furthermore, according to MIRAGE, the production of especially oil crops (e.g., rapeseed, sunflower) and also other first-generation bioenergy crops (e.g., wheat, maize, sugar beet) will increase. In the province of

Lublin, however, the current production of oil crops and maize is very small. Therefore, the total decline in the production of potatoes and cereals (except wheat and maize) is larger than the total growth in the production of wheat, sugar beet and other crops (see Data S2). But the resulting reduction in land use is not in line with recent developments in Lublin and Poland (CSO, 2014a). In addition, other competitive uses for released land, such as afforestation, exist. These are not taken into account in this analysis. It is recommended to further assess the potential pathways for crop production and land use and specifying under which conditions each scenario could be realized.

Legally available and suitable area

Although the surplus land area available in 2020 is potentially very large, a limited area of 203–269 thousand hectare (12–15% of the total agricultural area) is considered to be legally available and biophysically suitable for miscanthus production based on the criteria for protecting high conservation areas and minimum requirements for land suitability. As a result, in all scenarios, the amount of surplus land that could be used for miscanthus production is restricted. The limitation on land use is mainly caused by the suitability criteria and especially the sensitivity of miscanthus to water stress. However, this study only assessed the land suitability based on a few simple criteria like the minimum ground water level. It did not take into account other parameters such as soil characteristics (see, e.g., van der Hilst *et al.*, 2010) or the influence of the current vegetation and the conversion to miscanthus on the water balance and water availability. It is therefore recommended to further investigate how these factors affect the land suitability and the potential yield for miscanthus. The insights can be used to set a maximum area for growing miscanthus. In addition, lands that are only very marginally suitable for miscanthus may be suitable for other crops that have a higher tolerance to water stress, for example, reed canary grass and switchgrass (Lewandowski *et al.*, 2000, 2003; Richter *et al.*, 2008). Agro-ecological zoning data (FAO and IIASA, 2014) shows that in Poland, the soil suitability for reed canary grass, and to a lesser extent also switchgrass, is considerably higher than the suitability for miscanthus. Fischer *et al.* (Fischer *et al.*, 2010) found that a total of 61% of the agricultural land in Poland is moderately to very suitable for reed canary grass, miscanthus and/or switchgrass. Thus, selection of the most appropriate crop for each area could significantly increase the use of the surplus land area and raise the total biomass production potential.

Low-ILUC-risk bioethanol potential

Depending on the productivity of the bioethanol value chain, the low-ILUC-risk bioethanol production potential ranges from 12 to 35 PJ per year (522–1479 million liter per year). For comparison, the national Polish target for second-generation bioethanol consumption is almost 9 PJ. This means that the province of Lublin could play a key role in achieving this target and help Poland even become an exporter of second-generation bioethanol. This potential, however, is the technical potential that accounts only for key environmental aspects such as the protection of high conservation value areas. However, the (sustainable) implementation potential may be lower than the technical potential. The implementation potential is the fraction of the technical potential that can be produced at economically profitable levels and implemented within the considered timeframe, taking into account local constraints and policies (Smeets *et al.*, 2007). For Lublin, several factors are identified that could significantly affect the implementation potential. First, the agricultural sector in Lublin is characterized by a large number of small farms and low average management levels compared with regions such as Western Poland and Germany (see Table S2 in Data S1). To realize above-baseline yield increases, scaling up, modernization and intensification of agricultural production are needed. However, farmers have little capital to invest, and land prices are considered too low for selling or leasing land. Second, when the ILUC mitigation measures are implemented and land is made available for biomass production, several hurdles exist for farmers to start cultivating bioenergy crops. For example, in recent years, the production and trade of biomass for heat and electricity in Lublin province have been constrained by the lack of a stable market. Large amounts of biomass were imported from the Ukraine and, according to local experts, biomass prices offered to farmers in Lublin were too low (Faber, 2014; Galczynska, 2014; Gradziuk, 2014). With regard to miscanthus, a potential additional hurdle may be the high establishment costs compared with other energy crops (Lewandowski *et al.*, 2003). The sustainable biofuel potential is the fraction of the technical potential that can be implemented while delivering positive environmental, social and economic impacts. To assess the sustainability of biofuels, sustainability criteria and indicators have been developed [see, e.g., Cramer *et al.* (2007), Franke *et al.* (2012), McBride *et al.* (2011) and Dale *et al.* (2013)]. It is unknown yet what will be the environmental and socioeconomic impacts of implementing ILUC mitigation measures in Lublin. These aspects should be addressed in future research.

To maximize the implementation potential, governance and policies are considered vital. First, this could, for example, include financial support to farmers to facilitate improved production practices. Such support is already included in European and Polish agricultural and rural development policies (European Commission, 2014; Ministry of Agriculture and Rural Development), but should be increased to realize the full potential. Second, in the medium and high scenarios, it was assumed that miscanthus production in some protected areas with high conservation value may actually lead to improved biodiversity. Therefore, land-use policies should clearly define which areas are allowed to take into production for biomass. Third, it is recommended to further assess the potential barriers for implementing ILUC mitigation measures and producing bioenergy crops and biofuels at large scale. In addition, it should be investigated how these hurdles could be addressed. Fourth, monitoring ILUC risks and the implementation of ILUC mitigation measures is important. This case study identified several parameters that are useful for monitoring, for example, land use, land cover and annual yields. However, the availability and quality of the data required for monitoring vary for the different parameters; especially, data about losses in the food supply chain and underutilized lands should be improved. Finally, the assessment of the ILUC mitigation measures and the miscanthus-based bioethanol production potential with low ILUC risk in Lublin province in Poland shows that the mitigation or prevention of ILUC from bioenergy is only possible when the close link between the agricultural and bioenergy sectors is recognized. Therefore, an integrated perspective on these sectors in planning and implementing policies on ILUC prevention specifically (as well as on land use in general) is essential. Doing so would allow realizing a significant bioenergy potential with a low risk of causing ILUC while boosting the performance of the agricultural sectors as a whole.

Acknowledgements

The authors would like to thank David Laborde for sharing data from the MIRAGE model. The research presented in this paper was conducted within the 'ILUC prevention project' which was funded by the Netherlands Enterprise Agency, the Dutch Ministry of Infrastructure and the Environment, the Dutch Sustainable Biomass Commission and the Rotterdam Climate Initiative/Port of Rotterdam. This case study on miscanthus-based ethanol was funded by Shell. The funder contributed to the data collection and commented on the original report, but the authors take complete responsibility for the integrity of the data and the accuracy of the data analysis. The views expressed in this paper are those of the authors and do not necessarily reflect those of the funding agency. Magdalena Borzęcka-Walker and Rafał Pudełko were financed by the S2Biom project under grant agreement No FP7-608622.

References

Abengoa Bioenergy (2014) Abengoa celebrates grand opening of its first commercial-scale next generation biofuels plant. Available at: http://www.abengoabioenergy.com/web/en/prensa/noticias/historico/2014/abg_20141017.html (accessed 19 December 2014).

Aden A, Ruth M, Ibsen K, Jechura J, Neeves K, Sheehan J, Wallace B (2002) *Lignocellulosic Biomass to Ethanol Process Design and Economics Utilizing Co-Current Dilute Acid Prehydrolysis and Enzymatic Hydrolysis for Corn Stover*. National Renewable Energy Laboratory, Golden, CO, USA.

Agricultural Sustainability Institute, UCDAVIS. What is sustainable agriculture? Available at: http://www.sarep.ucdavis.edu/about-sarep/def (accessed 7 May 2015).

Al-Riffai P, Dimaranan B, Laborde D (2010) *Global Trade and Environmental Impact Study of the EU Biofuels Mandate*. International Food Policy Research Institute, Washington, DC, USA.

Bansal A, Illukpitiya P, Singh SP, Tegegne F (2013) Economic competitiveness of ethanol production from cellulosic feedstock in tennessee. *Renewable Energy*, **59**, 53–57.

Borkowska H, Molas R (2013) Yield comparison of four lignocellulosic perennial energy crop species. *Biomass and Bioenergy*, **51**, 145–153.

Brinkman M, Wicke B, Gerssen-Gondelach S, van der Laan C, Faaij A (2014) *Methodology for Assessing ILUC Prevention*. Copernicus Institute of Sustainable Development, Utrecht University, Utrecht, The Netherlands.

Brinkman MLJ, Wicke B, Gerssen-Gondelach SJ, van der Laan C, Faaij APC (2015) *Methodology for Assessing and Quantifying ILUC Prevention Options*. Copernicus Institute of Sustainable Development, Utrecht University, Utrecht, The Netherlands.

Caslin B, Finnan J, Easson L (2011) 1 crop production. In: *Miscanthus Best Practice Guidelines* (eds Caslin B, Finnan J, Easson L), pp. 5–26. Teagasc and Agri-Food and Bioscience Institute, Carlow and Hillsborough, Ireland.

Cramer J, Wissema E, deBruijne M et al. (2007) *Testing framework for sustainable biomass, final report from the project group "Sustainable production of biomass"*. Project group Sustainable Production of Biomass, The Hague, The Netherlands.

CSO (2013) *Rural Areas in Poland - National Agricultural Census 2010*. Central Statistical Office (CSO), Warsaw, Poland.

CSO (2014a) Local data bank. Available at: http://www.stat.gov.pl/bdlen/app/strona.html?p_name=indeks (accessed 2013, 2014).

CSO (2014b) *Statistical Yearbook of Agriculture 2013*. Central Statistical Office (CSO), Agricultural Department, Warsaw, Poland.

Dale VH, Efroymson RA, Kline KL et al. (2013) Indicators for assessing socioeconomic sustainability of bioenergy systems: a short list of practical measures. *Ecological Indicators*, **26**, 87–102.

van Dam J, Faaij APC, Lewandowski I, Fischer G (2007) Biomass production potentials in Central and Eastern Europe under different scenarios. *Biomass and Bioenergy*, **31**, 345–366.

Dauber J, Jones MB, Stout JC (2010) The impact of biomass crop cultivation on temperate biodiversity. *GCB Bioenergy*, **2**, 289–309.

Donnelly A, Styles D, Fitzgerald J, Finnan J (2011) A proposed framework for determining the environmental impact of replacing agricultural grassland with miscanthus in Ireland. *GCB Bioenergy*, **3**, 247–263.

Ekvall T, Weidema BP (2004) System boundaries and input data in consequential life cycle inventory analysis. *The International Journal of Life Cycle Assessment*, **9**, 161–171.

EPA (2010) *Renewable Fuel Standard Program (RFS2) Regulatory Impact Analysis*. Environmental Protection Agency, Washington, DC, USA. EPA-420-R-10-006.

European Commission (1992) Directive 92/43/EEC of 21 may 1992 on the conservation of natural habitats and of wild fauna and flora. *Official Journal of the European Communities*, **206**, 7–50.

European Commission (2014) Rural development 2014-2020 - agriculture and rural development. Available at: http://ec.europa.eu/agriculture/rural-development-2014-2020/index_en.htm (accessed 29 September 2014).

European Parliament, and Council of the European Union (2003) Directive 2003/30/EC of the european parliament and of the council of 8 may 2003 on the promotion of the use of biofuels or other renewable fuels for transport. *Official Journal of the European Union L* 123, **46**, 42–46.

European Parliament, and Council of the European Union (2009) Directive 2009/28/EC of the european parliament and of the council of 23 April 2009 on the promotion of the use of energy from renewable sources and amending and subsequently repealing directives 2001/77/EC and 2003/30/EC. *Official Journal of the European Union*, **140**, 16–62.

Eurostat (2012) Eurostat - data explorer. Fallow land and set-aside land: Number of farms and areas by size of farm (UAA) and size of arable area. Available at: http://appsso.eurostat.ec.europa.eu/nui/show.do?dataset=ef_lu_ofsetasid&lang=en (accessed 8/15/2014).

Eurostat (2013) Eurostat - data explorer. Key farm variables: Area, livestock (LSU), labour force and standard output (SO) by agricultural size of farm (UAA), legal status of holding and NUTS 2 regions. Available at: http://appsso.eurostat.ec.europa.eu/nui/show.do?dataset=ef_kvaareg&lang=en (accessed 29 July 2014).

Faber A (2014) *Personal Communication*. Institute of Soil Science and Plant Cultivation (IUNG), Pulawy, Poland.

Faber A, Pudełko R, Borek R, Borzecka-Walker M, Syp A, Krasuska E, Mathiou P (2012) Economic potential of perennial energy crops in poland. *Journal of Food, Agriculture & Environment*, **10**, 1178–1182.

FAO (2003) Chapter 1. introduction. In: *Fertilizer Use by Crop in Poland*. Food and Agriculture Organisation of the United Nations (FAO), Rome, Italy.

FAO (2014) FAOSTAT. Available at:http://faostat.fao.org (accessed 2014).

FAO, and IIASA (2014) GAEZ global agri-ecological zones. Available at: http://gaez.fao.org/Main.html# (accessed 19 June 2014).

Finnveden G, Hauschild MZ, Ekvall T et al. (2009) Recent developments in life cycle assessment. *Journal of Environmental Management*, **91**, 1–21.

Fischer G, Sylvia P, van Velthuizen H, Lensink SM, Londo M, de Wit M (2010) Biofuel production potentials in europe: sustainable use of cultivated land and pastures. part I: land productivity potentials. *Biomass and Bioenergy*, **34**, 159–172.

Franke B, Reinhardt G, Malavelle J, Faaij A, Fritsche U (2012) *Global assessments and guidelines for sustainable liquid biofuel production in developing countries*. A GEF targeted research project. Heidelberg/Paris/Utrecht/Darmstadt.

Galczynska M (2014) Personal communication. Foundation for Lubelskie Development.

Gerssen-Gondelach S, Wicke B, Faaij A (2014) *ILUC Prevention Strategies for Sustainable Biofuels. Case Study on the Bioethanol Production Potential from Miscanthus with Low ILUC Risk in the Province of Lublin, Poland*. Copernicus Institute of Sustainable Development, Utrecht University, Utrecht, The Netherlands.

Gerssen-Gondelach S, Wicke B, Faaij A (2015) Assessment of driving factors for yield and productivity developments in crop and cattle production as key to increasing sustainable biomass potentials. *Food and Energy Security*, **4**, 36–75.

Górski T, Zaliwski A (2002) Agroclimate model of Poland (in Polish). *Pamiętnik Puławski*, **130**, 251–260.

Gradziuk P (2014) Personal communication. Polish Chamber of Biomass.

Hamelinck CN, Faaij APC (2006) Outlook for advanced biofuels. *Energy Policy*, **34**, 3268–3283.

Hertel TW, Golub AA, Jones AD, O'Hare M, Plevin RJ, Kammen DM (2010) Effects of US maize ethanol on global land use and greenhouse gas emissions: estimating market-mediated responses. *BioScience*, **60**, 223–231.

van der Hilst F, Dornburg V, Sanders JPM et al. (2010) Potential, spatial distribution and economic performance of regional biomass chains: the north of the Netherlands as example. *Agricultural Systems*, **103**, 403–417.

Institute of Geology (1957) *Hydrogeological map of Poland. Scale 1:300,000*. Institute of Geology, Warsaw, Poland.

IUNG (1974) *The Agricultural soil Suitability Map of Poland. Scale 1:100,000*. Institute of Soil Science and Plant Cultivation (IUNG), Puławy, Poland.

Laborde D (2011) *Assessing the Land use Change Consequences of European Biofuels Policies*. International Food Policy Research Institute, Washington, DC, USA.

Lewandowski I, Clifton-Brown JC, Scurlock JMO, Huisman W (2000) Miscanthus: European experience with a novel energy crop. *Biomass and Bioenergy*, **19**, 209–227.

Lewandowski I, Scurlock JMO, Lindvall E, Christou M (2003) The development and current status of perennial rhizomatous grasses as energy crops in the US and Europe. *Biomass and Bioenergy*, **25**, 335–361.

Matyka M, Kus J (2011) Yielding and biometric characteristics of selected miscanthus genotypes. *Problemy Inżynierii Rolniczej*, **2**, 157–163.

McBride AC, Dale VH, Baskaran LM et al. (2011) Indicators to support environmental sustainability of bioenergy systems. *Ecological Indicators*, **11**, 1277–1289.

Ministry of Agriculture and Rural Development. Program rozwoju obszarów wiejskich 2014-2020. Available at: https://www.minrol.gov.pl/Wsparcie-rolnictwa-i-rybolowstwa/PROW-2014-2020 (accessed 26 September 2014).

Ministry of Economy (2010) *National Renewable Energy Action Plan*. Ministry of Economy, Warsaw, Poland.

Ministry of the Environment (2003) *National System of Protected Areas (KSOCH) 2003*. Ministry of the Environment, Poland.

Mioduszewski W (2014) Small (natural) water retention in rural areas. *Journal of Water and Land Development*, **20**, 19–29.

Monti A, Fazio S, Venturi G (2009) The discrepancy between plot and field yields: harvest and storage losses of switchgrass. *Biomass and Bioenergy*, **33**, 841–847.

Nunes de Lima LimaMV (2005) *Image 2000 and CLC 2000 – Products and Methods. CORINE land Cover Updating for the Year 2000*. JRC, Ispra, Italy.

Observ'ER (2011) Biofuels barometer. *Systémes Solaires Le Journal Des Énergies Renouvelables* **204**: 68–93.

Observ'ER (2012) Biofuels barometer. *Systémes Solaires Le Journal Des Énergies Renouvelables* **210**: 42–62.

Observ'ER (2013) Biofuels barometer. *Systémes Solaires Le Journal Des Énergies Renouvelables* **216**: 48–63.

Observ'ER (2015) *Biofuels barometer*. Observ'ER, Observatoire des énergies renouvelables.

Patton J (n.d.) Value-Added Coproducts from the Production of Cellulosic Ethanol. North Dakota State University.

Plevin RJ, O'Hare M, Jones AD, Torn MS, Gibbs HK (2010) Greenhouse gas emissions from biofuels' indirect land use change are uncertain but may be much greater than previously estimated. *Environmental Science & Technology*, **44**, 8015–8021.

Plevin RJ, Beckman J, Golub AA, Witcover J, O'Hare M (2015) Carbon accounting and economic model uncertainty of emissions from biofuels-induced land use change. *Environmental Science & Technology*, **49**, 2656–2664.

POET-DSM Advanced Biofuels (2014) First commercial-scale cellulosic ethanol plant in the U.S. opens for business - POET-DSM advanced biofuels. Available at: http://poetdsm.com/pr/first-commercial-scale-cellulosic-plant (accessed 19 December 2014).

Pudełko R, Borzęcka-Walker M, Faber A, Borek R, Jarosz Z, Syp A (2012) The technical potential of perennial energy crops in Poland. *Journal of Food, Agriculture & Environment*, **10**, 781–784.

Ray DK, Ramankutty N, Mueller ND, West PC, Foley JA (2012) Recent patterns of crop yield growth and stagnation. *Nature Communications*, **3**, 1293.

Reinhard J, Zah R (2011) Consequential life cycle assessment of the environmental impacts of an increased rapemethylester (RME) production in Switzerland. *Biomass and Bioenergy*, **35**, 2361–2373.

Richter GM, Riche AB, Dailey AG, Gezan SA, Powlson DS (2008) Is UK biofuel supply from miscanthus water-limited? *Soil use and Management*, **24**, 235–245.

Rutten M, Nowicki P, Boogaardt MJ, Aramyan L (2013) *Reducing Food Waste by Households and in Retail in the EU: A Prioritisation Using Economic, Land Use and Food Security Impacts*. LEI Wageningen UR, The Hague, The Netherlands. 2013–2035.

Searchinger T, Heimlich R, Houghton RA *et al.* (2008) Use of U.S. croplands for biofuels increases greenhouse gases through emissions from land-use change. *Science*, **319**, 1238–1240.

Shinners KJ, Boettcher GC, Muck RE, Weimer PJ, Casler MD (2010) Harvest and storage of two perennial grasses as biomass feedstocks. *Transactions of the ASABE*, **53**, 359–370.

Sliz-Szkliniarz B (2013) 4 methodological approach for bioenergy potential assessment. In: *Energy Planning in Selected European Regions - Methods for Evaluating the Potential of Renewable Energy Sources* (ed. Sliz-Szkliniarz B). Karlsruher Institut für Technologie (KIT). Available at: http://www.ksp.kit.edu/9783866449510 (accessed 12 December 2013).

Smeets EMW, Faaij APC, Lewandowski IM, Turkenburg WC (2007) A bottom-up assessment and review of global bio-energy potentials to 2050. *Progress in Energy and Combustion Science*, **33**, 56–106.

Smeets EMW, Lewandowski IM, Faaij APC (2009) The economical and environmental performance of miscanthus and switchgrass production and supply chains in a European setting. *Renewable and Sustainable Energy Reviews*, **13**, 1230–1245.

van de Staaij J, Peters D, Dehue B *et al.* (2012) *Low indirect impact biofuel (LIIB) methodology - version zero*.

Stampfl PF, Clifton-Brown JC, Jones MB (2007) European-wide GIS-based modelling system for quantifying the feedstock from miscanthus and the potential contribution to renewable energy targets. *Global Change Biology*, **13**, 2283–2295.

Szymańska D, Chodkowska-Miszczuk J (2011) Endogenous resources utilization of rural areas in shaping sustainable development in Poland. *Renewable and Sustainable Energy Reviews*, **15**, 1497–1501.

Tao L, Aden A (2009) The economics of current and future biofuels. *Vitro Cellular & Developmental Biology - Plant*, **45**, 199–217.

Terelak H, Witek T (1995) Poland. In: *Soil Survey: Perspectives and Strategies for the 21st Century* (ed. Zinck JA), pp. 100–103. Land and Water Development Division, FAO, Rome, Italy.

Tyner WE, Taheripour F, Zhuang Q, Birur DK, Baldos U (2010) *Land use Changes and Consequent CO2 Emissions Due to US Corn Ethanol Production: A Comprehensive Analysis*. Center for Global Trade Analysis, Purdue University, West Lafayette, IN, USA.

Warner E, Zhang Y, Inman D, Heath G (2014) Challenges in the estimation of greenhouse gas emissions from biofuel-induced global land-use change. *Biofuels, Bioproducts and Biorefining*, **8**, 114–125.

Wicke B, Verweij P, van Meijl H, van Vuuren D, Faaij A (2012) Indirect land use change: review of existing models and strategies for mitigation. *Biofuels*, **3**, 87–100.

Wicke B, Brinkman M, Gerssen-Gondelach S, van der Laan C, Faaij A (2015) *ILUC Prevention Strategies for Sustainable Biofuels: Synthesis Report from the ILUC Prevention Project*. Copernicus Institute of Sustainable Development, Utrecht University, Utrecht, The Netherlands.

de Wit M, Faaij A (2010) European biomass resource potential and costs. *Biomass and Bioenergy*, **34**, 188–202.

de Wit M, Londo M, Faaij A (2011) Productivity developments in European agriculture: relations to and opportunities for biomass production. *Renewable and Sustainable Energy Reviews*, **15**, 2397–2412.

Witcover J, Yeh S, Sperling D (2013) Policy options to address global land use change from biofuels. *Energy Policy*, **56**, 63–74.

Woods J, Lynd LR, Laser M, Batistella M, deCastro VictoriaD, Kline K, Faaij A (2015) Chapter 9. land and bioenergy. In: *Bioenergy & Sustainability: Bridging the Gaps* (eds Souza GM, Victoria RL, Joly CA, Verdade LM), pp. 258–300. Scientific Committee on Problems of the Environment (SCOPE), Paris, France.

15

Radiation capture and conversion efficiencies of *Miscanthus sacchariflorus, M. sinensis* and their naturally occurring hybrid *M. × giganteus*

CHRISTOPHER LYNDON DAVEY[1], LAURENCE EDMUND JONES[1], MICHAEL SQUANCE[1], SARAH JANE PURDY[1], ANNE LOUISE MADDISON[1], JENNIFER CUNNIFF[2], IAIN DONNISON[1] and JOHN CLIFTON-BROWN[1]

[1]*Institute of Biological, Environmental and Rural Sciences (IBERS), Aberystwyth University, Gogerddan, Aberystwyth, Ceredigion SY23 3EE, UK,* [2]*Rothamsted Research, Harpenden, Hertfordshire AL5 2JQ, UK*

Abstract

Miscanthus is a rhizomatous C4 grass of great interest as a biofuel crop because it has the potential to produce high yields over a wide geographical area with low agricultural inputs on marginal land less suitable for food production. At the moment, a clonal interspecific hybrid *Miscanthus × giganteus* is the most widely cultivated and studied in Europe and the United States, but breeding programmes are developing newer more productive varieties. Here, we quantified the physiological processes relating to whole season yield in a replicated plot trial in Wales, UK. Light capture and conversion efficiency were parameterized for four carefully selected genotypes (*M. sinensis, M. sacchariflorus* and *Miscanthus × giganteus*). Differences in the canopy architecture in mature stands as measured by the extinction coefficient (k) were small (0.55–0.65). Sensitivity analysis on a mathematical model of *Miscanthus* was performed to quantify the accumulative intercepted photosynthetically active radiation (iPAR) in the growing season using (i) k, (ii) variation in the thermal responses of leaf expansion rate, (iii) base temperature for degree days and (iv) date start of canopy expansion. A 10% increase in k or leaf area per degree day both had a minimal effect on iPAR (3%). Decreasing base temperature from 10 to 9 °C gave an 8% increase in iPAR. If the starting date for canopy expansion was the same as shoot emergence date, then the iPAR increases by 12.5%. In *M. × giganteus*, the whole season above ground and total (including below ground) radiation-use efficiency (RUE) ranged from 45% to 37% higher than the noninterspecific hybrid genotypes. The greater yields in the interspecific hybrid *M. × giganteus* are explained by the higher RUE and not by differences in iPAR or partitioning effects. Studying the mechanisms underlying this complex trait could have wide benefits for both fuel and food production.

Keywords: diversity, extinction coefficient, *Miscanthus*, radiation-use efficiency, sensitivity analysis, yield

Introduction

Miscanthus is a rhizomatous C4 grass of interest as a potential biofuel crop (Visser & Pignatelli, 2001; Hastings *et al.*, 2009a; Somerville *et al.*, 2010; Zub & Brancourt-Hulmel, 2010). This is because it has the potential to produce high yields (Clifton-Brown *et al.*, 2001) over a wide geographical area with low agricultural inputs (Beale & Long, 1997; Zub & Brancourt-Hulmel, 2010) and can be grown on marginal land not cultivated for food production. At the moment, a *Miscanthus × giganteus* genotype is the one most often grown and studied in Europe because of its high yields. *M. × giganteus* is a naturally occurring hybrid of *M. sinensis* and *M. sacchar-*

iflorus (Greef & Deuter, 1993; Hodkinson & Renvoize, 2001). As *Miscanthus* is an undomesticated plant, there is scope to increase yields over *M. × giganteus* and so breeding programmes are ongoing to achieve this by utilizing the considerable phenotypic diversity found across the genotypes (Robson *et al.*, 2013). The range of this diversity has not been systematically modelled at the level of descriptions of the physiological processes relating to yield. Nor has the range of the variation been used to inform which features of *Miscanthus* are most amenable to giving increases in yield above those produced by *M. × giganteus*.

To address both these issues, a physiologically based model of yield is needed which can be easily parameterized for different genotypes. The range of parameter values can then be used in 'what if' simulations using the model to access their impact on yield using

Correspondence: Christopher Lyndon Davey
e-mail: cdd@aber.ac.uk

$M. \times giganteus$ as a reference. Such a study using different genotypes would only give an indication of the variation in key yield parameters currently expressed in the breeding populations. In addition, genetic segregation and recombination during breeding would be expected to produce further variation especially in a genetically diverse and nondomesticated plant like *Miscanthus* (Hartl & Clark, 2007).

A fundamental model of yield based on the work of Monteith (Monteith, 1977) is as follows:

$$Yield = \text{Sum incident PAR over current growing season length} * \text{Proportion PAR intercepted} * \text{RUE} \quad (1)$$

For *Miscanthus,* yield is the above-ground dry matter at harvest. PAR is photosynthetic active radiation and RUE is the radiation-use efficiency which quantifies the amount of dry matter created for each MJ of PAR the canopy intercepts.

The proportion of PAR intercepted by the canopy in Eqn (1) has two components. The first is the innate ability of the canopy architecture to capture light: this is quantified by the canopy's extinction coefficient (k). The second component is the development of the canopy leaf area index (LAI in m^2 leaf area m^{-2} ground) over the growing season. This depends on the start date and rate of canopy expansion. Canopy expansion is largely driven by temperature above a base level (T_b) below which the leaf expansion ceases. This affect is quantified by the leaf expansion rate (LER) in LAI $°C\ day^{-1}$, where the denominator is the degree days above T_b. Thus, there is a complex interaction in yield production of the canopy intercepting the ambient PAR but with the canopy expansion (and hence its ability to intercept light) being driven by temperature. To combine the influences of PAR levels and temperature on yield over a growing season, one needs to convert Eqn (1) to a model which can be stepped through time using real met data. The model could then be made genotype specific using the appropriate values of the parameters T_b, LER, k and RUE.

The genotypes investigated in this study include two *M. sinensis* a *M. sacchariflorus* and the hybrid $M. \times giganteus$. They were selected based on their large variations in canopy architecture and also variation in flowering time which is known to affect yield (Jensen *et al.,* 2013). *Miscanthus sinensis* genotypes (including those used in this trial) tend to flower prolifically, whereas *M. sacchariflorus* does not flower in the field in the United Kingdom and $M. \times giganteus$ only flowers on exceptionally warm years (Jensen *et al.,* 2013).

In this study, field data are first used to estimate T_b, LER, k and RUE for the four genotypes. This provides an estimate of the cross-genotype (species) variation that can be expected in these key values that determine light interception and yield. In particular, the impact of the widely different canopy architectures of the plants on light interception (via k) can be accessed. In addition, the RUEs (calculated with the aid of the model in Eqn 1) can be used to help understand why $M. \times giganteus$ is such a productive genotype. The values of T_b, LER, k and RUE are of direct use in reparameterizing complex models of *Miscanthus* to emulate different genotypes, but this would require additional work to access the by-genotype values for other model parameters. To avoid this, the simpler model in Eqn (1) is used to investigate by simulation the light capturing ability and yield of a potential new hybrid created by incorporating into it the variation in values of T_b, LER, k and RUE seen in this study. This is achieved by inserting these values into the model of $M. \times giganteus$ and accessing if this increases its performance. Thus, suggestions on potential breeding targets can also be made.

Materials and methods

Plant material and trial configuration

Four genotypes were selected that included the extremes of *Miscanthus* canopy morphology (Fig. 1). Two were *M. sinensis* types: Sin-11 (also known as Emi-11) a diploid and Goliath a triploid. Sac-5 was a tetraploid *M. sacchariflorus* genotype and $M. \times giganteus$ (clone Gig-311) a triploid hybrid of *M. sinensis* and *M. sacchariflorus*. The trial was situated on former grassland at IBERS on the West Wales (UK) coast (52.4139′N, −4.014′W). It consisted of a randomized block design with four blocks each containing a replicate plot of each genotype. All the plots contained 121 plants at a density of 2 plants m^{-2} and were established using rhizome grown plantlets in May 2009. An automated meteorological station (Campbell Scientific) fitted with a CR1000 data logger at the trial site recorded soil and air temperature, PAR levels and soil moisture content. Additional weather data came from met stations at nearby field sites. A replicated trial was run at Rothamsted Research (Harpenden, Hertfordshire, England, UK) which also carried out all the measurements taken at IBERS except for LAI. Unless otherwise stated, data are for IBERS.

Emergence

The above-ground biomass of the *Miscanthus* was removed on 20 February in 2011. Emergence of new buds from the rhizomes was then scored at weekly intervals on four randomly selected (pseudo-replicate) plants in each plot. Stage 'NEB' indicated the presence of at least one newly emerged bud (shoot). There was no frost causing damage to the emerged shoots in 2011. A bud from each plant was then monitored, and its progressive development recorded. Stage 'FLL' was scored when its first leaf with a ligule was observed. The plot

Fig. 1 Photographs showing the different canopy architectures of the four genotypes used in the trial (taken in September 2013): (a) Gig-311, (b) Sac-5, (c) Sin-11 and (d) Goliath.

was designated as being at stage NEB or FLL when two of the four plants reached the given stage.

Destructive harvests

Destructive harvests were carried out periodically through the growing season starting on the 21 February 2011 (before emergence) until January 2012. A single plant per experimental plot was randomly selected and the above-ground material cut at 10 cm height. A quadrat was used to demarcate the plant's 0.5 m^2 ground area and the rhizome completely excavated and then washed. Samples of the above- and below-ground material were dried in a drying oven until constant mass and the total above- and below-ground dry weights were then calculated.

Leaf area index (LAI) measurements

LAI was estimated from repeat measurements of leaf area on marked stems and from counts of stem numbers m^{-2}. Stem counts were repeatedly carried out on at least three randomly selected plants per experimental plot. This was made at weekly or 2-weekly intervals on the same plants at 60% of the canopy height (Clifton-Brown et al., 2000). This height was chosen as it included all the light capturing leaf area but excluded later emerging stems with small leaf areas from the stem count. Including such stems in the stem counts would artificially elevate the LAI if multiplied by the large leaf areas on the older leaf area measurement stems which formed the canopy itself. The pseudo-replicate stem counts were averaged to give a single stem number m^{-2} value for each experimental plot on a given day.

On each experimental plot, three plants were picked at random and a single stem from each randomly selected and marked for repeat leaf area measurements. Measurements were made weekly or every 2 weeks. As each leaf unfurled, it was numbered and its length and maximum width were measured until it gained a ligule at which point its area no longer increased. The highest leaf with a ligule in the last measurements was remeasured to ensure its dimensions were unchanged. All leaves were checked and if they had died or badly fragmented, they were given a leaf area of zero. The leaf area measurements ceased before the onset of canopy senescence and so green leaf area and green LAI were estimated.

To convert field leaf dimension data to actual areas, a calibration data set was produced. For each genotype, a single mature stem was selected at random from each replicate experimental plot. Every undamaged leaf was measured as in the field and its real area estimated using image analysis. This was performed twice, once at the beginning of July 2011 to include the small early season leaves before they died. The second set was in mid-August 2012 so that the very large late season leaves were included. The data for each genotype were combined and plotted as actual leaf area (cm^2) vs. leaf length * maximum leaf width (cm^2). Genotypes with significantly different straight line fits to their data were found by ANCOVA using R's lm() function (R Core Team, 2013) as described in Crawley (2007). A 5% significance level was used for all statistical tests in this article.

The slopes of the significantly different straight lines fitted to the leaf area calibration data were then used to convert all the field leaf dimension data to actual leaf areas and hence to leaf areas per stem. On each measurement day, the leaf areas of the pseudo-replicate stems on each experimental plot were averaged to give a single value of m^2 leaf area $stem^{-1}$ for that plot. Typically, the stem counts were estimated on different days to the leaf area measurements and so they were adjusted to give the counts on the leaf measurement days using linear interpolation. Once this had been performed, multiplying the experimental plot's average m^2 leaf area $stem^{-1}$ by the equivalent average stem number m^{-2} ground gave the LAI (m^2 leaf area m^{-2} ground).

Canopy PAR transmission measurements and extinction coefficient (k) estimations

For each experimental plot, three plants were selected at random for repeat transmission measurements using a SunScan SS1 (Delta T Devices Ltd, Cambridge, UK: www.delta-t.co.uk/) at weekly or 2-weekly intervals. The instrument consisted of hand-held device with a 1 metre long probe with 64 diodes that measure PAR intensities and a separate station that measures the PAR incidence on the canopy top. The measurements were made in accordance with the manufacturer's instructions. In particular, all readings were taken within 2 h either side of solar noon and not on rainy or dark cloudy days. Stakes were placed around each plant so that measurements could be made with the probe centred at the plant's middle at ground level. They were situated so that a pair of measurements could be made, one in a south-westerly and the other in a south-easterly

direction. For each plant on a plot, one probe measurement was taken in each direction and then converted to transmission values by dividing the mean diode PAR readings by the PAR incident on the canopy top. The two transmissions were then averaged to give the plant's transmission. The pseudo-rep plant transmissions were averaged to give a single value for the plot's transmission. Linear interpolation was then used to find the plot's transmission on the days equivalent to the LAIs. These transmissions were then averaged across the replicate plots to give the mean transmissions for each genotype over time. Loss of transmission data in 2011 meant some additional measurements of transmission, and LAI was made in 2012 to allow the estimation of the canopy extinction coefficient (k).

For each genotype, the mean transmission values were plotted against the mean LAIs. For Gig-311, Sac-5 and Goliath, the 2011 and 2012 data sets were combined. However, for Sin-11, the data from the 2 years were very different from each other and therefore treated separately. The k values were estimated by fitting the formula transmission = $e^{-k * LAI}$ to the data using nonlinear regression with grouped data as described in Ritz & Streibig (2008). Pairs of genotypes were compared using the R nls() function, first using a separate k for each genotype and then using a model with the same k for both. An F-test on the two fits was then used to determine whether the k values were significantly different.

Estimating the leaf expansion rate (LER, LAI °C day^{-1}) values

Initially, the strategy used in Clifton-Brown et al. (2000) to estimate the LER of Gig-311 was adopted. This was to plot LAI vs. cumulative degree days (°C days) calculated using a given T_b value and then to fit the data by linear regression. This was repeated for a range of T_b values and the fit with the lowest R^2 gave both the LER from its slope and the optimal T_b. However, during this study, it was found that some of the genotypes had more than one growth phase and that R^2 was very insensitive to T_b for the data used here. It was therefore decided to adopt a T_b of 0 °C as was used by Hastings et al. (2009b) for Miscanthus modelling.

The two growth patterns that occurred were as follows: (i) a single straight line and (ii) a change in growth rate that could be approximated by two straight lines with a single breakpoint between them. Linear regression using the R lm() function was used for the 1st growth pattern data. Growth pattern 2 was fitted using segmented regression using the R 'Segmented' library (Muggeo, 2008) which gave both straight lines and the breakpoint in one fit. Each experimental field plot was fitted separately using the minimal number of lines that gave a reasonable fit to the data so that between-plot variation could be accessed. In all cases, the LAI and cumulative degree days were set as zero on the actual day that particular experimental plot first produced leaves with ligules (stage FLL) and the late season steady-state LAIs when shading effects occurred were excluded.

The slopes of the two possible growth phases on the LAI vs. degree day graphs gave the LERs. The replicate LER values for each genotype (one per replicate field plot) were then compared to the other genotypes by ANOVA. This was performed using the R aov() function and the Duncan's multiple range test from the R library 'Agricolae' (Mendiburu, 2013). Where the mean genotype LERs were not significantly different, the mean LER across the genotypes was used as the final LER value, otherwise the significantly different genotype mean was used.

For use in simulations, the mean LAI values for Gig-311 across the four replicate plots were plotted against degree days using the known base temperature of this genotype of 10 °C (Clifton-Brown *et al.*, 2000). The data were fitted to a single straight line as above to give the equivalent LER.

Simulations

Simulations were executed by running the model in Fig. 2 in the modelling program Simile (Simulistics, Edinburgh, UK: www.simulistics.co.uk). The model consisted of four coupled ordinary differential equations in which Simile were represented graphically in System Dynamics notation (Haefner,

1996). These equations form the core of the *Miscanthus* model MiscanMod (Clifton-Brown *et al.*, 2000) which can be downloaded from the PLASMO web-portal (www.plasmo.ed.ac.uk/plasmo). The model was stepped through time in Simile using an Euler numerical integrator with 1-day time steps using the genotype specific values for the parameters T_b, LER, k and RUE derived from the field data. In Simile, the representation of the model and the integration algorithm are kept strictly separate (Muetzelfeldt, 2004), and the version of the model in Fig. 2 is a combination of the model and the Euler integrator.

For the simulations and RUE calculations, it was necessary to confirm that the mean genotype values for the LERs derived from fitting each individual field plot, and the mean breakpoint degree days for those with two growth phases still enabled the model to predict the actual mean LAIs of each genotype. This was performed by running the model parameterized for each genotype: starting with zero LAI and cumulative degree days on the actual stage FLL day and using the 2011 met data to drive the model (see Fig. 2). The simulated LAI values were

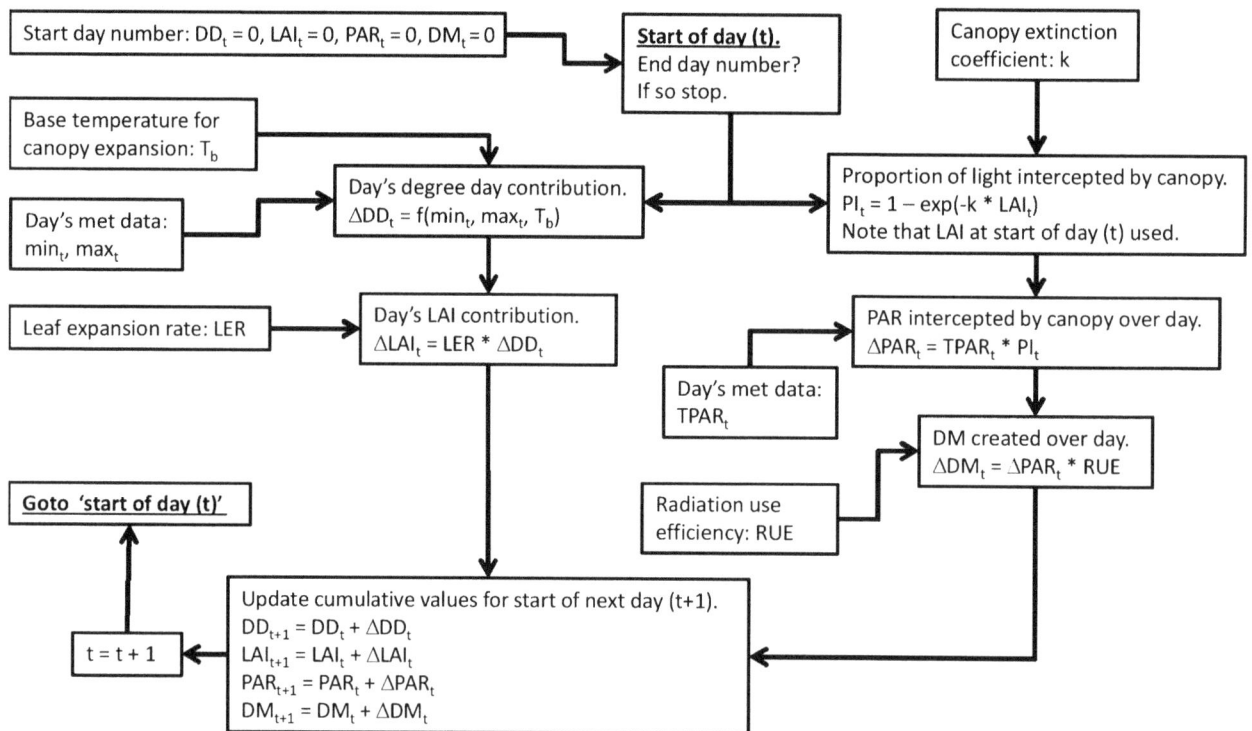

Fig. 2 The mathematical model of *Miscanthus* canopy development and yield represented as a flowchart. One cycle around the chart steps the model through one day using that day's met data. Δ: value (change) over given day t. DD: thermal time (degree days in °C days). DM: dry matter (gDM m^{-2} ground). f(min$_t$, max$_t$, T_b): the formula for calculating DD using the daily min and max air °C values and the base temperature for canopy expansion (T_b) (Clifton-Brown *et al.* (2000)). k: canopy extinction coefficient (m^2 ground m^{-2} leaf, i.e. dimensionless). LAI: leaf area index (m^2 leaf m^{-2} ground, i.e. dimensionless). LER: leaf expansion rate which is the degree day to LAI conversion factor (LAI °C day^{-1}, or °C day^{-1}). If a genotype has more than one growth phase, then this value may be switched to a different one once a threshold cumulative DD value has been reached. Max$_t$: maximum air temperature on day t (°C). Min$_t$: minimum air temperature on day t (° C). PAR: photosynthetically active radiation (MJ m^{-2} ground). PI$_t$: proportion of the PAR hitting the top of the canopy that is intercepted by the canopy on a given day (dimensionless). RUE: radiation-use efficiency (gDM MJ^{-1} intercepted PAR). t (subscript): day number in year. T_b: base temperature for canopy expansion (°C). TPAR$_t$: total PAR hitting the top of the canopy over day t (MJ m^{-2} ground).

then plotted with the equivalent real mean genotype LAIs (across the four replicate experimental field plots) against time. As with all the graphs in this article, this was carried out in R using the error bar function from the library 'Plotrix' (Lemon, 2006) as required.

Simulation was then used to estimate the LAIs of each genotype over the growing season, including for the missing values just after the canopy started to expand, and using the fitted k values, the cumulative PAR on each day could also be found. These simulations were used for two purposes. Firstly, they gave the cumulative PAR values on the days when destructive harvests gave the equivalent plant dry weights so enabling the calculation of the RUEs. Later in the season, the plants achieved a steady-state LAI which the model fails to predict. Over this time range, the simulated LAIs were replaced by values found from linear interpolation between the real LAIs and the met PAR data then used to increment the cumulative PAR interception accordingly. These adjusted cumulative PAR values were then used in the estimation of the genotype RUEs although the correction made only a few percentage points difference to the cumulated PAR because of the canopy intercepting 'all' of the light from quiet early in the growing season.

The second use of simulation was to access the effect of the variation in the fitted parameter values on potential cumulative PAR interception. As Gig-311 was the genotype to improve over, the effects were investigated by looking at the changes in cumulative PAR that they would produce in Gig-311 (i.e. in an equivalent new hybrid). Gig-311 reached steady-state LAIs by day 230, and as the model does not account for this, the simulations were run from its stage FLL day until day 230. By this stage, all the plants were intercepting light at the same rate (due to their high LAIs) and so the few weeks until the first plants started to senesce wont materially affect the conclusions expressed relative to Gig-311. For each simulation, a reference run was carried out with Gig-311 having only its fitted parameter values. Percentage changes in cumulative PAR on day 230 for the additional simulations were expressed relative to the equivalent value for the reference simulation using: 100 (simulation –ref simulation)/ref simulation. The simulations on the variations in fitted k and LER used the model parameterized for a T_b of 0 °C. For Gig-311, the T_b is known to be 10 °C, and for simulations of the effect of T_b changes on the cumulative PAR intercepted, the LER estimated with this base temperature was used. All the simulations used the 2011 met data set.

Estimating the radiation-use efficiencies (RUE)

The destructive harvests gave the plant dry matter values on particular days. The simulations above also gave the cumulative PAR interception values for each genotype on the same days up to the 5 September harvest. The destructive harvest before the plants emerged (21 February) was assumed to give the baseline below-ground biomass with the above-ground mass as zero because of the removal of the previous year's growth. These values corresponded to zero cumulative light interception by the plants. The slopes of straight lines fitted to plots of dry matter vs. cumulative PAR interception gave the RUEs in g dry matter MJ^{-1} PAR intercepted. For each

genotype, three such plots were done with the below-ground, above-ground and total dry matter on the y-axis. These gave the below-ground, above-ground and total RUEs, respectively. The straight lines were fitted either to each genotype's data by linear regression, or ANCOVA was used to find across genotype values as described above.

Results

Climate in 2011 and 2012

Miscanthus yield can be affected by drought (Clifton-Brown et al., 2002; Richter et al., 2008). There was a period of slightly lowered rainfall in March/April 2011, but the vegetation cover was low and the soil water was at winter levels before it. Monitoring of stem extension rate did not indicate drought effects during the growing seasons.

Emergence

The days in the year when new buds from the rhizome first appeared above the ground (stage NEB) and when these produced their first leaves with ligules (stage FLL) are recorded in Table 1. Gig-311 produced its first buds 2 weeks before Sac-5 which in turn had buds between 1 and 2 weeks before the other two genotypes. Despite the differences in bud emergence day, all the genotypes produced their first leaves with ligules from these buds within 1 week of each other. The buds of Sin-11 and Goliath emerged and produced ligule leaves within 1 week. In addition, some emergence from buds that over-wintered above ground and regreening from stems cut during harvesting was also observed.

Figure 3 shows the mean daily air temperature and maximum daily soil temperature (at 10 cm depth) for the period leading up to bud and ligule leaf emergence in 2011. Marked on the figure are the dates when the buds and first ligule leaves of Gig-311 appeared. The temperatures at which the buds first appeared above

Table 1 The mean day numbers in the year (in 2011) when newly emerged buds from the rhizomes first appeared above ground (stage NEB day) and when the first leaf with a ligule was recorded (stage FLL day)

Genotype	Day newly emerged buds appeared (stage NEB day)	Day first leaf with a ligule observed (stage FLL day)
Gig-311	76	102
Sac-5	92	106
Sin-11	102	109
Goliath	100	107

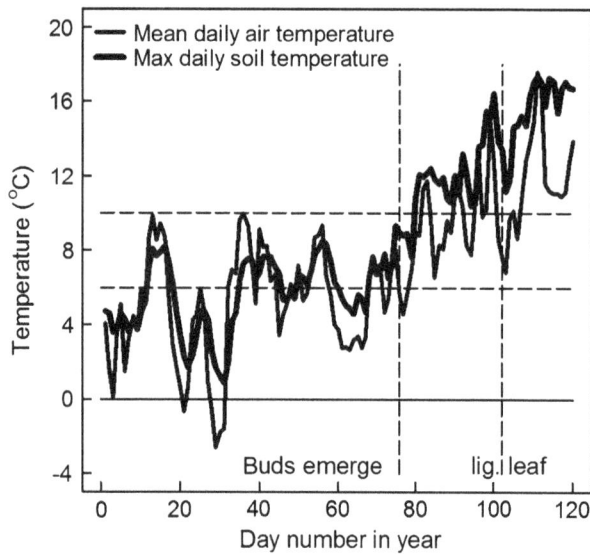

Fig. 3 The mean daily air temperature (at 1.5 m) and maximum daily soil temperature (at -10 cm) for 2011 up to the dates of the appearance of the first leaves with ligules for the four genotypes. For Gig-311, new buds from the rhizome emerged above the ground (stage NEB) on day 76 but did not produce the first leaves with ligules (stage FLL) until day 102 (vertical lines on the figure). The horizontal lines on the figure are at 0, 6 and 10 °C. 10 °C is thought to be the base temperature for canopy expansion (T_b) of Gig-311 whilst 6 °C is the lowest T_b currently recorded for *Miscanthus*.

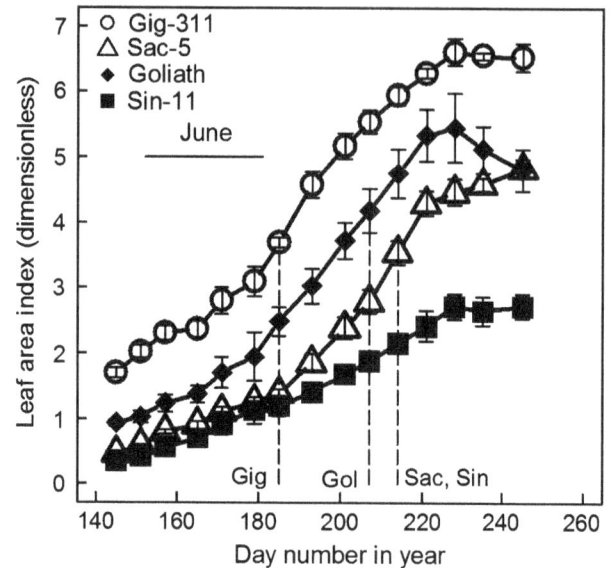

Fig. 4 The mean LAIs across the replicate trial plots for the four genotypes (in 2011) vs. day number in the year. The error bars are plus-and-minus one standard error of the means. The vertical lines show when the canopies first intercepted 90% or more of the incident PAR based on the LAIs and the fitted k values. The horizontal bar indicates June which has the peak in the annual PAR levels.

the soil (stage NEB, day 76) were not different from those that had already occurred several times earlier in the year and were below the base temperature (T_b) for canopy expansion of 10 °C. Once the buds emerged, the air temperatures were at times above the 10 °C T_b value but leaves with ligules were not unfurled from the developing stem (stage FLL) until day 102.

LAI and canopy development

The field measurements of leaf length and width were converted to real areas using the results from the calibration graph of actual leaf area (cm²) vs. leaf length * maximum width (cm²). The data gave two significantly different straight line fits both with a zero y-intercept. Gig-311 had a conversion factor (slope) of 0.745, whilst the other three genotypes all had factors of 0.684 (R^2 of the ANCOVA was 0.99).

The leaf areas on each stem were then combined with the equivalent stem numbers m⁻² to give the LAIs. Figure 4 shows the mean LAIs across the four replicate field plots of each genotype over the 2011 growing season. The maximum LAIs were in the order Gig-311 > Goliath > Sac-5 > Sin-11. Gig-311 achieved 90% PAR interception first, then Goliath and finally Sac-5

and Sin-11, and in all the genotypes, this point was reached after the peak PAR levels in June. This ordering was also reflected in the final above-ground yields in January 2012 which using two-way ANOVA gave the following significantly different values (all in tonnes dry matter hectare⁻¹): Gig-311 15.27; Goliath 8.81; Sac-5 4.97 and Sin-11 4.34 (the last two were not significantly different). There were no significant block effects. The plateaux (steady state) LAIs on Fig. 4 were due to shading effects resulting in the death of lower stem leaves, whilst new leaves were still being produced.

Calculation of the canopy extinction coefficients (k)

The significantly different curves fitted to the transmission vs. LAI data are shown on Fig. 5 and the k values from the fits on Table 2. Gig-311 and Sac-5 have the same k value which is slightly larger than the value for Goliath and Sin-11 (using Sin-11's 2012 data only). Sin-11 in 2011 had a very high k value, but by the following year, its k was the same as for Goliath.

Leaf expansion rate (LER)

Figure 6 shows examples of the LAI vs. cumulative degree days graphs used to estimate the LERs of the different growth phases. Example data showing the two

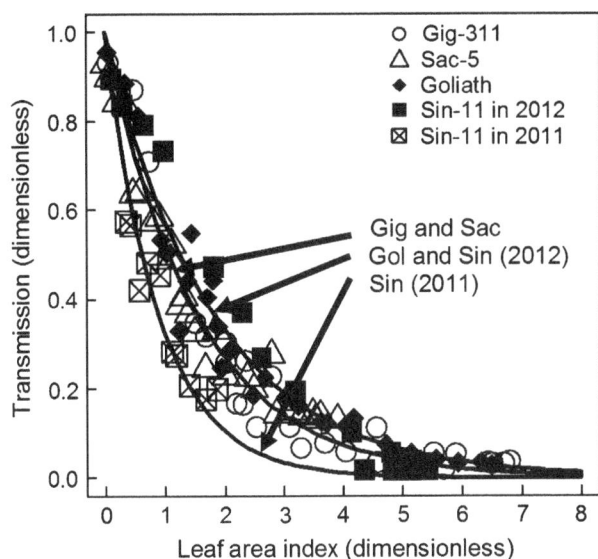

Fig. 5 Transmission vs. leaf area index showing the three significantly different fitted lines used to estimate the extinction coefficients (*k*). Apart for Sin-11, the 2011 and 2012 data have been combined. Thus, Sin-11 has separate fits and *k* values for both years. For clarity, the standard errors are not shown.

Table 2 The significantly different canopy extinction coefficient values (*k*, m^2 ground m^{-2} leaf) from the fitted lines shown in Fig. 5. The year values show the data sets combined to give the data fitted. Thus, only Sin-11 had separate fits and hence *k* values for 2011 and 2012

Genotype	*k*-Value	Standard error
Gig-311 and Sac-5 (both 2011 and 2012)	0.6539	0.01637
Goliath (2011 and 2012) and Sin-11 (2012 only)	0.5533	0.01832
Sin-11 (2011 only)	1.129	0.07686

methods used to fit straight lines to the data are shown (Fig. 6a, b). Two canopy expansion phases were identified; the initial phase 1 of growth was present in all the genotypes whilst Sac-5 and Goliath showed the later second phase of LAI expansion. The mean LER values for each phase for each genotype after comparison by ANOVA are given in Table 3. Only Sin-11 and Goliath flowered in this trial. The former produced its flag leaves on day 196 whilst the latter did so on day 215. The secondary phase of growth and also the continued LAI increase of Sin-11 even after flowering were due to a flush of new stems reaching the upper canopy and contributing to the leaf area capturing light.

That the mean LER data in Table 3 would give a good estimate of the actual mean LAI data shown on Fig. 4 was checked on Fig. 7 by simulation. The simulated

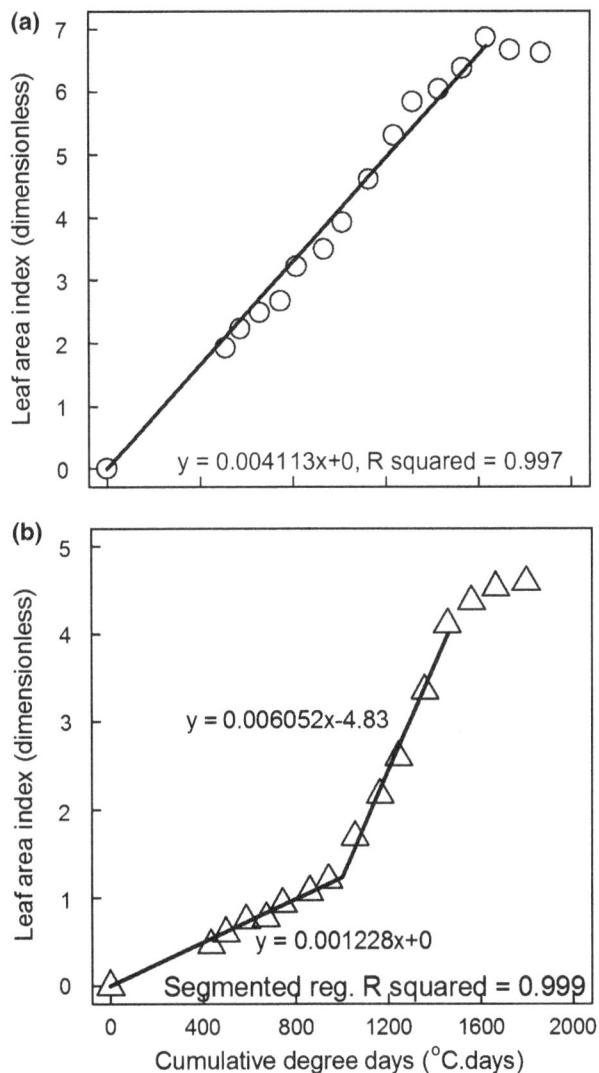

Fig. 6 Measured LAI vs. cumulative degree days calculated using a T_b of 0 °C and with the LAI and degree days equal to zero on the day when the first leaves with ligules appeared (stage FLL day). Example data from individual trial plots are shown for two genotypes demonstrating the fits to the two models used: a single linear regression line ((a) Gig-311) and segmented regression with two linear sections and one breakpoint ((b) Sac-5). The slopes of the lines were used to estimate the LAI °C day^{-1} (leaf expansion rate: LER) values. The later season 'plateau' LAI values were not used in the fits.

LAIs were a good fit to the data up to the plateaux (steady state) LAI points apart from Sin-11 where the simulation was noticeably less consistent with the real data than for the other genotypes.

Radiation-use efficiencies (RUE)

The cumulative PAR estimates equivalent to the dry matter values from the destructive harvests were calculated

Table 3 The significantly different leaf expansion rates (LER, LAI °C day^{-1}) values (T_b = 0 °C) for the 1st and 2nd (later) phases of growth (in 2011). For genotypes with both growth phases, the breakpoint between the two phases is given as cumulative °C days (T_b = 0 °C) after the 1st ligule leaves unfurled (see Table 1)

Genotype	Mean LAI °C day^{-1} for 1st growth phase	Std. error of mean LAI °C day^{-1} for 1st growth phase	Mean LAI °C day^{-1} for 2nd growth phase	Std. error of mean LAI °C day^{-1} for 2nd growth phase	Mean breakpoint cumulative °C day^{-1}
Gig-311	0.003931	0.0001141	–	–	–
Sac-5	0.001395	0.00005638	0.006225	0.0003270	1029
Sin-11	0.001395	0.00005638	–	–	–
Goliath	0.002276	0.0002752	0.006225	0.0003270	866

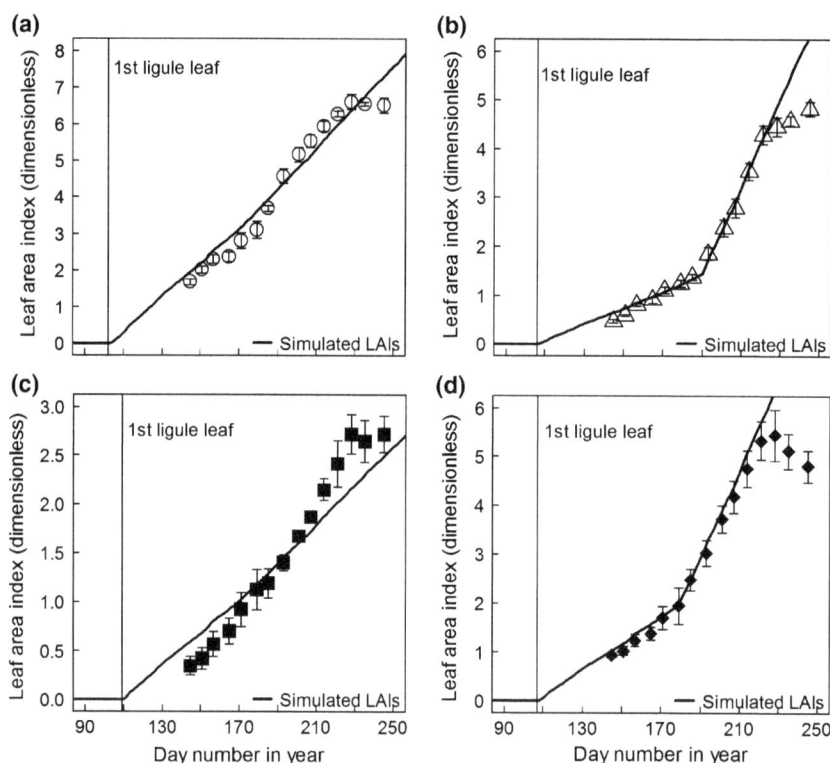

Fig. 7 The mean LAIs across the replicate trial plots for the four genotypes vs. day number in the year in 2011: (a) Gig-311, (b) Sac-5, (c) Sin-11 and (d) Goliath. The error bars are plus-and-minus one standard error of the mean. These are the data plotted on Fig. 4. The LAI values were assumed to be zero on the day when the first leaves with ligules appeared (stage FLL day, vertical lines on the plots). The lines through the data points are simulations with the model on Fig. 2 using the 2011 met data. The model parameterization for each genotype came from Tables 1, 2 and 3.

as in the methods apart from for Sin-11. The relatively poor simulation of LAI for this genotype warranted all the LAIs being derived by linear interpolation between the actual LAIs.

Figure 8 shows the below-ground, above-ground and total dry matter vs. cumulative PAR interception for the four genotypes. Early season remobilization from the rhizome to support the start of the new season growth meant that the below-ground data was expected to show physiological meaningful deviations from linear-

ity. Hence the below ground and total biomass were tentatively fitted using linear regression to give the RUEs (slopes) in Table 4. The above-ground data were expected to increase reliably with increased PAR interception (Beale & Long, 1995; Clifton-Brown *et al.*, 2000) and so were analysed using ANCOVA (see Table 4). This showed that the above-ground RUE for Gig-311 was 2.40 gDM MJ^{-1} PAR intercepted and was significantly higher than the other three genotypes which were not significantly different from each other with

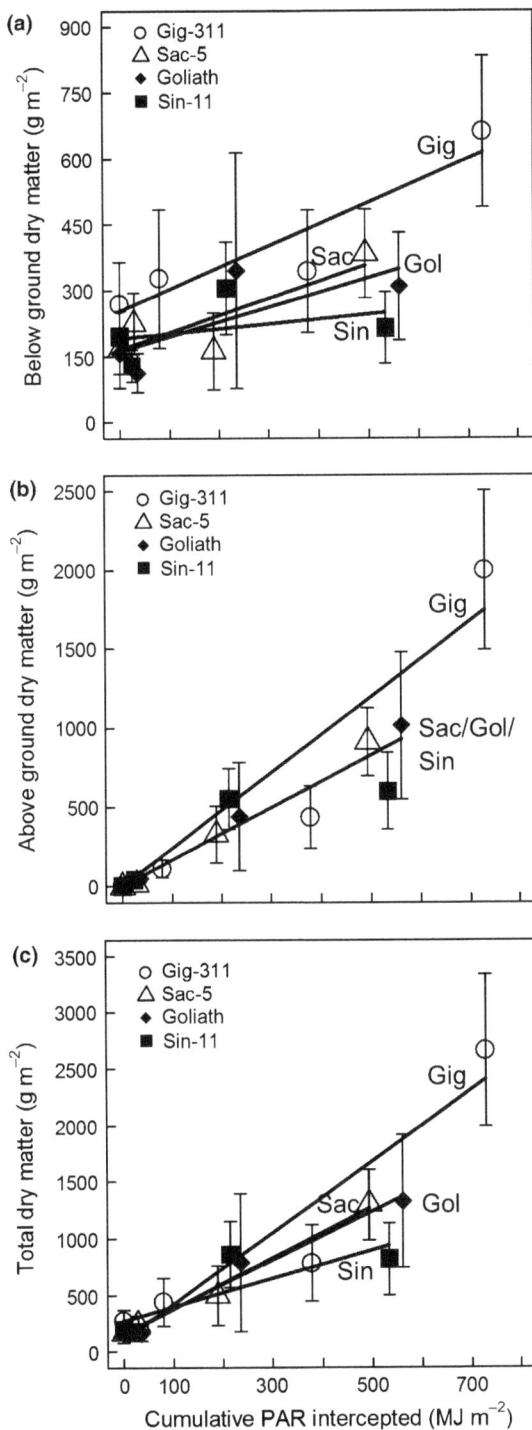

Fig. 8 The dry matter accumulated by the plants vs. the cumulative PAR intercepted by their canopies as estimated by simulation and interpolation. The error bars are plus-and-minus one standard error of the mean dry matter values. (a) Below-ground dry matter, (b) above-ground and (c) total above- and below-ground dry matter. The straight lines on (a) and (c) are linear regression fits to each genotype. The two significantly different lines on (b) are the minimal adequate model fitted by ANCOVA. The slopes of the fitted lines gave the RUEs on Table 4.

Table 4 The RUEs estimated from the slopes of the linear regression or ANCOVA fits to the data in Fig. 8 (in 2011)

Genotype	Biomass used in fit	Slope (RUE) gDM MJ^{-1} PAR	y-intercept gDM m^{-2}	R^2
Gig-311	Below ground	0.49	253.04	0.86
	Above ground*	2.40	0.00	
	Total	3.16	99.26	0.90
Sac-5	Below ground	0.39	164.65	0.72
	Above ground*	1.66	0.00	
	Total	2.27	143.89	0.99
Sin-11	Below ground	0.11	189.66	0.14
	Above ground*	1.66	0.00	
	Total	1.26	265.75	0.68
Goliath	Below ground	0.34	160.86	0.58
	Above ground*	1.66	0.00	
	Total	2.16	157.90	0.98

*From minimal adequate model fitted using ANCOVA (R^2 of fit: 0.94).

1.66 gDM MJ^{-1} PAR. Thus, Gig-311's above-ground RUE is 45% higher than the other genotypes. The data may also indicate that the early flowering of Sin-11 might have reduced its yield in line with the findings of Jensen *et al.* (2013). The cumulative PAR in the model in Fig. 2 can be converted to dry matter yield by direct multiplication by the RUE. Therefore, any increases in RUE would give directly proportional increases in yield.

Simulations

Figure 9a shows the effect on the cumulative PAR interception of Gig-311 (reference simulation) of changing its k value to that of the other genotypes or of increasing its k value by 10%. Although the Sin-11 k value in 2011 (1.13) was 73% higher than Gig-311's k value (0.65) when the Gig-311 simulation was run using this value, it only gave a 14.1% increase in cumulative PAR by day 230. Likewise, a 10% increase in Gig-311's k only gave a 2.9% increase in the day 230 cumulative PAR compared to the reference simulation.

Figure 9b plots the proportion of PAR intercepted vs. LAI for the k values used in Fig. 9a. Gig-311 can intercept 90% of the incident PAR at the fairly low LAI of 3.5 (Clifton-Brown *et al.*, 2000) which it reached at the beginning of July in 2011 (Fig. 4). By the end of that growth season, Gig-311 had almost doubled this LAI. The 90% interception dates are shown for each genotype on Fig. 4.

Increases in k made surprisingly modest changes in the cumulative PAR interception because a 10% increase in k (relative to the k of Gig-311) only gives a 10%

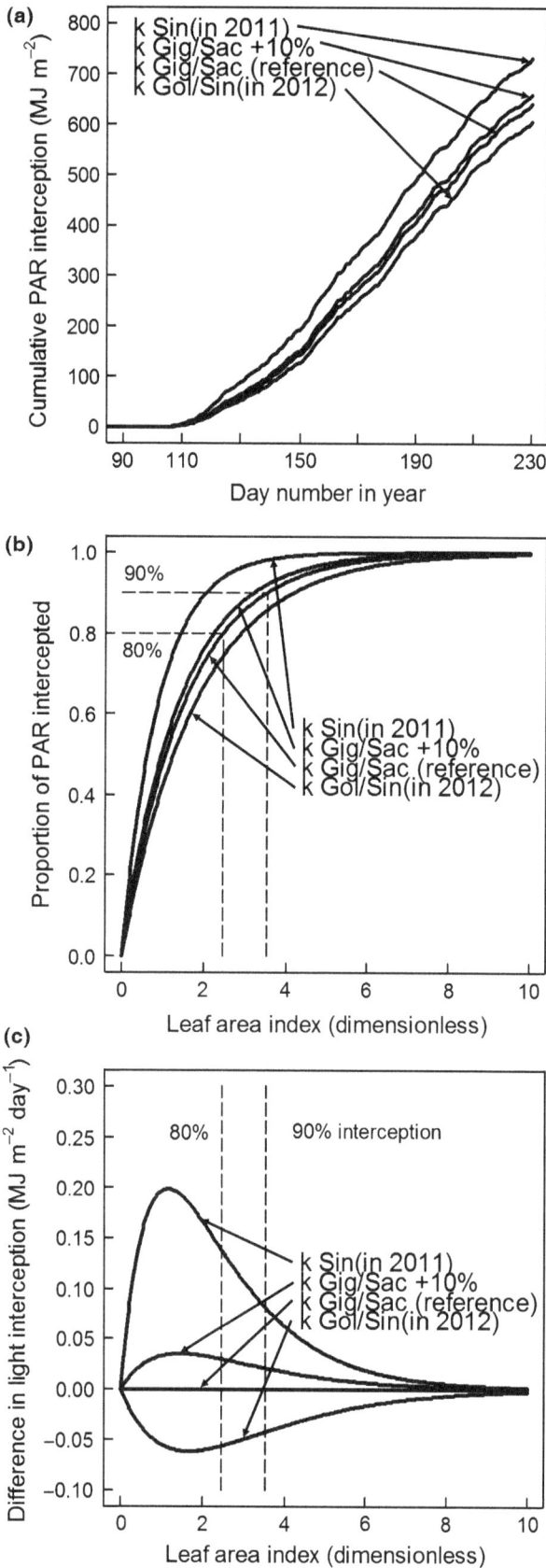

(a)

(b)

(c)

Fig. 9 (a) The effect of changes in canopy extinction coefficient (k) of Gig-311 on cumulative PAR interception vs. day number. The plots were produced by simulation starting at the stage FLL day using the model on Fig. 2 and parameterized using the values for Gig-311 on Tables 1, 2 and 3. The reference line was for Gig-311 using its k value. Also shown are the effects on cumulative PAR interception of increasing Gig-311's k value by 10% and from using the k values for Goliath/Sin-11 (in 2012) and Sin-11 (in 2011) from Table 2 (all other parameterization remained the same as for the reference simulation). Note that Sac-5 has the same k value as Gig-311. (b) The proportion of PAR intercepted by the canopy vs. LAI for the k values used in (a). The dashed lines show when Gig-311 achieves 80% and 90% interception. (c) At each of the LAIs and k values in (b), the proportion of PAR intercepted was used to calculate the daily PAR interception if the canopies had been illuminated with an intensity of 1 MJ PAR m^{-2} ground day^{-1}. These interceptions were then expressed as differences relative to that of Gig-311 with its own k value (reference lines on (a), (b) and (c)) and plotted vs. LAI.

increase in the light interception as LAI approaches 0: when the leaf area is so small that virtually no light is intercepted. As the LAI increases the percentage, increase in interception rapidly falls and even at quite low LAIs essentially all the light is intercepted (see Fig. 9b) and so there is then no difference in PAR captured. This point is made clearer if the PAR intercepted over a day, which gives the daily yield, is considered. Figure 9c shows the changes in the (normalized daily) PAR interception ability of a *Miscanthus* canopy relative to Gig-311's at a given LAI caused by changes in k value. Only over a restricted range of quite low LAI values does an increase in k allow the canopy to outperform Gig-311's at the same LAI, and then, the increase is lower than the percentage rise in k value. For instance, on Fig. 9c, when Gig-311 is intercepting 80% of the light (e.g. 0.80 MJ PAR m^{-2} day^{-1} if the incident radiation is 1.0 MJ PAR m^{-2} day^{-1}), a canopy with the k value of Sin-11 in 2011 would only be collecting an additional 0.14 MJ PAR m^{-2} day^{-1} despite the 73% higher k value. In any case, the low LAIs that give the peak increase in canopy performance are exceeded by Gig-311 early in the growing season and it gets to those levels far sooner than the other genotypes (Fig. 4).

There are two aspects to the rate of canopy expansion: the leaf expansion rate (LER) in LAI °C day^{-1} and the actual rate of canopy expansion on a given day in LAI increase day^{-1} ($\triangle LAI_t$ on Fig. 2) which results from multiplying the degree days contributing to canopy expansion by LER. The magnitude of the degree day contribution on a given day is in turn controlled by the base temperature for canopy expansion (T_b). For Gig-

311, T_b is thought to be 10 °C (Clifton-Brown *et al.*, 2000) although values as low as 6°C have also been estimated for *Miscanthus* (Price *et al.*, 2003; Farrell *et al.*, 2006). The potential effect of decreasing T_b from 10 to 6 °C on degree days can be seen on the temperature profile above these temperatures on Fig. 3. The effect of changes in leaf expansion rate (LER) on cumulative PAR is shown on Fig. 10. A 10% increase in Gig-311's LAI °C day^{-1} only gave a 2.9% increase in the cumulative PAR intercepted by day 230.

The simulations on Fig. 11 use Gig-311's actual base temperature for canopy expansion (T_b) of 10 °C. The reference simulation is for Gig-311 using the T_b of 10 °C and starting at the actual start day for canopy expansion (stage FLL day). Shown are the effects on cumulative PAR interception of reducing the T_b by 10% to 9 °C and then to 6 °C. The effect on the reference simulation of allowing canopy expansion as soon as the buds appeared (stage NEB day) and of also doing this with a T_b of 6 °C is also shown. The percentage increases in cumulative PAR intercepted by day 230 relative to the reference simulation are as follows: T_b of 9 °C 8.1%, T_b

of 6 °C 21.1%, reference starting at NEB day 12.4% and the NEB day start with T_b 6°C 39.0%. The simulation starting at the NEB day with a T_b of 10 °C had a LAI of 0.48 by the real day on which the canopy started to expand (stage FLL day). This LAI is some fifteen times larger than that given by the actual leaf areas of all the first ligule leaves and indicates inhibition of leaf production early in the growing season.

Replicated trial data

Yield data collected from the replicated Rothamsted Research trial were compared to the results from the model. The model was run using the meteorological and emergence data collected from the Rothamsted trial. The model over predicted yield compared to the observed data in both 2011 and 2012. A potential cause

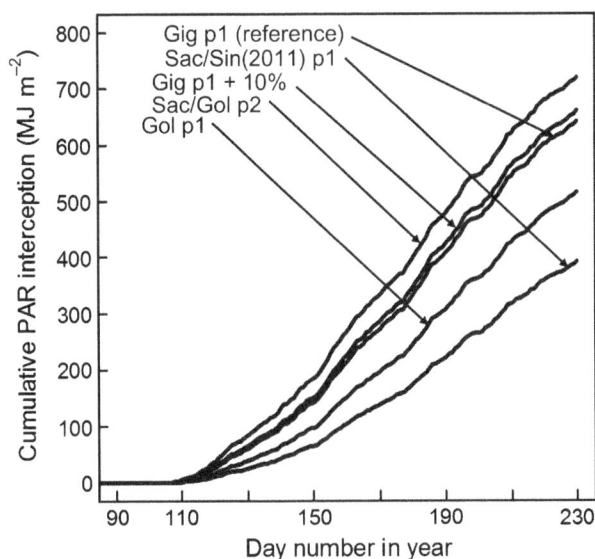

Fig. 10 The effect of changes in LER (LAI °C day^{-1}) on the cumulative PAR interception of Gig-311 vs. day number. The plots were produced by simulation using the model on Fig. 2 and parameterized using the values for Gig-311 on Tables 1, 2 and 3 (starting at the stage FLL day). The reference line was for Gig-311 using its actual LAI °C day^{-1} value for growth phase 1 (p1 on figure). Also shown are the effects on the cumulative PAR interception of increasing Gig-311's value by 10% and from using the LAI °C day^{-1} values for phase 1 Sac-5/Sin-11 (in 2011); Goliath phase 1; and Sac-5/Goliath phase 2 (p2 on figure). The LER values were from Table 3, but all other parameterization remained the same as for the reference simulation.

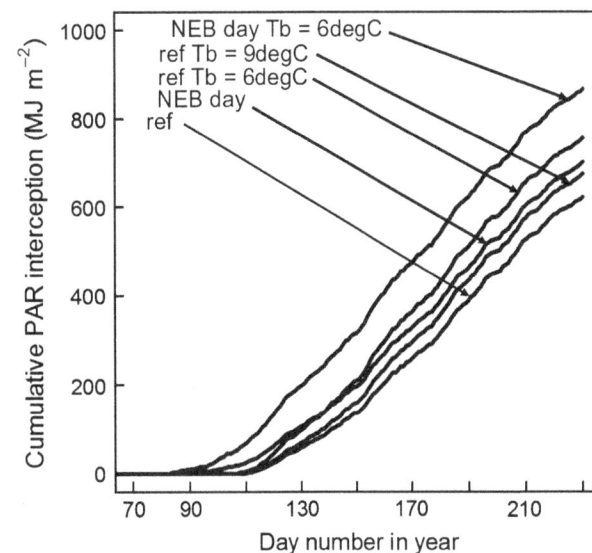

Fig. 11 The effect of changes in the start day for canopy expansion and of changes in base temperature for canopy expansion (T_b) on the cumulative PAR interception of Gig-311. The plots were produced by simulation using the model on Fig. 2 but with a T_b of 10 °C and its equivalent fitted LAI °C day^{-1} for Gig-311 of 0.01668. The other parameters were for Gig-311 from Tables 1 and 2. The reference line is for Gig-311 starting at the actual day when the canopy started to expand (stage FLL). The effect of decreasing the T_b value by 10% to 9 °C and to 6 °C but keeping all the remaining reference line parameterization the same are shown. The impact on the cumulative PAR interception of allowing the canopy to start expanding from the day when the buds first emerged (stage NEB) is also illustrated. Two such simulations are shown as follows: the first with the reference line parameterization but starting at the bud emergence day, and the second doing the same but with a T_b of 6 °C as well. The former shows that the leaf production should have been possible earlier in the year than was observed in the field.

for the reduction in yield was drought; however, attempts to introduce drought into the model by reducing RUE with respect to soil moisture deficit proved ineffective, but did reduce the error. This implies that other factors are effecting the yield at Rothamsted. A reduction in LER due to droughting (Clifton-Brown et al., 2000) is a likely contributor, but there are also possible differences in crop maturity between the sites due to their different climates.

Discussion

Effect of the canopy extinction coefficient k on cumulative PAR interception

Canopy architecture is crucial to intercepting light and hence producing yield. Increases in the canopy extinction coefficient (k) should result in higher cumulative PAR interception and hence give a larger yield. This study used genotypes with the extremes of Miscanthus canopy architecture and estimated their k values. The Gig-311 k value of 0.65 is very close to the 0.68 previously found for this genotype in Clifton-Brown et al. (2000). However, Cosentino et al. (2007) found a k of 0.56 for M. × giganteus which is close to the values for the sinensis types (Goliath and Sin-11) found here. Vargas et al. (2002) gave the Goliath k value as 0.66. Thus, despite the large differences in canopy architecture, the k values are very similar. The exception in this study was for Sin-11 in 2011 which may have still been immature in that year as its high k value became very similar to the other genotypes in the following year. Sin-11 matured more slowly than the other genotypes due to slower growth rates, which delayed canopy closure in Aberystwyth at the planting density of 2 m^{-2}. The high extinction coefficient does suggest that the genetic flexibility for high k exists in Miscanthus and that stand maturity may affect it in some genotypes.

Gig-311 is a superior performing genotype (Naidu et al., 2003; Wang et al., 2008; Dohleman & Long, 2009; Dohleman et al., 2009) and currently the only one commercially available. Therefore, the aim of breeding research is to outperform M. × giganteus (of which Gig-311 is an example). The effect on cumulative PAR interception of introducing the k values observed in Sin-11 and Goliath into Gig-311 to simulate the creation of a new hybrid made surprisingly modest changes in PAR interception.

The time period in which increases in the k value had the greatest impact on light interception was at the beginning of the growing season before LAI ≥4. After this stage, most of the light is intercepted in all genotypes and so increases in k had little effect. Therefore, to exceed the performance of Gig-311, the duration of max-imum canopy interception must be extended beyond the current length of the growing season. Peak PAR levels at IBERS are in June, and by September, light levels are already approximately the same as March. Thus, there is considerable under-utilized PAR in the spring when the canopies are still developing. However, temperatures in this period are usually still cold which is inhibitory to growth, risks destruction of the early canopy from frost and could reduce RUE (Hastings et al., 2009b) and so breeding for cold tolerance is important. In the warmer climate of south-east England (compared to IBERS), Beale & Long (1995) found that M. × giganteus achieved 90% light interception in early June, whilst the cooler conditions at IBERS meant that the June PAR peak was nonoptimally intercepted. This emphasizes the need to include the effect of temperature on canopy expansion in any consideration of light interception. To accelerate canopy closure either stem emergence along with the start of canopy expansion must be brought forward and/or the early spring canopy growth rate must be increased.

Emergence and the start and rate of canopy expansion

The base temperature for canopy expansion (T_b), photoperiod or a combination of these with a threshold level of accumulated degree days has been thought to control when new buds first appear above ground at the start of a growing season (stage NEB day) (Clifton-Brown et al., 2000; Farrell et al., 2006; Hastings et al., 2009b). The emergence of Gig-311 in this study is consistent with the 12-h photoperiod emergence criteria used Hastings et al. (2009b). However, if the temperature of the rhizome controls emergence of new buds/stems, then the T_b of 10 °C is too high as we observed stem emergence at temperatures below this value. We also observed that the new buds of each of the genotypes emerged at different times and the processes controlling this and regreening from cut stems are unknown. The wide geographic distribution of Miscanthus may mean that different genotypes have different emergence mechanisms attuned to the requirements of their original locations.

For Gig-311 and Sac-5, the emergence of the stems did not coincide with the production of unfurled leaves with ligules (see also Beale & Long, 1995) even though the temperatures were conducive to leaf production in Gig-311. The young buds of Gig-311 are particularly susceptible to frost damage and yet its stems appeared above ground well before its canopy started to expand (Zub et al., 2012). Frost damage to the stems would result in more resources being drawn from the rhizome to restart growth and repeated episodes could eventually kill the plants with insufficient rhizome resources to drive another flush of shoots. Inhibition of leaf production could be due to photoperiod, temperature or accumulated degree day thresholds or to cold stress in the mornings inhibiting leaf expansion later in the day.

Cumulative PAR interception was not very sensitive to increases in LER over Gig-311's already high rate. The role of the T_b value in controlling the canopy expansion, especially in the early part of the growing season, is complex. If the daily average air temperatures are about equal to T_b, then some days will not contribute degree days whilst many others only contribute for part of the day. If the T_b is lowered, then substantial increases in the daily degree day contributions are possible in the spring. The higher degree day values would result in the LAI expanding rapidly with concomitant substantial increases in early season PAR interception.

The large genetic variability in Miscanthus T_b values combined with the potential gains in yield that lowering it could make means this is a good breeding target especially if the start of canopy expansion could be moved to earlier in the season. The beneficial impact of early canopy establishment on yield has been demonstrated in a study of 244 Miscanthus genotypes (Robson et al., 2013). A lower T_b than Gig-311's might be less beneficial in a warmer climate where the canopies would reach levels intercepting most of the light earlier in the year anyway. In addition, to gain the full advantage from a lower T_b, the LER should not decrease appreciably (Farrell et al., 2006). Increases in LER could have useful synergistic effects when combined with a lower T_b. To be useful in a breeding programme, simple and quick methods need to be developed to measure these traits and their narrow sense heritabilities (h^2) in the actual breeding populations needs to be sufficiently high to make them useful for selection (Falconer, 1989).

Radiation-use efficiencies (RUE)

The PAR accumulated over a growing season is not the only factor dictating final yield, the efficiency with which that energy has been converted to dry matter (i.e. RUE) is also important. For Miscanthus where the above-ground material is harvested, it is the above-ground RUE that is most important. For Gig-311, this was estimated as 2.40 g dry matter MJ^{-1} PAR intercepted which was close to the 2.35 gDM MJ^{-1} found by Clifton-Brown et al. (2000) in Ireland. The Gig-311 above-ground RUE was 45% higher than for the other three genotypes (which included both its nominal parent species) which in part explains why it is such a productive hybrid. The physiological mechanism causing Gig-311's RUE being higher than the other genotypes in this trial is not known although heterosis is the likely genetic cause.

RUE estimated using harvested dry matter is net of the photosynthetic rate offset by many factors that remove dry matter after photosynthesis has created it. Thus, it is not surprising that there is variation in the RUE values for Miscanthus in the literature (Beale & Long, 1995; Clifton-Brown et al., 2000; Zub & Brancourt-Hulmel, 2010; Kiniry et al., 2012). Such variation in RUE is also a complicating factor in studies attempting to relate canopy duration with yield (Robson et al., 2013). The RUE estimated here of 2.40 gDM MJ^{-1} is much lower than the 3.7 gDM MJ^{-1} found for M. × giganteus in a trial in the USA (Kiniry et al., 2012). The probable cause of this is the comparatively cool summers in West Wales (U.K.) which would limit biomass accumulation and the RUEs seen. Thus, the RUE values in this paper are strictly only applicable to areas of similar climate and probably explains why the value for M. × giganteus grown in Ireland (Clifton-Brown et al., 2000) is so close to that in Wales. Some biofuel crops such as switchgrass can have very high RUEs of 4–5 gDM MJ^{-1} under favourable climates (Kiniry et al., 1999). This emphasizes the need for breeding for increased biomass and RUE in Miscanthus if full advantage is to be taken of its ability to grow on marginal land with low agricultural inputs.

Work for the next loop of the modelling cycle

From modelling Miscanthus yield across genotypes using the model in this manuscript and using a more detailed model as well, it became clear that several processes important in yield production are far from well understood. The current understanding of emergence failed to predict the field outcomes and even the emergence of M. × giganteus needs clarification. Despite evidence that flowering decreases yield in Miscanthus (Jensen et al., 2013), there is no comprehensive understanding of what triggers flowering or its potential impact on RUE and senescence. Senescence processes other than those triggered by frost are also in need of further investigation. Any potential effect of plant maturity on yield could not be modelled and meaningfully parameterized in a short-term trial such as this one, but the fact that the estimated k and RUE values for M. × giganteus were so close to those previously measured on older stands (Clifton-Brown et al., 2000) suggests that these key model parameters do not vary much with maturity (unless the plants are very immature). Thus, for future progress towards detailed models that can incorporate the diversity seen in Miscanthus, research is needed to produce full processes descriptions of several key stages in its growth and development which are based on solid physiological knowledge. Experiments are then needed under controlled (and field) conditions to parameterize these process descriptions.

In summary, this study has demonstrated that extending canopy duration at the start of the growing season has the potential to increase yields to a greater extent than improving k (i.e. changing canopy architecture). From the results, it is clear that Gig-311 is such a high yielding hybrid (under nondrought conditions) because it has a slightly higher k value, more rapid leaf expansion and significantly higher above-ground RUE compared to the other genotypes studied. If a frost resistant hybrid could be bred that combined the high RUE and k of Gig-311 with a lower T_b and earlier canopy expansion, then significant increases in yield are achievable.

Acknowledgements

This work funded as part of the BSBEC BioMASS Programme (http://www.bsbec-biomass.org.uk/) under the BBSRC Sustainable Bioenergy Centre (BSBEC) grant (BB/G016216/1), the IBERS component of the BBSRC Energy Grasses & Biorefining Institute Strategic Programme (BBS/E/W/00003134) and also the Cropping Carbon Institute Strategic Programme at RRes and Ceres Inc.

References

Beale CV, Long SP (1995) Can perennial C4 grasses attain high efficiencies of radiant energy conversion in cool climates? *Plant, Cell and Environment*, **18**, 641–650.

Beale CV, Long SP (1997) Seasonal dynamics of nutrient accumulation and partitioning in the perennial C4 grasses *Miscanthus × giganteus* and *Spartina cynosuroides*. *Biomass and Bioenergy*, **12**, 419–428.

Clifton-Brown JC, Neilson BM, Lewandowski I, Jones MB (2000) The modelled productivity of *Miscanthus × giganteus* (GREEF et DEU) in Ireland. *Industrial Crops and Products*, **12**, 97–109.

Clifton-Brown JC, Lewandowski I, Andersson B, Basch G, Christian DG, Bonderup-Kjeldsen J, Teixeira F (2001) Performance of 15 *Miscanthus* genotypes at five sites in Europe. *Agronomy Journal*, **93**, 1013–1019.

Clifton-Brown JC, Lewandowski I, Bangerth F, Jones MB (2002) Comparative responses to water stress in stay-green, rapid- and slow senescing genotypes of the biomass crop, *Miscanthus*. *New Phytologist*, **154**, 335–345.

Cosentino SL, Patane C, Sanzone E, Copani V, Salvatore F (2007) Effects of soil water content and nitrogen supply on the productivity of *Miscanthus × giganteus* Greef et Deu. in a Mediterranean environment. *Industrial Crops and Products*, **25**, 75–88.

Crawley MJ (2007) *The R Book*. John Wiley and Sons Ltd, Chichester.

Dohleman FG, Long SP (2009) More productive than Maize in the Midwest: how does miscanthus do it? *Plant Physiology*, **150**, 2104–2115.

Dohleman FG, Heaton EA, Leakey AD, Long SP (2009) Does greater leaf-level photosynthesis explain the larger solar energy conversion efficiency of *Miscanthus* relative to switchgrass? *Plant, Cell and Environment*, **32**, 1525–1537.

Falconer D (1989) *Introduction to Quantitative Genetics*. Longman, Essex, UK.

Farrell AD, Clifton-Brown JC, Lewandowski I, Jones MB (2006) Genotypic variation in cold tolerance influences yield of *Miscanthus*. *Annals of Applied Biology*, **149**, 337–345.

Greef JM, Deuter M (1993) Syntaxonomy of *Miscanthus × giganteus* GREEF et DEU. *Angewandte Botanik*, **67**, 87–90.

Haefner JW (1996) *Modeling Biological Systems*. Chapman and Hall, London.

Hartl DL, Clark AG (2007) *Principles of Population Genetics*. Sinauer Associates, Sunderland.

Hastings A, Clifton-Brown J, Wattenbach M, Mitchell CP, Stampfl P, Smith P (2009a) Future energy potential of *Miscanthus* in Europe. *Global Change Biology Bioenergy*, **1**, 180–196.

Hastings A, Clifton-Brown JC, Wattenbach M, Mitchell CP, Smith P (2009b) The development of MISCANFOR, a new *Miscanthus* crop growth model: towards more robust yield predictions under different climatic and soil conditions. *Global Change Biology Bioenergy*, **1**, 154–170.

Hodkinson TR, Renvoize S (2001) Nomenclature of *Miscanthus × giganteus* (Poaceae). *Kew Bulletin*, **56**, 759–760.

Jensen E, Robson P, Norris J, Cookson A, Farrar K, Donnison I, Clifton-Brown J (2013) Flowering induction in the bioenergy grass *Miscanthus sacchariflorus* is a quantitative short-day response, whilst delayed flowering under long days increases biomass accumulation. *Journal of Experimental Botany*, **64**, 541–552.

Kiniry JR, Tischler CR, Esbroeck GA (1999) Radiation use efficiency and leaf CO2 exchange for diverse C4 grasses. *Biomass and Bioenergy*, **17**, 95–112.

Kiniry JR, Johnson MVV, Bruckerhoff SB, Kaiser JU, Cordsiemon RL, Harmel RD (2012) Clash of the titans: comparing productivity via radiation use efficiency for two grass giants of the biofuel field. *BioEnergy Research*, **5**, 41–48.

Lemon J (2006) Plotrix: a package in the red light district of R. In: *R News*.

Mendiburu F (2013) Agricolae: statistical procedures for agricultural research in R.

Monteith JL (1977) Climate and the efficiency of crop production in Britain. *Philosophical Transactions of the Royal Society*, **B281**, 277–294.

Muetzelfeldt R (2004) *Declarative Modelling in Ecological and Environmental Research*. European Commission, Luxembourg.

Muggeo VMR (2008) Segmented: an R package to fit regression models with broken-line relationships. In: *R News*.

Naidu SL, Moose SP, Al-Shoaibi AK, Raines CA, Long SP (2003) Cold tolerance of C_4 photosynthesis in *Miscanthus × giganteus*: adaptation in amounts and sequence of C_4 photosynthetic enzymes. *Plant Physiology*, **132**, 1688–1697.

Price L, Bullard M, Lyons H, Anthony S, Nixon P (2003) Identifying the yield potential of *Miscanthus × giganteus*: an assessment of the spatial and temporal variability of *M. × giganteus* biomass productivity across England and Wales. *Biomass and Bioenergy*, **26**, 3–13.

R Core Team (2013) *R: A Language and Environment for Statistical Computing*. R Foundation for Statistical Computing, Vienna.

Richter GM, Riche AB, Dailey AG, Gezan SA, Powlson DS (2008) Is UK biofuel supply from *Miscanthus* water-limited? *Soil Use and Management*, **24**, 235–245.

Ritz C, Streibig JC (2008) *Nonlinear Regression with R*. Springer, New York.

Robson P, Farrar K, Gay A, Jensen E, Clifton-Brown J, Donnison I (2013) Variation in canopy duration in the perennial biofuel crop *Miscanthus* reveals complex associations with yield. *Journal of Experimental Botany*, **64**, 2373–2383.

Somerville C, Youngs H, Taylor C, Davis SC, Long SP (2010) Feedstocks for lignocellulosic biofuels. *Science*, **329**, 790–792.

Vargas LA, Andersen MN, Jensen CR, Jorgensen U (2002) Estimation of leaf area index, light interception and biomass accumulation of *Miscanthus sinensis* 'Goliath' from radiation measurements. *Biomass and Bioenergy*, **22**, 1–14.

Visser P, Pignatelli V (2001) Utilisation of *Miscanthus*. In: *Miscanthus - for Energy and Fibre* (eds Jones MB, Walsh M), pp. 109–154. James and James (Science Publishers), London.

Wang DF, Portis AR, Moose SP, Long SP (2008) Cool C-4 photosynthesis: Pyruvate P-i dikinase expression and activity corresponds to the exceptional cold tolerance of carbon assimilation in *Miscanthus × giganteus*. *Plant Physiology*, **148**, 557–567.

Zub HW, Brancourt-Hulmel M (2010) Agronomic and physiological performances of different species of *Miscanthus*, a major energy crop. A review. *Agronomy for Sustainable Development*, **30**, 201–214.

Zub HW, Arnoult S, Younous J, Lejeune-Henaut I, Brancourt-Hulmel M (2012) The frost tolerance of *Miscanthus* at the juvenile stage: differences between clones are influenced by leaf-stage and acclimation. *European Journal of Agronomy*, **36**, 32–40.

Permissions

List of Contributors

Vamini Bansal, Vianney Tumwesige and Jo U. Smith
Institute of Biological and Environmental Science, University of Aberdeen, 23 St Machar Drive, Aberdeen AB24 3UU, UK

Andreas Kiesel and Iris Lewandowski
Department of Biobased Products and Energy Crops, Institute of Crop Science, University of Hohenheim, Stuttgart, Germany

Stephen D. Leduc and Christopher M. Clark
National Center for Environmental Assessment, U.S. Environmental Protection Agency, 1200 Pennsylvania Ave., NW, 8623P, Washington, DC 20460, USA

Xuesong Zhang
Joint Global Change Research Institute, Pacific Northwest National Lab, 5825 University Research Court, Suite 1200, College Park, MD 20740, USA
Great Lakes Bioenergy Research Center, Michigan State University, East Lansing, MI 48824, USA

R. César izaurralde
Great Lakes Bioenergy Research Center, Michigan State
University, East Lansing, MI 48824, USA
Department of Geographical Sciences, University of Maryland, College Park, MD 20742, USA
Texas AgriLife Research, Texas A&M University, Temple, TX 76502, USA

Chae-In Na, Jeffrey R. Fedenko, Lynn E. Sollenberger and John E. Erickson
Agronomy Department, University of Florida, Gainesville, FL 32611, USA

Mattias De Hollander and Eiko E. Kuramae
Department of Microbial Ecology, Netherlands Institute of Ecology (NIOO/KNAW), Droevendaalsesteeg 10, 6708 PB Wageningen, The Netherlands

Heitor Cantarella
Soils and Environmental Resources Center, Agronomic Institute of Campinas (IAC), Av. Barão de Itapura 1481, 13020-902 Campinas, SP, Brazil

Leonardo M. Pitombo
Department of Microbial Ecology, Netherlands Institute of Ecology (NIOO/KNAW), Droevendaalsesteeg 10, 6708 PB Wageningen, The Netherlands
Soils and Environmental Resources Center, Agronomic Institute of Campinas (IAC), Av. Barão de Itapura 1481, 13020-902 Campinas, SP, Brazil
Department of Environmental Sciences, Federal University of São Carlos (UFSCar), Rod. João Leme dos Santos Km 110, 18052-780 Sorocaba, SP, Brazil

Janaína B. Do Carmo
Department of Environmental Sciences, Federal University of São Carlos (UFSCar), Rod. João Leme dos Santos Km 110, 18052-780 Sorocaba, SP, Brazil

Raffaella Rossetto
Polo Piracicaba, Agência Paulista de Tecnologia (APTA), Rodovia SP 127 km 30, 13400-970 Piracicaba, SP, Brazil

Maryeimy V. L Opez
Department of Soil Science, University of São Paulo (ESALQ/USP), Avenida Pádua Dias 11, CEP 13418-260 Piracicaba, SP, Brazil

Aaron J. Sindelar and Marty R. Schmer
Agroecosystem Management Research Unit, USDA-ARS, 251 Food Industry Complex, UNL-East Campus, Lincoln, NE 68583, USA

Russell W. Gesch, Frank Forcella, Carrie A. Eberle and Matthew D. Thom
North Central Soil Conservation Research Lab, USDA-ARS, 803 Iowa Ave, Morris, MN 56267, USA

David W. Archer
Northern Great Plains Research Laboratory, USDA-ARS, 1701 10th Ave SW, Mandan, ND 58554, USA

Richard G. F. Visser and Luisa M. Trindade
Wageningen UR Plant Breeding, Wageningen University and Research Centre, Wageningen 6700 AJ, the Netherlands

Andres F. Torres
Wageningen UR Plant Breeding, Wageningen University and Research Centre, Wageningen 6700 AJ, the Netherlands
Graduate School Experimental Plant Sciences, Wageningen University, Droevendaalsesteeg 1, 6708 PB Wageningen, the Netherlands

Moritz Wagner and Iris Lewandowski
Department of Biobased Products and Energy Crops, Institute of Crop Science, University of Hohenheim (340 b), 70593 Stuttgart, Germany

Qirong Shen
Jiangsu Collaborative Innovation Center for Solid Organic Waste Utilization and National Engineering Research Center for Organic-based Fertilizers, Department of Plant Nutrition, Nanjing Agricultural University, Nanjing, Jiangsu 210095, China

James M. Tiedje
Center of Microbial Ecology, Michigan State University, East Lansing, MI 48824, USA

Chao Xue
Jiangsu Collaborative Innovation Center for Solid Organic Waste Utilization and National Engineering Research Center for Organic-based Fertilizers, Department of Plant Nutrition, Nanjing Agricultural University, Nanjing, Jiangsu 210095, China
Center of Microbial Ecology, Michigan State University, East Lansing, MI 48824, USA

Christopher Ryan Penton
Center of Microbial Ecology, Michigan State University, East Lansing, MI 48824, USA
School of Letters and Sciences, Faculty of Science and Mathematics, Arizona State University, Mesa, AZ 85212, USA

Bangzhou Zhang
Center of Microbial Ecology, Michigan State University, East Lansing, MI 48824, USA
State Key Laboratory of Marine Environmental Science and Key Laboratory of the Ministry of Education for Coast and Wetland Ecosystems, School of Life Sciences, Xiamen University, Xiamen, Fujian 360102, China

Mengxin Zhao
Center of Microbial Ecology, Michigan State University, East Lansing, MI 48824, USA

State Key Joint Laboratory of Environment Simulation and Pollution Control, School of Environment, Tsinghua University, Beijing 100084, China

David E. Rothstein
Department of Forestry, Michigan State University, East Lansing, MI 48824, USA

David J. Mladenoff and Jodi A. Forrester
Department of Forest and Wildlife Ecology, University of Wisconsin, Madison, WI 53706, USA

Rosalie Van Zelm, Patience A. N. Muchada and Mark A. J . Huijbregts
Department of Environmental Science, Institute for Water and Wetland Research, Radboud University, Nijmegen, GL 6500, the Netherlands

Marijn Van Der Velde, Georg Kindermann and Michael Obersteiner
International Institute of Applied Systems Analysis, Ecosystem Services and Management Program, Laxenburg A-2361, Austria

Peng Zhu and Qianlai Zhuang
Department of Earth, Atmospheric, and Planetary Sciences, Purdue University, West Lafayette, IN 47907, USA

Joo Eva
Department of Plant Biology, University of Illinois, Urbana, IL, USA

Carl Bernacchi
Department of Plant Biology, University of Illinois, Urbana, IL, USA
Global Change and Photosynthesis Research Unit, USDA-ARS, Urbana, IL, USA

Danilo A. Ferreira and Henrique C. J. Franco
Brazilian Bioethanol Science and Technology Laboratory – CTBE/CNPEM, Rua Giuseppe Máximo Scolfaro, 10.000, CP. 13083-970 Campinas, SP, Brazil

Rafael Otto
Soil Science department – ESALQ/USP, University of Sao Paulo, Av. Pádua Dias, 11, CP. 13418-900 Piracicaba, SP, Brazil

Andréc. Vitti
Sao Paulo's Agency for Agribusiness Technology, Center South Pole, Rodovia SP 127, Km 30, CP. 13400-970 Piracicaba, SP, Brazil

Caio Fortes, Carlos E. Faroni and Paulo C. O. Trivelin
Stable Isotopes Laboratory - CENA/USP, University of Sao Paulo, Av.
Centenário, 303, CP. 13400-970 Piracicaba, SP, Brazil

Alan l. Garside
Visiting Researcher, Brazilian Bioethanol Science and Technology Laboratory – CTBE/CNPEM, Rua Giuseppe Máximo Scolfaro, 10.000, CP. 13083-970 Campinas, SP, Brazil

Ernie Marx and Mark Easter
Natural Resource Ecology Laboratory, Colorado State University, 1499 Campus Delivery, Fort Collins, CO 80523, USA

John L . Field
Natural Resource Ecology Laboratory, Colorado State University, 1499 Campus Delivery, Fort Collins, CO 80523, USA
Department of Mechanical Engineering, Colorado State University, 1374 Campus Delivery, Fort Collins, CO 80523, USA

Paul R. Adler
Pasture Systems and Watershed Management Research Unit, United States Department of Agriculture-Agricultural Research Service, University Park, PA 16802, USA

Keith Paustian
Natural Resource Ecology Laboratory, Colorado State University, 1499 Campus Delivery, Fort Collins, CO 80523, USA

Department of Soil and Crop Science, Colorado State University, 1170 Campus Delivery, Fort Collins, CO 80523, USA

Sarah J. Gerssen-Gondelach and Birka Wicke
Copernicus Institute of Sustainable Development, Utrecht University, Heidelberglaan 2, 3584 CS Utrecht, The Netherlands

Magdalena Borzezckawalker and Rafał Pudełko
Department of Agrometeorology and Applied Informatics, Institute of Soil Science and Plant Cultivation State Research Institute, 8 Czartoryskich Str., 24-100 Puławy, Poland

Andre P. C. Faaij
Energy and Sustainability Research Institute, University of Groningen, Blauwborgje 6, 9747 AC Groningen, The Netherlands

Christopher Lyndon Davey, Laurence Edmund Jones, Michael Squance, Sarah Jane Purdy, Anne Louise Maddison, Lain Donnison and John Clifton-Brown
Institute of Biological, Environmental and Rural Sciences (IBERS), Aberystwyth University, Gogerddan, Aberystwyth, Ceredigion SY23 3EE, UK

Jennifer Cunniff
Rothamsted Research, Harpenden, Hertfordshire AL5 2JQ, UK

Index

www.ingramcontent.com/pod-product-compliance
Lightning Source LLC
Chambersburg PA
CBHW082040190326
41458CB00010B/3418